D1259200

DISCARD

History of the Space Shuttle
Volume 2

Development of the Shuttle, 1972–1981

CENTRALIA HIGH SCHOOL LIBRARY
CENTRALIA, WASHINGTON 98531

History of the Space Shuttle
Volume 2

Development of the Shuttle, 1972–1981

T 37836

T. A. HEPPENHEIMER

SMITHSONIAN INSTITUTION PRESS
WASHINGTON AND LONDON

By special arrangement with the National Aeronautics and
Space Administration, History Office, this publication is being offered for sale
by the Smithsonian Institution Press, Washington, D.C. 20560-0950.

Copy editor: Karin Kaufman
Production editor: Ruth W. Spiegel
Designer: Janice Wheeler

Library of Congress Cataloging-in-Publication Data
Heppenheimer, T. A., 1947–
History of the space shuttle / T. A. Heppenheimer.
p.   cm.
Includes biographical references and index.
Contents: v.1. Development of the space shuttle, 1965–1972 –
v. 2. Development of the space shuttle, 1972–1981.
ISBN 1-58834-014-7 (v. 1 : alk paper) – ISBN 1-28834-009-0 (v. 2 : alk. paper)
1. Space shuttles–United States–History.   I. Title.
TL795.5 .H4697 2002
629.44′1′0973–dc21   2001049233

British Library Cataloguing-in-Publication Data available

Manufactured in the United States of America
09 08 07 06 05 04 03 02   5 4 3 2 1

∞ The paper used in this publication meets the minimum requirements of the American
National Standard for Information Sciences—Permanence of Paper
for Printed Library Materials ANSI Z39.48-1984.

For permission to reproduce illustrations appearing in this book, please correspond
directly with the owners of the works, as listed in the individual captions.
The Smithsonian Institution Press does not retain reproduction rights
for these illustrations individually, or maintain
a file of addresses for photo sources.

To Connie and to Raizel, who look toward the future

Contents

Illustrations

Tables

Acknowledgments

The development of the space shuttle is a large topic, and no single author can hope to cover the pertinent technologies without considerable support. I indeed received such assistance in writing this book, and it is a pleasure to note the people who helped me.

The management of Rocketdyne was particularly supportive. My interviewees there included Program Manager John Plowden, Fred Jue, Bob Biggs, and Maynard "Joe" Stangeland. Several key people who now are retired gave interviews as well: Paul Castenholz, Ed Larson, and Willy Wilhelm. Bob Biggs also reviewed my draft chapter on the Space Shuttle Main Engine (SSME) and helped to improve its accuracy.

I received additional support within NASA. Dave Geiger and Boyce Mix hosted a visit of several days at the Stennis Space Center. There I saw two SSMEs in tests at full duration while receiving good help in the archives from historian Mack Herring. J. R. Thompson, former SSME project manager at NASA-Marshall, gave an interview as well. Other key interviewees included Dale Myers and Hans Mark.

Dennis R. Jenkins, author of *Space Shuttle* (Stillwater, Minn.: Voyageur Press, 2001), provided valuable help with this book. He gave a lengthy and critical review of the text. He generously shared many line drawings and photos that he had used in his own book and that now appear in this one. In addition, he shared his considerable knowledge of the Launch Processing System (LPS). In this fashion he greatly improved my initial treatment of the LPS, which was both sketchy and inadequate.

Dill Hunley, historian at NASA-Dryden, gave a review of the text that was very detailed and therefore very useful. Other people also reviewed this book in draft form: Leroy Day, Charles Donlan, Tim Kyger, Robert F. Thompson, and John Yardley. Together they noted a number of errors. They also encouraged me to expand my range of topics.

Within the History Office at NASA Headquarters, the chief historian, Roger Launius, deserves particular note. His pointed critiques led me to broaden this book's scope, as by treating issues of safety and program management. He also steered me through extensive archival holdings. Members of his staff helped as well: Nadine Andreassen, Colin Fries, Mark Kahn, Stephen Garber, and Louise Alstork.

I also received good help from archivists, librarians, and center historians at other NASA libraries. At Marshall Space Flight Center, key people included Mike Wright, Alan Grady, and Laura Ballentine. At Kennedy Space Center, I relied on Donna Atkins and Elaine Liston. At Johnson Space Center, I obtained good help from Janine Bolton and Sharon Halprin. The late Mack Herring, at Stennis Space Center, also gave valuable guidance.

The Air Force maintains its own archive at Maxwell Air Force Base, which enabled me to present aspects of this narrative from a military perspective. Ann Webb, the librarian, offered valuable assistance. A security officer, Archie DiFante, worked with classified materials and provided the highly useful service of releasing some of their unclassified sections for my use.

At the European Space Agency, Jane Mellors sent me an extensive series of in-house historical reviews. At Rocketdyne, Haroldeane Snell helped me find my way through that company's photo archive and library. This made it possible for me to work with original project documents.

At Rensselaer Polytechnic Institute, Tammy Golbert of the Folsom Library performed a similar service. She gave me access to the papers of George Low that this library holds within its special collections. At the University of Utah–Salt Lake City, Stanley Larson arranged similar access to the papers of James Fletcher, which are on deposit at that university's main library. John Logsdon of George Washington University, who holds a major collection of memos and correspondence, allowed me to work with this material freely.

The source material for this book is on deposit at the NASA History Office at NASA Headquarters in Washington, and is available for use by researchers. This material includes memos, letters, and correspondence; reports from the General Accounting Office; technical papers; program documents, particularly from Rocketdyne (including original audiotapes of personal interviews); and Air Force historical summaries. All these documents are arranged by topic or subject.

In acknowledging those who contributed, it is appropriate to note a man who expected to but could not: the artist Dan Gauthier. He had prepared original line art for two of my earlier books and was ready to do the same for the present one. However, he died suddenly of a heart attack in September 1999. Dennis Jenkins stepped in and provided me with his own drawings as replacements. He and I nevertheless were able to use some of Gauthier's earlier drawings, which continue to set a standard for quality.

The present book complements my earlier one, *The Space Shuttle Decision* (NASA SP-4221, 1999). In this book as in the previous one, it is appropriate to note my former wife, Phyllis LaVietes. On both projects, she has taken care of my word processing, thereby freeing me to continue to work with my IBM Correcting Selectric typewriter.

It is a pleasure to note that two new grandchildren joined my family in March 2001. My daughter Connie had one of them; the other came from my daughter-in-law Raizel, the wife of my son Alex. Many people have viewed the space shuttle as a new beginning, and it therefore is appropriate to dedicate this book to these two young women, who are going forward with beginnings of their own.

Prologue

Some activities go forward more by indirection than by frontal attack. The space shuttle was one of these, during its formative years. The concept of a shuttle first emerged during the mid-1960s, when Apollo activity was passing its peak and the question was at hand: what would NASA do next?

George Mueller, associate administrator for manned space flight, set his eyes on a space station. He wanted a big one, to fly to orbit atop a Saturn V. He also wanted low cost for the frequent space flights he believed it would attract. Existing launch vehicles, such as the Saturn I-B, looked unpromising in this respect. Thus, anticipating that an all-new design would do better, Mueller decided that a reusable space shuttle would serve this purpose. It was to offer the low logistics costs that he wanted.

His space station plans took root as the Apollo Applications program and proved sufficiently viable to yield the Skylab program. But his space station failed to win support, for there was little demand for it outside of NASA itself.[1] However, during 1969 the new Nixon administration carried through a high-level planning effort. NASA now had a bold visionary at its helm: its administrator, Thomas Paine. Flushed by the success of Apollo, he accepted Mueller's plans and took them several steps farther.

The goal now was Mars. En route to that destination, Paine proposed to sow the Earth-moon system thickly with space stations, and to follow with much larger space bases. Nuclear rockets were to provide transportation. Space station modules would be adapted for the Mars voyage, with nuclear engines pushing the Mars ship on its way. And—oh yes—NASA would need a space shuttle for routine flight to orbit.[2]

The Tom Paine of 1776 had helped to spark a revolution in government. His namesake, two centuries later, hoped to achieve his own revolution in space flight. But Paine the Younger was a liberal Democrat—and the White House was full of Republicans. Nixon presided over the successful moon landings but shared a general view that having reached the moon, there was no need for the nation to go farther. He was not about to grant NASA more giant leaps for mankind.[3]

Within a year and a half, NASA sustained three important setbacks. The first came from the Office of Management and Budget (OMB), which drew up the annual federal budget that the president sent to Congress. In November 1969 the OMB chopped Paine's request by a billion dollars. With this, Mars went out the window, along with space bases and nuclear rockets.

Paine responded by falling back on Mueller's earlier plans, as he called for a "shuttle/station." This concept continued to focus on the space station, with the shuttle providing low-cost logistics. NASA officials described the shuttle/station as a single interrelated program. This approach was straightforward enough, but it nearly killed both. During 1970 critics in Congress attacked NASA harshly, asserting that the shuttle/station was really a back door to

Mars. A Mars program would be quite costly, and funding for the shuttle/station survived by the narrowest of margins in the recorded votes.

This was the second setback. George Low, the acting administrator, following the resignation of Paine, responded by changing his agency's approach. He knew that the station needed the shuttle, but the shuttle might find plenty to do even without a station. The shuttle had been planned as an element of a far-reaching program of *piloted* space flight. Low and his colleagues now argued that it could earn its way by launching and servicing the nation's *automated* spacecraft.[4] To buttress this assertion, and to win support in Congress, Low and his colleagues turned to the Air Force. Its generals now found themselves in an unexpectedly delightful position, for they had tried and failed to win a position in piloted space flight through their Dyna-Soar and Manned Orbiting Laboratory programs. In 1969 both lay abandoned. Yet here now was NASA, offering a shuttle for Air Force use—and promising to pay for its development.

One sometimes hears that when two people are in a relationship, the one that wants it more is the weaker one. NASA needed the Air Force, but the Air Force did not need NASA, for it was happy with current launch vehicles such as the Titan III.[5] Air Force officials therefore declared that to win their support, shuttle advocates were to go back to the drawing board and start with a fresh design.

NASA had planned its shuttle to provide resupply flights to its putative space station, a task that demanded twenty-five thousand pounds in payload. The Air Force was in the business of launching heavy reconnaissance satellites, and insisted on sixty-five thousand. At a stroke, the shuttle became far larger and more costly—with no prospect that the Pentagon would share the expense. It might offer political support, but nothing more.

In addition, Air Force officials expected to launch the shuttle into polar orbit from Vandenberg Air Force Base on the California coast. There was particular interest in single-orbit missions, to conduct quick reconnaissance (and perhaps to snare a Soviet spacecraft before anyone in Moscow knew about it). However, following the ninety-minute duration of that orbit, Vandenberg would have moved to the east by eleven hundred nautical miles, due to the rotation of the earth. To avoid ditching in the Pacific, and to return to this base, the shuttle then would need "crossrange," the ability to fly to the side of its track during reentry. The crossrange was to be eleven hundred nautical miles.[6]

NASA had little interest in either crossrange or polar orbits. Its designers therefore had planned to build a shuttle with a simple straight wing, like that of a World War II fighter. But the Air Force insisted on crossrange, and there was only one way to get it. This was to change to a delta wing. Such a wing would be larger, heavier, and more demanding of thermal protection. Hence it would drive up costs anew. Nevertheless, the Air Force insisted on it.

In January 1971, at a meeting in Williamsburg, Virginia, NASA responded by giving the Air Force most of what it wanted. The gift was a set of conceptual designs that met their requirements. NASA's ploy worked: the Air Force secretary, Robert Seamans, soon was telling Congress that such a shuttle indeed could prove attractive. This addressed the political problem, for opposition waned on Capitol Hill, never to reawaken.[7] The OMB, however, now launched a renewed assault. In May 1971 Low learned that its budget examiners expected to hold NASA to strict financial limits during future years. These limits particularly covered the annual cost of a shuttle program—and were barely half as generous as NASA had hoped for. That was the third setback.[8]

A new administrator, James Fletcher, responded by initiating a sweeping program of studies that sought a low-cost redesign. Within months, these studies turned out new shuttle concepts that apparently could do everything promised by the earlier designs—while costing far less to develop. This gave encouragement to NASA's critics at the OMB. They took the view that because NASA had done so well in response to their recent budget cuts, the agency might do even more if squeezed harder. By the end of 1971 Fletcher was on the verge of asking for no more than a mini-shuttle, too small for the Air Force.[9]

However, his position within the White House was stronger than he thought. George Shultz, director of the OMB, worked closely with Nixon and shared a view that piloted space flight was something that the nation would continue to pursue. Then, with this basic policy in hand, Shultz was not about to quibble over the modest cost savings that might result by building a smaller shuttle. With Nixon's assent, Shultz gave support to a Fletcher request for funding that would allow NASA to build a shuttle of full size. It had the payload bay that the Air Force wanted, fifteen by sixty feet in size.[10]

Why, finally, did the shuttle prevail whereas Mueller's space stations did not? NASA operated a national program, one that had to serve a community of users that included the Air Force, the builders of commercial communications satellites, and the scientific community. These people had turned thumbs down on the space station, finding no compelling reason to want it. But the shuttle was something different, for it might serve this entire community by providing both low launch costs and the renovation of satellites in orbit. Russell Drew, a White House staffer, emphasized this as early as 1969. The Hubble Space Telescope, more than twenty years later, benefited from such refurbishment.[11]

Nixon publicly came out for the shuttle early in 1972, at the start of an election year. But a question remained: What would it look like? NASA by then had a good concept for an orbiter, which closely resembled the design that was actually built. This orbiter needed a booster, however, and here the matter was not so well defined. Booster designs included both liquid- and solid-propellant concepts, and NASA Headquarters held a definite preference for the liquids. Big liquid-fueled rockets were the specialty of the house at NASA's Marshall Space Flight Center. By contrast, the agency had little experience with big solids.

Again, though, budget issues forced NASA's hand. Hardly had the ink dried on its shuttle program cost estimate, a few weeks after Nixon's endorsement, than problems with a liquid booster pointed to a substantial overrun. By contrast, solid booster costs promised to remain well controlled. This was important, for the OMB was continuing to emphasize low cost. NASA responded by going over to solids. Such boosters were already in use, and this technology held ample background. The Air Force, builder of the Titan III, had been flying it since the mid-1960s with solid boosters that were close to shuttle-sized. Hence there was a reserve of experience that NASA could tap. NASA-Marshall adapted to this choice of solids by taking responsibility for those of the shuttle.

By March 1972 the shuttle configuration was in its final form, except for details of its specific dimensions. It was NASA's largest and most important new program since Apollo. It had been crafted to suit the Air Force, and budget pressures from the OMB had led to the choice of boosters. Conceived by this mixed parentage, the shuttle program now faced detailed design and engineering development, with flight to orbit as the goal.[12]

Abbreviations and Acronyms

2-D	two dimensional
3-D	three dimensional
A	ampere
AAS	American Astronautical Society
ABES	Airbreathing Engine System (jet engines)
ABPS	Airbreathing Propulsion System (jet engines)
ACE	Acceptance Checkout Equipment
ACS	attitude control system
AFB	air force base
AFL-CIO	American Federation of Labor–Congress of Industrial Organizations
AFRSI	Advanced Flexible Reusable Surface Insulation (TPS)
AIAA	American Institute of Aeronautics and Astronautics
ALT	Approach and Landing Tests
APS	Aft Propulsion Subsystem
APU	Auxiliary Power Unit
ARMSEF	Atmospheric Reentry Materials and Structures Evaluation Facility (NASA-JSC)
ASI	Augmented Sparkplug Igniter (SSME)
ASRM	Abort Solid Rocket Motor
ASSET	Aerothermodynamic/elastic Structural Systems Environmental Tests
ATOLL	Acceptance Test or Launch Language
AT&T	American Telephone and Telegraph Company
BCWP	budgeted cost of work performed
BCWS	budgeted cost of work scheduled
BTU	British thermal unit
CASI	Center for Aerospace Information
CCMS	Checkout, Control and Monitor Subsystem (LPS)
CCTV	closed-circuit television
CDS	Central Data Subsystem (LPS)
CEI	contract end item specification
CETS	Conference Européene des Telecommunications par Satellites
CG	center of gravity
CNES	Centre National d'Études Spatiales (French space agency)
COSMIC	Computer Software Management and Information Center
CPU	central processor unit
DDR&E	director of defense research and engineering
DDT&E	design, development, test, and evaluation
DM	Demonstration Motor (Thiokol)
DOD	Department of Defense
ELDO	European Launcher Development Organization
ENIAC	Electronic Numerical Integrator and Computer
ERNO	Entwicklungsring Nord
ESA	European Space Agency
ESRO	European Space Research Organization
ET	external tank
F, °F	Fahrenheit degrees
FPB	fuel preburner (SSME)
FRC	Flight Research Center (NASA)
FRCI	Fibrous Refractory Composite Insulation (TPS)

FRF	Flight Readiness Firing (SSME)
ft.	foot/feet
FY	fiscal year
g	force of gravity
GAO	General Accounting Office
GE	General Electric Company
GOAL	Ground Operations Aerospace Language
GOES	Geostationary Operational Environmental Satellite
GSE	ground-support equipment
GTS	Geostationary Technology Satellite
HiMAT	Highly Maneuverable Aircraft Technology
HMX	cyclotetramethylene tetranitramine (solid-propellant oxidizer)
HPFTP	high-pressure fuel turbopump (SSME)
HPOTP	high-pressure oxygen turbopump (SSME)
HPU	Hydraulic Power Unit
HQ	Headquarters
HRSI	High-temperature Reusable Surface Insulation (TPS)
HSR	History Study Report (ESA)
Hz	hertz (frequency)
IAF	International Astronautical Federation
IBM	International Business Machines Corporation
ICAO	International Civil Aviation Organization
ICBM	intercontinental ballistic missile
ICD	Interface Control Document
IEEE	Institute of Electrical and Electronic Engineers
ILS	Instrument Landing System
IMU	inertial measurement unit
I/O	input/output
ISS	International Space Station
ISTB	Integrated Subsystem Test Bed (SSME)
IUS	Interim Upper Stage; after 1977, Inertial Upper Stage
JSC	Johnson Space Center
JSLWG	Joint Spacelab Working Group
K	Kelvin degrees
KSC	Kennedy Space Center
KW	Kilowatt
kword	1,024 words in computer memory
L3S	Lanceur 3ème Génération Substitut (early name of Ariane)
lb.	pound
LC	Launch Complex (KSC)
LCC	Launch Control Center (KSC)
L/D	lift-to-drag ratio
LDEF	Long Duration Exposure Facility
LEM	Lunar Excursion Module (Apollo)
LO−2−, LOX	liquid oxygen
LPS	Launch Processing System
LRSI	Low-temperature Reusable Surface Insulation (TPS)
LST	Large Space Telescope
MBB	Messerschmitt Boelkow Blohm
MFV	main fuel valve (SSME)

MIC	Management Information Center
MIL-STD	Military Standard
MIT	Massachusetts Institute of Technology
MOL	Manned Orbiting Laboratory
MOU	memorandum of understanding
MOV	main oxygen valve (SSME)
MPTA	Main Propulsion Test Article
MSC	Manned Spacecraft Center (NASA)
MSFC	Marshall Space Flight Center (NASA)
MST	Mobile Service Tower (Vandenberg AFB)
MW	megawatts
MX	Missile Experimental
NACA	National Advisory Committee for Aeronautics
NAE	National Aeronautical Establishment (Canada)
NAR	North American Rockwell
NASA	National Aeronautics and Space Administration
NASTRAN	NASA Structural Analysis
NMI	NASA Management Instruction
NOA	new obligational authority (OMB)
NOAA	National Oceanic and Atmospheric Administration
NORAD	North American Air Defense Command
NRC	National Research Council
NSAM	National Security Action Memorandum
NSC	National Security Council
N_2O_4	nitrogen tetroxide (storable liquid oxidizer)
OMB	Office of Management and Budget
OMCF	Orbiter Maintenance and Checkout Facility (Vandenberg AFB)
OMS	Orbital Maneuvering System
OMSF	Office of Manned Space Flight (NASA)
OPF	Orbiter Processing Facility (KSC)
OSTP	Office of Science and Technology Policy (White House)
OTS	Orbital Test Satellite
OV	Orbiter Vehicle
PAM	Payload Assist Module
PASS	Primary Avionics System Software
PBAN	polybutadiene-acrylic acid-acrylonitrile
PCR	Payload Changeout Room
PDL	Process Design Language
PERT	Program Evaluation and Review Technique
PIO	pilot-induced oscillation
ppm	parts per million
PRR	Preliminary Requirements Review; also Program Readiness Review
psi	pounds per square inch
P&W	Pratt and Whitney
PRCB	Program Requirements Control Board
RAM	Research and Applications Module
RCA	Radio Corporation of America
RCC	reinforced carbon-carbon (TPS)
RCS	Reaction Control System
R&D	research and development
RFP	Request for Proposal
RHC	Rotational Hand Controller

Abbreviations and Acronyms

RMS	Remote Manipulator System (Canadian robot arm)
rpm	revolutions per minute
RP-1	Rocket Propellant 1 (high-grade kerosene)
RPS	Record and Playback Subsystem (LPS)
RSI	Reusable Surface Insulation
RSS	Rotating Service Structure (KSC)
SAB	Shuttle Assembly Building (Vandenberg AFB)
SAE	Society of Automotive Engineers
SAIL	Shuttle Avionics Integration Laboratory (JSC)
SALT	Strategic Arms Limitation Talks
SAMPE	Society for the Advancement of Material and Process Engineering
SAMSO	Space and Missile Systems Organization (USAF)
SAMTEC	Space and Missile Test Center (Vandenberg AFB)
SBS	Satellite Business Systems
s/c	spacecraft
SCA	Shuttle Carrier Aircraft (Boeing 747)
sec	second
SEP	Société Européene de Propulsion
SiC	silicon carbide
S-IC	first stage of Saturn V
S-II	second stage of Saturn V
SIP	Strain Isolator Pad (TPS)
S-IV	second stage of Saturn I
S-IVB	third stage of Saturn V
SLC	Space Launch Complex (Vandenberg AFB)
SMAB	Solid Motor Assembly Building (Cape Canaveral)
SMS	Shuttle Mission Simulator
SOAR	Shuttle Orbital Applications Requirements
SP	Special Publication (NASA)
SPS	Service Propulsion System (Apollo)
SRAM	Short-Range Attack Missile
SRB	Solid Rocket Booster
SRM	Solid Rocket Motor
SSME	Space Shuttle Main Engine
SST	supersonic transport
SSUS	Spinning Solid Upper Stage
STA	Structural Test Article
STS	Space Transportation System
STS-1	first flight of the shuttle to orbit
T	scheduled launch time
TACAN	Tactical Air Navigation
TDRS	Tracking and Data Relay Satellite
TM	Technical Memorandum (NASA)
TPS	Thermal Protection System
TRW	Thompson Ramo Wooldridge, Inc.
TVC	thrust vector control
UDMH	unsymmetrical dimethyl hydrazine (storable liquid fuel)
USAF	United States Air Force
UTC	United Technologies Corporation
VAB	Vehicle Assembly Building (KSC)
VIB	Vertical Integration Building (Cape Canaveral)
WBS	Work Breakdown Structure

CHAPTER ONE

Launching the Program

In mid-1972 the shuttle program was NASA's, and fairly won. President Nixon himself had made it so, having held a well-publicized meeting with NASA administrator James Fletcher in January. Nixon's eyes lit up like a young boy's when Fletcher presented him with a model of this new launch vehicle. His support had ensured that NASA's budget request for fiscal year 1973, $200 million for the shuttle proper and $28 million for facilities, would appear in full within the overall federal budget that went to Congress later that month.

In March George Shultz, director of the powerful Office of Management and Budget (OMB), gave written assent to NASA's decision to select solid-propellant rockets as boosters and to set the dimensions of the payload bay at a generous fifteen by sixty feet, enough to serve Air Force needs in full. Then in July, a NASA source evaluation board chose Rockwell International as the winner of the program's most important award, a $2.6 billion contract for the orbiter. Under the firm's earlier names, North American Aviation and North American Rockwell, its Space Division had won laurels for its work on Apollo. Now the division would distinguish itself anew.[1]

Nevertheless, not all was well within Fletcher's world. Nixon indeed had given the shuttle a crucial measure of support, but this did not mean that he would give it strong personal commitment. Throughout 1972, and in subsequent years, he left the program in the hands of Shultz and other OMB officials. The nation faced economic difficulties, which led the OMB to impose budget cuts and program stretch-outs. Fletcher, having no strong supporter in the White House on whom he could rely, found himself virtually alone in facing the ongoing budget battles to determine the program's pace and schedule.

NASA Fights for Its Budget

Early in 1972, in the immediate wake of Nixon's endorsement, the shuttle's position appeared to hold some strength. The request of $228 million for fiscal year 1973 continued a rapidly rising curve of support: $12.5 million in fiscal year 1970, $78.5 million in fiscal year 1971, and $118.5 million in fiscal year 1972.[2] Fletcher also had reason to believe that he had trimmed the cost of the shuttle program to a level that the OMB could endorse.

A year earlier, in 1971, shuttle-design concepts had envisioned a two-stage, fully reusable configuration, with an enormous rocket-powered airplane as the first stage. The projected development cost had been $9.92 billion, which would have required an annual budget exceeding $2 billion during the peak years. The OMB had rejected this out of hand, advising NASA that it had to anticipate stringent limits on future appropriations. NASA responded with a determined effort to craft a shuttle concept that would keep its program cost below $1 billion per year. Its effort succeeded.

A succession of design studies, extending through much of 1971, deleted the big flyback first stage and substituted a pair of booster rockets that were to be recovered after falling into the ocean. The program cost estimate fell to $5.5 billion, but the shuttle retained the payload capacity of the 1971 version: sixty-five thousand pounds in a fifteen-by-sixty-foot bay. Even so, amid continuing pressure from the OMB, NASA sponsored studies of smaller shuttles that promised still lower costs, though with considerable compromise in their ability to carry large and heavy payloads.[3]

On 3 January 1972 Fletcher had presented Shultz with NASA's full-size shuttle concept in addition to its smaller shuttle designs, anticipating that Shultz would choose one of the latter. Shultz, however, proved willing to accept the full-size concept, taking the view that the proper shuttle design was one that would best serve the national interest. A full-size payload bay, Shultz reasoned, would serve the widest community of users at the lowest pro-rata program cost. Faced with Fletcher's alternatives, he said, "If we're going to do it, let's do it right; let's do the big shuttle and forget about the Bureau of the Budget shuttle."[4] Two days later, Fletcher met with Nixon and gained his endorsement as well.

Still, in dealing with the OMB, NASA officials wanted everything in writing. Indeed, an early budget exercise showed that verbal agreements with Shultz's staff members might not be worth the paper they were not written on. Discussions with the OMB had led NASA to the hopeful view that Shultz might accept a cost overrun of as much as $1 billion. A NASA fact sheet, released at Nixon's Western White House at the time of the meeting with Fletcher, referred to "a contingency of 20 percent above the $5.5 billion (R&D) figure." But in mid-February John Sullivan, an OMB economist who dealt particularly closely with the shuttle, wrote a memo to Shultz's deputy, Caspar Weinberger, stating flatly, "Such a contingency has not been recognized by the OMB." It quickly dropped from sight, never to be revived.[5]

Meanwhile, NASA was continuing to refine its shuttle concepts. By March, the agency had chosen solid-propellant rockets for the shuttle boosters, cutting the program cost to $5.15 billion.[6] Yet in the year-to-year world of federal budgets, what counted was not this "runout cost," which was spread over several years. Rather, attention focused on the cost for particular years, such as fiscal year 1973 or 1974, which had to be accommodated within NASA's annual budgets. Since the mid-1960s, the OMB had shown an unnerving tendency to cut this budget repeatedly.

Moreover, in 1972 it was clear that NASA would have to swallow further cuts. The entire federal budget now was under heavy pressure. NASA offered one of the few areas that could be cut extensively. The pressure stemmed from a sudden and strong need to protect the value of the U.S. dollar, amid what the historian William Manchester has called "the worst monetary crisis since the Depression."[7]

This crisis involved nothing less than a collapse of postwar arrangements for international finance. Those arrangements dated to 1944, when a conference of economists at Bretton Woods, New Hampshire, established the dollar as the reserve of value lying behind every major noncommunist national currency. Pounds, francs, and subsequently the deutschmark and Japanese yen were quoted as a fixed number per dollar. In turn, the dollar was valued at thirty-five dollars per ounce of gold, redeemable in this precious metal upon demand. Ordinary citizens could not purchase gold in this fashion, but governments could do so.

During the 1950s and 1960s, as the economies of Europe and Japan recovered and grew, those nations traded with the United States and obtained dollars in exchange for their goods. Central banks used these dollars to back additional issues of national currency, expanding their nations' money supplies and promoting further economic growth. West Germany in particular prospered, which tended to make the deutschmark worth more in dollars. Its central bankers followed the Bretton Woods rules and fought this rise by selling deutschmarks in exchange for dollars, making these marks less valuable. Other international transactions transferred gold out of the United States, as overseas nations redeemed some of their dollars. Their governments built up reserves of this metal, which gave additional backing to their currencies.

In this fashion, other countries accumulated large quantities of dollars, called eurodollars, while the amount of gold at Fort Knox steadily diminished.[8] The situation nevertheless remained controllable, for the United States held a favorable balance of trade; it exported more than it imported. American bankers were able to demand payment in dollars, stemming their overseas flow. During 1971, however, this balance turned negative.

In Bonn, the deutschmark was already under strong upward pressure, tending to rise anew against the dollar. The change in the American trade balance boosted this pressure sharply. West Germany's Zentralbank fought the pressure by selling marks and purchasing many more eurodollars, but that did not work. Early in May, German bankers notified the International Monetary Fund that they would buy no more. Instead, they would allow the deutschmark to "float," seeking its natural exchange rate.

With this, speculators sensed an opportunity. With the dollar losing value, they used euro-dollars to buy more French and Swiss francs, which were rising. Central banks in Paris and Zurich, unable to maintain their fixed exchange rates, followed West Germany's lead and al-lowed their currencies to float as well. Those nations' banks, holding eurodollars that were steadily becoming cheaper, sought refuge by redeeming some of these dollars for gold. This caused U.S. gold reserves to drop perilously close to $10 billion, a minimum level set by law to maintain backing for U.S. currency.

Why was this important? The fixed exchange rates of Bretton Woods had promoted inter-national trade. For example, auto manufacturers in Wolfsburg or Nagoya could know the cost in dollars of every car they sold in the United States. In addition, gold had been the measure of value for thousands of years. But if currencies rose when they floated against the dollar, that would have the same effect as high tariffs, discouraging trade and perhaps promoting a reces-sion. If the U.S. Treasury could not redeem dollars in gold, the dollar would have only such value as the government said it possessed. And if people lost confidence in such promises—European speculators had already done so—the value of the dollar might decline sharply, in a burst of inflation. That would strike a heavy blow at the prospects for domestic prosperity.

Nixon faced the implications of the new global monetary order, and he did not flinch. On 15 August 1971, in a nationwide television address, he announced a bold program. He halted the free conversion of dollars into gold. To fight inflation, he announced a ninety-day freeze on wages and prices, to buy time for more permanent measures. Nixon also called for a $4.7 billion cut in federal spending.[9]

In 1970 the inflation rate had approached 6 percent. In addition, the country was in a re-cession, which cut tax revenues. Nixon had projected a balanced budget for fiscal year 1971, which ended in mid-1971, but the budget actually showed a deficit of $23 billion, which was appallingly large. To defend the dollar, the spending cut was essential, to show that Uncle Sam was not a spendthrift but could live within his means.

At $3.3 billion, NASA's fiscal year 1971 budget covered little more than 1 percent of all federal outlays. Yet it was inevitable that NASA would have to take far more than 1 percent of the overall cuts. The reason was that much federal spending was "uncontrollable," not sub-ject to restriction. Social Security was an example; retirees were entitled by law to receive their benefits, even as the deficit ballooned. On 12 August, in a memo to Nixon, OMB deputy director Weinberger had noted similar budget items: "Welfare, interest on the National Debt, unemployment compensation, Medicare." Only 28 percent of the federal budget was con-trollable, capable of being cut, and NASA was in this category. Its outlays would be trimmed accordingly.[10]

Thus, in 1972 the shuttle stood amid two Nixon decisions. One, made the previous August, sought to defend the dollar and to redraw the terms of international trade and finance. The second, dating to early January, had endorsed the shuttle as a new program for NASA. No one could doubt that of these policy choices, the first was by far the more significant.

During 1972, in the wake of these decisions, the immediate issue was whether NASA could halt its budgetary retreat and hold the line. Fletcher, submitting his fiscal year 1973 request, had called for $3.385 billion in budget authority and $3.225 billion in outlays. (Budget authority, formally known as "new obligational authority," represented the requested congressional appropriation. Outlays represented the amount to be spent during that fiscal year. The difference constituted appropriated funds to be spent in a subsequent year.) He had described this as his "minimum recommended program," but Shultz's minimum was lower still. The Nixon budget invited Congress to provide $3.379 billion, while holding NASA to $3.192 billion in outlays. In constant or uninflated dollars, this budget would still be a cut when compared to the request for fiscal year 1972. But it would hold the line on outlays, which had been $3.181 billion, and would add $92.2 million, some 3 percent, in budget authority.[11]

Much now depended on whether NASA could persuade the OMB to stay at that level during subsequent years, preferably with allowance for inflation. Fletcher's deputy, George Low, who had seen NASA receive far harsher treatment in recent years, described the fiscal year 1973 dollar amounts as "essentially the same but slightly higher than our FY 1972 actuals. They are also well-above our budget guidelines." On 9 February Weinberger sent a letter to Fletcher that proposed to continue such funding: "*Five Year Planning*—For planning purposes an annual spending level of $3.2 billion should be assumed for the foreseeable future. NASA is requested to develop and submit by mid-April 1972 a five year Agency Program Plan in which the peak budget authority and outlays for FY 1974–1978 does not exceed $3.2 billion, including new starts. For the purposes of this plan, price levels should be assumed to be the same as those included in the FY 1973 budget."[12]

Weinberger appeared willing to accept a budget that would remain constant at the fiscal year 1973 level in outlays, and in constant 1973 dollars. Shultz endorsed this plan a week later, within a letter of his own: "I would add our specific understanding that NASA's peak annual spending during the period of development of the shuttle will not exceed *$3.2* billion of outlays in the dollars of the 1973 budget." The time would shortly come when such outlays would appear as distant as the moon, but Fletcher wanted more. Early in March, he and Low met with Weinberger and with Jonathan Rose, a White House staffer. The NASA men stated that their understanding differed from that of the OMB, for they were thinking in terms of the fiscal 1973 level of $3.379 billion in new obligational authority (NOA), not $3.2 billion in outlays. Fletcher emphasized this point a few days later in a letter to Weinberger.[13]

This was no mere game in which Fletcher would try to nail down his $3.2 billion by asking for more. It reflected his agency's internal planning. In a memo to Fletcher, Low reviewed this work:

All of our planning has been based on NOA and not on outlays. The outlays to go with this planned NOA would reach $3.334 billion in 1974; $3.397 billion in 1975; $3.332 billion in 1976; and would finally come down to the $3.2 billion level by 1977. These numbers do not include any future year new starts. . . . This means that to meet the OMB objective as stated in George Shultz's many letters, we would have to cut out existing programs to decrease the outlays by $134 million in 1974; $197 million in 1975, etc.; and we would, of course, not be able to have any new starts."[14]

Meanwhile, Weinberger was replying to Fletcher with his own letter:

I understand we had a general agreement that future NASA budgets would total approximately in the $3.2 billion range. I do not recall that there was any specific discussion as to whether this was outlays or budget authority, although when I speak of budget totals, I nearly always mean outlays, since that is the point with which I am almost always immediately concerned. . . . I am sure we will be able to resolve this matter amicably when we come to fix the 1974 ceilings.

Shultz, in turn, had his own thoughts, emphasizing "the importance of developing the shuttle system . . . within a peak annual NASA spending rate of $3.2 billion (in the dollars of the FY 1973 budget)."[15]

Was there room for compromise? Fletcher met with Weinberger at the end of March. Weinberger invited him to write another letter, stating that NASA would "try" to maintain average outlays of $3.2 billion during the next ten years, while going above this figure during fiscal year 1974 and 1975. After that, NASA would work to bring the outlays to that desired level. "In accordance with our discussions," Fletcher wrote in late April, "I have considered the possibility of an *alternate baseline budget plan* which would preserve the option of holding total NASA outlays to the target of $3.2 billion (in 1971 dollars) you have suggested. . . . I believe that the minimum alternate baseline plan that is worth considering is one which would reach the $3.2 billion (1971 dollars) suggested target in FY 1978, in budget authority, with outlays reaching that level in FY 1979."[16] Weinberger replied a month later, noting that the $3.2 billion of outlays was "in the dollars of the FY 1973 budget," not the dollars of 1971. The difference was significant, for inflation had reduced the worth of the dollar by about 4 percent during the previous year, thereby cutting the value of $3.2 billion by some $130 million. Apart from this, Weinberger was noncommittal, writing that "in this business, nothing is certain."[17]

Soon afterward, Fletcher had the opportunity to raise the matter of his budget with Nixon personally, as the president invited the crew of Apollo 16 to a state dinner. Nixon was his usual ebullient self, as he greeted the astronauts warmly and asked Fletcher whether there

6

might be life on one of the moons of Jupiter—a point on which Fletcher had to demur. Nixon proved to be well aware of NASA's hope for a constant budget. He suggested that the agency's troubles would only last for the duration of a current recession. He expected considerably more interest in space once the economy picked up—or if the Soviets were to make another breakthrough. It was as much as to say that NASA would see its problems resolve themselves in good time, and without Nixon lifting a finger in direct support.[18]

Then, in July 1972, Weinberger sent Fletcher a letter intended to provide "policy guidance for the preparation of [NASA's] fiscal year 1974 budget submission." Weinberger had succeeded Shultz as director of the OMB, and his proposed ceilings were definitely on the low side: $2.9 billion in new budget authority, $3.03 billion in outlays. He was not merely lowballing; he meant it. The two men met a few days later, and Fletcher offered a ploy of his own, suggesting that in December he might have to announce a cancellation of the shuttle, as a result of the severe cuts. Weinberger said that he would vigorously oppose this and that Nixon would as well, for they all agreed that the shuttle was important. Somehow it would have to go forward, even if the program was to be stretched out.[19]

A week later, Fletcher met with Jonathan Rose. Rose offered no help in challenging the OMB budget levels, saying, "Well, we've got to do something to reduce the deficit." He added that the general White House view was negative toward NASA—and stated that Weinberger actually was the best source of support on avoiding permanent damage to the agency. Indeed, Weinberger had already told Fletcher that he would not favor eliminating the planetary program, a major NASA function.[20] Yet Viking, a significant set of missions that aimed at exploring Mars, was in trouble. Events soon showed that it might face outright cancellation.

At an in-house budget meeting, Fletcher asked Dale Myers, the associate administrator for manned space flight, whether he would prefer to cut Skylab or the shuttle, should such a choice became necessary. Myers responded that "the Shuttle Program would have to take any additional cuts" because Skylab was nearly ready for launch—it would fly in less than a year—whereas the shuttle program was just getting under way. He nevertheless added, "I strongly believe that any further cut in the Shuttle Program will invite its cancellation."[21]

Fletcher, to be sure, was under no obligation to accept Weinberger's budget marks. He was free to respond with his own, and would do so by submitting his own budget request at the end of September. The point of departure lay in the actual congressional appropriations for fiscal year 1973, beginning on 1 July 1972. Those appropriations matched in full the OMB request of January; George Low wrote that his budget was "at exactly the level that NASA had originally requested from the Congress. To the best of my knowledge, this is the first time in NASA's history where we got 100% of our request."[22]

The appropriations included the full $200 million for shuttle research and development, along with the extra $28 million for construction of facilities. It was another matter, however, whether NASA would get to spend this largesse, at least during fiscal year 1973. The OMB was pressing hard for deferral of expenditures, which would shift some outlays into fiscal year 1974 and reduce the need for fiscal year 1974 appropriations as well. Moreover, there was reason to defer some of the shuttle spending. The $200 million mark dated to September 1971, a time when NASA still hoped to build the booster as a winged and fully reusable rocket stage that would, in fact, be the world's largest airplane. With the booster now about to take shape as simple solid rocket motors, that $200 billion was out of date, and NASA could live with a smaller amount.

NASA's fiscal year 1973 operating plan thus called for outlays of only $144 million for the shuttle and $11 million for facilities. Even so, this plan was to maintain the existing schedule, which called for first flight to orbit on 1 March 1978.[23] Robert Thompson, shuttle program manager at the Johnson Space Center in Houston, was responsible for preparing budget estimates. Early in September he presented them in a letter to NASA Headquarters. To keep the shuttle on schedule, he wrote, the program would need $721 million in fiscal year 1974, a fivefold increase in just one year. Lesser sums would lead to a program stretch-out; a delay of twelve months in that first flight would permit expenditures of only $540 million in that fiscal year. He knew that the high figure was out of the question; he even doubted whether NASA could spend such a sudden windfall effectively. He therefore recommended a target of $602 million, describing this as "a realistic objective for balancing the resources and expending effort as realistically as possible" while "attempting" to hold to the first flight date of March 1978.[24]

In Washington this target, $602 million, transmogrified into a formal request for $560 million. This meant that even at the level of Fletcher's recommended budget, which amounted to a counteroffer to Weinberger, NASA already was prepared to slip the schedule by close to a year. Within the overall budget request, Fletcher stuck to his guns by doing what he had promised, as he requested $3.54 billion in new budget authority. This was the total budget authority of fiscal year 1973, $3.408 billion (the original $3.379 billion, plus an additional $29 million in new budget authority, in the area of program management) plus an adjustment for inflation. It also was $640 million above Weinberger's mark.

Fletcher also proposed $3.495 billion in outlays, $470 million above the OMB figure. He went on to warn that Weinberger's budget would bring disaster, forcing a choice between "termination of the Space Shuttle Program," "termination of Skylab and Viking," or "going out of business" in important areas such as aeronautics, space science, or space applications. The first alternative was clearly unacceptable; the second would cancel Viking, with an unmanned orbiter and lander that stood as the linchpin of the planetary program.[25]

Clearly, it would not be easy to negotiate a compromise. Indeed, it quickly became more difficult still, for within days of Fletcher's budget letter, he received a letter from Weinberger that built on previous correspondence: "Unfortunately, the 1973 deficit picture has further deteriorated. . . . Most agencies will have to make further program cuts that will result in greater 1973 outlay reductions from currently planned levels. I regret that it may therefore be necessary to ask you not only to make the program reductions outlined in your letter, but also to find additional reductions. . . . The President is determined that we shall not fail the test of fiscal responsibility."[26] In this spirit, NASA's Dale Myers met with counterparts at OMB, who suggested a fiscal year 1974 funding level for the shuttle of $440 million. As Myers wrote in a memo to Fletcher, "I responded that *I would recommend that the program be canceled.*"[27] Subsequent discussions did not go well for NASA, and by early November George Low was writing that OMB would "most likely recommend that either Viking or Skylab or the Shuttle be completely eliminated." He added, "We will lose the shuttle if Viking gets canceled."

This reflected more than a general gloominess; it drew on Low's understanding of Congress. During prior years NASA had presented OMB with well-considered justifications for Viking and had received appropriations for that program. Indeed, Low noted that "more than half of the $800 to $900 million on Viking has already been spent." But if Viking were to be canceled, in the words of NASA's J. F. Malaga, Low believed that such action "would be totally unacceptable in all of NASA's Congressional Committees and would expose NASA and the Administration to severe criticism for the huge waste." In such an atmosphere, Low anticipated that Congress would refuse to throw good money after bad and would cancel the shuttle preemptively.[28]

Moreover, Viking held the support of the nation's scientists, who were eager to learn about Mars. They stood as a key NASA constituency, part of a community of users of spacecraft that the agency hoped to serve. But, in Low's words, "it would be almost impossible to sustain the support of the scientific community for the rest of the NASA program if Viking were canceled." This loss of support would weaken the case for the shuttle by diminishing its constituency.[29]

In December 1972, in a meeting at the OMB, Low and Fletcher received the bad news. Nixon was insisting that the fiscal year 1973 federal budget show total outlays of no more than $250 billion, with a similar ceiling on all federal expenditures for fiscal year 1974. NASA would have to surrender its share. It had already done so, preparing draft budgets at several levels, including a low or "marginal" level and a lower or "submarginal" one. But in Low's words, "OMB had accepted the 'submarginal submission' and made drastic cuts below that level."

NASA's submarginal budget had already proposed to eliminate funding for a quiet aircraft-engine program and an Orbiting Solar Observatory spacecraft while terminating most work on a nuclear rocket engine and on nuclear power for use in space. The OMB zeroed out NASA's funding request for a short-takeoff-and-landing aircraft program. This budget also delayed the shuttle—and zeroed out Viking. In particular, the Viking decision was a policy option approved by Nixon himself and was not NASA's to change.

These OMB acts did not constitute outright cancellation. Congress still held the power of the purse under the Constitution. The OMB might include zero dollars in the president's budget that would soon go to Capitol Hill, but the House and Senate could vote to restore funding. Still, this procedure was certainly worth avoiding, for Congress rarely overrode White House decisions by adding money in such a fashion. Fletcher and Low thus were highly interested in means whereby they might win restoration of funds in the OMB draft budget.[30]

To save Viking, these officials hoped to free up some leeway within the overall NASA budget by cutting or eliminating requests in other areas. The affected programs thus were to be sacrificed to protect the funding for Viking, which had high priority. Low suggested that the time was right for NASA to phase out its involvement in advanced communications satellites and to turn over this work to commercial industry. Fletcher also elected to "suspend" the High Energy Astronomical Observatory program, keeping it alive with a skeleton staff while working on redesigns that would cut costs. He also terminated almost all of the nuclear work and shut down NASA's nuclear research facility at Plum Brook, Ohio.

Just before Christmas, Low wrote, "Our budget submission, as revised, was only approximately $50 million over the OMB mark for both fiscal year 1973 and fiscal year 1974. Weinberger was apparently quite pleased with our proposals, and it is quite probable that they will be accepted."[31] The Nixon budget for fiscal year 1974, which went to Congress early in 1973, continued a pattern of cutbacks that had already made themselves felt within the fiscal year 1973 budget, even after its passage and enactment as law (table 1.1).

The fiscal year 1973 budget showed a drop of $91 million in research and development; these funds were carried over to fiscal year 1974. New obligational authority—the total, not just research and development (R&D)—fell accordingly, to only $3.016 billion. The cited total, $3.107 billion, included this $91 million carryover.[32] But this budget included $201.2 million for Viking, which indeed survived this near-debacle and went on to fly. What of the shuttle? Fletcher had requested $560 million; the final mark was $475 million, which meant a delay in the first orbital flight to December 1978. He nevertheless stated that total research and development costs for the shuttle would remain at $5.15 billion—in 1971 dollars.[33]

For NASA as a whole, this was the nadir. Its budget never went lower, at least in year-to-year dollars; indeed, this budget rose during subsequent years. But this did not reflect growth; it stemmed from inflation, with NASA's budget holding steady in purchasing power. George Low had sought a constant budget, and events would prove that he now had it, but at a level markedly lower than he had wished.[34] In the words of Senator Frank Moss, chairman of the Senate space committee, "The Congress endorsed last year the concept of a 'constant level' NASA budget of $3.4 billion in 1971 dollars. If any of you can't remember what a 1971 dollar was, take my word that it looks a lot bigger now than it did then. But NASA is down to about $3 billion in current dollars, well more than half a billion below the 'constant' budget in 1971 dollars."[35]

Nor was OMB finished in its treatment of the shuttle. During preparation of the fiscal year 1975 budget, NASA sought $889 million, a sum that would have kept the program on

Table 1.1. NASA Budget Summary (in Thousands of Dollars)

	Fiscal Year		
Obligational Authority	**1973** **(Early 1972)**	**1973** **(Early 1973)**	**1974** **(Early 1974)**
Research and development	2, 600,900	2, 509,900	2,288,000
Construction of facilities	77,300	77,300	112,000
Program management	700,800	715,000	707,000
Total obligational authority	3,379,000	3,302,200	3,107,000
Total outlays	3,192,400	3,062,100	3,136,000

Sources: Data from "NASA's Budget, *Aviation Week,* 31 Jan. 1972, 5; and Shumann, "No Projects Axed," *Aviation Week,* 5 Feb. 1973, 23.

its delayed schedule.[36] This request went to OMB late in September 1973, and Dan Taft, a staffer who dealt with NASA's budget, responded with a list of questions. George Low met with Taft early in October and summarized some of his responses:

> *I was quite firm with Taft this year, much more so than I had been in the past. I told him that we would not be in a position this year to sit still for the kind of treatment we received from the OMB last year.*
>
> *In response to the question concerning a $3.3 B constant budget (without adjustment for inflation), I said that if this were to be the President's decision our most likely reaction would be to cancel the Shuttle now rather than see it go down the drain in a few years after we had spent significant funds on it.*
>
> *On the question concerning a 12- to 18-month slip of the Shuttle, I said that I was not at all sure that we could get the Shuttle program back under control if such a slip was imposed this year after last year's slip. We would fight very hard against any budget that would require such a slip.*[37]

Nevertheless, such a slip proved unavoidable. In mid-November John Sawhill, the OMB's deputy director, invited Fletcher and Low to a meeting. Sawhill agreed that the program needed a firm commitment, and he was willing to discuss its terms. In exchange for such a commitment, however, he stated that NASA would have to accept one more delay.

Sawhill discussed an alternative: delay all other new starts. Low and Fletcher suggested another possibility: the OMB might give them a little more money in the next two or three years. Sawhill raised the possibility that Nixon might personally commit to a flight date for the shuttle, perhaps in 1980. In Low's words, "We answered that a Presidential commitment would certainly be better than an OMB commitment because OMB commitments don't seem to be worth anything."[38]

In the end the OMB granted $800 million for fiscal year 1975, which postponed the date of the first orbital flight to mid-1979. This meant that in the two years since Nixon approved the program, in January 1972, the date of this flight slipped by fifteen months. But in exchange for this delay, Fletcher told *Aviation Week* that he now had a "firm commitment" from OMB that "there will be no more slips in the shuttle schedule for budgetary reasons." He added that this assurance was in "a piece of paper from OMB saying it is a commitment on the runout of the shuttle."[39]

For NASA, this marked a consolidation of its prospects, and it is worth noting how this happened. The 1971 monetary crisis had eased. The nation had been in recession early in the 1970s, but prosperity had returned in time for the 1972 elections. Inflation had eased as well, falling from 5.9 percent in 1970, as measured by the Consumer Price Index, to 3.1 percent during 1972.[40] Clearly, the move to floating exchange rates had not brought a new economic downturn, while cutting the link between dollars and gold had not sparked renewed inflation.

Cuts in NASA's budget had forced Fletcher to curtail or abandon programs of lesser priority and to concentrate funds on programs such as Viking, the importance of which was considerably greater. Moreover, the OMB was well aware that Nixon had endorsed the shuttle; it was not a program that could be delayed indefinitely. The rising curve of its funding reflected the fact that the OMB dealt with management as well as with budgets. It was generally understood that as with any major procurement, the shuttle effort had to build up in a prudent fashion and proceed at a measured pace.

The first orbital flight now was set for mid-1979. The program encountered further delays; this flight did not take place until April 1981. But these delays stemmed largely from technical difficulties, which particularly beset the main engine and the thermal-protection system. The shuttle never received lavish support, at least during the 1970s, but it proved to have enough to move forward.

Congress and the General Accounting Office

The shuttle program was shaped by seasoned professionals. At the OMB these included John Young (not the astronaut), who headed its Economics, Science, and Technology Programs Division, which dealt with NASA's budget. His boss, William Morrill, was an OMB assistant director.[41] Their budgets went to Congress for enactment, where their proposed allocations were treated with due respect.

Congress generally viewed the shuttle as a national launch system, intended to serve the Air Force as well as NASA. Some opposition existed; in the House, for instance, Bella Abzug of New York declared, "I think the Space Shuttle will be so stuffed with armaments that there may be no room for people." However, her motion to eliminate shuttle funding went down to defeat by vote of ninety-five to twenty.[42] Much the same happened to a leading Senate opponent, Minnesota's Walter Mondale. He introduced similar budget amendments that sought to strike out funding for the shuttle, but his quest proved somewhat lonely. In recorded floor votes, his motions lost by votes of sixty-four to twenty-two in 1971 and by sixty-one to twenty-one in 1972.[43]

Some Mondale allies switched sides. Senator Clifford Case of New Jersey, for one, opposed the shuttle in 1971 but changed his mind a year later. His legislative assistant told *National Journal* that "the reduced development costs and the cost-benefit studies convinced the Senator that the program was worth pursuing." Other opposition proved soft. George Low wrote,

> *I went to visit Senator Pastore, complete with my Shuttle model. I explained the workings of the Shuttle to him and he was most interested. Apparently, until now he had no idea of how the Shuttle worked, what it was used for, what its purpose was, etc.; all this in spite of the fact that he voted against it in the fiscal year 1972 budget Mondale amendment. However, after I talked to him, he kept the model, had it on his desk on the floor of the Senate during the debate, briefed all interested Senators on it, then took it back to his office, used it there as well in some other Committee markup sessions during the same day. Apparently we have won him over even though I lost my Shuttle model in the process.*[44]

Nevertheless, there was at least one potentially dangerous adversary: Senator William Proxmire of Wisconsin. During 1970 and 1971 he had led the Senate debates that had killed the supersonic transport (SST). He then had worked closely with Mondale, supporting the latter's attempt to do the same to the shuttle. He and his legislative assistant, Richard Wegman, held long-established contacts with dissident space scientists and engineers, including James Van Allen, discoverer of the world-circling belts of trapped radiation that bore his name; Thomas Gold, a leading astrophysicist at Cornell University; and Ralph Lapp, a distinguished nuclear physicist.

Proxmire's opportunity came in August 1972, when the death of Senator Allen Ellender brought a shakeup in Senate committee assignments. Very quickly, Proxmire won the chairmanship of the Appropriations Committee's subcommittee that dealt with NASA's budget. He soon sought to wield his power, asking Fletcher to consider the effect of cuts as large as 30 percent in the agency's request for $2.288 billion in new obligational authority for research and development.

Fletcher resisted and declined to respond. Proxmire complained, "They were totally unresponsive to this request, despite the fact that I warned them that some cuts might be coming and without their advice we'd be less equipped to do this intelligently. They spend six to eight months working up their budget within the budget bureau, and then refuse to allow Congress the benefit of any of these detailed budgetary analyses and negotiations. The questions they refused to supply to us are given routinely every year to the OMB."[45]

Proxmire's opposition, like that of Mondale, proved ineffective. Indeed, during 1972 the Senate actually increased the NASA fiscal year 1973 appropriation by $24 million over the Nixon request. This modest boost was deleted when the final bill went to conference committee with the House, but the enacted measure matched this request in full. In particular, it included $200 million for the shuttle and $28 million for construction of facilities, just as Fletcher had sought. During 1973 the fiscal year 1974 request called for $475 million in shuttle funding, and again the House and Senate appropriated this sum in full.[46]

How did this happen? It helped that there was little significant opposition to the shuttle among the public at large. This encouraged Congress to weigh the program on its merits, amid awareness that the budgets that came from the OMB had already been "pre-shrunk." As Proxmire complained in 1973, "The trouble is that the shuttle program doesn't directly cut across the goals of any powerful public interest group. For instance, the environmental movement, which did yeoman work on the SST, isn't interested at all in the shuttle."

By contrast, another powerful group definitely was interested and was in favor of the shuttle: organized labor. Floyd Smith, president of the International Association of Machinists and Aerospace Workers, took the lead in rallying support within the labor movement. In February 1972 he persuaded the entire AFL-CIO Executive Council to issue a strong declaration of support for the shuttle. This statement noted particularly that the shuttle program would "provide 50,000 jobs in the United States." With the nation emerging from a recession, this was nothing to sneeze at.[47]

Nevertheless, Mondale and Proxmire sought support for their views where they could find it, and the economics of the shuttle represented a promising source. The OMB had insisted that NASA justify this program on economic grounds, with the alternative to the shuttle being the continued use of expendable launch vehicles. Because these expendables already existed, as the Delta, Atlas and Titan III families, they would incur no development cost.

By contrast, the shuttle would require a rather hefty cost during the 1970s, both for development and for construction of an operational fleet. However, NASA expected to recover these outlays through savings in the costs of launch and of spacecraft development, during the 1980s. The potential stumbling block lay in OMB's insistence that NASA calculate these savings using a 10 percent discount rate, equivalent to a rate of interest. In essence, NASA was to view the shuttle program as borrowing money at this interest rate to pay for development and construction and then to pay back these borrowings, along with interest accruing year by year, out of operational savings.

NASA's economic analysts, led by Klaus Heiss and Oskar Morgenstern of Mathematica in Princeton, New Jersey, had set to work with a will during 1970 and 1971. At that time, NASA's designers hoped to build a costly two-stage, fully reusable shuttle with a cost-effectiveness that indeed was open to serious doubt. The Mathematica analysts nevertheless succeeded in arguing that this configuration could meet the OMB economic goal—by making at least thirty-nine flights per year between 1978 and 1990, and with a generous allowance for savings in the projected cost of future spacecraft.

But NASA's planned budget lacked funds for a two-stage, fully reusable shuttle. Pressed by fiscal stringency, NASA responded with a succession of design studies during 1971 and 1972, which cut the development cost nearly in half and evolved a design closely similar to the one that would be built. There was good reason to believe that such a shuttle would be even more economically attractive than the earlier one, and Heiss and Morgenstern sought to show this with a new report in January 1972. They concluded that the new shuttle configuration would not merely break even under the OMB ground rules but would actually produce a sizable net saving.

The break-even level of activity now fell as low as 25 flights per year, or 300 flights between 1979 and 1990. A 514-flight schedule, closely resembling the earlier baseline of 39 missions per year, would return savings of $10.2 billion in "discounted dollars" (constant dollars of 1970, reduced in present value in consideration of the interest these dollars were to earn at 10 percent). A 624-flight program, reflecting NASA and Defense Department plans of 1971, would return savings of $13.9 billion, again in discounted dollars.[48]

In mid-February 1972, less than three weeks after Mathematica issued this second report, Mondale wrote a letter to Elmer Staats, the comptroller general and head of the General Accounting Office (GAO). Like the OMB, the GAO retained a staff of economists and analysts; it sought to provide members of Congress with independent reviews of federal programs. The GAO did not participate directly in shaping the federal budget, a role Congress reserved for its committees. Nevertheless, its reports carried weight. Accordingly, Mondale asked Staats to conduct a review of Mathematica's findings.

Mathematica had worked with Lockheed, which had given close attention to prospective methods of low-cost spacecraft design. The Princeton analysts had also received cost estimates from the Aerospace Corporation, which in turn had compiled explicit lists of shuttle payloads based on information supplied by NASA and the Air Force. The GAO made no attempt to carry through an independent study at such depth; in the words of its report, "We did not conduct an independent cost-benefit analysis of the Space Shuttle Program. We worked with estimates received from Mathematica but did not analyze NASA's March 1972 estimate."

Nevertheless, the GAO made its own contribution by conducting a sensitivity analysis, determining levels of cost overrun at which the program would fail to meet the OMB's criterion for cost-effectiveness. (Again, this criterion called for the shuttle program to recover its outlays through cost savings during the 1980s, in dollars discounted at 10 percent per year.) Working mostly with 1970 dollars, the GAO then summarized the permissible overruns for NASA's selected shuttle configuration (table 1.2). These findings did not permit much leeway. Each putative overrun assumed all other factors held constant, but it was only too likely that more than one would deteriorate simultaneously. In addition, the GAO noted that the Rand Corporation had reviewed a number of other aerospace programs and had found an average cost growth of 40 percent.[49]

George Low read a draft version of the report in mid-May, and he did not like it:

> *It turns out that the analysis presented in the report is not bad, and, in effect, supports the Mathematica analysis. However, the conclusions reached from that analysis are extremely bad. Specifically, the GAO report concludes that the Shuttle will not be economical if we have overruns of more than 20% and further points out that 20% overruns are not unusual. However, the GAO's definition of economical is extremely misleading: GAO calls the Shuttle uneconomical if it earns less than 10% return on investment. Also, GAO fails to point out that we have already included in our Space Shuttle estimates major provisions for cost growth so that we don't expect a very large overrun.[50]*

Low phoned Staats and sent him a letter, while Fletcher followed and sent one of his own. The latter letter quoted Mathematica's report: "The 10% rate of discount is among the highest rate used for the evaluation of public investment projects in large scale research and development programs." Low viewed this OMB criterion as "extremely misleading" because he felt the 10 percent rate was artificially high.

This correspondence brought results. In mid-June, following release of the final version, Low wrote that "the report itself was quite favorable. For this reason it received almost no publicity and did not lead to a renewed Mondale outburst."[51] But Mondale was not through with NASA. Arguing that no cost-benefit analysis performed by NASA contractors could be wholly credible, he urged the GAO to conduct an entirely new review of the shuttle's economics. In July, in a heavily detailed letter, he outlined the full-scale analysis that he wanted and invited the GAO to complete this study by March 1973.[52]

It took a few weeks longer, but during April draft chapters from the GAO were available at NASA for internal review. As Low soon wrote, "The news is very bad. The GAO tried to shoot us down on all points of the economic analysis. Furthermore, in their report they indicate that they have succeeded." Low summarized the results in a table (see table 1.3).[53] In his words, "The GAO had turned what, in our own economic assessment, was a $5 billion advantage of the Shuttle over conventional launch vehicles into a $5 billion disadvantage—a

Table 1.2.　GAO's　Permissible Overruns, 1972

Area Considered	Important Conditions	Permitted Overrun (%)
Launch system life-cycle costs	514 flights 624 flights	25 30
NASA program changes	Space tug and space station in 1985 (not 1979), 581 flights (not 514), 1971 dollars	20
Cost per launch	514 flights	75
Payload retrieval or refurbishment	514 flights	150

Source: Staats, *Cost-Benefit,* 4.

swing of $10 billion." He added that "this is their first attempt at such an analysis. Thus, they were selective in the subjects they undertook to examine and tilted the entire Shuttle analysis out of balance. More than that, they were wrong in many of the things that they did."[54]

Late that month, Low and his colleagues spent an entire day at the GAO. The NASA contingent included Fletcher, Myers, the comptroller William Lilly, and other senior officials. In turn, the GAO's Staats stayed for much of the day as well. Low came with a paper that strongly criticized the draft report. "In effect," he wrote, "we told the GAO not only that they were wrong but also that they were biased. This must have come as quite a blow to Staats."

"We were weak on one or two points," he admitted. "Dale Myers tried to bluff his way out of these, which, in my opinion, was a mistake." On the whole, however, the NASA group was loaded for bear. For example, the GAO had determined that the shuttle's external tank would cost two and a half times as much as the NASA estimate, based on a GAO analysis of the S-II, the second stage of the Saturn V. As Low put it, "It turns out that our Shuttle tank manager happened to have been chief engineer of the S-II stage at MSFC for six or eight years. He did an outstanding job in presenting the differences between the S-II and the Shuttle external tank. In a similar vein, GAO indicated that our reliability numbers for the Titan-class launch vehicle were too low and for the Shuttle were too high. It turned out that our Shuttle reliability man (Guy Cohen) had years of experience on the Titan before coming to NASA. Again, he told a very good story."[55]

In mid-May the GAO delivered a revised draft of its report. Low wrote, "Although they had made some attempt to incorporate our comments, especially in terms of the tone of the report, they did not, by any stretch of the imagination, make all of the required changes. In

17

Table 1.3. Program Cost Estimates

	NASA (1972)	GAO Report (1973)
Current expendables	$48 billion	$43 billion
New expendables	$45 billion	$40 billion
Shuttle	$43 billion	$48 billion

Source: NASA Deputy Administrator to NASA Administrator, memo, 23 Apr. 1973.

---

fact, the report is still extremely biased and unfair." Clearly, further discussions and meetings were in order. NASA continued to disagree over such issues as reliability and the prospect of cost overruns, whereas Low was prepared to argue that external-tank costs might actually be lower than NASA had estimated.[56]

Another draft came on 20 May. Low described it as "still very bad and obviously quite biased, especially in the words used, but the GAO had deleted from that version the table which showed a disbenefit of $5 billion. . . . I think we have made reasonable headway. The GAO report, as it now stands, is very weak and not all that negative. However, the very negative bias of the GAO auditors against the Shuttle still shows in several places, but it is quite clear that the GAO report by itself can no longer be used to kill the Shuttle in the Senate."[57]

The published report came out on 1 June and highlighted a table that presented NASA's own cost estimates, prepared a year apart (table 1.4). What had happened? With the shuttle actually increasing in cost, how could the cost of expendables increase so much more, in only a year, as to triple the estimated savings? From the start, in 1970, NASA and Mathematica had been well aware that the way to win the game of economic analysis was to project the highest possible number of shuttle flights, thus augmenting the advantage of its reusability. The 1972 program had anticipated 581 shuttle missions between 1979 and 1990. The new program projected 779 flights, carrying 1,031 payloads. The operational fleet would demand seven orbiters, which were to fly 86 missions—one every four days—in 1990.[58]

Late in April Low had described this ambitious program as offering "one possible way out," suggesting that "the GAO could use this model as an out in that it represents new information to them." It did more; it convinced the GAO that cost-benefit analysis was little more than a fool's game. After all, NASA could inflate the anticipated benefits of the shuttle merely by swelling the roster of spacecraft that did not yet exist even on paper and that would not fly for fifteen years or more. In its report, the GAO admitted defeat: "GAO is not convinced that the choice of a launch system should be based primarily on cost comparisons."[59]

A year earlier, in a letter to Staats that had accompanied the previous GAO review, Fletcher had written, "NASA's position is that development of the shuttle would be justified

Table 1.4. NASA's Total Program Cost Estimates

	15 March 1972	27 April 1973
Current expendables	$48.3 billion	$66.2 billion
New expendables	$45.7 billion	(Not estimated)
Space shuttle	$43.1 billion	$50.2 billion
Estimated savings	$5.2 billion	$16.0 billion

Source: Staats, *Analysis,* 1.

even if we had not been able to demonstrate at this time that it will have a substantial economic return." Mondale's legislative assistant responded, "It's incredible. They've defended the shuttle all along on economic grounds, and now Fletcher says in effect cost savings don't mean a damn thing." But Fletcher's position was now the GAO's. To Low, the published report "probably was as good a write-up as we could expect."[60]

Still, this evaluation of the shuttle using cost-benefit analysis was somewhat out of the ordinary for a large federal procurement, and one may ask, Why was it necessary? Few other such projects have had to prove themselves in such a fashion, and there are reasons.

Cost-benefit analysis, virtually by definition, concerns itself with estimates of benefits that can be quantified in dollars. Such analysis is routine within the private sector, as when a company decides whether to build a new plant. The reason it is rare within the federal government is that Washington has only limited involvement with procurements that are designed specifically to pursue economic benefit. It leaves most such initiatives to private industry. The government has procurement programs aplenty, but they usually aim to provide services rather than quantifiable benefits. The Defense Department, for one, does not expect its fighters and bombers to turn a profit as if they were commercial airliners. Instead, the Pentagon has the goal of maintaining military readiness, which is hard to quantify in economic terms. What is the dollar value of a war averted, or won?

The analyst Philip Klass of *Aviation Week* notes that weapons must show "survivability," the ability to stay alive in a modern battle zone. Recalling the F-80 fighter of the Korean War, he states, "We can't build an air force with a hundred thousand F-80s," from the 1950s. "They aren't survivable." The pursuit of this goal often drives up the cost of new weapons, amid understanding that in combat money saved may mean aircraft lost. The B-2 bomber program carried this to an extreme, spending $45 billion to procure twenty-one aircraft.[61]

Other federal programs also have goals that do not lend themselves to cost-benefit analysis. The Federal Aviation Administration (FAA) promotes safety in commercial aviation. It has repeatedly carried through upgrades to its radars and its computers used in air traffic control. These have gone forward under contracts secured through competitive bidding, amid close scrutiny of the expenditures. Yet the FAA does not rely on cost-benefit analysis. Instead it considers what the nation's airways and airports need, if they are to be safe.[62]

Still, cost-benefit analysis is not unknown within governments. It flourished at the state level during the 1950s, when a number of states built toll roads of advanced design. These included the New Jersey Turnpike and the New York Thruway. The costs of construction were defrayed through issuances of bonds, which were to earn interest through cash paid by motorists at the toll booths. Hence it was necessary to show that there would be enough traffic to cover this expense. This led to cost-benefit analysis conducted at the state level.[63]

In Washington, the Corps of Engineers has been required by law to conduct such analysis before it can proceed with a major construction project, such as a dam or waterway. This had led to some egregious instances wherein the Corps, like NASA, has tailored its analysis to fit its hopes. Among the more noteworthy examples has been the Tennessee-Tombigbee Waterway, which runs along the border between Alabama and Mississippi, connecting two major rivers. Its proponents faced a difficulty, when arguing in its favor, for an entirely adequate waterway ran parallel to the proposed Tenn-Tom, only a short distance to the west: the Mississippi River. Still, economic analysis gave support to the project, as a 1971 study found a ratio of benefits to cost of 1.6. The benefits included cost savings to shippers, who could use the new route to transport coal and other bulk commodities. The estimated cost came to $323 million.

In 1976 the ratio of benefits to costs was down to 1.08. Still the project went ahead, excavating enough rock and dirt to build a two-lane highway reaching to the moon. A 1981 GAO review found that the benefits had been overstated, for many prospective shippers no longer were interested in using it. The costs had been badly underestimated. The GAO set the total at $2 billion, with the largest element of the difference, $960 million, being the cost of improvements to an existing channel that was to be part of the completed waterway. The Corps had ignored such costs within its analyses on the ground that its internal rules did not require their consideration. With the cost-benefit argument having evaporated, critics in Congress moved to kill the project. Proponents asserted that cancellation would waste the funds already spent, which totaled more than a billion dollars, while leaving what one supporter called "the largest swamp in America." On a key vote in the House, the advocates of the Tenn-Tom prevailed—by the slender margin of 208 to 198.[64] The project went to completion, opening to customers in 1985.[65]

NASA came late to cost-benefit analysis. Even Apollo, the most costly of its programs, had not been required to win justification in such terms. It had gone forward because the view in Washington was that it would serve the national interest in non-economic ways, as by enhancing American prestige.[66] But the advent of the shuttle changed the rules. Beginning in 1969 the OMB took the view that the purpose of the shuttle was to save money by lowering the cost of launching satellites. Ergo, justification through cost-benefit analysis was in order.

20

The shuttle was to incur its costs, principally for development, during the 1970s. The benefits to users were to come only later, during the 1980s. The OMB then treated the shuttle as if it were a toll road—a turnpike into space, as it were. Then, although it was to have its program costs covered through federal appropriations, the OMB approached the shuttle as if it were to be paid for with highway bonds. The benefits then amounted to revenue at the toll booths that was to repay both principal and interest.

The OMB interest rate, again, was 10 percent in constant or uninflated dollars. A toll-way built with 10 percent bonds would need a great deal of traffic to service its debt, and the same was true of the shuttle. This high interest reflected the low priority that the budget makers assigned to this program. Amid other national needs, the shuttle would justify itself only if its proponents could show prospective cost advantages that were large indeed.

These putative advantages included low-cost methods of satellite design. However, success with such methods proved to be out of reach, for existing design approaches gave reliability, whereas low-cost design methods did not. Then, during the 1980s, operational use of the shuttle showed that it was entirely incapable of achieving the low cost and high flight rate that its advocates had projected.[67]

Trimming the Weight, Farming Out the Work

Amid NASA's mixed budgetary prospects, Rockwell International initiated a buildup as it proceeded with detailed design of the shuttle orbiter, while preparing for the day when its workers could begin to cut metal. The firm received notice of its contract award on 26 July 1972, following a competition that had matched it against three rivals: Lockheed, McDonnell Douglas, and Grumman.

Contractor selection was an exercise in competitive bidding, each firm preparing a lengthy proposal that covered management as well as technical design. These proposals went to NASA, where a source selection board prepared detailed evaluations, scoring them on points. The two top contenders, Grumman and Rockwell, showed different strengths. Grumman's technical concept was rated the best, winning the highest score. Rockwell, however, outshone Grumman in management, particularly by arguing convincingly that it could offer the lowest cost of the four contenders. This, plus a technical proposal that was nearly as good as Grumman's, gave the prize to Rockwell.[68]

Following announcement of this award, company officials declared that they would begin hiring immediately and would boost the employment within their Space Division from the existing level of sixty-two hundred to as many as sixteen thousand by 1975. This news drew cheers within the industry, which lately had been in a slump. *Aviation Week* noted that "at $2.6 billion, the contract was the juiciest financial plum in the space program for the past ten years and probably for the next ten."

The award led quickly to a flurry of activity. Negotiations on the formal contract were expected to take as long as eight months, but within days after announcement of the award, NASA issued a letter contract that authorized the company to proceed. Bastian Hello, the project manager, launched an extensive program of subcontracting. He initiated negotiations aimed at awarding sole-source contracts to three firms that had supported his company in preparing the winning proposal: American Airlines, for maintenance and refurbishment of the orbiter; Honeywell, for flight controls; and IBM, for data management.

Some four hundred engineers and technicians already were part of the company's shuttle team, having worked on the proposal. These formed the nucleus of the project staff. Already, the personnel office had sent questionnaires to over five thousand former employees, asking if they cared to return if Rockwell were to win the contract. Most of these people responded, with more than forty-three hundred stating that they indeed would return if invited. Hello stated that he would give priority to rehiring individuals who had worked on Apollo and the Saturn V and who would be transferred into the shuttle program based on their backgrounds as specialists.[69]

The outlook was not nearly so brilliant at Lockheed, Grumman, and McDonnell Douglas, which had lost in the competition. Grumman had employed a staff of some three hundred in preparing its proposal. A company spokesman stated that Grumman would need a commitment as a subcontractor, "forthcoming within a month," or these people would face layoff. "We are overstaffed," he said, "and we will have to know something soon. Without some shuttle work, we are out of the space business in December."[70]

McDonnell Douglas also faced layoffs, not only because it had lost its own shuttle bid but because it had nearly finished its work as prime contractor on Skylab, and because sales of its DC-9 airliners were in a downturn. A formal company statement declared, "It is anticipated that a total of 6,000 layoffs will be required by the end of 1972 and an additional 5,000 in 1973. Had McDonnell Douglas been selected to build the shuttle it is anticipated that a significant portion of this reduction would not have been necessary."[71]

But Rockwell had no intention of keeping the orbiter to itself. Instead, it was prepared to farm out the work. To a degree, this reflected limits on its in-house expertise. The firm had little background in advanced thermal protection, for instance, whereas Lockheed was strong in this area and went on to win a separate contract of its own. This contract provided thermal protection for the orbiter.[72]

Politics, too, was part of the picture; the shuttle program drew in subcontractors with a broad net that covered the nation. Every new award of this type put a few more jobs in some congressman's district and widened the base of support in Washington. Yet the use of subcontracts did more, for it helped preserve strong existing engineering groups within the industry.

Aerospace contracting was not ordinarily a matter of corporate life and death, with a winner prospering and a loser going out of business. By issuing subcontracts, the winner could steer some work to competitors, enabling them to contribute and to remain active. In turn, when the present winner became the loser in some future high-stakes competition, it could hope for similar support from its successful rival.

In mid-September George Low wrote of these matters:

Even before the Space Shuttle selection was made, Fletcher, McCurdy [NASA's associate administrator for organization and management], and I decided that it would be essential that a good share of the business go out on subcontract in order to get the best possible talent in American industry and also in order to get the most widespread support for the continuing development of the Shuttle. We discussed this immediately after the selection with Bob Anderson [president of Rockwell International], and with the leaders of the losing competitors. At the same time we began to receive pressure from the White House through Jon Rose asking us to put an immediate subcontract on Long Island with Grumman. Our response was that we could not do this on a sole source basis, that it wasn't right to go to Grumman only, but that we could justify going to both Grumman and McDonnell Douglas to maintain their sound manned space flight capabilities. However, to do this we would need an extra $20 million in FY 1973 expenditure limits. This was more or less on track before I went on vacation, derailed completely while I was on vacation, and Fletcher asked me to get into it again when I returned. The outcome was a white paper and two alternative press releases. . . . On September 14 the White House selected Option B and gave us the additional $20 million.[73]

Low had reason to feel the pain of Grumman and McDonnell Douglas. Those companies had been mainstays of all the piloted programs to date: Mercury, Gemini, Apollo, and Skylab. Yet in the highly competitive world of aerospace, a corporation was only as good as its latest proposal, and in seeking the contract for the shuttle orbiter, their proposals had come up short.

Nevertheless, Rockwell and NASA expected to help these firms. Rockwell intended to subcontract some 53 percent of the work, with its initial sole-source contracts accounting for less than 6 percent of the whole. In particular, Rockwell's Space Division would build the orbiter's nose and crew compartment, the forward fuselage section, and the aft fuselage that would accommodate the main rocket engines. But the firm planned to subcontract the design and fabrication of the wing, mid-fuselage, vertical fin, and orbital maneuvering system that would employ auxiliary rocket engines and propellant tanks.

"We want to avoid the illusion of activity by subcontracting development of subsystems too soon, then having to redesign them later," Charles Donlan, NASA's acting space shuttle program director, told *Aviation Week* in August. "We won't push subcontracting for subsystems faster than we push the design of the spacecraft." Even so, early in November he gave assent as Rockwell issued Requests for Proposal on three major structural components: wing, mid-fuselage, and vertical stabilizer.[74]

Nevertheless, there was wisdom in Donlan's words, for by then it was clear that the orbiter and shuttle designs were coming in badly overweight. Weight growth is common as detailed configurations mature; the design concept that wins a contract is often merely an outline of the complete working drawings, explicit in all particulars, that define the project when approved for construction. In turn, that final set of specifications addresses a host of engineering problems that arise during the design process, with the solutions to these problems adding weight. But weight costs money, and with the shuttle being held to a stringent budget, weight growth was unacceptable.

In its proposal, Rockwell had presented an orbiter concept with gross weight of 253,000 pounds. By November this was up to 277,500. The weight of the complete shuttle, including solid rocket boosters and external tank with propellants, went from 4,696,000 pounds to 5,410,500. The shuttle had to go on a diet.[75]

At the outset, attention focused on a pair of small solid rockets for use in abort. In an emergency, immediately after liftoff, the orbiter was to separate from its external tank and solid boosters. These Abort Solid Rocket Motors (ASRMs), thirty feet long and each with 386,000 pounds of thrust, would remain attached to the orbiter and were to push it to a safe altitude, allowing it to glide to a landing. Could the shuttle lose weight by deleting the ASRMs? This was no simple matter, for its implications ramified throughout the entire shuttle system. The ASRMs indeed added weight, but they promised to provide thrust that could offset this weight. To do this, the shuttle would fly with less propellant, shutting down its main engines after reaching a very low orbit and relying on the ASRMs for a boost to a somewhat higher orbit.

Related issues arose within the main solid boosters. Initial concepts lacked the ability to change the direction of thrust; that is, they lacked thrust vector control (TVC). This simplified their design. If the ASRMs were to be deleted, then the big solids would have to serve for abort. They then indeed would need TVC, complicating their design. Yet this offered advantage, for adding TVC to those solids would make the overall shuttle more controllable. Lacking TVC on the big solids, the shuttle could steer only with its liquid-fueled main engines. Adding TVC would allow the shuttle to steer with the muscle of those solids, which were far more powerful. They also were set well apart, giving very strong control moments.[76]

NASA performed extensive analyses of the merits of keeping or deleting the ASRMs. A NASA official, E. W. Land, noted subsequently that these studies concluded that those rockets were "not required":

> *The ARMs were useful for only about the first 30 seconds of flight and a "gap" existed in that period during the first few seconds while on, and just above, the launch pad.*

> *The ASRMs created complexities in overall vehicle design, operations tech-
> niques, abort mode selection logic, flight stability, mechanical systems, and soft-
> ware requirements which made the system marginal at best.*
>
> *Failure modes existed in the ASRM system itself which could cause loss of an
> otherwise nominal mission (e.g., failure to separate, inadvertent ignition) and pos-
> sible loss of the Orbiter (e.g., uncontrollable CG in the landing phase).*

Weighing heavily in the balance was an appreciation that the main solid motors indeed could serve for abort, even at low altitude. Land described them as "extremely highly reliable." He added that using these solids "to gain the necessary altitude and time to establish a flight trajectory back to the landing site for the Orbiter is considered a safer situation than the time critical, complex, and demanding flight control and systems requirements associated with the earlier abort solid rocket concept."[77]

The decision to delete the ASRMs came in November 1972. This promised to cut the orbiter weight by 7,000 pounds and reduce the weight of the complete shuttle by 99,000 pounds. The reason was that a lighter orbiter meant less weight in the solid boosters and external tank. This also brought an anticipated drop of $20 million in shuttle development costs, while reducing the cost per flight by $800,000. Deletion of the abort rockets thus was a good decision, for it cut cost as well as weight, without compromising safety, while introducing TVC that improved the system overall. But it still left the shuttle with an overly heavy orbiter and a total weight of 5,246,000 pounds. There was considerable interest in further weight reduction, and the key to this proved to be the wing.[78]

Early shuttle studies, in 1970, had emphasized a simple straight wing like that of a World War II fighter. It would have been light indeed, and would have solved the weight problem in a trice. From the start, however, the Air Force had insisted on a delta wing, to allow the orbiter to fly to the side of a straight-ahead track during reentry. Such "crossrange" was necessary to return to the Air Force space center at Vandenberg Air Force Base from a once-around polar orbit, which could conduct quick reconnaissance—or grab a Soviet satellite while in space.[79]

A delta wing was much larger and heavier than a straight wing. It also spanned more area, requiring far more thermal protection; this added weight as well. However, there are many ways to design a delta wing, and some approaches offered useful reductions in weight. Rockwell, in turn, made an early commitment to an extensive program of wind-tunnel testing, running to fifteen hundred test hours, with scale models of orbiter configurations that mounted various types of wing. This work defined an initial wing, which for a time seemed light enough.[80]

Four requirements defined the aerodynamics of an acceptable wing: to provide satisfactory control throughout the entire flight regime, from hypersonic reentry to landing; to minimize aerodynamic heating; to provide enough hypersonic lift-to-drag ratio to meet Air Force

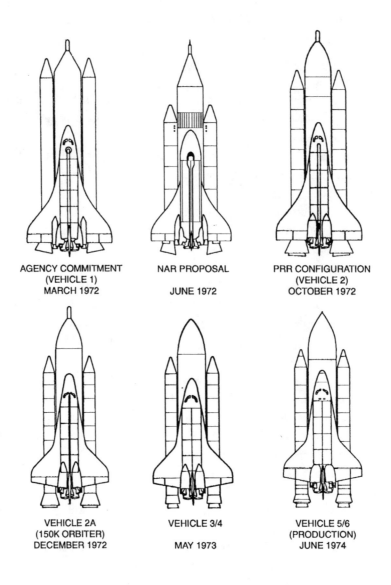

AGENCY COMMITMENT
(VEHICLE 1)
MARCH 1972

NAR PROPOSAL

JUNE 1972

PRR CONFIGURATION
(VEHICLE 2)
OCTOBER 1972

VEHICLE 2A
(150K ORBITER)
DECEMBER 1972

VEHICLE 3/4

MAY 1973

VEHICLE 5/6
(PRODUCTION)
JUNE 1974

Fig. 1. *Evolution of the space shuttle configuration during the first two years. (Art by Dennis R. Jenkins)*

demands for crossrange of eleven hundred nautical miles; and to provide low-speed lift for good performance during final approach and landing.

Landing requirements dictated the wing size. Excessive landing speed was unsafe and unacceptable; hence, the wing had to permit a design touchdown speed of 150 knots at sea level. In addition, concern for safety mandated some margin between landing speed and stall speed; hence, the latter could be no higher than 136 knots. The wing that resulted had an area of 3,220 square feet, with an orbiter landing weight of 170,000 pounds.

Landing considerations also determined the sweep of the wing's leading edge. Airplanes with swept wings land with the nose high and with marked angle of attack; this angle increases with leading-edge sweep. To prevent the tail from dragging on the runway, the aircraft needs long landing gear, which adds weight. For the shuttle, the maximum angle of attack at landing was limited to seventeen degrees. The wing weight minimized for a leading-edge sweep of sixty degrees, but this would have violated the angle-of-attack limit. However, the wing incurred only a modest weight penalty for a sweep of fifty degrees, with this penalty being offset by the permitted use of shorter and lighter landing gear. This sweep angle was the one chosen.

The wing that resulted was described as a "50° blended delta." It had a straight trailing edge with elevons, along with a gently curving leading edge that faired or blended smoothly into the fuselage. This minimized aerodynamic heating due to interference between the wing and body. The orbiter retained this wing through Program Requirements Review, a major design review in November 1972 that confirmed the need for a new wing that would cut the weight still further.

Rockwell and NASA reduced this weight by opting to accept a higher landing speed of 165 knots, as this increased speed permitted a smaller wing. The angle of attack at landing went down to fifteen degrees, which saved additional weight by further shortening the landing gear. The wing's basic shape, or planform, changed as well: from a blended delta to a "double delta," with a highly swept forward extension of the main wing. The sweep of the main wing diminished to forty-five degrees, which provided better thermal protection for the leading edge while also accommodating the reduced angle of attack at landing. The forward extension, a delta wing in its own right, was swept seventy-five (later, eighty-one) degrees.

The change in planform amounted to taking a bite out of the leading edge, thereby cutting the wing area. With the wing and the overall orbiter now reduced in weight, the wing could shrink further, for additional savings in poundage. The final wing had an area of 2,690 square feet, making it five-sixths the size of the earlier one. The orbiter dry weight fell from 170,000 pounds to 150,000.[81]

These changes brought a large measure of control to an overall shuttle design that had been getting badly out of hand. The solid boosters, for instance, had been baselined with a diameter of 156 inches, close to the limit of what could be transported via rail or interstate highway. Rockwell's proposal had called for these boosters to have a length of 150 feet. By

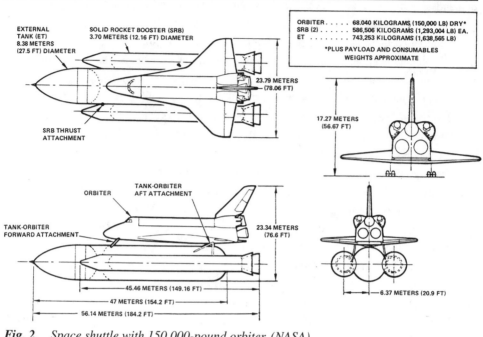

Fig. 2. *Space shuttle with 150,000-pound orbiter. (NASA)*

autumn, with the weights at their peak, this length was up to 185 feet, making them longer than the external tank. For a time there was serious discussion of increasing the diameter of the solid boosters to 162 inches, which was as wide as they could get and still fit, barely, under the nation's overpasses.[82]

But the lightweight orbiter brought a shuttle that was considerably more compact. Total weight, which had peaked at 5.41 million pounds, now went to 4.101 million, a reduction of nearly one-fourth and a considerable saving over the 4.7 million of Rockwell's proposal. The solid boosters shrank to a diameter of 142 inches and length of 145 feet, for a saving of close to one-third of their maximum. Cost per flight, which had been nudging upward toward $12 million, now fell back to $10.5 million, in line with NASA estimates of a year earlier.[83]

This weight-trimming exercise had amounted to a true redesign, assessing the merits of alternatives. Thus, in considering planforms other than the blended delta, NASA and Rockwell had considered adding canards to the orbiter, small wings placed well forward on the orbiter fuselage. This approach was rejected; canards promised performance advantages, but the double delta gave similar advantage and was simpler. In the wake of this work, there was good reason to believe that the shuttle was approaching its definitive form.[84]

"We are confident we will be able to meet the 150,000-pound gross weight for the orbiter which is a key factor in achieving the 4.101-million-pound gross liftoff weight for the total system," Rockwell's Bastian Hello told *Aviation Week* in April 1973. "Actually, we are below the 150,000-pound gross weight target for the orbiter and have a margin for the normal weight growth encountered in a development program of this type."[85]

The design saw further changes, notably the deletion of a docking hatch and airlock, which shortened the orbiter by three feet. Specified lengths and diameters changed slightly during subsequent years, whereas weight growth proved unavoidable; in 1978 the gross liftoff weight was up again, to 4.457 million pounds. But by August 1973, when Rockwell conducted its System Requirements Review as a second major design evaluation, the shuttle was virtually indistinguishable in appearance from the configuration that in time would fly.[86]

These technical developments went forward as Rockwell prepared to issue its principal subcontracts. In its enthusiasm, the company at first sought to conduct a series of subcontracting seminars as early as September 1972, less than two months after contract award. These conferences actually took place around Thanksgiving, following the Program Requirements Review, with this firm hosting symposia in Los Angeles, Boston, and Fort Worth, Texas. The meetings included detailed briefings on the latest shuttle configuration, as well as information on the business opportunities available in the program.

These seminars were timely, for the firm had already issued its requests for proposals on the orbiter's main structural elements. Rockwell had also set deadlines for submission of proposals: 15 December for the vertical fin, 10 January for the wing, and 24 January for the mid-fuselage. The symposium that followed, held at the Space Division on 21 and 22 November, thus amounted to a bidders' conference.[87]

Rockwell selected the winners by early March 1973. It then submitted its recommendations to NASA's Space Shuttle Program Office at the Johnson Space Center (JSC). Following reviews with Rockwell officials, the recommendations went to NASA Headquarters on 26 March for approval by Fletcher. His assent followed swiftly; three days later, the new subcontractors learned of their good fortune.

Grumman indeed had gone out of the space business following the final Apollo flight in December 1972, but this firm now was back in satisfying fashion. It took a subcontract, with initial value of $40 million, to design and fabricate the wing. *Aviation Week* stated that this was "expected to involve the main technical challenge." McDonnell Douglas, facing significant layoffs, found relief in a $50 million subcontract for the orbital maneuvering system. *Aviation Week* noted that this dollar award was the largest "because of the advanced technology involved in this system." It required its own rocket engines, which were to burn toxic and corrosive propellants; yet it had to be safe enough for use with a flight crew. There were other awards as well: $13 million to the Republic Division of Fairchild Industries for the vertical fin and $40 million to the Convair Division of General Dynamics for the mid-fuselage. This was the orbiter's largest structural element, for it included the payload bay.[88]

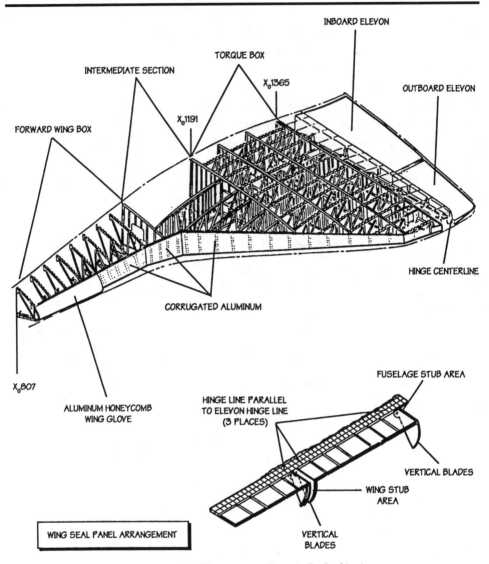

Fig. 3. *Orbiter wing structure. (NASA, courtesy Dennis R. Jenkins)*

Meanwhile, Rockwell was proceeding with its own orbiter elements: the nose, crew cabin and forward fuselage, and aft fuselage. These belonged to the company's Space Division; another company division, Rocketdyne, held a separate contract for the main engine. These arrangements reflected the experience of Apollo, wherein the Space Division had built the Apollo spacecraft as well as the second stage of the Saturn V, with Rocketdyne building the Saturn V's engines.

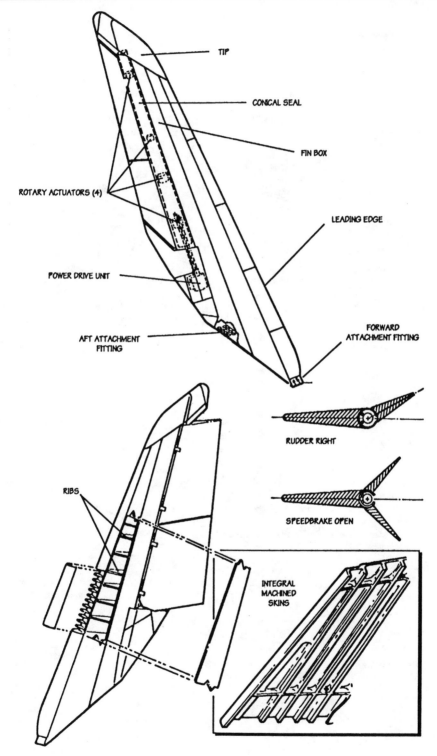

Fig. 4. *Vertical stabilizer. (NASA, courtesy Dennis R. Jenkins)*

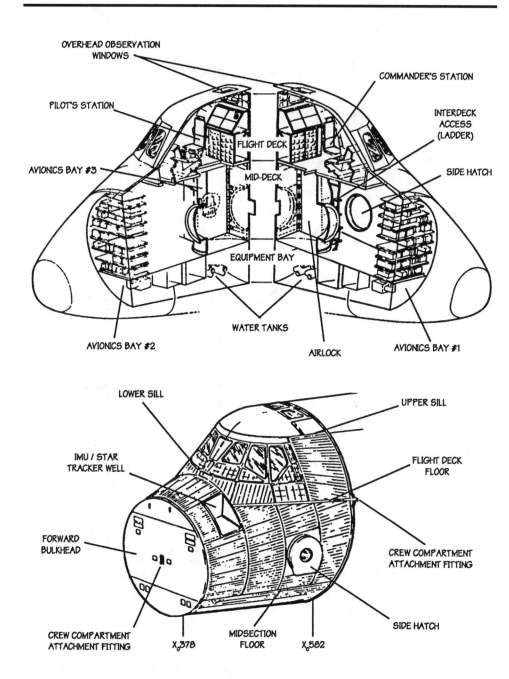

Fig. 5. *Orbiter crew cabin and forward fuselage. (NASA, courtesy Dennis R. Jenkins)*

The subcontract awards also reflected other firms' experience with Apollo and with earlier programs, along with their work on the design studies of space shuttles during recent years. Grumman had experience with piloted lunar flight to rival Rockwell's, for it had built the lunar module that actually carried astronauts to the moon's surface. Grumman also had shown outstanding originality in its shuttle designs, taking the initiative during 1971 as it recommended the use of expendable external tanks. This firm had actually come out ahead of Rockwell during the bid for the main contract, at least in its technical design, with Rockwell winning on points due to a superior management plan. Its wing subcontract was little more than a consolation prize, but it certainly was worth having.

McDonnell Douglas had its own splendid record: Delta launch vehicles, the Mercury and Gemini spacecraft; Aerothermodynamic/elastic Structural Systems Environmental Tests (ASSET), for hypersonic boost-glide reentry; the third stage of the Saturn V; and the Skylab space station. This firm also matched Rockwell in holding the best contracts for shuttle design studies during 1970 and 1971, Grumman's study contract being of a lesser character. But in competing for the main shuttle contract during 1972, McDonnell Douglas had found itself outclassed by both Rockwell and Grumman and came in third. Its subcontract award thus testified to the company's continuing strength.

General Dynamics was more of an also-ran. It had built the Atlas as the nation's first intercontinental ballistic missile (ICBM), with this missile finding continuing use as a launch vehicle. But it had never crafted a piloted spacecraft and had played no significant role in Apollo. It had been among the first to pursue space shuttle studies, but had failed to win significant study contracts after 1970. Its subcontract thus meant that this corporate division had made something of a comeback.[89]

During the subsequent two years, Rockwell continued to expand its roster of subcontractors. As of February 1975, eighteen projects were under subcontract, building elements of the orbiter and having value of $10 million or more (table 1.5).[90] Midway through 1975, some thirty-four thousand workers in forty-seven states were employed within the space shuttle program, working for NASA, the prime contractors, and the subcontractors as well. The buildup continued, reaching a peak of forty-seven thousand during 1977.[91]

Managing the Program

Pacing this activity on the corporate side, NASA already had a management structure in place that was providing supervision over the entire shuttle program. It had evolved over several years, growing out of the arrangements that had served Apollo.

Early in the 1960s, NASA entered the Apollo effort with a blank managerial slate. The agency had never taken responsibility for any program remotely as extensive or demanding,

---

Table 1.5.   Subcontracted Elements of the Shuttle Program

Element	Subcontractor
Flight control systems	Honeywell
Data processing and software	IBM
On-board computers	IBM
Mass Memory and Display System	IBM
Orbital Maneuvering System Engine	Aerojet
Auxiliary Power Unit	Sundstrand
Reaction Control System and APU propellant tanks	Martin Marietta
OMS/RCS Aft Integrated Module	McDonnell Douglas
Vertical Stabilizer	Republic
Wing	Grumman
Mid-fuselage	General Dynamics
Reusable Surface Insulation	Lockheed
Leading-Edge Thermal Protection	LTV
Cabin atmosphere revitalization and thermal heat transport	Hamilton Standard
Fuel-cell reactant storage	Beech Aircraft
Multiplexer-Demultiplexer	Sperry
Structural Test Program	Lockheed
Carrier aircraft modification	Boeing

Source: CASI 76A-18650: Peters, *NASA Management,* 3.

---

and responded with a major buildup. New field centers came to the forefront, notably Houston's Manned Spacecraft Center (MSC; renamed the Johnson Space Center in 1973) and the Marshall Space Flight Center (MSFC). They had different histories and vastly different levels of experience.

MSC grew out of the Space Task Group at NASA-Langley, which managed Project Mercury, the nation's first man-in-space program. The group's director, Robert Gilruth, became the director of MSC during the Apollo years. MSC went on to manage development of that program's piloted moonship, including the lunar landing module built by Grumman as well as the spacecraft built by North American.[92] By contrast, MSFC was virtually an empire in its own right. Its task was the development of rockets, big ones, and its director was the redoubtable Wernher von Braun. He and his rocket team had been crafting successively larger missiles and launch vehicles since the late 1930s and had served three masters: the wartime German army, the postwar U.S. Army, and NASA.[93] Historian Stephen B. Johnson notes that the managers at MSFC "understood rocketry and each other so well as to make formal coordination mechanisms redundant."[94]

Stated succinctly, the task of Apollo was to develop very large liquid-fueled rockets—the Saturn I-B, the Saturn V—and use them to launch the piloted spacecraft. This work brought about a natural division of NASA authority between MSC and MSFC. NASA Headquarters had overall responsibility for budgets, schedules, and broad program requirements. However, NASA administrator James Webb, who headed the agency during most of the 1960s, gave considerable leeway to Gilruth and von Braun within their respective domains.

With MSFC and MSC both ranking as powers holding authority second only to that of NASA's headquarters itself, post-Apollo planning envisioned a managerial future in which both centers would continue to play similar roles. In NASA parlance, each was to continue to serve as a "lead center." Then, following a tendency that is common to federal bureaucracies, post-Apollo programs showed a strong propensity for two-part programs that would serve these institutional needs.

Initial plans, during the late 1960s, focused on a space station with a space shuttle to provide its logistics. MSFC expected to build on current experience with Skylab, which had grown out of the S-IVB stage of the Saturn V, and to manage work on this station. MSC was to take responsibility for the shuttle.[95]

During 1969, a preferred shuttle concept emerged. It called for a two-stage, fully reusable configuration, each stage being piloted, winged, and powered by the liquid-fueled Space Shuttle Main Engine. In a series of meetings during September and October managers from the two centers worked out a joint agreement that extended the two-part approach to the shuttle itself. Gilruth and von Braun agreed that development of the two shuttle stages, an orbiter and a booster, should be handled by separate contractors. Separate centers then could manage each contract, with Marshall taking the booster and MSC handling the orbiter.[96]

At NASA Headquarters, shuttle activity fell within the domain of the Office of Manned Space Flight (OMSF). In 1970 its director, Dale Myers, set up a shuttle program office within OMSF, naming Charles Donlan as its manager. Donlan had spent his career at NASA-Langley; his experience with piloted space programs dated to the early days of Project Mercury. He asserted that MSC should not only hold responsibility for the orbiter but should serve as the lead center for the entire shuttle program as well.[97]

This drew on two lines of argument, as background. The first was a change in the role of Headquarters. Leroy Day, Donlan's deputy, recalls a strong intention to reduce the staffing in Washington by giving more responsibility to centers such as MSC:

There was a ground rule laid down that said we didn't—we weren't going to have any support contractors. So whatever we had in Washington, at least the program office there would be quite small; we would not be large. The comparison is that, at its peak, the Apollo program office in Washington, including all the support contractors, was about five hundred people, if you can imagine that. This included Bellcomm, General Electric, and a large number of Boeing people. We were not going to have any support contractors; we were going to have a lean program office in Washington. We would rely on the centers.[98]

Such an approach promised to draw on the experience that these centers had gained during Apollo, which qualified them to take more responsibility in program management.

The second argument stemmed from the nature of the shuttle as a flight vehicle. Donlan used this as his point of departure:

> *My argument for the management structure was this: the Shuttle is a system. It's a system that's composed of these elements: the Orbiter, the tanks, the boosters, and everything else, and ought to be designed not only as a system but managed as a system. This means that, at all times, the status of the Orbiter should be the principal consideration, i.e. what happens to it. It's the element of the system that's going to pay off. Anything that affects its weight or its performance has got to be weighed very carefully. That means if, for whatever reason, the Orbiter's size is to change and requires a change in the tank, that would have to be done, and so on, right through the system.*
>
> *So how do you manage something like that? My concept was to have prime management where the Orbiter was the prime element; that the boosters and the solids were subordinate elements. I structured a way in which Johnson [Space Center], as the Orbiter manager, would retain the prime field management for the system. And that Marshall would be delegated to carry out the management of the tank and solids, would indeed do that independently, but would report through the Orbiter office of the Shuttle Program Office, and similarly for KSC [Kennedy Space Center], where all the requirements for launch facilities were being delegated. That created quite a furor, initially. Finally, Dale bought the idea. And he and I went over and sold George Low on it. Then, Dale and I went and met with the directors and deputy directors of each of the three Centers and discussed the proposed management plan with them.*[99]

Within NASA as a whole, KSC and MSFC continued to stand on an equal footing with JSC, as all three centers—Kennedy, Marshall, and Johnson—remained within the purview of Myers and the OMSF. But under Donlan's proposed lead-center approach, shuttle project offices at the three centers were not to report to a central office in Washington. Instead, this central office was to be located at Johnson. JSC then would hold two levels of management: one for the orbiter alone and a separate one for the entire shuttle program. The latter, holding greater responsibility, would deal with the complete shuttle, including the booster, as well as with launch facilities.[100]

In George Orwell's *Animal Farm* all animals were equal, but some were more equal than others. Officials at MSFC did not want JSC to benefit from such pseudo-equality; they had no desire to stand subordinate to their counterparts in Houston. They shared their concerns with Senator John Sparkman, who represented their home state of Alabama. The final management plan, announced on 10 June 1971, mollified their criticisms.

Marshall received responsibility for both the booster stage and the main engine, which was to power the orbiter as well. In turn, JSC was confirmed as lead center for the shuttle as a whole. Myers stated that in this role, JSC would have "program management responsibility for program control, overall systems engineering and systems integration, and overall responsibility and authority for definition of those elements of the total system which interact with other elements." In addition, JSC held separate responsibility for the orbiter.[101]

These arrangements crystallized within a NASA Management Instruction, NMI 8020.18, issued in July 1971. At that time, NASA's preferred shuttle configuration called for a large liquid-fueled flyback booster, a concept that fell by the wayside in subsequent months. Still, the basic management patterns persisted amid the design changes. The shuttle continued to use an orbiter, for which responsibility remained within its Houston project office. The fall of the big flyback booster still left MSFC with responsibility for the main engine. When this reusable first stage gave way to the external tank flanked by a pair of solid rocket boosters, Marshall took on both new elements as well.[102]

Within NASA, the words "program" and "project" are not synonymous. A program is a coordinated major undertaking, within which two or more projects form the building blocks. Thus, the shuttle as a program embraced its solid and liquid rockets, the tank, the orbiter, and the launch facilities, all of which stood as distinct projects. The overall management structure reflected this, organizing this management at three levels.

Level I, at Headquarters in Washington, provided program direction within the OMSF. It dealt with overall issues of budget, schedule, and performance requirements, while allocating funds to the NASA field centers. The choice of solid boosters was a Level I decision, as was the commitment to a payload bay with dimensions of fifteen by sixty feet. The key positions were the associate administrator for manned space flight, who headed OMSF, and the space shuttle program director. Dale Myers held the first position, succeeded by John Yardley; the latter post was held successively by Donlan and by Myron Malkin.

Level II provided overall program management, within guidelines set at Level I. It concerned itself with integration, both of systems and of project managements, to ensure that the major components of the shuttle would go forward in parallel. For example, the main engine had to be ready when the orbiter needed it or the overall program would face a delay. Level II management resided not in Washington but at JSC, with Robert F. Thompson holding the top post as space shuttle program manager. Project managers, at Level III, reported to him.

Level III provided project management for each major element of the shuttle, such as the orbiter and the main engine. At this level, JSC had a project office for the orbiter, with Aaron Cohen as the project manager. Marshall held similar Level III responsibility for all major propulsion systems: the main engine, the solid boosters, and the external tank. NASA-Kennedy had its own Level III office, covering shuttle launch and landing.[103]

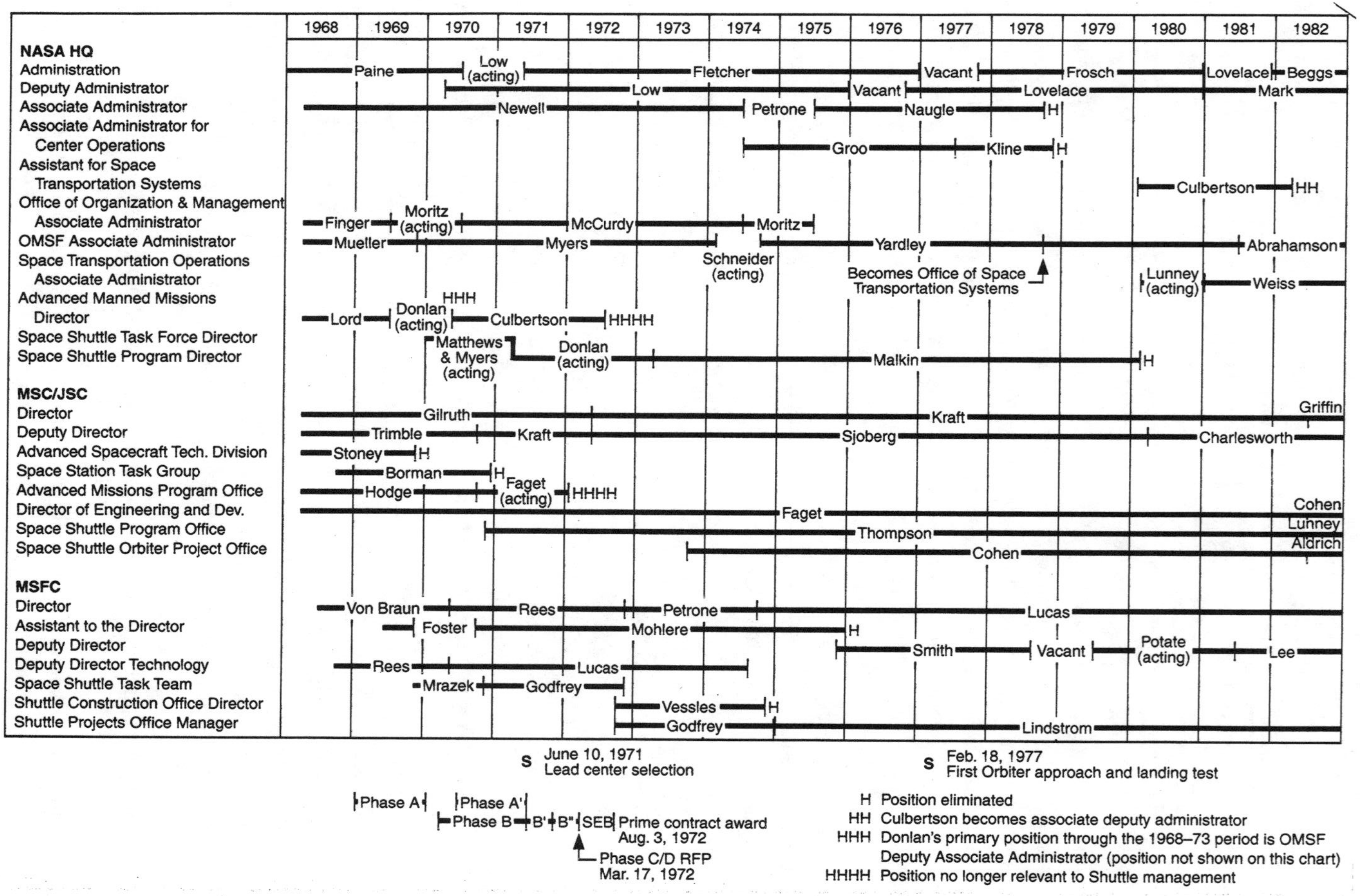

Fig. 6. NASA management during the post-Apollo and early space shuttle eras. (NASA)

Who were these people? Myers had spent much of his career at North American Aviation, the future Rockwell International, having joined this company fresh out of college in 1943. He worked in aerodynamics; he recalls using his high school German to translate captured documents on swept-wing aircraft. He rose to become project manager of Navaho, a Mach 3 cruise missile powered by ramjets. As early as 1956 he was a corporate vice president, responsible for Hound Dog, an air-launched cruise missile. In 1964 he became project manager for the Apollo manned spacecraft, for which North American held the contract. He came to NASA Headquarters in 1970 as head of OMSF, holding this position until 1974. He then returned to Rockwell, this time as president of its aircraft group.[104]

Yardley, his successor at OMSF, spent much of his career at McDonnell Aircraft in St. Louis. Following wartime service in the Navy, he joined that firm in 1946 and worked with aircraft structures. He directed design of the Mercury capsule, the first American spacecraft to carry an astronaut, and then did the same for Gemini. He managed launch operations for both programs between 1960 and 1964, then served as Gemini technical director during the following three years. A corporate merger with Douglas Aircraft formed the enlarged firm of McDonnell Douglas and brought him to a vice-presidency, with responsibility for Skylab. He took charge of OMSF in 1974, remaining in that post at NASA until 1981, then returned to McDonnell Douglas as its president.[105]

Donlan had spent his career within NASA and its predecessor, the National Advisory Committee for Aeronautics (NACA). He had joined its Langley research center in 1938, following graduation from Massachusetts Institute of Technology. In the realm of space flight he was present at the creation, participating in early studies of the X-15. He played active roles in Mercury, Gemini, and Apollo, joining OMSF in 1973 as its deputy associate administrator–technical. When the shuttle became a program in its own right and qualified for a program office at Headquarters, late in 1969, Donlan became its first director. He continued as space shuttle program director until 1973.[106]

Malkin, his successor, was another war veteran; he had served in the Marines. He matched his courage with a penetrating mind, earning master's and doctorate degrees in nuclear physics at Yale. He then joined General Electric, working there during the 1960s on missiles and launch vehicles, as well as on the Manned Orbiting Laboratory (MOL). When he came to Washington, in 1972, he returned anew to the military world, becoming deputy secretary of defense for intelligence. He transferred to NASA the following year, continuing to direct the shuttle program at Headquarters until 1980.[107]

Thompson, another wartime Navy man, also was a career NASA man. Like Donlan, he started his career at NACA-Langley, which he joined in 1947. He took aircraft stability and control as his area of specialty, winning progressively greater responsibility within Langley's Stability Research Division. In 1959 he became a founding member of Langley's Space Task Group, the nucleus both of Project Mercury and of MSC. In 1966 he became manager of

Apollo Applications, which developed into Skylab, remaining in this position until April 1970. In that month Robert Gilruth, director of MSC, selected him to head that center's Space Shuttle Program Office, at Level II. He remained there until he retired from NASA in 1981.[108]

Cohen, managing the orbiter project at Level III, had been with NASA since 1962. His academic background included work in mathematics and mathematical physics, along with a bachelor's degree in mechanical engineering. When he joined MSC in that year it still was rising from the Texas prairie. Between 1969 and 1972 he was MSC's manager for the Apollo piloted spacecraft, then headed JSC's Space Shuttle Orbiter Project Office from 1972 to 1982. Then, in 1986, NASA lost the Challenger. This brought a particularly strong need for seasoned managers, and Cohen responded by taking the directorship of NASA-Johnson.[109]

Also at JSC, separate from the formal shuttle organization, was a man no one could overlook: Maxime Faget. His early career followed a familiar path: wartime Navy service, in submarines; NACA-Langley, beginning in 1946; membership in the Space Task Group. Throughout the 1970s he was JSC's director of engineering and development. In fact, he was much more. He had spent much of his life crafting shapes for piloted flight into space.

He had started this work in 1954, as a member of a five-man study group that set forth the first concept for the X-15. Then in 1958 he presented the basic design of the Mercury capsule, with its blunt forward end covered with thermal protection for reentry. This shape, a cone with blunt base, served as well for Gemini and Apollo. In 1969 he gave a focus for studies of the nascent space shuttle, as he presented a highly influential concept for a two-stage, fully reusable system. It served as the point of departure for much work during the subsequent two years. He remained in the forefront of shuttle conceptual design between 1970 and 1972, turning out a succession of new concepts that guided the work of contractors.[110]

This was the organization and these were the men that managed the shuttle program. After 1971, the question was how well this system would work. There was an obvious prospect that the JSC Level II office would root for the home team, JSC Level III. At times this happened, with Houston approving its own facility requirements and disapproving those of Marshall. Still, it took several years of experience before an extensive management review was in order.

Yardley directed it, soon after taking over OMSF in 1974. Just after Christmas he wrote a letter to his superiors, Deputy Administrator George Low and Associate Administrator Rocco Petrone. He took particular note of discussions with Thompson and with Christopher Kraft, director of JSC: "Management surveillance and evaluation of sister center activities are very weak. R. L. Thompson doesn't believe it should be beefed up, but, rather, feels that once the technical direction has been given, it's clearly the other center's job to produce, and he feels little, if any, responsibility if problems arise. Chris Kraft weighed the question very carefully and concluded that JSC could not direct MSFC as a practical matter. This problem is further complicated by the fact that the two centers have entirely different management styles." The latter point reflected the continuing influence of von Braun's rocket team, in contrast to more formal procedures at JSC.

Yardley continued: "Lead center allocation of funds has a number of problems. . . . The sister centers are always suspicious of the objectivity of the lead center. The different management styles create problems here, also. In addition, the matter of fund recommendations for program support puts the program managers in a difficult spot between the center director and the program director, and required detailed headquarter review to assure objectivity." He recommended that "no major organizational changes be made, but that headquarters be reorganized to perform the missing management surveillance function and to play a more in-depth role in the fiscal area. This essentially means the 'lead center' becomes the 'lead technical center' which really more closely describes the role they play today."[111]

The changes that followed brought a partial return to the Apollo management system. The center directors, both at MSFC and KSC, already reported to the OMSF as a matter of course. They now won formal authority over their Level III projects, with their direct relation to OMSF allowing them to bypass the Houston office. Headquarters also strengthened its hand, for while existing Level II and III responsibilities remained largely intact at the centers, OMSF took a stronger role both in day-to-day project management and in reviewing budget recommendations from Level II at JSC. Under these revised arrangements, the space shuttle program went forward.[112]

There also was Level IV: the prime contractors. Though organization charts placed them low on the totem pole, they did far more than merely execute commands that came from on high. NASA's managerial structure had only a loose resemblance to a military chain of command, for there was a good deal of interplay between program offices and the contractors. In particular, Level I specifications were broad and general, leaving these companies with considerable leeway in how to comply. To find the most appropriate way, engineering groups engaged in design studies, with these studies' findings guiding decisions made at higher levels. When this happened, the relationship between NASA and its contractors could amount to a dialogue.

For example, the basic layout of the orbiter featured delta wings, with all propellants being carried in the external tank. That configuration dated to May 1971, when Maxime Faget of NASA-Johnson had drawn it up as a design concept and had offered it to contractors for study and evaluation. Lockheed, McDonnell Douglas, and North American Rockwell all gave it strong endorsements, and Grumman declared it was at least as good as any concept seen to date. With this, that configuration ceased to be merely another design option and became mandatory. Faget crafted a variant, designated MSC-040. In September NASA officials formally instructed these firms to adopt it and to use it as a basis for their ongoing studies.[113]

Subsequently, the shuttle's weight-trimming exercise expressed this dialogue anew. The pertinent directive called for a design with reduced weight, but the contractor, Rockwell International, was welcome to use its ingenuity in finding a solution. The redesigned wing became the key, and after its merits were clear, NASA again issued an instruction, directing that this was the design to use.

As the shuttle approached its final form, the program was broadening its reach. In addition to the orbiter, the shuttle needed its external tank and solid rocket motors, which were the subject of separate procurements. The Air Force showed an increasing involvement, particularly in preparing to construct its own shuttleport at Vandenberg Air Force Base. Concepts for shuttle upper stages also drew attention. In Europe, space program leaders prepared to cooperate with NASA in developing Spacelab, to conduct research within the shuttle's payload bay. Moreover, the fiercely independent French were preparing to develop their own launch vehicle, Ariane. It was to supplement the shuttle—and perhaps to supplant it if the Yankees were to falter.

CHAPTER  TWO

The Expanding World of the Shuttle

Midway through 1969, as astronauts Neil Armstrong and Buzz Aldrin were carrying the flag to the surface of the moon, NASA administrator Thomas Paine was looking ahead to bolder leaps for mankind. The dazzling success of Apollo 11 coincided with a high-level exercise in planning NASA's future, led by Vice President Spiro Agnew. Its report, issued in mid-September, sought nothing less than to build on the achievements of Apollo by setting sail for Mars. A piloted mission to that planet was to stand as NASA's long-term goal for the 1980s.

En route to Mars, space stations were to represent a major theme. A space station module, launched by Saturn V, would be "the basic element of future manned activities." Variants of this element were to provide stations in low Earth orbit, in geosynchronous orbit, and in lunar orbit, the latter providing a base for expeditions to the moon's surface. An assembly of several such modules was to form a space base with a staff of up to a hundred people, while "the space station module would be the prototype of a mission module for manned expeditions to the planets."

These stations and bases would demand logistics. Hence there was to be a space shuttle, reaching low orbit. A space tug would serve as the shuttle's upper stage "for moving men and equipment to different earth orbits. This same tug could also be used as a transfer vehicle between the lunar-orbit base and the lunar surface." In addition, a nuclear-powered rocket stage was to push space station modules into geosynchronous orbit and lunar orbit and, in time, to Mars.[1]

Paine expected that international cooperation would highlight these post-Apollo plans. In mid-October, flush with Apollo's success, he visited London, Paris, Bonn, and Rome, inviting leaders of those nations to work with NASA.[2] Europe indeed had a presence in space, but whereas the United States already was landing men on the moon, the Europeans were struggling to get off the ground.

Europe's Background in Space

No single European nation had the economic strength for a serious space program, at least during the 1960s. However, by pooling their limited contributions, nations of western Europe succeeded in pursuing several cooperative ventures. NASA helped as well, by providing launch services.

The European effort was small by American standards, for it avoided costly initiatives in piloted space flight. At a time when NASA officials looked forward with dread to budgets that might drop as low as $3 billion, European spacefarers were pushing forward with total annual expenditures of barely one-tenth that level. This sum of $300 million covered $90 million for the European Launcher Development Organization (ELDO), which was developing a new launch vehicle; $50 million for the European Space Research Organization (ESRO), which built scientific satellites; and $160 million for national programs. The largest such program, that of West Germany, was spending some $100 million annually.[3]

Both within ESRO and the national programs, small budgets supported development of a number of small spacecraft, which flew aboard U.S. launchers. Britain had been the first; its Ariel 1 satellite reached orbit aboard a Delta launch vehicle in April 1962. Most subsequent European payloads used the Scout, America's smallest launcher. The largest Scout-launched European satellite, Italy's San Marco 1, weighed only 660 pounds and flew in a low orbit. Still, from 1962 to 1972, twenty-one satellite launches succeeded for ESRO and the national programs, six of which used the more capable Delta.[4]

These projects enabled Europe to stake a foothold in the new field of space research. At international conferences, its scientists could do more than merely sit in the audiences and watch presentations by speakers from America and the Soviet Union. These specialists could give papers of their own. This hands-on experience made them attractive to the space superpowers, who at times included European instruments on their own spacecraft. In addition, Europe's home-grown scientific satellites encouraged thoughts of orbiting spacecraft that might provide direct benefits in telecommunications.[5]

During 1962 the United States took the lead in establishing the International Telecommunications Satellite Organization (Intelsat), a global consortium for satellite communications. From the start, Intelsat was powerful enough to bypass American Telephone and Telegraph Company (AT&T), which operated most of America's telephones. In 1962 AT&T had become the first corporation to launch a satellite, purchasing a Delta and placing its Telstar into orbit. Nevertheless, Intelsat acquired its spacecraft through competitive bidding, awarding its contracts first to Hughes Aircraft and later to the firm of Thompson Ramo Wooldridge (TRW).[6]

European representatives responded in May 1963, when they met in Paris and founded CETS, Conference Européen des Telecommunications par Satellites. This organization proceeded to represent Europe in negotiations with Intelsat. But in those discussions, and in subsequent dealings with the United States, CETS found that its hand was weak indeed. As one observer put it, "The only thing Europe had was the other end of the line."[7]

An American company, the Communications Satellite Corporation (Comsat), had been chartered by law to control U.S. work in this area. Comsat also served as Intelsat's operating manager, and by law it had the explicit responsibility to uphold American interests. Comsat controlled the procurements for Intelsat and was required to make them in ways that would promote the growth of American industry. In 1967 a committee of Parliament found that of $32 million awarded for the Intelsat III spacecraft, Britain's share was only $500,000. A year later the journal *Nature* found that Europe was obtaining only 4 percent of the value of Intelsat's contracts.

The obvious riposte was for Europe to build its own communications satellites. No single nation in Europe was large enough to need them, but such spacecraft could easily provide regional service, covering Europe as a whole. Television transmission drew particular interest, for a satellite, transmitting to receiving stations across Europe, could replace a cumbersome system of coaxial cables. The coverage could readily include northern Africa, where French influence remained strong.[8]

Europe, however, lacked its own launch vehicles. The United States also squelched hope of using NASA's by making it clear that any such European satellite could not compete with those of Intelsat. This became official American policy in September 1965, as President Lyndon Johnson approved National Security Action Memorandum NSAM-338, "Policy Concerning U.S. Assistance in the Development of Foreign Communications Satellite Capabilities":

> *It is the policy of the United States to support development of a single global commercial communication satellite system to provide common carrier and public service communications.*
>
> *The United States should refrain from providing direct assistance to other countries which would significantly promote, stimulate or encourage proliferation of communications satellite systems.*
>
> *The United States should not consider requests for launch services or other assistance in the development of communications satellites or other assistance in the development of communications satellites for commercial purposes except for use in connection with the single global system established under the 1964 Agreements.*

The memorandum also stated that any offer of launch services "should be conditioned upon express (written) assurances" that such launch support would be used only within the framework of Intelsat.[9] This policy statement loomed in the background during April 1967, as France and West Germany signed an agreement to develop their own communications satellite, Symphonie. It counted as an experimental system, which could qualify for American launch. It also had strong support in France because it would cover northern Africa.[10]

Would NASA launch Symphonie? The project directors wrote to NASA in October 1968, requesting launch services for two such spacecraft and emphasizing that they were to serve experimental purposes only. NASA's Paine did not believe it; his sources led him to believe that Symphonie had commercial objectives as well. Three weeks later, he replied, "We (NASA) would launch the two Symphonie satellites on a reimbursable basis if we could agree at a mutual understanding of the experimental character of the project." To qualify, Symphonie was to be "used exclusively for experimental and demonstration purposes, not for the transmission of regular commercial or governmental traffic or broadcasts." Arnold Frutkin, NASA's assistant administrator for international affairs, emphasized that Europe had to guarantee that Symphonie would never be used against Intelsat.[11]

This drove home the point that America would refuse to launch operational communications satellites for Europe and would launch spacecraft such as Symphonie only under tight restrictions. To the proud French, this was intolerable. "Symphonie made it clear to everyone that Europe needed its own launcher," Frederic d'Allest, a leading French space official, later told *Air and Space.* "We were not about to remain dependent on anyone for launch services. The issue was clearly sovereignty."[12]

France already had an active program of rocket development, sponsored by CNES (Centre National d'Études Spatiales), the national space agency. Beginning in 1960 it had carried through the design and test of a series of rocket stages named for precious stones: Emeraude, Topaze, and Rubis ("ruby"). Together, they formed the launch vehicle Diamant. It was small; even Scout outstripped its capabilities, for even in an upgraded version, Diamant could barely place three hundred pounds into a low orbit. Still, it made France the third nation, after the Soviet Union and the United States, to place a satellite in orbit. The first such launch took place in November 1965, with six more successes being tallied during the next six years.[13]

Diamant gave a useful education to French engineers, but the mainstream European launch-vehicle effort again was cooperative. Its point of departure was a British liquid-fueled missile, Blue Streak. With a planned range of twenty-five hundred miles, its performance stood midway between that of America's intermediate-range Thor and its intercontinental Atlas. In size, Blue Streak stood midway between those two Yankee missiles as well. It never achieved deployment as a weapon but won new life as the first stage for the three-stage Europa launcher. Development of Europa was the principal aim of the ELDO consortium.[14]

Blue Streak first flew in 1964, racking up an impressive run of successes in flight tests. For Europa, it mounted two upper stages: Coralie, built by France, and Astris, from West Germany. Both burned storable propellants, which simplified the designs and prevented propellant boil-off during long holds on the launch pad.[15] However, Blue Streak and Europa had been projects of the Tory government. The Tories, led by Prime Minister Harold Macmillan, strongly supported ELDO, but they lost control of Parliament to Labour in a general election in October 1964. The new technology minister, Anthony Wedgwood Benn, broke with the

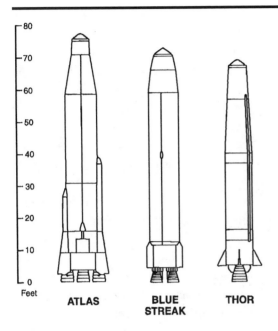

Fig. 7. *Britain's Blue Streak missile was intermediate between the United States' Atlas and Thor in size, range, and number of engines. (Art by Dennis R. Jenkins)*

Tory practice of favoring costly efforts in space and nuclear energy. With support from the new prime minister, Harold Wilson, he proposed to rely on the United States for such technology and to redirect the funding into social programs.

Within this general policy, Wedgwood Benn had specific reasons to swing the budget axe at ELDO. Its costs were escalating rapidly; an initial estimate of 1961 for development of Europa, $196 million, had grown to $400 million by early 1965. Yet with Blue Streak showing success in flight test, British industry had already reaped whatever technical advances it was likely to win. Nor did Wedgwood Benn fear a Yankee monopoly over satellite telecommunications, for the British were prepared to work within Intelsat. At the same time they had their own communications project, Skynet, which qualified for American launch because it was a military program. It thus bypassed the restrictions of NSAM 338, which applied only to civilian activities.

In June 1966 the Foreign Office announced that it was the intention of Her Majesty's Government to pull out of ELDO. This brought a storm of criticism, particularly from France. The French were eager to develop an uprated launch vehicle that could carry a communications satellite, and they declared that a pullout from ELDO could harm British prospects for membership in the Common Market. The British responded by accepting that they would remain within ELDO, under new and more favorable terms.[16]

Meetings at the ministerial level, during June and July, hammered out a compromise. France won a commitment to develop the new launcher, Europa 2. The new launch vehicle placed a solid-fueled fourth stage atop the existing three-stage Europa design and was to carry 150 kilograms, the weight of the Intelsat III satellite, to geosynchronous orbit. Britain won an important concession, reducing its share of the contributions that supported ELDO from 38.79 percent of its budget to 27 percent. As a further sweetener, Britain was awarded the contract for the inertial guidance system of Europa 2.[17]

Then, in November 1967, the British devalued the pound from $2.80 to $2.40. A new application to join the Common Market was currently pending, though people remembered that the highly nationalistic French premier, Charles de Gaulle, had vetoed an earlier attempt in 1963. Now he did so again, citing British economic problems as the reason. This new veto destroyed the argument that Whitehall should support ELDO for the sake of broader membership in the European community.

Following devaluation, Prime Minister Wilson was committed to large cuts in public expenditures. In April 1968 the British foreign secretary advised his French counterpart that Britain would accept no further financial obligations to ELDO, particularly as regarded Europa 2. Britain would do no more than to maintain existing levels of spending under the ELDO charter, which had the force of a treaty.[18]

A year later, after NASA had refused to launch Symphonie, a new conference of ministers made further concessions to the British. Their contribution to ELDO fell from 27 percent to 5.245 percent, payable in a lump sum in 1970. Italy, which had established good relations with NASA, left ELDO as well. France and West Germany covered most of the financial shortfall that resulted from the British and Italian decisions. Together, Paris and Bonn agreed to pay 87.9 percent of the cost of Europa 2. This was a prelude to those nations' subsequent collaboration that built Ariane.[19]

There already was new interest in a next-generation launcher. A 1967 review had looked ahead to the continued use of Blue Streak far into the future, flying it with increasingly capable upper stages. If Britain had accepted that report and had made it a basis for policy, then it would have had orders galore for this first stage, on a silver platter. But with the British on their way out of ELDO, they would play no role in defining future launch vehicles. They were willing to continue building Blue Streaks and to sell them to ELDO, but even this was equivocal. Whitehall would guarantee the continued delivery of Blue Streak only "for a limited number of years." In November 1968, pressed strongly by French delegates, a British minister promised no more than to supply Blue Streak and its components "at least up to 1976."[20]

Why did Whitehall not understand its value? The Labour Party was in a position to continue to build it for export, but saw no reason to anticipate large demand. NASA's launch vehicles were already available for most purposes, particularly when launching scientific payloads. The standard Europa had already lost its prime payload, ESRO's Large Astronomy

Satellite, which had been canceled due to budget pressures. Europa 2 had only two orders, for Symphonie. Subsequent decades brought a surge in demand for launch services, along with the rise of a commercial launch-vehicle industry. But in 1968 it took an act of faith to believe in a future for Blue Streak, within a government that was now in massive retreat from rocketry.[21]

As Blue Streak faltered within the government, it succeeded in further flight tests of Europa. In August 1967, having performed well in five earlier launches, Blue Streak carried a live Coralie second stage. Again this first stage flew properly—and Coralie refused to ignite. ELDO tried anew in December, again with live first and second stages. The reliable Blue Streak flew as planned, but Coralie once more failed to fire.

ELDO tried again a year later. The third stage, Astris, now was ready, and with it, a complete Europa flew for the first time with all three stages live. Blue Streak worked properly, again. Coralie also separated and ignited at altitude, winning its own success. Astris also ignited—and shut down after only seven seconds. Much the same happened on the next flight in July 1969. Again the first two stages operated as planned, but this time Astris did not fire at all.[22]

This was the situation in the fall of 1969, when NASA's Paine came to Europe. His visit raised hope of a cooperative program that could span the Atlantic, carrying prospects for broadened access to American launch services. Europe then might launch operational communications satellites after all, even if its own rocket-development efforts continued to falter.

Europe and the Shuttle

Paine pursued his expansive views on the space program all through 1969. This expansiveness extended to hope of greater international collaboration. In February, having learned that Nixon was interested in this topic, Paine sent him a six-page letter, single-spaced, on European efforts in space.[23] A month later the new secretary of state, William Rogers, prepared a memorandum for the president that included a staff report on international cooperation in space.[24]

Then in July Nixon flew out to the Pacific to greet the returning Apollo 11 astronauts, immediately following splashdown. Paine accompanied him, thus gaining a rare opportunity to talk at length with the president. They particularly discussed the prospects for international cooperation in space flight, and Paine came away believing that Nixon would support new initiatives in this area.[25]

Three weeks later, Paine met with the physicist Hermann Bondi, who was director general of ESRO. In a letter to Nixon written that same day, Paine stated that Bondi wanted "top NASA personnel" to make "a series of presentations . . . to senior space officials in Europe within the next few months to raise their sights to more advanced projects of greater mutual value." Paine's letter then outlined his own plans for such briefings and gave an update on current issues in European space policy.[26]

During Paine's visit to Europe in October he addressed the European Space Conference, a minister-level body embracing ELDO, ESRO, and CETS. He laid out the plans of Spiro Agnew's Space Task Group: space shuttle, space station, nuclear rockets. He emphasized "the desire of America not only to continue but to expand the cooperation," adding that "we will welcome your suggestions as to new means whereby we can achieve a greater degree of cooperation between our proposed space programs and your own plans for European programs."[27] With this, the historian Kevin Madders writes that "a Europe that had serious hesitations about taking on basic satellite applications was therefore invited to join NASA on a road signposted towards the Moon and Mars."[28]

Paine made his presentations at a time when studies of post-Europa launch vehicles were blossoming. France was in the forefront. Naturally, the French favored their own industry; they had no wish to rely on perfidious Albion. An initial concept retained the upper stages of Europa 2 but replaced Blue Streak with a new first stage, the L-95, holding ninety-five metric tons of propellant. It mounted four engines designated M-40, each with 40 metric tons, or 88,000 pounds, of thrust. This engine was already entering test.

The M-40 was another French product, built by the Société Européene de Propulsion (SEP). This engine contrasted sharply with that of Blue Streak, built by Britain's Rolls Royce. That engine had drawn on technology from Rocketdyne, purchased under license, and used Rocketdyne's standard propellants: liquid oxygen (LOX) and RP-1, a highly pure form of kerosene. The SEP M-40 echoed the engine of Coralie by using storables: unsymmetrical dimethyl hydrazine (UDMH) and nitrogen tetroxide, N_2O_4. American manufacturers had already made this change, with the Titan I, powered by LOX and RP-1, giving way to the more powerful Titan II, which burned similar storables.[29]

A conference in April 1969 took a sharp turn in the direction of this new approach. Government ministers, attending this meeting, set up the new and powerful Directorate of Future Activities. Its head, Jean-Pierre Causse, headed a field center of the French space agency; he had conducted the 1967 review that had looked ahead to continuing use of Blue Streak. He now took the post of deputy to the ELDO secretary general with full design authority over future projects. He set to work on studies of new-generation launch vehicles, both with and without Blue Streak. He retained a strong preference for this British booster, but in May 1970, ELDO's governing council dropped such concepts in favor of a new one that broke all ties to the existing Europa program. This was Europa 3. It was intended to rival NASA's Atlas-Centaur, placing 1,500 pounds in geosynchronous orbit.[30]

Europa 3 was a two-stage rocket. Its first stage resembled the L-95 but was to carry over 50 percent more propellant. It also was to mount four engines, from SEP, each with 132,000 pounds of thrust. The second stage would burn hydrogen; its engine, with 44,000 pounds of thrust, was to result from a partnership between SEP and West Germany's Messerschmitt-Boelkow-Blohm (MBB). The payload, a communications satellite, included its own engine

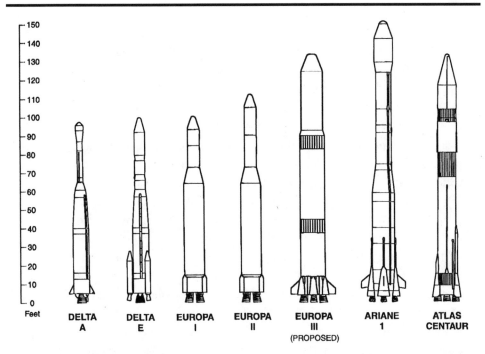

Fig. 8. *The Europa 1 and Europa 2 launch vehicles were comparable to NASA's Delta launchers. The proposed Europa 3 and the subsequent Ariane 1 compared with the Atlas-Centaur. (Art by Dennis R. Jenkins)*

as a third stage. Here was truly a leap into the future, for a version of that first stage would power Ariane.[31]

Meanwhile, Europa was still in development, and on 12 June came the flight that people would remember. Blue Streak fired for 157 seconds, as planned. Coralie then scored its third straight success as it burned for 105 seconds, again as planned. Now it was the turn of Astris—and this stage worked, firing for a full 366 seconds. (One pictures the engineers in the control center, watching their instruments with mounting hope as these minutes ticked by.) All three stages operated properly, which should have placed the 573-pound payload in orbit.

It did not. A 661-pound payload shroud had failed to separate when scheduled, 200 seconds into the flight, and had ridden with the rocket. It was much as if this test had attempted to carry a payload of double the rated weight. It fell short of orbital velocity by twenty-six hundred feet per second. With this, the Europa program indeed had demonstrated all the elements of a successful ascent to orbit—though not on the same flight. This latest attempt had come close, but full success continued to elude their grasp.

51

This was the last launch of the standard Europa, also called Europa 1.[32] Attention now turned anew to Europa 2 and 3, providing a background against which Europe's space leaders shaped a response to Paine's initiative. ELDO and ESRO started by setting up a joint working group to examine the possibilities with Jean-Pierre Causse as a co-chairman.

The space tug drew interest. Paine had not emphasized it in his presentations of the previous October, but it was very much a part of NASA's post-Apollo plan. It was to burn liquid hydrogen; hence it might provide experience that would be useful in designing the second stage of Europa 3. In July 1970 ELDO awarded contracts totaling $500,000 for early space-tug studies. These went to Britain's Hawker Siddeley and to the partnership of MBB and SEP, the same association that was studying the Europa 3 second stage.[33]

West German firms also participated actively in detailed studies of the space shuttle. In Munich, MBB teamed with North American Rockwell and examined attitude-control systems. Dornier, in Friedrichshafen, worked with Grumman and addressed several items: rocket-exhaust impingement on structures, separation of external propellant tanks during ascent, and the attitude-control problems of using jets with aerodynamic controls during reentry. In Bremen, ERNO Raumfahrttechnik collaborated with McDonnell Douglas by building models of that firm's space shuttle configurations and testing them in wind tunnels.[34]

Paine kept Nixon informed of these developments. In June he wrote another letter to the president: "ESRO has committed funds to technical studies of a space station module which might constitute an integral element of a future U.S. space station. ELDO has similarly committed funds to studies of an orbit-to-orbit 'space tug' which would supplement the space shuttle. . . . Nearly one million dollars of European funds are involved in these two programs which directly follow up our proposals." Paine, however, cautioned that Europe wanted something in return: "We must clear the way for assurances to our prospective partners that the United States is prepared to agree to their purchase, for peaceful purposes consistent with international agreements, of the launch vehicles or launch services that they will have helped to develop (and, until such launch services become available, the purchase of existing NASA launch vehicles or services.)"[35]

By then Paine's grand plans were receiving rude shocks, with both the OMB and Congress erecting large stop signs on the road to Mars. Pressed by budget cuts, NASA scaled back its plans to a single space station served by the shuttle. Moreover, this station would not be launched in one piece. Instead it was to be assembled from modules carried aloft by the shuttle. However, even this modular station lay well off in the future; the shuttle would come first.

As the classic space station faded, attention turned to a variant of these modules, which could ride in the shuttle's payload bay. Within NASA and its contractors, this was called a Research and Applications Module (RAM), a Sortie Can, or, more formally, a Sortie Laboratory. Europeans, who eventually built it, gave the name by which it would become known: Spacelab.

52

Initial thoughts, in 1969, called for small laboratories that could ride within the shuttle and extend its usefulness. Such concepts soon drew on studies of modular space stations. As elements of a working station, the modules were to rely on a central core for attitude control, power, storage of consumables, crew quarters, and a control center. The shuttle's orbiter itself could provide all these services. This raised the prospect that individual laboratory modules, carried within the shuttle, could function as small and temporary space stations.[36]

Work at NASA's Ames Research Center gave impetus to this idea. The center operated a fleet of aircraft that served as observation laboratories, including a Lear Jet and a Convair CV-990 airliner. The latter found use in weather surveys, infrared astronomy, and land-use studies conducted using multispectral photography. People expected the shuttle to fly frequently and routinely. Experience with that 990 encouraged thoughts of using the shuttle as a similar flying research post.

NASA's Marshall Space Flight Center took the lead in studying such concepts. It was known as a center for the development of large rocket stages, having managed the design and flight test of the Saturn I-B and Saturn V, together with their engines. During the late 1960s its purview broadened to include space stations, for it took responsibility for Skylab, with an orbiting workshop that grew out of a spent S-IVB stage. When Apollo and Skylab reached completion, when the big space station did not materialize, and as the shuttle came to the forefront, Marshall was left with little more than components of the shuttle: its main engine, its solid boosters and external tank. Spacelab became a welcome addition to this center's portfolio. The new project kept it more nearly on a par with NASA-Johnson, while drawing on Marshall's experience with Skylab and with space-station studies.[37]

As preludes to Spacelab, several preliminary-design efforts set the pace. A study at Convair, directed by Marshall, looked at experiment modules for a space station and went on to examine the feasibility of operating such modules from the shuttle, in the absence of a station. Convair studied such modules in greater detail during 1971 and 1972, placing equal emphasis on shuttle-based operations and on use with a space station. This work envisioned an evolutionary approach, with the shuttle being available first.

McDonnell Douglas carried out a related study called SOAR, Shuttle Orbital Applications Requirements. This was a broad investigation of a number of types of shuttle payloads: modules for on-board scientists, instrument-carrying pallets or platforms, spacecraft, and upper stages. At Marshall's Preliminary Design Office, another important study took place during the autumn of 1971. It focused entirely on shuttle missions with a crew-tended laboratory in the cargo bay. This effort, called Sortie Can, drew on the SOAR and RAM work and sought conceptual designs that could accommodate a wide range of experimental objectives. The study team concluded that the cylindrical "can" should have a length of twenty-five feet, less than half that of the payload bay, and should carry a pallet in the form of an open truss, to mount instruments outside the pressurized volume.[38]

Meanwhile, both in Europe and within the White House, people were learning that the path of international collaboration might not run smoothly. In seeking transatlantic cooperation, Paine hoped to broaden the base of political support for his post-Apollo program. He also hoped that Europe would share the expense, putting up as much as $1 billion within a $10 billion program. But in February 1971 a memo from the presidential assistant Peter Flanigan to the senior presidential advisor John Ehrlichman raised questions: "We have not yet decided what we want our post-Apollo program to be or how fast it will go, but if NASA successfully gets a European commitment of $1 billion, the President and the Congress will have been locked into NASA's grand plans because the political cost of reneging would be too high." Flanigan's memo also raised the issue of "technology outflow," whereby the United States might sell its crown jewels cheaply: "Finally, the U.S. trade advantage in the future will increasingly depend on our technological know-how. The kind of cooperation now being talked up will have the effect of giving away our space launch, space operations, and related know-how at 10 cents on the dollar."[39]

Moreover, people in Europe were learning that working with NASA might not necessarily make their dreams come true. They had entered the post-Apollo effort amid high hopes of access to U.S. launch services and to American technology. But as historian Lorenza Sebesta notes, they found themselves obtaining much less (see table 2.1).[40]

During 1972, decisions within the White House and State Department substantially reduced the scope of possible European involvement. NASA had nurtured thoughts of having Europe build structural components of the orbiter itself, offering five possibilities: elevons, vertical fin with rudder, landing gear, nose cap, cargo bay doors. John Walsh, a staff member of the National Security Council, set up an interagency subcommittee that examined these prospects.[41] In mid-February he wrote a memo to Herman Pollack, the State Department's top official for science and technology:

> *That simply subcontracting to Europe, on a "no transfer of funds" basis, the design, development and fabrication of major pieces of the orbiter entailed great risks of contention which would, in fact, be counterproductive to improved foreign relations. Such contention normally develops between prime and subcontractors because of disputes over responsibilities for delays, performance shortfalls, overruns, changes of scope and the like. It is a fact of life of U.S. business, but may not be recognized as such by European businessmen. Moreover, since such prime-subcontractor arrangements would take place under government-to-government umbrella arrangements, any disputes would soon be escalated to the government level. Therefore it was agreed that, ideally, European development participation should be limited to clearly separable systems such as the tug, sortie cans, and research application modules.*[42]

Table 2.1. European and American Objectives in the Space Shuttle Program

European Objectives	U.S. Objectives
Development of elements critical to the shuttle	Avoidance of reliance on Europe for success of the shuttle
Participation in decision making at all levels of management; shared management	Final decisions retained by the United States
Access to all the technology developed within the post-Apollo program	Detailed access to technology necessary for European contribution; general access for the rest
Commitment by the United States to procure from Europe the hardware developed on a permanent basis	Freedom from any formal commitment, linked to financial constraint coming from Congress
Privileged access to the post-Apollo system, with special pricing; unrestricted availability	Normal access on a commercial basis
Each prime contractor to be responsible not only for technical management and direction of subcontractors but for their funding	Participants, at governmental level, would bear full financial responsibility for their part of the work
Exchange of funds	No exchange of funds

Source: Sebesta, "Blueprints," 52.

The memo continued to hold open the option of European development of the space tug. A month later, a memo from the National Aeronautics and Space Council, addressed to NASA's Fletcher, argued against the tug:

> *Any advantages to the U.S. of a European tug project seem to be more than offset by several disadvantages: the probability of Europe producing an unacceptably low performance system, the likelihood of technology outflow, the enhancement of their own booster capability, the dollar outflow to buy production units (perhaps up to $500 million), and the difficulty in accommodating DOD's unwillingness to rely on a foreign supplier.*
>
> *Concerning the other side of this option, the advantages of Europe developing the sortie can version of RAM is that the task clearly can be within their capabilities, has minimum risk of technology transfer, could contribute a useful element to the post-Apollo program, and has no military implication.*[43]

Late in April Secretary of State William Rogers wrote a memo to Nixon:

Within the last several months U.S. views that we should minimize European participation have begun to harden. These views hold that we should not permit European participation in development of the Shuttle because of domestic economic considerations and the difficulties of sharing such a task with foreign governments and subcontractors. With respect to the Tug they hold that the development task will be too difficult technically to rely on European performance. European participation would thus be limited to development of one or more of the RAMS.

Commenting on Rogers's memo, Fletcher concurred: "Our preferred objective is to obtain European agreement to develop a specified type of sortie module for use with the shuttle, reserving other types for our own development. . . . We agree with the Secretary's letter that the tug requires further study. It is, therefore, a distinctly second choice, and much less desirable."[44]

Fletcher presented this response to Henry Kissinger, the national security advisor, on 5 May. On 1 June Kissinger replied with a memo of his own, and George Low summarized its content: "After more than a year of indecision, we finally received a memo from Kissinger concerning the extent of European post-Apollo cooperation. Specifically, the memorandum indicated that we should seek participation in the Sortie Module, should deny the tug to the Europeans, and should discourage but allow essential participation in bits and pieces of the Shuttle. This is the package that Jim Fletcher and I had hoped for, and with the exception of the bits and pieces of the Shuttle, it is exactly how we would like to handle it."[45] Two weeks later, Pollack hosted a meeting in Washington with European space officials and delivered the news: no tug for Europe, no participation in the shuttle orbiter. The Europeans were told, diplomatically, that the United States had not decided whether to proceed with a tug and that, therefore, they could not share in such a project. This was true as far as it went, but there were other reasons as well.[46]

The tug was to be a high-performance upper stage, somewhat resembling Astris, but would burn hydrogen where Astris had used storable propellants. Though storable rocket technology was relatively straightforward, Astris had not exactly covered Europe with glory, and liquid hydrogen would be considerably more demanding. Hence to succeed with the tug, Europe was likely to require technology outflow, wherein the United States would share its expertise in hydrogen. This expertise was proprietary. NASA was using it to good advantage in the Centaur upper stage and expected to use it again in the shuttle, and the State Department did not want to give it away.[47]

Objections from the Pentagon also doomed a European tug. "There was no way that was going to happen, not from NASA's standpoint but from the military's standpoint," notes William Lucas, a director of NASA-Marshall. "That tug was to serve both NASA's interest and the military's payload interests. The military certainly would not have been willing to have a foreign entity that they had no control over to be in the loop as far as their payloads were concerned."[48]

Many of the Europeans had regarded the tug as their most attractive choice, for it was technically challenging and promised to stretch the talents of their industries. Its sudden withdrawal came as a highly unpleasant surprise.[49] But when the sortie lab remained as their only option, the Europeans accepted it with a will.

The firms of MBB and ERNO, along with Sweden's Saab-Scania, already had some background, for they had participated in the studies of Convair. Now, late in June, a NASA technical team visited ESRO's technology center in the Netherlands and shared the findings of American studies. This launched an active collaboration between NASA and ESRO. In mid-1972 ESRO awarded contracts for feasibility studies, at $250,000 each, to industrial groups led by MBB, by ERNO, and by British Aerospace. Late in the year, ESRO raised the ante tenfold, awarding $2.5 million contracts to the three consortia. These funded more detailed efforts that defined their respective designs.[50]

Spacelab, Ariane, and ESA

Studies of sortie labs offered prospects for the future, but the ongoing present was more concerned with communications satellites and with European launch vehicle development. In December 1970 France had announced that it intended to leave ESRO. The only way that it would remain a member would be if this consortium were to switch its emphasis from science to applications. Giampietro Puppi, chairman of ESRO's governing council, responded by searching for a compromise that indeed would broaden ESRO's role.

In December 1971 this search led to a package deal whereby ESRO undertook three new projects: a communications satellite, a weather satellite, and a spacecraft for use in air traffic control over the Atlantic. This compromise kept France within the fold while pointing ESRO toward a new future. Its plans now called for its budget to double by the mid-1970s, with applications rising to take two-thirds of the expenditures. ESRO's science programs already were in decline, for those of the national space programs were better funded. This shift in emphasis gave ESRO new activities and a new purpose.[51]

It also coincided with a softening in U.S. launch policy. This followed negotiation of the definitive Intelsat international agreement in August 1971. Article XIV stated that any nation within Intelsat, seeking to set up satellite communications separate from those of Intelsat, was obliged to "consult with the Assembly of Parties, through the Board of Governors . . . to avoid significant harm to the global system of INTELSAT."

On 1 September Under-Secretary of State U. Alexis Johnson wrote a letter to Theo Lefevre, chairman of the European Space Conference. This letter was noteworthy; a year later, parts of it appeared almost verbatim in a White House policy statement. He wrote that even if Intelsat refused to support the launch of a competing system, the United States could nevertheless provide launch services if the requesting organization "considers in good faith that it has met its relevant obligations under Article XIV." In addition, Johnson opened a loophole in Article XIV. He wrote that a "possible operational system of European communication satellites," which he had already discussed with Lefevre, "would appear to cause measurable, but not significant, economic harm to Intelsat." Therefore, "we would expect to support it in Intelsat."[52]

What brought this change of heart? Within Washington, there was growing acceptance that America could not maintain a monopoly by denying launch services to Europe. Not only were European launchers in development; such a heavy-handed use of export controls would actually work against U.S. interests. In the words of a NASA paper, "The health of Intelsat is assured in part by the feeling of the major Intelsat partners that they are indeed partners and not puppets in an organization dominated by the U.S." Another official predicted that if Washington fought too hard against Europe, "the international system will be the one which breaks up and fails."[53]

Still, if Europe hoped for more than the kindness of strangers, it was well advised to continue to develop its own launchers. Much hope rested on Europa 2, which made its first flight in November 1971. One minute and forty-seven seconds after liftoff, the guidance failed. The rocket pitched over and broke apart due to aerodynamic forces. Tracking films, taken through telescopes, showed a white cloud that blossomed from the Blue Streak as it exploded, accompanied by a red-orange fireball from the Coralie and Astris.[54]

Like all such failures, this one was given careful attention and was the subject of a formal investigation. The chairman of the inquiry, Robert Aubinière, had headed the French space agency, CNES, and had just taken over as director general of ELDO. His panel took a close look at the guidance system, which was new. They also reexamined the three liquid-fueled stages. They were the stages of the standard Europa and had been built respectively by Great Britain, France, and West Germany.

The report from this group pulled no punches. It stated that the problems of both Europas stemmed largely from the policy that gave each of these stages to a separate nation. Hence there was no centralized program management that could coordinate the work of companies in different countries. Many ELDO contracts were deficient in such areas as completeness of specification and procedures for freezing a design.

The loss of that Europa 2 had resulted from failure of the guidance system computer in the third stage. This computer was a prototype that was used nowhere else; lack of operational experience meant inevitably that it was prone to defects. The third stage had electrical systems from two German manufacturers; their connections obeyed "none of the elementary rules concerning separation of high and low level signals, separation of signals and electrical power supply, screening, grounding, bonding, etc." No one took responsibility for this. Aubinière's commission concluded that "in its current configuration," Europa 2 was "unflightworthy." He nevertheless was confident that the program could succeed—if ELDO were to institute centralized management and if Europa received more funds.[55]

While Aubinière was attacking both the design and the management of Europa 2, the new American communications satellite launch policy undercut its rationale—along with that of Europa 3—by holding out the prospect of easier access to NASA's launch services. The new policy left France unmoved. Its officials wanted to place Symphonie over the Atlantic so they could use it to transmit French-language programs to Quebec and French Guiana. Intelsat was willing to lease channels on its own satellites for this purpose, but it wanted Symphonie placed over Africa, serving only the Eastern Hemisphere, and with regular broadcasts ruled out.[56]

France had broader reasons to distrust the Yankees, for Paris now had been left holding the bag on two separate occasions. The first, in 1968, had denied NASA's launch services for Symphonie, which now appeared to be available only under unfavorable restrictions. The second and more recent such event had ruled out participation in developing the shuttle or tug.

But the Germans, partners in Symphonie, found much of interest in the new policy from Washington. Despite the objections of Intelsat, they believed, correctly, that NASA indeed would launch Symphonie. (NASA went on to launch two such spacecraft, which reached geosynchronous orbit in December 1974 and August 1975.)[57] More broadly, by late 1972 the Germans were arguing that America was likely to refuse to launch only a very small number of satellites, if any. At a ministerial meeting in November, the German delegate, strongly supported by Italy, declared that his government saw little reason to proceed with Europa 2, particularly in the light of the Aubinière report. Further, Europa 3 now did not appear worth the effort.

The Netherlands had already joined Italy and Britain in declining to support Europa 3. Now West Germany was also ready to decline, leaving only France and Belgium as active proponents of independent European access to space.[58] Yet as Europa 3 faltered, France stepped forward with a new concept called L3S, Lanceur 3ème Génération Substitut, "third-generation substitute launcher." The world would know it as Ariane.

L3S retained the first stage of Europa 3, with its four engines from SEP. Designated Viking 2, this engine was already on the test stand as a long-lead item. It burned the standard French cocktail, the storable propellants UDMH and N_2O_4. The Europa 3 second stage would have burned liquid hydrogen, but L3S dispensed with it. Instead, its design called for a new second stage, also using storable propellants, powered by a single Viking 2 engine. L3S now needed a third stage, which indeed was to burn hydrogen. However, it was to be smaller than the planned Europa 3 second stage and have considerably less thrust. That promised to make it easier to develop.

Studies of L3S had begun in June 1972, through a collaboration between the space agency CNES and the missile-development branch of the French Defense Ministry. These studies accelerated during subsequent months, as Europa 2 and 3 lost backing. L3S promised the capability of Europa 3, placing seventeen hundred pounds in geosynchronous orbit but at a cost of development of only $440 million, two-thirds that of Europa 3. This saving would result from the large degree of commonality in its first two stages and from development of a small rather than a large hydrogen-fueled upper stage.[59]

On 19 December Jean Charbonnel, France's minister of industrial and scientific development, met in Paris with his West German counterpart, Klaus von Dohnanyi. Charbonnel hoped to proceed with L3S and asked if the Germans might share the cost of development. Dohnanyi had a request of his own: West Germany was preparing to collaborate with NASA; would France wish to share its cost in return?[60]

This nascent collaboration grew out of activities during the 1960s wherein the Germans had pursued increasingly demanding efforts in partnership with the United States. The most ambitious was Helios, the subject of a memorandum of understanding in June 1969. This joint program provided for launching two German-built probes into heliocentric orbit to reach within thirty million miles of the sun. West Germans designed, assembled, and integrated the two spacecraft while providing seven of the ten instruments. They flew to deep space in December 1974 and January 1976. NASA provided two Titan III-E–Centaur launch vehicles, the nation's most powerful, and supported the missions with its network used for tracking planetary probes. West Germans operated and controlled the spacecraft from a German control center. Here was a major step beyond the small Earth-orbiting payloads and the Scout launchers of earlier years.[61]

When Charbonnel met Dohnanyi in December 1972, each had plans for a new program. Charbonnel carried material on the L3S launch vehicle. Dohnanyi, with studies from his country's MBB and ERNO as background, expected to work with NASA on the sortie lab. The two men agreed that France would finance some 60 percent of the development cost of L3S, that West Germany would take on a similar percentage of the cost of the sortie lab, and that each nation would obtain the other 40 percent from other countries in Europe.

Charbonnel and Dohnanyi also were prepared to discuss the future of inter-European space cooperation and to look beyond ELDO and ESRO. ESRO was flourishing, taking on new work with applications satellites while sponsoring the studies of sortie labs. By contrast, ELDO was on its last legs. The British, West Germans, Dutch and Italians all had declined to support Europa 3, leaving only France and Belgium. This suggested that Europe's future in space could feature a single unified international agency, which might grow out of ESRO.[62]

Proposals for unification were not new. Causse's report in 1967 had proposed to merge ELDO and ESRO, but British rejection of that report had doomed this plan. Another plan obtained preliminary agreement at a conference of ministers in mid-1970, but it too failed to advance. Then in 1970 Britain took a significant step in this direction. The Tories defeated Labour in a general election, bringing in a new aerospace minister, Michael Heseltine. In November he proposed that Europe should set up a new and unified space agency—and that his government was prepared to put a large portion of its space budget into it.[63]

Two years later, Charbonnel and Dohnanyi had their initial discussion in Paris. They then traveled to Brussels to participate in another ministerial meeting. Charbonnel unveiled the L3S, making clear that French financial support for the sortie lab was contingent on acceptance by other nations of either this launcher or Europa 3. Dohnanyi had no love for the latter; he regarded it as costly and unlikely to see much operational use. He was willing to support the less expensive L3S. However, L3S would not be an ELDO project; it was to go forward within a "common European framework" that was still to be defined. With this decision, Europa 3 was dead.

Working late into the night, the ministers also accepted Heseltine's proposal for a European Space Agency (ESA), to form through a fusion of ELDO and ESRO. It was to grow out of ESRO, the charter of which would serve as a model. The merger was to be completed by 1 January 1974, barely a year away, and individual national space projects were to go over to this new agency as rapidly as possible.[64]

This meeting made it clear that West Germany was the principal supporter of the sortie lab. Hence it was appropriate to ensure that German firms would take the lead in its development. ESRO accordingly dropped British Aerospace from the current round of studies, leaving MBB and ERNO to continue their work.[65] During 1973 those two companies went on to compete for the prime contract.

Europa 2 remained as a loose end, for the Brussels conference had kept its fate unsettled. West Germany had no wish to pursue it. Horst Ehmke, the minister for technology and research, declared that it was "not sufficiently reliable" to launch Symphonie and that ELDO had "tried to do too much at once to close the technology gap with the Americans." Indeed, the Germans stated that abandonment of Europa 2 was part of their price for participation in the L3S. France held on to Europa 2 for a while, but when West Germany indeed agreed to share the expense of L3S, the French relented.

In April 1973, at a ministerial meeting in Paris, envoys of West Germany and France agreed jointly to cancel Europa 2, effective on 1 May. This knocked the final prop from under ELDO; it had been prepared to take on both Europa 3 and the space tug, but now had lost both. ELDO was given another month to finish its official activities. It then went out of business, its assets and some staff members going over to ESRO. With this, ELDO too was dead.[66]

But ESA was very much alive, if not yet born, and the British were ready to assert their own preferences. They expected to rely on the United States for launch services and regarded L3S as a warmed-over Europa 3, believing that Europa would never have enough space traffic to justify its development. However, the British held a strong interest in satellite communications. Hawker Siddeley was studying a Geostationary Technology Satellite (GTS). Reflecting Britain's centuries-long involvement with the sea, this was to provide links from ship to shore. It neatly fitted an existing ESRO initiative in this field.

Late in 1971 ESRO had accepted a package deal that took it strongly into satellite applications, with three projects in the forefront: a communications satellite, a weather satellite, and a spacecraft for use in air traffic control. The first of these now was taking conceptual form as the Orbital Test Satellite (OTS). It was a prototype of a class of spacecraft that were to accept different electronic packages with ease, to serve various communications functions. Maritime communications was one of them. This raised the prospect that the British might drop OTS and accept prime responsibility for Marots, an OTS for mariners.[67]

By mid-1973, then, three new projects were in play: L3S, Marots, and the sortie lab. The time was ripe for a second package deal, to parcel out the nation-by-nation financial shares and establish the managerial arrangements. This came together at a ministerial meeting in Brussels on 31 July. The main negotiating session lasted fourteen hours and ran past midnight. When the chairman failed to obtain a firm definition of national positions, he adjourned the general session and met with each delegation in private. This produced the desired agreement.

ESRO was to initiate the new activities pending final approval of the ESA Convention. But France's CNES won management authority over L3S. Here indeed was a break with the practices of ELDO, for whereas that consortium had accepted a different project manager for each participating nation and for each stage of Europa, L3S was to have centralized program management within a single agency: CNES. In NASA parlance, it was to serve as the "lead center" for L3S. Other nations would contribute their shares, but France would make the decisions.[68] This conference also defined the respective national contributions (table 2.2).[69]

With France now responsible for L3S, that nation's officials exercised a prerogative and gave it a name. A classic name was in order, for Europa, predecessor of L3S, had been a goddess and one of Zeus's lovers. These officials considered "Phoenix," but this suggested that L3S had risen from the ashes of ELDO and would repeat its fate. "Penelope" drew attention, as the faithful wife of Ulysses who waited for his return, but this suggested program delays. For a time the name "Vega" was in vogue, but this was well known as a brand of Belgian beer.

The final choice, Ariane, was the French form of Ariadne. Using a spool of thread, this

Table 2.2. Percentage of European National Contributions, 1973

	L3S ($445 million)	Sortie Lab ($370 million)	Marots ($90 million)
France	62.5	10.0	12.5
West Germany	20.12	52.55	20.0
Britain	2.47	6.3	58.5
Italy	1.75	18.0	2.3
Belgium	5.0	4.2	1.0
Netherlands	2.0	2.1	0
Spain	2.0	2.8	1.0
Switzerland	1.2	1.0	0
Sweden	1.1	0	0
Denmark	0.5	1.5	0

Sources: Data from Krige and Russo, *Europe,* 112; and Madders, *New Force,* 167–68.

woman guided Theseus out of the Labyrinth, a maze inhabited by the dreaded Minotaur. Similarly, Ariane was to lead Europe out of its own maze. To the nationalistic French, the name also brought thoughts of Marianne, a young woman who stood as a symbol of the nation and who thus was a counterpart of John Bull and Uncle Sam. She had appeared in such works as the painting by Delacroix, *Liberty Leading the People,* and had inspired the Statue of Liberty, a gift from France to America.[70]

The sortie lab also received a new name. In America a sortie was an individual flight or mission. But in France, the word's country of origin, *sortie* meant "exit," which was inappropriate. The Europeans preferred the name "Spacelab." It invited confusion with NASA's existing Skylab program, but Europe was paying for its development, which gave them their own prerogative. On 24 September Fletcher issued a memo: "The Sortie Lab is officially renamed 'Spacelab,' which designation will henceforth be used in all official correspondence and public releases."[71]

That same day brought a ceremony at the State Department, as NASA's James Fletcher and his ESRO counterpart, Director General Alexander Hocker, signed the memorandum of understanding that implemented their transatlantic cooperation. A coordinating body already was in place, managing work both in America and in Europe. It now took the name of Joint Spacelab Working Group, or JSLWG (pronounced "jizzlewig"). At NASA Headquarters, Douglas Lord was the Spacelab program manager; he co-chaired this group. ESRO's head of the Spacelab program was the other co-chairman. He was, again, Jean-Paul Causse.[72]

At MBB and ERNO, system definition studies continued through the whole of 1973.

Major design reviews took place between October 1973 and February 1974; then, early in March, ESRO issued requests for proposal to both contractors. The proposals reached ESRO in mid-April, as NASA stepped aside and gave full responsibility to the Europeans for their evaluation. A month later, the source evaluation board presented its report to a panel headed by Director General Hocker.

The board put MBB on top by a score of 662.5 to 650.1; MBB won this advantage in the area of management. But as NASA's Douglas Lord would write, "ERNO's technical concept was superior and employed low-cost design features, ERNO's depth of design was better for the immediate implementation of [project development], the suitability of ERNO's concept to users' needs was superior, ERNO's proposal showed particular strength in the top management aspects, the shortcomings of the ERNO proposal could be repaired either more easily or later in the program, and ERNO's price was better." MBB had won on points, but Hocker's preference was for ERNO, and there was no form of appeal for the loser. On 5 June ESRO's Administrative and Finance Committee confirmed the choice of MBB. ERNO now held a six-year, $226-million prime contract with delivery of a fully qualified Spacelab flight unit set for April 1979.[73]

With this, and with Ariane, Europe now was set on a course that would offer NASA both support and defiance. NASA's post-Apollo planning, several years earlier, had centered on space stations. Even the shuttle was conceived initially as merely a logistics vehicle to support such stations. But those concepts faded, and although President Reagan went on to commit NASA to a space-station program in 1984, no such facility was built during that century. In America and Europe, for more than twenty years after 1974, cooperation with Moscow provided the only means to gain access to such a station.[74]

While Spacelab supported the shuttle, Ariane challenged it. NASA expected that the shuttle would be all things to all people and would replace its expendable launch vehicles: Delta, Atlas, and Titan III. Ariane amounted to a French assertion that NASA was wrong, that the world would continue to need expendables, and that Europe would provide them. Ariane was not conceived as an alternative to the shuttle; it grew out of France's prickly determination not to depend for launch services on the unreliable Yankees. But when the shuttle faltered with the loss of Challenger in 1986, and when American expendables were not immediately available, Ariane picked up the slack.[75]

Blue Streak now lay abandoned, though other flights of Europa 2 had been scheduled. When ELDO canceled that program in April 1973, another first stage was in the harbor of Kourou, French Guiana, which served the nearby launch site. It never flew but was left to rust. Some years later a farmer bought it. "He made a home of it for his chickens," the historian Kevin Madders notes, "where they may still be laying their eggs today."[76] Still, this was better than an alternative. Like the program that built it, this rocket stage might have sunk into French Guiana's swamp and jungle without a trace.

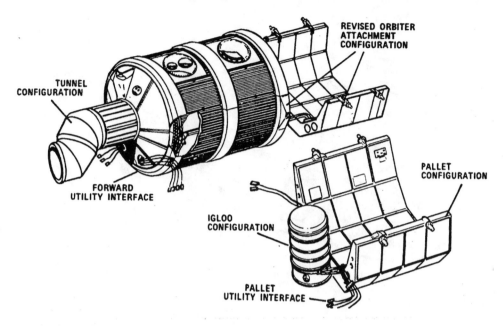

Fig. 9. *General layout of Spacelab. The pallet and igloo carried experiments and instruments. (ESA)*

The External Tank

In 1973, although Ariane was in the news, it existed largely on paper. The world of the shuttle was far from sunny, for storm clouds were flashing from the OMB. Nevertheless, Ariane amounted to a cloud no larger than a man's hand. The shuttle's builders had plenty of work right in front of them, particularly because major elements of the complete space shuttle were not yet under contract.

The complete system was to include the external tank (ET) and two solid rocket boosters (SRBs). They appeared prominently in all conceptual drawings, but this was not the same as preparing to build them. These components did not go immediately to contract, for they were the subject of phased procurements. This permitted the mature definition of one shuttle element before proceeding to the next. Thus, the weight of the orbiter had to be known to size the ET. Both of these major components had to be sized before anyone could truly design the SRB. The SRB and ET had their own contracts for procurement, competition for which occurred during 1973.[77]

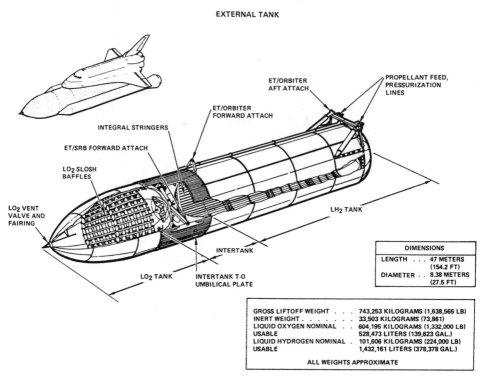

EXTERNAL TANK

ET/ORBITER
AFT ATTACH

PROPELLANT FEED,
PRESSURIZATION
LINES

ET/ORBITER
FORWARD ATTACH

INTEGRAL STRINGERS

ET/SRB FORWARD ATTACH

LO₂ SLOSH
BAFFLES

LO₂ VENT
VALVE AND
FAIRING

LH₂ TANK

INTERTANK

LO₂ TANK   INTERTANK T-O
UMBILICAL PLATE

DIMENSIONS	
LENGTH . . .	47 METERS (154.2 FT)
DIAMETER . .	8.38 METERS (27.5 FT)

GROSS LIFTOFF WEIGHT . . .	743,253 KILOGRAMS (1,638,565 LB)
INERT WEIGHT	33,503 KILOGRAMS (73,861)
LIQUID OXYGEN NOMINAL . .	604,195 KILOGRAMS (1,332,000 LB)
USABLE	528,473 LITERS (139,623 GAL.)
LIQUID HYDROGEN NOMINAL .	101,606 KILOGRAMS (224,000 LB)
USABLE	1,432,161 LITERS (378,378 GAL.)

ALL WEIGHTS APPROXIMATE

Fig. 10.   *External Tank. (NASA)*

The ET was slated for production at the Michoud Assembly Facility, a large manufacturing plant near New Orleans. Its site first entered history in 1763, through a plantation grant from France's King Louis XV. It took its name from one of the land's owners, Antoine Michoud, an eccentric New Orleans junk dealer and recluse. The plantation produced some lumber from swampland, but it was never successful. The only lasting relic of those days was a pair of chimneys in front of the Michoud administration building.

This plant had been built during World War II to assemble Liberty cargo ships. Problems with contracts brought a shift in purpose to the assembly of C-46 cargo aircraft, but only two of these planes rolled out by war's end. The facility, owned by the government, remained largely inactive until the Korean War, when Chrysler Corporation used it to build engines for Army tanks. It then relapsed into dormancy. Before long, it was a mess.

In 1961, during the buildup for Apollo, it passed into the hands of NASA. Its main manufacturing area spanned two million square feet, and the aviation writer Robert Serling later described it as "rat-infested" and "decaying," with "cobwebs, dust, dirt, cracked floors, rust-ing rafters, and corroded overhead cranes." Cleanup crews also faced detritus from a recent flood that had left high-water marks eight feet off the floor.

Within the Apollo program, not the least of the accomplishments was that Boeing, the contractor, cleaned up the facility and turned it into a working factory for construction of rocket stages. These were S-ICs, first stages for the Saturn V, each of them standing 138 feet tall with a diameter of 33 feet. When people watched Apollo missions on television and saw that first stage lift off with 7.5 million pounds of thrust, they were watching a product of Michoud. Now, with Apollo having run its course, this cavernous plant was to serve for production of shuttle external tanks.[78]

These would be huge. With diameter of 27.5 feet, each tank was to be more than 6 feet wider than the fuselage of a Boeing 747. Total internal volume was nearly six times that of Skylab. In *Spaceflight,* science writer David Baker described it as "the tallest single structure yet designed for a launch (or payload) system; in fact, one of the two propellant tanks it contains is the tallest and most voluminous vessel ever built for a rocket stage. So large is the ET that if it were a hollow structure of the same dimension (and 2 m taller), it could entirely contain early models of the Saturn 1 launch vehicle!"

NASA's Apollo experience had provided background in the transportation of such outsize containers. Like the S-IC, it was to ride a barge along the Intracoastal Waterway from Michoud to Cape Canaveral. This requirement for waterborne transport had shaped Apollo's choice of Michoud, with its riverfront location, and carried over to the use of the facility for the shuttle program. Plans were afoot to launch the shuttle from California's Vandenberg Air Force Base, which would require the barging of external tanks through the Panama Canal. NASA had done this, too. Its S-II stage, built on the coast near Los Angeles, had traveled in this fashion to reach the Cape as well.[79]

The stages of the Saturn V also gave technical background for the external tank. The most directly pertinent were the second and third stage, the S-II and S-IVB, which had mounted insulated internal tankage for liquid hydrogen and liquid oxygen, while combining structural strength with low weight. This strength allowed those stages to withstand the acceleration of rocket thrust, while carrying their payloads. But whereas the Saturn stages had been simple cylinders that took their loads from the top and bottom, the ET received its loads from SRBs mounted at the sides. A transverse beam, spanning the ET interior, provided hard points for mounting the SRBs and included structural fittings that distributed their loads into the ET.[80] The orbiter, with powerful engines of its own, rode this tank piggyback.

The S-II was the more nearly comparable of the two stages, and it is appropriate to compare and contrast them (table 2.3).[81] The external tank was to be nearly twice as long and more than half again as voluminous. Propellant mass fraction, the ratio of propellant weight to fully fueled weight, was to be as high as possible. The S-II mounted five rocket engines, which added inert weight and diminished this value, but the external tank was to carry no engines and would hold more than twenty times its empty weight in propellant. In addition, NASA intended to use low-cost manufacturing techniques such as spray-on insulation and electron-beam welding, while achieving economies of scale through a large production run. In 1973 cost estimates per tank were $2.1 to $2.3 million once production was well under way.[82]

The tank, together with the solid boosters that flanked it, had a marked resemblance to the Titan III. This point was not lost on the builder of the Titan, Martin Marietta. Its president, Tom Pownall, met with Fletcher in September 1972 and gave him a letter that sought to remind him of this company's unique qualifications: "Since 1962 [Martin Marietta] has been involved in all phases of integrating and launching large solid rocket motors with large liquid stages in the Titan III program. We were the first and still are the only contractor currently engaged in this type of integration, which closely parallels the shuttle integration requirements. During these ten years we have developed an in-depth understanding of the problems and their solutions to successfully launch large SRMs with large liquid tanks." Pownall noted particularly that his firm had been responsible for "providing the design, development, qualification and installation of the electrical/structural and ordnance hardware required to structurally attach and electrically interconnect the SRM to the Titan III core and to laterally separate the SRM from the Titan III core during the ascent phase. This hardware has performed consistently and flawlessly for all 20 Titan IIIC and D flights." He concluded by describing Martin Marietta as "the only contractor that has successfully integrated large SRMs with large liquid propellant tanks."[83]

Competition for the external-tank contract got under way early in April 1973, as NASA issued a Request for Proposal (RFP) to interested bidders. Rockwell, which had built the S-II, would have been a competitor par excellence, but NASA was not about to put its eggs in one basket; it did not even invite that company to bid. But McDonnell Douglas, builder of the S-IVB, received an invitation. So did Martin Marietta. Boeing, which had worked at Michoud and had built the S-IC, got one as well. A fourth RFP went to Chrysler, which had operated as Wernher von Braun's manufacturing arm and had constructed Saturn I and Saturn I-B first stages, within its own section of Michoud.[84]

Few people were surprised when the award went to Martin, in mid-August. George Low described the decision:

> The [Source Evaluation Board] presented us with a rather clear-cut case for selecting the Martin Marietta Company, which we did. Technically, Martin Marietta was just slightly better than the next company, McDonnell Douglas, but their costs both for [design, development, test, and engineering] and for production work was considerably lower. Although we recognized that the Martin Marietta Company was partially "buying in" with their lowest costs, we nevertheless strongly felt that in the end the Martin Marietta costs would, indeed, be lower than those of any of the other contenders.

Table 2.3. Comparison of S-II and External Tank

	S-II	External Tank
Dimensions (ft.)	82 x 33	154 x 27.5
Propellant weight (lbs.)	982,200	1,556,000
Empty weight (lbs.)	81,800	73,861
Propellant mass fraction	92.3%	95.5%

Sources: Data from "Saturn V" (NASA-MSFC poster); Powers, *Shuttle,* 240; and Ley, *Rockets,* 636.

McDonnell Douglas had a different kind of buy-in, in that they shaved the weights below a reasonable value. Boeing's major weakness was their selection of General Dynamics as a partner in areas where they really did not need this company. This presented an awkward marriage and probably would have caused major problems during the years. Chrysler was weak all around. In selecting the Martin Company we felt that they will undoubtedly do the best cost management job in that they have clearly proposed doing this "the new way" (design to cost). The others, in effect, were proposing to do business as usual.[85]

Martin now had two major programs that drew on its specialized experience in building launch vehicles that mated large solid motors to large propellant tanks. This was not the same as showing leadership in manufacturing automobiles or in producing oil or steel, not by a long shot. It was, however, an industrial niche for which a market existed, providing the company with a continuing flow of business. Moreover, in serving that niche, Martin stood alone. McDonnell Douglas was building versions of its Delta that placed small solid boosters alongside large propellant tanks, but General Dynamics with its Atlas did not even go that far. Atlas was built as a thin pressurized shell that lacked hard points for attaching large solid rockets. By contrast, the Titan III had been designed with such hard points.[86]

As with the orbiter, the external tank of mid-1973 differed noticeably from the design concepts of a year earlier, and the differences reflected weight reduction. The new tank was not only shorter but also simpler, reflecting a change in the method selected for its disposal. Though it was to accompany the orbiter and provide propellants for its main engines, the tank itself would not enter orbit. It was to fly halfway around the world and reenter the atmosphere near Australia, in a region of the Indian Ocean where few ships would note its flaming entry—or face the risk of impact.

Initial studies, within both NASA and Rockwell, designed the tank with a retro-rocket mounted to its nose. This gave the ET the appearance of an elongated artillery shell with a fuse at its tip. Such a tank indeed was to reach orbit, then fire the rocket to reduce its velocity and fall

69

from the sky. The new configuration deleted the retro-rocket. The orbiter, its ET attached, now was to accelerate to just below orbital velocity, with the tank dropping away. The orbiter then would gain the final small increment of velocity by using its orbital maneuvering engines.[87]

In the words of Myron Malkin, who succeeded Charles Donlan as director of the space-shuttle program in April 1973, this change in mode "made it possible to delete the retro-rockets and associated electrical systems (such as power, sequencers, and pyrotechnic devices). Eliminating the need for remote retrofiring simplifies operating procedures, improves reliability, and reduces the crew's workload."[88]

The Solid Rocket Boosters

In 1973 one could look at a flight-ready S-II stage, such as the one that launched Skylab in May of that year, and envision the shuttle's external tank as a much longer counterpart. But in the matter of the solid rocket motors, there was far less need for imagination. At Cape Canaveral, solid rockets of similar size were matters for standard industrial practice, within the Titan III launch facilities.

The boosters arrived by rail from the manufacturer, United Technology Center, in the form of drum-shaped segments filled with rubbery propellant. These segments, ten feet across, were stored following arrival, inspected, then moved to ready storage. Meanwhile the liquid-fueled Titan III core vehicle, complete with payload, was being assembled and checked out within one of four bays of the towering Vertical Integration Building (VIB). The core vehicle stood atop a platform 44 feet across, fitted with wheeled trucks and riding on a double set of railroad tracks.

The twin solid boosters came together within the Solid Motor Assembly Building (SMAB), which stood more than two hundred feet tall and showed a windowless structure that resembled a pharaoh's tomb. Here a crane lifted the components and stacked them vertically: a nozzle assembly at the bottom, surmounted by five propellant-filled segments, with a nose section at the top. Completed boosters stood eighty-five feet tall.

Now, on both the VIB and SMAB, enormous doors opened as diesel locomotives prepared to move the wheeled platform, holding the core vehicle, between the two buildings. Within the SMAB, a three-hundred-ton crane lifted and transported the finished solid boosters, placing them alongside the core. Technicians mated them, and following final checks, the Titan III was ready for flight. Still riding its platform, the locomotives moved it from the SMAB to Launch Complex 40 or 41, to be counted down and launched. The solid motors ignited on the launch pad. Then, their propellants exhausted, small stage-separation rockets pushed them away from the core, which continued onward toward orbit.[89]

The nation supported at least four major builders of solid rockets: Lockheed Propulsion, Thiokol Chemical, Aerojet-General, and United Technology. The last of these had won its laurels with the Titan III, and the others also had stellar records. During the 1960s they had vied to build the largest possible test versions. All three firms fired experimental solids of 156-inch diameter, while Aerojet and Thiokol went on to design behemoths that measured 260 inches across. Aerojet went further and conducted two successful test firings.[90]

These companies also had distinguished themselves in building operational solid-fuel missiles for the Air Force and Navy. Lockheed built the Navy's Polaris, with Aerojet supplying the solid-fuel motor. Aerojet also built the second stage of Minuteman. The energetic Thiokol constructed the Minuteman first stage and went on to provide propulsion for the Poseidon, the Navy's successor to Polaris, through a joint venture with the firm of Hercules in Utah.[91] Lockheed, Aerojet, and Thiokol, along with United Technology, all were well acquainted with the requirements of the space shuttle. Early in 1972 NASA's Marshall Space Flight Center, which had responsibility for shuttle solid boosters but held little background in this area, awarded contracts to these four firms to study the use of 120-inch and 156-inch solids.[92]

During July 1973 NASA-Marshall issued a Request for Proposal for shuttle solid booster development, inviting all four companies to submit their bids by late August. A source evaluation board placed these proposals under close review, with some 289 people participating in the examination. They did more than read the documents with care; they conducted independent analyses and design studies.[93]

Thus, although standard practice called for fabricating large solids from segments, Aerojet drew on its experience with the 260-inch program and proposed to build a single monolithic case, free of joints, and to transport it by water rather than rail. Thirteen years later, the failure of such a joint would doom the shuttle Challenger, but the review board proved quite willing to accept segmented designs. However, the Aerojet concept failed on its technical merits: "The strength of the case was found inadequate for the prelaunch bending moment loads and was not designed with an adequate safety factor for water impact loads." Aerojet's proposal held other deficiencies as well, and the board ranked it fourth in the competition.[94]

The other three bids were more evenly matched. Lockheed submitted a very strong technical proposal but was weak on the management side, offering little promise of controlling costs. Thiokol was deficient in some technical areas, but its proposal gave good reason to anticipate costs that would be both low and well controlled. United Technology fell generally between these two, with a bid that was good but not outstanding in either area. This put the choice between Lockheed and Thiokol.

Fletcher, summarizing the board's review, pointed to "the thoroughness and detail that Lockheed utilized in designing for low cost." An innovative case design offered ease of assembly and check-out, with no severe safety restrictions. Detailed stress analyses, included with the proposal, showed that Lockheed had optimized the structural design of the case to meet all load requirements. The nozzle was simple and used conventional materials for moderate cost. Fletcher wrote that Lockheed had "clearly focused on a highly reliable, low risk, conservative design, using all proven materials."

But Lockheed flopped on the management side. It proposed a rapid and early buildup of both staff and new facilities, which was not the way to control costs. Fletcher wrote that this "resulted in high early year funding which is contrary to one of the program goals." An anticipated relocation of personnel to the New Orleans area "was not adequately planned or described." The program office "was heavily staffed and showed a tendency toward overcontrol and duplication." The project manager would have no deputy and hence would be "spread quite thin." Most managers and key technical people "had no experience on a solid rocket motor program comparable in size or scope" to that of the space shuttle."[95]

Thiokol's technical proposal showed weaknesses. Its nozzle called for use of materials "not currently developed or characterized," with "attendant technical and program risk." The nozzle design "was insufficient to meet required safety factors" and "could require a redesign. . . . The design was complex and would contribute to difficulty in manufacturing." The proposal showed "major weakness . . . in the area of case fabrication," which "could require extensive redesign."

Thiokol, however, sparkled in its management. It anticipated no buildup of new facilities; rather, it expected to use an existing plant in Utah that had constructed first stages for Minuteman. Nor would Thiokol hire people in droves; it would shift people over to the shuttle as current programs wound down. Fletcher noted that "Thiokol structured the development program so that all major costs were deferred to the latest practicable date. This resulted in low early year funding, which is a key program objective."

Fletcher also saw strength in Thiokol's people. Its supervisors "were experienced and had worked together as a team" on Minuteman and Poseidon. The project manager "was considered exceptionally strong and . . . is widely known for his excellent performance." His deputy "had important and successful engineering management roles in previous major motor programs and has an excellent reputation in the trade." Fletcher further noted the "depth of experience available" within the company's operating departments.[96]

Which would it be, good management or good design? The board's report ranked these top two contenders among the four competing firms (table 2.4).[97] More particularly, the board scored the proposals on points.[98] Lockheed received 714 points; Thiokol, 710; United Technology, 710; and Aerojet, 655.

Board members were not entranced with Thiokol's technical proposal, but they were even less enthralled with Lockheed's management. Lockheed anticipated high cost, the worst of the four contenders, in carrying through the design and development during the next sev-

Table 2.4. Rankings of Thiokol and Lockheed

	Thiokol	Lockheed
Design, development, verification	4	1
Manufacturing, refurbishment, product support	2	1
Management	1	3
Program cost	1	2
Design and development	1	4
Operational production	1	2
Cost per flight	1	2

Source: Fletcher, "Selection of Contractor."
Note: Rankings ran from 1 to 4 because four companies were competing.

eral years. NASA had picked solid boosters in preference to liquids early in 1972 because solids promised costs that could be both low and well-controlled. Selecting Lockheed then would have tended to vitiate that earlier decision. Moreover, as Low noted, the board "decided that Lockheed had written an outstanding proposal, but that this did not mean that they could also produce a good rocket; and that the differences, which were purposely amplified by the Source Selection Board, were all in easily correctable areas."

The board was willing to give leeway to Thiokol in these "easily correctable [*sic*]" technical matters, for this company's senior managers were highly experienced and ready to learn. But board members took the view that Lockheed's management deficiencies reflected inexperience, stemming particularly from a lack of corporate facilities and staff. The board thus decided to award the contract to Thiokol, and Fletcher, Low, and McCurdy concurred.[99]

The Thiokol solid booster was the one that eventually failed in flight in 1986, causing loss of the Challenger. This gave grist for the mills of investigative writers, who noted a curious set of facts: Fletcher was from Utah; Thiokol was located in Utah; Utah had a senator, Frank Moss, who chaired the Senate space committee and who put strong pressure on Fletcher to select Thiokol during the bidding. Had a network of good old boys foisted a compromised design upon the nation, merely to help the folks back home?[100]

Fletcher's links to Utah were undeniable. He had spent part of his youth there, with his father on the faculty of Brigham Young University. He attended that university himself during part of his undergraduate years. Decades later he returned to Salt Lake City and took over the presidency of the University of Utah. He also cherished strong ties to the Mormon Church. His parents were Mormons. He not only practiced this religion but also rose to prominence within its hierarchy. Still, to conclude from this that Fletcher was a Mormon puppet smacks of saying that because President Kennedy was Catholic, he was a tool of the Pope.

73

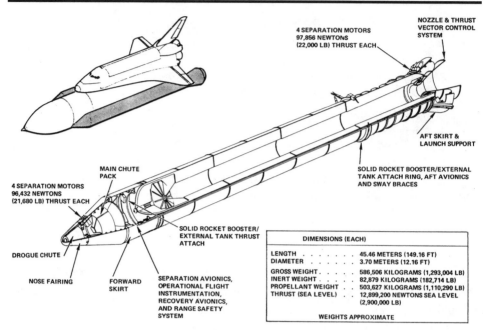

Fig. 11. *Solid Rocket Booster. (NASA)*

Fletcher was his own man, and he let people know it. Like other Mormons he neither smoked nor drank, but when he took his university presidency in 1964, he placed an ashtray on his desk. This was his way of stating that although Utah was a society dominated by Mormons, he would not force their views on those who disagreed.[101]

As NASA administrator, he indeed fell under strong pressure in favor of Thiokol, not only from Senator Moss but also from Mormon elders, including the president of this church, Nathan Eldon Tanner. Fletcher, however, was well aware that he was in Washington to serve not only secular interests but also the national interest. In exchanges of letters with Moss, he repeatedly insisted that he would wield a free hand in the SRB procurement. Thus, in January 1973 he wrote to Moss, "I know that President Tanner and various of your state officials have manifested an unusual zeal in hopes that NASA would send some of our business your way. As you probably already know several firms have recently approached me in person, which I suspect was through your persuasions no doubt. . . . But it would be comforting to know that amicable solutions can be reached without any undue pressure on our part or the intervention of politics as such."[102] Several weeks later, under continuing pressure, Fletcher responded more personally. Bypassing his secretary, he typed his own letter and sent it to Moss, disregarding its errors in spelling and grammar:

Dear Mr. Chairman:

I feel an obligation to respond to the numerous efforts made by your office of late to have this Agency, and, in particular myself, look with considerable favor at the placing of some of our business in your State. Not only would it be highly irregular to say the least, but might provoke the kinds of inquiries we are not prepared at this time to handle. . . .

But the fact remains, Mr. Chairman, that my hands are tied for the time being. In my present position here at this particular Agency, it would be extremely difficult if not somewhat unethical for me to channel any more of our contracts towards your State without arousing further suspicion. . . .

I would also like to call your attention to another matter along these same lines. One of your staff . . . went so far as to insinuate sometime ago that I had a moral, if not a spiritual obligation to acquiese [sic] on some of business issues previously raised by President Tanner. This person voiced an unthinkable opinion to the effect that my Church membership took precedent over my Government responsibilities.

Knowing that you share similar sentiments with me in the clear separation of Church and State, I would like to request that you take this unpleasant matter under advisement with the individual in question and explain just how serious and unconscionable those inferences were. . . .

But for right now I must pursue a course that, at least, seems to be equitable to all parties concerned. Sometimes substantive actions don't count as much as how others perceive them to be.[103]

Yet Fletcher gave the contract to Thiokol. Kem C. Gardner, Moss's former top aide, later stated, "There's no question that one of the main reasons Thiokol got the award was because Senator Moss was chairman of the Aeronautical and Space Sciences Committee and Jim Fletcher was the administrator of NASA." Moss's position "gave us major clout in lobbying for it." Gardner later backed away, asserting that Moss in fact lacked the necessary clout.[104] Yet one certainly must not accept that Thiokol could sit back, knowing that the fix was in.

Its proposal held strength, particularly in the key areas of cost and management. As events showed, it would stand on its merits. Further, the quality of this proposal placed Fletcher in a position to help Utah in a way that would also help NASA. Moss faced re-election in 1976; Fletcher's decision enabled him to tell the people of his state that he had brought them jobs. Fletcher thus was in a position to ask Moss for a favor. Pressed as he was by the OMB as well as by Moss, he knew he would need all the help he could get.

Following the loss of Challenger, Fletcher might have been denounced and made a scapegoat, based on his preference for Thiokol. Instead, he returned anew as NASA administrator, holding this post from 1986 to 1989 amid a widespread view that he was a man of integrity who could rescue the shuttle. With Washington investigative reporters being plentiful, it is significant that his 1973 acts did not lead to a scandal.[105]

NASA announced its decision for Thiokol on 20 November 1973. It took people by surprise, as the president of United Technology declared, "With our record of achievement in the large solid motor field, it seems incomprehensible."[106] Lockheed did more; on 5 December it notified the GAO that it intended to protest NASA's decision. Its case appeared strong, for it had won the scoring outright. It filed the formal appeal on 8 January, arguing that NASA had "arbitrarily and improperly increased" Lockheed's cost proposal and had introduced an "unprecedented and improper design correction process" in giving Thiokol the benefit of the doubt in its engineering work. This was as much as to say that the source evaluation board had been biased against Lockheed and in favor of Thiokol.

The first of these causes of complaint was particularly significant. Just as the board's reviewers had challenged Aerojet's stress analyses and had critiqued that firm's technical design in the light of their own calculations, so independent financial analysts took issue with some of Lockheed's cost figures. These analysts had substituted their own—and then had graded Lockheed's proposal on the basis of those modified figures, not Lockheed's. This was both legal and prudent; Lockheed had a rather odorous reputation for buying in, winning contracts by quoting unrealistically low cost estimates. It had done this particularly in winning a 1965 Air Force contract for the C-5A transport. But in seeking the solid-booster award, A. H. von der Esch, a Lockheed vice president, argued that his firm had the lowest proposed cost, "before the improper adjustment and normalization by NASA."[107]

This appeal put the SRB on hold, for NASA could not prejudge the issue by confirming its contract award to Thiokol. The most it could do was to allocate small driblets of money until the GAO made up its mind. NASA granted $864,000, but this covered only the most pressing matters during the next ninety days. These funds ran out with Lockheed's protest still unresolved. NASA gave Thiokol another $500,000 to hold things together during the subsequent forty-five days, with this money slated to run out at the end of June. A NASA official met with his GAO counterparts on 23 April and urged them to hurry. The GAO replied with a letter to Fletcher, writing that "we firmly intend to issue our decision no later than June 24, 1974."[108]

On that date the GAO delivered its response to Lockheed in a ninety-eight-page report, providing copies to NASA as well. On the whole, it was surprisingly favorable to NASA, endorsing most of the procedures and findings of the review board. In one significant area, however, the GAO disagreed: the production of ammonium perchlorate, the oxidizer that was to be mixed within the solid propellant itself.

Lockheed expected to build its solid motors using NASA's Michoud and Mississippi Test Facility sites, the latter being a major center for rocket engine testing. Lockheed planned to transport the motors' segments by barge, along with external tanks from Michoud. They would need plenty of perchlorate, some twelve thousand tons per year for a shuttle fleet in routine operation. This substance is commonly used within the chemical industry, with the firm of Kerr-

McGee as the nation's largest supplier. The other three bidders—Thiokol, Aerojet, and United Technology—expected to meet their own needs for perchlorate by relying on an existing plant in Henderson, Nevada. But Lockheed took the view that ongoing demand within the chemical industry was likely to tax this plant's capacity and proposed to build its own perchlorate factory, also in Mississippi. This new facility would be costly, and the source evaluation board had determined that relative to Thiokol, Lockheed faced a cost disadvantage of $122 million.

The GAO took the view that the nation faced no shortage of capacity to produce perchlorate, that Lockheed did not need its new plant and would not receive permission from NASA to build it, that Lockheed could purchase its perchlorate in Nevada—and that on this basis, it would save some $68 million. This cut Thiokol's cost advantage from $122 million to $54 million. The GAO had other quibbles—$6 million in transportation costs, $15 million in labor costs—that put the final estimate of cost difference between $48 million and $63 million. Still, this placed Lockheed's costs much closer to Thiokol's. Far from buying in, Lockheed actually had overestimated its expenses. Accordingly, the GAO invited NASA to "determine whether . . . the selection decision should be reconsidered." Significantly, this was not a direct order to reconsider the choice of Thiokol, which would have reopened the competition. It gave NASA a good deal of leeway in shaping a response.[109]

George Low called a meeting of his colleagues on 26 June. They quickly agreed to stand on the existing decision. Even at $48 million, Thiokol's cost advantage still was large enough to justify its selection. Nor did anyone want a recompetition; that would bring further delay, at a time when everyone was eager to get the show on the road. They wrote a memo to Fletcher, advising him that he had ample cause to reconfirm the selection of Thiokol. Signers of the memo included Low, Yardley, Malkin, McCurdy (who now was a consultant), two other associate administrators, two assistant administrators, NASA's general counsel, and two other senior attorneys.[110]

This was not quite the end of the matter, for politics entered anew. Senator John Stennis was a member of his chamber's space committee and wanted that Lockheed plant for his home state of Mississippi. Senator Russell Long, from nearby Louisiana, felt the same way. These men saw ambiguities in the GAO report and wrote a letter to Staats that requested clarification, which might help Lockheed. They also asked Fletcher not to act until the GAO had time to respond. They could not push too hard, however, for they were well aware that Senator Moss was the committee chairman, outranking Stennis, and that Moss was from Utah and firmly supported Thiokol. Fletcher talked with Stennis on the phone, telling him that he was about to go ahead with a contract for Thiokol. In Low's words, "Stennis apparently accepted that decision, but not very graciously."

Fletcher then signed a letter to Staats, which read, "I have determined that there is no justification for reconsideration of the selection decision. . . . I have directed that a letter contract be awarded promptly to the Thiokol Chemical Corporation." The contract was signed the next day, at eight o'clock in the morning. With this, the last major piece of the shuttle program fell into place.[111]

It now was two and a half years since Nixon had announced his support for the space shuttle. Twenty-three months had passed since Rockwell had received the contract for the orbiter, with Lockheed's protest to the GAO having delayed the solid-booster program by no more than six months. It was the summer of Watergate, a moment when Nixon's own hold on the presidency had just six weeks to run. And within the Pentagon, Air Force leaders were shaping their own plans for space.

The Air Force and the Shuttle

There is a place along the California coast where the U.S. Navy suffered its worst peacetime disaster in history. Amid dense fog, on a night in September 1923, nine destroyers ran onto rocks with a loss of sixty men. To this day, visitors can look down into a small cove and see rusted steel that once had been part of a ship.[112] To the left, on a low rise less than a mile away, stands the site of the Air Force's shuttle launch facilities. In their heyday they showed an array of tall, stark, boxlike forms. A massive and windowless structure stood hundreds of feet tall. Another structure, a vast cube, loomed nearby. A third such tower, wide and with a high, overhanging top, completed the group.

The shuttle was a NASA program, but the Air Force had every intention of becoming a major user. That service was an important space power in its own right, with an active program and with plans to fly the shuttle in the service of national defense. The focus of the Air Force effort involved reconnaissance satellites, launched by military rockets on behalf of the Central Intelligence Agency.

The first successful reconnaissance spacecraft had flown as early as 1960. Built within the CIA's Corona program, they reached orbit aboard launch vehicles that used versions of the Thor as the first stage, with Agena as the second stage. Though small and limited in capability, these spies in the sky performed services of great value as they photographed much of the Soviet Union. The photos showed that Moscow was deploying far fewer ICBMs than had been feared. This success whetted appetites, which the Air Force satisfied for a time by launching increasingly capable Corona satellites. Larger and heavier spacecraft soon appeared, necessitating bigger launch vehicles. Gambit, which first flew in 1963, started by using the Atlas-Agena and then moved up to the Titan III-B, the core of a Titan III that lacked the large solid boosters. Then in 1971 the Air Force began launching Big Bird, which required the powerful Titan III-D.[113]

The Air Force anticipated that the space shuttle would continue this trend and used its clout to shape its design accordingly. Early shuttle concepts, in 1969, had been designed to meet NASA's needs only and to carry twenty-five thousand pounds to orbit. Air Force demands raised this to sixty-five thousand. The anticipated size of future reconnaissance satellites dictated a payload bay with length of sixty feet. In addition, related requirements led to the choice of a delta wing for the orbiter.[114]

The Air Force cooperated willingly with NASA but was independent in its charter and most of its activities. It had its own launch center at Vandenberg Air Force Base, with a location on the California coast that was well suited for polar orbits. These orbits gave complete coverage of the Soviet Union. NASA and the Pentagon shared a common stable of launch vehicles, but the Air Force had sponsored the development of the Titan III family and claimed these space boosters as its own.[115]

In addition to photo reconnaissance, Air Force and other military payloads provided secure communications, missile launch detection, interception of signals and other transmissions, and mapping of targets.[116] From this position of strength, the Air Force prepared for the shuttle. Much of its activity took place at Vandenberg, which now stood where that 1923 destroyer squadron had run aground.

Vandenberg was the West Coast counterpart of Cape Canaveral. It lay amid hilly country that adjoined the seacoast, 170 miles northwest of Los Angeles. It provided headquarters for the Air Force's SAMTEC, the Space and Missile Test Center, which operated twenty launch sites, mostly for missile launches down the Western Test Range. These included three Atlas launch pads, two Titan pads, fourteen active Minuteman ICBM sites, and a Scout facility. Two of the Atlas sites supported tests of advanced warheads; the third conducted space launches. SAMTEC also operated the Western Test Range, with instruments that included precision tracking radars, optical sensors, command transmitters, and telemetry receivers. The range facilities supported the Strategic Air Command, which launched Minuteman missiles in tests of readiness. SAMTEC also worked with NASA, which fired some of its own spacecraft into polar orbits, and with the Navy, which orbited navigation satellites.[117]

The Air Force also had extensive launch facilities at Cape Canaveral. However, these had not been suitable for satellite launches into polar orbit, for the geographical location of the Cape meant that such launches would have dropped spent rocket stages onto populated areas. Vandenberg avoided this problem; located on the seacoast, it provided unobstructed ocean extending for thousands of miles to the south. It had developed alongside Cape Canaveral, both serving military needs as major space centers.[118]

However, early shuttle design concepts spurred thoughts of a single national shuttle facility, accommodating both NASA and the Air Force while supporting all-azimuth launches that could include polar orbits. Indeed, the new facility might be located inland, in desert country where land was cheap to purchase and few people were around to complain of the rockets' roar. The pertinent shuttle concept had two winged and fully reusable stages, including a winged booster. It was to fly back to a landing on a runway following launch.

A number of existing Air Force bases and facilities seemed to offer good prospects. In April 1970 NASA administrator Thomas Paine set up an advisory panel, staffed jointly by NASA and military members, to weigh the merits of the alternatives. These included Edwards Air Force Base, New Mexico's White Sands Missile Range with its adjoining Holloman AFB, and a site near Wendover, Utah, on the Nevada border. The White Sands option appeared particularly promising. A community of missile specialists had worked there since World War II. An existing runway could accommodate orbiter landings. Its weather was virtually cloudless; its flying conditions approached the ideal. It also had support from its home-state senator, Clinton Anderson, who chaired the Senate space committee and was ready to fight vigorously for its selection.

However, selection of White Sands would have meant abandonment of NASA's Apollo launch facilities at Kennedy Space Center. These also offered all-azimuth capability for a flyback shuttle booster; they could be modified for use with the shuttle for $150 million, an acceptable cost. Moreover, KSC had its own champion in Congressman Olin Teague, chairman of the powerful Subcommittee on Manned Space Flight within the House space committee, of which he was a ranking member. In December 1970 Teague warned, "Unless I am convinced that NASA is making maximum use of existing facilities, I intend to oppose any money for the shuttle in every way, shape, or form."[119]

Changes in the shuttle design also changed the prospects for the contending sites. By late 1971 it was clear that the shuttle would not use a flyback booster. Instead, it was to fly with the ET and a pair of SRBs. The ET was far too bulky to travel by rail or highway; it needed a barge. The SRB casings were to come down by parachute and would be damaged if they fell on land; they had to descend into the sea. In addition, use of SRBs ruled out all-azimuth launches from KSC, for each SRB casing was very heavy and would cause a great deal of damage if it fell onto a building or into a town. Shuttle design changes thus ruled out all inland sites, restricting choices to the seacoast.[120]

These considerations gave new life to the prospects for continuing use of the existing launch centers: KSC for NASA, Vandenberg for the Air Force. One all-azimuth coastal site nevertheless remained in contention. It was in Matagorda County, Texas, near Galveston. It offered clear shots over the Gulf of Mexico, both to the east and to the south. It could accommodate polar orbits while permitting ocean recovery of SRBs for all directions of launch.

Matagorda, however, lacked development. It not only called for a complete set of shuttle launch facilities, built from scratch; it would need a new town for the work force, complete with highways, hospitals, and water and sewage systems. The two existing sites, KSC and Vandenberg, already had suitable nearby communities. The single Matagorda location promised operational cost savings, for it would avoid duplication. However, it would cost $300 million more than the price of achieving a shuttle launch capability at both KSC and Vandenberg. Such disadvantages proved compelling. In April 1972, NASA's George Low announced that KSC and Vandenberg constituted the choice.[121]

Vandenberg had no facilities that could compare with those of Apollo. Still, it indeed held a large launch pad that appeared suitable for conversion into a shuttle facility. This was Space Launch Complex 6, SLC-6 (pronounced "slick six"). It dated to the 1960s, when this complex was to have launched the Manned Orbiting Laboratory atop a Titan III-M. The combination of MOL and this launch vehicle had more than a casual resemblance to the eventual space shuttle. Both were to be piloted, carrying flight crews. Moreover, the Titan III-M flanked its core stages, heavy with liquid propellants, with two long solid motors. This arrangement closely resembled the ET with its twin SRBs.

Like its adjacent Titan launch sites, SLC-4E and SLC-4W, SLC-6 was to assemble a complete launch vehicle with payload right on the pad, atop the flame bucket. The complex included the Mobile Service Tower (MST), a tall structure with an overhang. It carried cranes for heavy lifting. This tower then was to roll away on railroad tracks, moving to a safe distance.

No orbiting laboratory ever flew from SLC-6, for that program was canceled in 1969. This left SLC-6 as a launch facility in search of a customer, and Air Force officials opted to revamp it for use with the shuttle. They also elected to retain the basic Titan III procedure for launch preparation, stacking a complete shuttle within the MST and then having this tower back away on its rails. This contrasted sharply with NASA's arrangements at KSC, which followed those of the Saturn V. NASA planned to stack its shuttle within the Vehicle Assembly Building and then move it to the pad atop a mobile platform. Hence at both launch centers, the existing facilities planned for reuse left legacies. At KSC, the assembly building was fixed and the shuttle moved. At Vandenberg, the shuttle was fixed and the MST moved.[122]

Seventeen miles from SLC-6, on the north side of the base, Vandenberg had an existing runway eight thousand feet long. To accommodate shuttle landings, it was to be lengthened to fifteen thousand feet and would receive a microwave landing system at each end. Orbiters then would land safely even in foggy weather. The Vandenberg area often was foggy indeed, sometimes providing close to zero visibility even at midday during the summer. Edwards Air Force Base offered the clear weather that pilots preferred, together with long runways galore. But Edwards was 150 miles away, and a 747 carrier aircraft would have to ferry the orbiter back to Vandenberg after every landing. Vandenberg also was considerably easier to reach in the event of an abort during flight to orbit. These concerns militated against Edwards as the landing site. The 17-mile distance at Vandenberg, from runway to SLC-6, contrasted with less than 2 miles at KSC. The Vandenberg shuttle therefore could not be towed, as at Kennedy; it was to ride a seventy-six-wheel transporter. This was inconvenient, but the hilly terrain ruled out a new and closer runway.

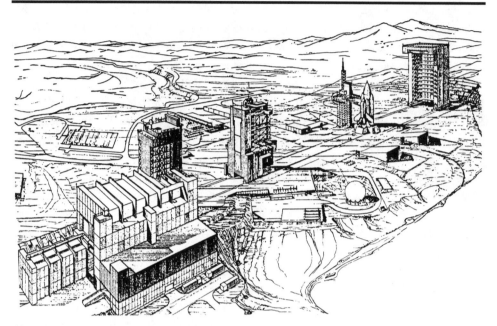

Fig. 12. *Vandenberg Air Force Base SLC-6 concept, c. 1980. Left to right: Payload Preparation Room, Payload Changeout Room, Access Tower with space shuttle, and Mobile Service Tower. (U.S. Air Force)*

In 1975 the Air Force's Space and Missile Systems Organization (SAMSO), the parent of SAMTEC, set forth a $650 million plan for construction of facilities that would turn Vandenberg into an active shuttleport. NASA's plans envisioned a first orbital flight in April 1979; SAMSO expected to begin building in 1978 and to fly the first Air Force shuttle in December 1982. The plane included a launch control center with NASA's Launch Processing System, a specialized propellant facility for the OMS and attitude-control thrusters, and a Mate-Demate Device, a steel framework with cranes to lift an orbiter on and off the back of its 747.

As at the Cape, elements of a shuttle were to arrive by land, sea, and air. The Southern Pacific Railroad had a coastal route with tracks that ran right through Vandenberg, providing connections to Thiokol in Utah. Filled SRB segments were to reach the center along this right-of-way, with empty casing segments returning to Thiokol along the same main line. Following launch, the spent SRBs would fall into the Pacific, to be recovered at sea and towed to nearby Port Hueneme. The ET required an oceangoing barge to transport it from NASA's Michoud Assembly Facility through the Panama Canal to the West Coast. These arrangements echoed those of KSC, which also received SRB segments by rail from Utah and its ETs by barge.

Both at the Cape and at Vandenberg, orbiters were to arrive by 747. The Vandenberg run-

Table 2.5. Weight in Orbit (in Pounds)

	Low Earth Orbit (Cape Canaveral)	Low Earth Orbit (Vandenberg AFB)	Geosynchronous Orbit
Titan III-C	29,200	22,500	3,600
Titan III-E	34,000	29,200	7,200
Titan 34D	34,000	27,000	4,200
Space shuttle	65,000	45,000	5,000

Sources: Data from Hanley et al., SAMSO, 1 Jan.–31 Dec. 1978, 71; and *Aviation Week,* 30 June 1975, 36; 15 May 1978, 48; and 16 July 1979, 53.

way was also to accept launches from Florida, which then could avoid the need for a ferry flight. After a landing, ground crews would shut down on-board systems for safety and tow the vehicle on its own gear to the Orbiter Maintenance and Checkout Facility, a counterpart of the Orbiter Processing Facility at the Cape. In California this would be the shuttle's hangar. Like any hangar, it adjoined the runway. The seventeen-mile trip was still ahead.

At SLC-6, with the MST rolled up to the launch pad, cranes were to stack the two SRBs, segment by segment, and then lift the ET to place it between them. The orbiter would arrive in its own good time, along a route with notches cut into the hillsides to provide clearance for its wings. Following its arrival, workers would raise it vertically and mate it to the ET. A second mobile facility then was to roll up on tracks and place itself against the orbiter's cargo bay, still protected within the broad MST. This was the Payload Changeout Room; it was to provide a clean room for check-out of a spacecraft, along with arrangements for its installation within the shuttle. Later, with the countdown under way, the two mobile buildings were to trundle off on their tracks, leaving the shuttle in a clear area and ready to launch.[123]

As it looked to the future the Air Force also kept a keen eye on its near-term needs. Together with NASA, its senior officials shared a strong interest in upgraded versions of the Titan III. This family of launch vehicles belonged to SAMSO, which had centralized the responsibility for its design and production within a single office. NASA was committed to the Titan III-E, with the Centaur as a high-energy upper stage. It was slated to launch a succession of planetary missions, which were to set sail for Mars, Jupiter, and Saturn. The Centaur was NASA's, but SAMSO owned the rest of the vehicle.

The Air Force had its own planned upgrade, the Titan 34D. It was to mount the standard Titan III second stage atop the first stage of the Titan III-B, which was elongated; longer solid boosters would give it extra payload. Like the Titan III-E, it promised a welcome improvement in launch capacity over the standard Titan III-C (table 2.5).[124]

Hence, amid the expanding world of the shuttle, this NASA program faced competition. It would not win the world's payloads automatically; it faced challenges not only from Ariane but also from the Air Force's advanced Titans. Indeed, the Air Force's situation resembled Europe's. Both were willing to cooperate with NASA, with the Pentagon building its shuttle-port at Vandenberg while the Europeans pursued Spacelab. But both had independent launch-vehicle programs. In time these surged to the forefront as the shuttle faltered.

CHAPTER THREE

Odyssey of the Enterprise

At Christmastime 1975, readers of the trade journal *Astronautics & Aeronautics* saw an arresting color photo on the cover of the January 1976 issue. It showed what was unmistakably an airplane in final assembly within a hangar—and which equally unmistakably was a space shuttle orbiter. "Space Shuttle 1976," read the cover's caption; "Into Mainstream Development." The wings and vertical fin were in place, along with much of the fuselage. This was OV-101, later to receive the name Enterprise. It never flew into space, but nonetheless had a useful career in flight and ground test.

Its most noteworthy moments came during 1976 and 1977. At first, amid considerable ceremony, it was rolled out for public display, thus showing dramatically that the shuttle program indeed was building hardware. It then served in the Approach and Landing Tests (ALT), soaring off the back of a Boeing 747 carrier aircraft and gliding to precision landings on runways at Edwards Air Force Base. Following the ALT flights, it continued to find useful roles, first in structural tests and then in exercising the shuttle's launch facilities.

The Shuttle Orbiter as a Glider

"The great bird will take its first flight upon the back of the great swan," wrote Leonardo da Vinci, "filling the world with wonder and all writings with renown, and bringing eternal glory to the nest where it was born."[1] Though written nearly five centuries before the ALT flights, these words offered a good description. Nevertheless, the shuttle's mode of flight test broke sharply with standard practice in the aerospace industry. Prototypes of new aircraft rarely take their initial flights as passengers, riding on the back of larger planes. Instead they conduct flight tests under their own power, beginning with undemanding excursions and gradually extending the envelope to the full range of speed and performance. Well into 1973, shuttle designers intended to fly their craft in this fashion, and the events that turned the orbiter into a glider took place over several years.

At the outset, in 1969, it was generally accepted that an operational shuttle would mount jet engines. These were to serve for flight tests, in the usual fashion, and would allow pilots to practice landings as preludes to return from orbit. The engines would also see use in ferry flights, flying the orbiter from a landing site to a launch site. Following a return from orbit, a shuttle could cruise in the atmosphere and might rendezvous with a tanker for in-flight refueling. This would provide crossrange while permitting use of a wing of simple design, optimized for subsonic flight.

Designers were reluctant to anticipate unpowered landings. Such landings gave a pilot only one opportunity for touchdown, with no possibility of a wave-off. They also took place at high speed. A 1962 survey had given reason to believe that at such landing speeds, aircraft would occasionally strike the runway at sink rates of up to four feet per second, twice the standard, and might damage the landing gear. But a powered approach would allow a pilot to fly slightly above stall speed, using the engines to maintain control. The approach would be conventional, familiar to pilots, and would use standard instrument-landing equipment. The orbiter was to have go-around capability and could abort a landing to come back for a second try. Coming in at a moderate approach angle, the pilot would flare immediately prior to landing, touching down at a low sink rate.[2]

Yet while a powered orbiter offered a host of advantages, it carried a strong disadvantage: weight. The engines were heavy; tanks with fuel added additional pounds. This would cut the payload, while demanding a larger and more costly shuttle to carry the extra mass to orbit. Still, while it was at least possible to envision an unpowered orbiter, there was little doubt that the shuttle would use a jet-powered booster. The configurations of 1970 and 1971 emphasized two-stage fully reusable designs with a winged first stage. It was to drop back into the atmosphere several hundred miles from the launch site and would have no alternative other than to fly back to that site under its own power. Design studies projected that this stage would be large indeed, with as many as twelve jet engines.[3]

Though the use of such engines appeared unavoidable, it at least appeared possible to reduce their number and the weight of their fuel. Studies showed that a flyback booster would require nearly 150,000 pounds of conventional jet fuel, a grade of kerosene, with another 45,000 pounds as the weight of the engines. It also appeared feasible to use hydrogen in the jets, which promised to save 100,000 pounds in weight of fuel. Moreover, the booster was lighter and hence would need fewer engines. Indeed, the move to hydrogen cut the booster's gross liftoff weight by as much as a million pounds.

In June 1970 NASA issued study contracts to General Electric (GE) and Pratt and Whitney (P&W) for nine-month studies of shuttle air-breathing engines using hydrogen. The work centered on existing turbofans: P&W's F401 and GE's F101, the latter being developed for use in the B-1 bomber. Mounted in a shuttle orbiter, such engines with fuel would weigh about 20,000 pounds. But the use of hydrogen cut this total by some 2,500 pounds, for a modest but welcome saving.[4] Meanwhile, work with test aircraft at Edwards Air Force Base was shedding new light on issues of unpowered landings in high-performance aircraft. This work went forward not only at Edwards proper but at NASA's Flight Research Center, which shared its facilities.

NASA-FRC was physically located within the built-up area of Edwards and bore somewhat the same relationship to this base as Monaco holds to France. Each was a kingdom within a kingdom. Monaco and FRC both stood as small enclaves surrounded by much larger territories but had their own separate administrations. NASA, which operated FRC, certainly was not the Air Force. In addition, while flight testing at Edwards itself generally addressed issues of aircraft development, FRC was a research center. Its pilots engaged in flight research, which called for them to push frontiers.[5]

At Edwards, unpowered landings were a specialty and a requirement for its test pilots. New graduates of the Air Force's Aerospace Research Pilot School were required to demonstrate proficiency by executing power-off approaches and landings in an F-104. Other test pilots made such landings in the X-15, gliding to predetermined touchdown points from speeds above Mach 6 and altitudes up to sixty-seven miles.

These men controlled the X-15 through a technique known as "energy management." Amid instructions from a control room, a pilot arrived at a "high key," some thirty-five thousand feet above the landing point. After reaching the high key, he could usually proceed by using his eyes and drawing on his experience with an F-104 that had been configured to simulate an X-15 when landing. He would make a 180-degree turn and proceed to a "low key" at around eighteen thousand feet in altitude, then turn through another 180 degrees and continue onward to a landing on Rogers Dry Lake. If his speed was excessive, he could bleed it off by deploying speed brakes that angled outward from the vertical fin or by making shallow or tight banks. In this fashion the X-15 landed safely as a matter of routine.[6]

NASA test pilots, at FRC, were dealing successfully with similar challenges. They were flying lifting bodies, bathtub-shaped craft that lacked wings and obtained lift from their fat and stubby fuselages. Lifting bodies also lacked jet engines; liquid-fueled rockets propelled them above Mach 1, but they glided to unpowered landings. They were considerably more demanding than a winged shuttle orbiter, but they proved tractable as pilots even landed them successfully at night. They exhibited good handling qualities, sometimes after only modest modification. The M2-F2 lifting body, for example, received a center fin, a redistribution of internal weights, and a roll-control system that used reaction jets. In the words of one of its test pilots, these changes turned "something I really did not enjoy flying at all into something that was quite pleasant to fly."[7]

Lifting-body pilots found that the best way to approach the runway was to dive at it rather steeply. Like dive bombers, these craft showed greater accuracy with steeper angles of descent. Approaches took place at high speed, which improved the stability and made the control surfaces more effective. Speed brakes were essential, to modify the drag and adjust the flight path, but pilots found that such landings were actually less demanding than conventional powered approaches. Lacking power, they no longer needed to watch their throttles. Nor were hard landings a problem. These craft showed acceptable sink rates at touchdown, even with landing speeds as high as two hundred knots.[8]

Major Jerauld Gentry, who was among the most experienced of the lifting-body pilots, presented an assessment to the Society of Experimental Test Pilots in September 1970. He reported tests of landing accuracy on a 10,000-foot strip. Thirty unpowered landings took place, from altitudes up to 90,000 feet and speeds as high as Mach 1.9. The average landing dispersion, using high-angle approaches, was less than 250 feet. "Many of us at Edwards feel that the requirement for the orbiter to have landing engines may be neither practical nor necessary," Gentry declared. He also reported results of unpowered instrument-flight approaches and landings in an F-111. It had variable-sweep wings and could fly with a range of gross weights, thus permitting flight at low lift-to-drag ratio and high wing loading. Pilots powered down their engines at Mach 2 and 50,000 feet, then navigated using on-board inertial navigation and a standard Instrument Landing System (ILS). They flew at night and under a hood, gliding to within a few hundred feet of the ground. The pilot then took over visually.

"All the pilots felt that these approaches were less demanding than flying a conventional low-speed powered ILS approach," Gentry reported. The approach speeds were sufficiently high that pilots did not have to worry about stalling if the speed fell off while maneuvering. In addition, they could adjust their glide slopes by making pitch changes only, rather than having to use both pitch and power.[9]

But at the Manned Spacecraft Center, "Deke" Slayton, head of the astronaut office, doubted that this experience would apply to the space shuttle: "We believe that their experience only indicates that, given the unique conditions of the Edwards environment, unpowered landings can be accomplished safely if the vehicle can be maneuvered through reentry to certain initial conditions relative to the desired landing point. It is not intended, however, to operate the orbiter under the same conditions as exist at Edwards, and the effects of these differences need to be assessed operationally before the decision is made to remove the orbiter engines." He added that at Edwards and FRC, "test pilots spend many practice hours in fixed base and free-flight simulators immediately prior to each flight. This provides a degree of proficiency that will not be available to the orbiter crew returning, perhaps, from seven days in orbit."[10]

Slayton's views carried weight, for in NASA the astronauts were kings. But during 1970, the Air Force emerged as a strong influence in favor of unpowered orbiter landings. This did not happen specifically because its commanders endorsed the views of Gentry and his fellow pilots but because they wanted the shuttle to lift heavier payloads. The shuttle concept had initially taken form as a logistics vehicle, supporting a space station. The shuttle and station were intimately linked; the shuttle's baseline mission was to carry twenty-five thousand pounds to a space-station orbit. That very intimacy, however, nearly became the shuttle's undoing, as it barely survived on close votes in both the House and Senate. New political support was vital, and the Air Force proved willing—if NASA would build a shuttle that it liked.[11]

The Air Force had its own basic mission: to launch heavy reconnaissance satellites. At a meeting with NASA officials on 2 October the Air Force insisted that it wanted forty thousand pounds in polar orbit, the equivalent of sixty-five thousand pounds launched due east from Cape Canaveral. NASA responded that its baseline of twenty-five thousand pounds was reduced due to the weight of jet engines carried aboard the orbiter. By removing them, the payload indeed would approach the Air Force's requirement.[12]

This design change did not turn the orbiter into a glider. The engines remained, but now they took the form of removable kits. They could be installed for flight test, for ferry, and for use in landing if the payload weight was modest. This raised the prospect of flying to orbit with these engines during initial missions, then removing them and accommodating full-weight payloads after operational experience was in hand. In this fashion, unpowered landings could become standard.[13]

During 1971 and 1972, the shuttle's configuration changed dramatically. The fully reusable winged first stage faded, to be replaced by a pair of solid rocket motors. The orbiter shrank in size as its propellants, which had been carried internally, went into an external tank. But amid the new concepts, the requirement for jet engines on the orbiter persisted with little change. Thus, a summary of Level I design and performance requirements, in April 1972, included:

Airbreathing Engine Subsystem (ABES)
The ABES shall provide propulsion for safe controlled vehicle operation during atmospheric flight when required for mission or ferry operations.

Loiter Time
The airbreathing propulsion system, when utilized for orbital missions, shall be required to provide 15 minutes loiter time at 10,000 feet standard day or higher to allow operational assessment of conditions existing at completion of reentry. This requirement shall not size the Space Shuttle System.

Ferry
The orbiter vehicle shall be capable of ferry flights between airports.

The only concession for unpowered landings was that the shuttle would not have to carry both the engines and a sixty-five-thousand-pound payload.[14]

Test pilots at FRC opposed the requirement for loiter, for they believed it would give a shuttle commander a false sense of security. They continued to favor unpowered shuttle landings, basing this preference on the success of some ten thousand unpowered approaches and landings in F-104s, as well as two hundred such landings in the X-15 and another hundred in lifting bodies. However, no one had yet shown that one could land a large airplane using this technique.

An initial series of tests, again at FRC, used a B-52. Its landing gear and flaps were extended to increase drag. Engines were throttled back to idle, except for a slight amount of power needed to run the plane's accessories on one engine. During the landing approaches, the B-52 dived at angles up to fifteen degrees. This ensured that the plane would have enough energy and speed to reach the runway and execute a successful flare immediately prior to touchdown.

During some tests, the pilot could see the runway through the windshield; in others, he was under a hood to four hundred feet in altitude and had to rely on instruments. All the attempts achieved success. One lifting-body pilot, who had never flown a multi-engine airplane, made a series of successful unpowered approaches. He subsequently tried to land the B-52 in the conventional fashion, using a powered low-angle approach, and was unable to do so. In addition, two pilots from United Airlines received permission to try a series of unpowered landings. In the words of Fred DeMeritte, the lifting-body project manager at NASA Headquarters, "They were amazed at the precision and ease involved with the unpowered approach concept." DeMeritte noted the accuracy of the touchdowns: "The landings seldom varied more than several hundred feet from the specified touchdown point." Moreover, "the pilot is never put in a position of requiring power for a successful landing."[15]

A complementary series of unpowered landings grew out of investigation of the shuttle's requirements for ground navigation aids. NASA's Ames Research Center began examining this issue in 1969. In 1972 it extended this work by using its Convair 990, which could simulate shuttle aerodynamics by deploying landing gear and speed brakes while throttling the engines back to idle. An on-board computer not only controlled the 990 during its descent but also conducted automated approaches and landings. At the FRC, these followed the B-52 tests.

"We've shown that—within the limitations of our tests—airbreathing engines are not necessary for landing the shuttle," Fred Drinkwater, a research pilot at NASA-Ames, told *Aviation Week*. "We've had a long history of unpowered landings with low L/D vehicles, but mostly with piloted landings of small aircraft such as the X-15 and the lifting bodies. The computer in the shuttle simulation avionics allows us to manage the energy of a larger and more sophisticated vehicle in a no-thrust situation."

This work contributed to a NASA decision to go to unpowered shuttle landings as a general standard, reserving the engines only for ferry. The engine layout changed accordingly. In July 1972, at the time of contract award, Rockwell's orbiter was to carry two mission engines, deployed from within the payload bay, along with two more engines for ferry that could be set on struts. By December the payload bay no longer held engines; these were to be mounted to the orbiter's flat undersurface, then removed for flight to orbit.[16]

Ferry range soon became an issue. The orbiter was similar in size to a DC-9 airliner, but was nearly twice as heavy in loaded weight. It also produced a good deal of drag. Hence it needed as many as five ferry engines, whereas the DC-9 flew nicely with only two.[17] The orbiter also lacked an airliner's capacious wing tanks; it had to get along with a single tank of limited capacity that could fit in the payload bay while remaining within structural weight limits. By late 1973 it was clear that with its numerous engines and its restricted fuel supply, the shuttle's range would be no more than five hundred miles.

This was unacceptable. NASA and the Air Force expected that shuttle orbiters would fly across the nation, landing after some missions at remote sites and returning either to Cape Canaveral or to Vandenberg AFB in California. Indeed, with one such contingency landing strip being at Hickam AFB in Hawaii, NASA expected the orbiter to fly twenty-five hundred miles to return to the mainland. Intermediate landing fields would be costly within the United States and out of the question within the Pacific. Aerial refueling demanded significant design modifications and appeared far less feasible than when suggested in 1969.[18] The only alternative was to drop the use of jet engines for ferry. This meant removing them entirely; they no longer would serve any purpose. Instead the orbiter was to ride atop a large carrier aircraft for ferry, an airplane with transcontinental range.[19]

Inevitably, test flights were to precede operational ferry flights, with this carrier-orbiter combination demonstrating safe takeoff, cruise, and landing. As a related issue, the orbiter required its own test flights to demonstrate safe unpowered landings. At NASA's Johnson Space Center a branch chief, John Kiker, took the lead in arguing that for this purpose the orbiter should glide from the back of its carrier aircraft.

Kiker supported his argument with flight tests of small radio-controlled models. His carrier airplane was a Sterling Gazariator, piston-powered and much appreciated by hobbyists. His shuttle orbiter was of plastic foam, at one-fortieth scale. In airborne tests, it separated cleanly.[20] Furthermore, it was not exactly news that a shuttle orbiter might fly from the back of a much larger carrier. Two-stage fully reusable shuttle configurations, which had received detailed attention prior to 1972, had called for the orbiter to do precisely that.

Aeronautical history also encouraged such thoughts. Within the shuttle community, many people saw presentations by the independent rocket specialist Robert Salkeld, who looked ahead to an era of advanced shuttle vehicles that could fly to orbit with a single stage. Salkeld liked to show photos of the Short-Mayo Composite of 1938, the first commercial airplane to cross the Atlantic. It had been a two-stage aircraft: a large flying boat carrying a smaller land plane, which separated in flight and flew on. Salkeld declared that the two-stage shuttle concepts of the day resembled the Short-Mayo, with his futuristic single-stage shuttle corresponding to transatlantic airliners that the world came to know.[21]

91

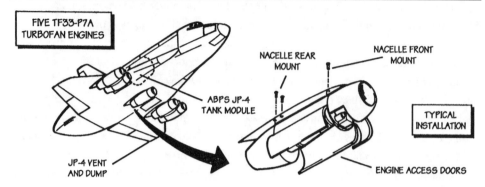

Fig. 13. *Concept for a five-engine installation on the orbiter. (NASA, courtesy Dennis R. Jenkins)*

In February 1974 NASA deleted its requirement for jet engines on the orbiter. A new review of overall shuttle requirements, issued the following month, made no mention of such engines whatsoever. The orbiter now was a glider and would remain so.[22]

Choosing the Carrier Aircraft

In anticipating the use of a carrier, NASA needed the largest airplane it could get, and its first thoughts were of the C-5A cargo aircraft. This appeared readily available because the Air Force might provide one from its existing fleet, as part of its cooperation with NASA on the space shuttle. By contrast, a Boeing 747 would have to be purchased, which would add cost. In August 1973 NASA awarded a $1 million study contract to Lockheed, builder of the C-5A. In October a similar contract went to Boeing, to examine the possible use of a 747. These preliminary studies, which went far beyond Kiker's models and Salkeld's trip down memory lane, shaped the NASA decision to delete the orbiter's jet engines.

Of particular concern was whether the orbiter could separate cleanly from the back of the carrier without striking its tail surfaces. The research included some six hundred hours of wind-tunnel testing. Boeing's facilities included a tunnel at the nearby University of Washington, while Lockheed used an in-house tunnel at its plant in Marietta, Georgia. In a supporting effort, McDonnell Douglas conducted some twenty-five hundred computer simulations of separation maneuvers with the orbiter and carrier moving and turning freely in all directions.[23] While this work was under way, John Conroy, a veteran of Apollo, stepped in with a proposal for an entirely new carrier aircraft. It was to drop the orbiter while in flight as if it were a bomb.

More than a dozen years earlier, Conroy had purchased several surplus Boeing Strato-cruiser airliners. He went on to fit them with enormously swollen and voluminous fuselages, crafting the Pregnant Guppy and, later, the more distended Super Guppy. These transported the S-IV and S-IVB upper stages of the Saturn-class rockets. Now he promised to build a similarly grandiose shuttle carrier called Virtus, Latin for "courage." It was to carry the orbiter between twin fuselages, beneath the wing. With a wingspan of 450 feet and length of 280, it would dwarf a 747, which had a span and length of 196 and 232 feet, respectively.

Conroy expected to use a cockpit and forward fuselage of a C-97, the Air Force's version of his beloved Stratocruiser, to eliminate the high cost of developing a new design. He also expected to use landing gear taken from surplus B-52 bombers. Even so, this craft would take two years for construction and six months for flight test and certification. As an aircraft of entirely novel type, it carried the risk of cost overruns and schedule delays. By contrast, the C-5A and 747 were known quantities, and NASA preferred to choose one or the other. Conroy would win attention only if there was excessive hazard in flying an orbiter from the backs of both.[24]

Analysis showed that this risk indeed existed for the C-5A. Its T-shaped tail, with the horizontal stabilizer high atop the vertical fin, produced aerodynamic effects that inherently would cause the aircraft to pitch toward the orbiter during separation. A C-5A pilot could prevent a collision with the tail with a large and carefully timed movement of the controls, but if this was not done properly, the collision indeed would take place—and could shear off the horizontal stabilizer. "There are no known instances of aircraft returning for a safe landing" following such a loss, noted John Yardley, NASA's associate administrator for manned space flight. The C-5A would tumble out of control and crash, allowing little chance for the crew to save themselves by ejecting.

The 747 was much safer. It lacked a T-shaped tail; its tail showed a conventional configuration, with horizontal stabilizers mounted to the fuselage, below the orbiter, and a vertical fin standing alone. Its aerodynamics were far more favorable for air launch. In Yardley's words, "The inherent characteristic of the 747 is to pull away from the Orbiter, thereby aiding separation." The pilot would not need to take sudden evasive action to prevent collision. Yet if the orbiter did collide with the fin (the horizontal stabilizers being below and out of the way), the consequences would not be catastrophic. The 747 could lose a large portion of this fin and still return safely.

The 747 had other advantages. Carrying its orbiter, it had an estimated range of 2,320 nautical miles, enough for a nonstop transcontinental flight. Some hopeful souls even argued that it could reach the mainland from Hawaii with this orbiter, though this proved not to be true. Still, the C-5A had much less range and would require in-flight refueling. The standard C-5A was equipped for this, but there was no experience in refueling it with an orbiter on its back. Hence it would be necessary to develop this experience by flying a C-5A with a dummy orbiter. The 747 avoided this problem.

The 747 had a further safety advantage in that the presence of the orbiter was most desta-bilizing while mated, allowing flight tests of this reduction in stability prior to flights with actual air launch. This was not possible with the C-5A, for its greatest destabilization oc-curred just after separation. Again, this difference resulted from the dissimilarity between the two airplanes' tail shapes.

The 747 could use shorter runways than the C-5A, if an engine were to fail during takeoff. The 747 could mount engines of greater power to increase the air-launch altitude, for better re-alism of the orbiter flights that were to simulate return from space. Carrier aircraft were ex-pected to see extensive use, at a time when people expected to fly the shuttle up to sixty times per year. This gave a further advantage to the 747, for it had a structural life of sixty thousand hours while the lifetime of the C-5A wing was no more than twelve thousand. This reflected the fact that commercial 747s flew every day, whereas the military C-5A flew less frequently.[25]

The decision, the choice of carrier aircraft, came from within NASA's center in Hous-ton. In May 1974 the center director, Christopher Kraft, wrote to William Schneider in Wash-ington, the acting associate administrator for manned space flight: "Dear Bill: This letter re-quests authorization for NASA Johnson Space Center to purchase a Boeing 747 aircraft."[26]

On 13 June the Space Shuttle Program Office gave a briefing to the NASA/DOD Space Transportation System Committee, comparing the C-5A and 747 and recommending the lat-ter. This committee concurred in the choice. On the next day George Low saw this briefing, and Low gave approval to Kraft's request of a month earlier. The procurement went through quickly, with NASA paying $15.6 million for a used Boeing 747-123 of American Airlines. NASA did not name it but merely gave it a new registration number, N905A. Its commercial markings remained plainly visible during subsequent years in the space program.[27]

During August, soon after NASA took ownership, this 747 contributed to aviation safety by conducting thirty wake vortex research flights. Wake vortices are narrow zones of extremely severe turbulence that trail for miles behind an airplane's wingtips. They are particularly strong when the airplane is large. Smoke generators, attached to the 747's wingtips and aft fuselage, made the vortices visible. A Lear Jet and an Air Force T-37 jet trainer, flying as chase aircraft, flew close to the danger zones. During one flight, the T-37 flew more than three miles behind the 747, but nevertheless rolled violently. A research pilot suggested that a safe separation be-tween such aircraft, when landing, would be as much as three times that distance.[28]

Because the orbiter was to be mounted in front of the vertical fin, it reduced that fin's ef-fectiveness. To restore the diminished stability of the 747, designers crafted rectangular fins, ten by twenty feet, to fit on the ends of the horizontal stabilizers. Wind-tunnel tests at the Uni-versity of Washington, close to Boeing's main plants, verified their usefulness. In addition, Boeing won a $30 million contract from Rockwell to carry through the 747's physical mod-ifications. The work took place at the 747's production facilities near Everett, Washington, between August and December 1976.[29]

Significant structural modifications also were part of the work. The commercial 747 had been built to carry passengers or air cargo on a strong deck within the fuselage. The fuselage proper withstood internal cabin pressure and external aerodynamic forces, but it most certainly had not been built to support the weight of a 150,000-pound orbiter. The fuselage of NASA's new jet therefore had additional bulkheads installed, for extra strength. In accordance with design principles dating to the 1930s, the 747 used "stressed skin" construction. Its loads, weights, and stresses were not only borne by the internal framework but were carried in part by the aircraft skin, which served as a major structural element in its own right. This skin was reinforced in critical areas, with overlays of sheet aluminum being riveted into place.

Other fittings, mounted atop the fuselage, supported the orbiter itself. These had the form of struts, mounted in locations matching the socket fittings that mated the orbiter to the external tank. The forward struts had the shape of an inverted V, poking into the underside of the orbiter, and came in two varieties. Those intended for the Approach and Landing Tests could telescope in length, to raise the nose of Enterprise for easy separation; the second version, for use in operational ferry missions, were fixed in length. Rear orbiter supports were mounted atop the aft fuselage. These structural modifications added some 11,500 pounds to the empty weight of NASA's 747. It therefore was given a stronger and more capable rudder control. In addition, the extra weight demanded more power from the plane's engines, which therefore went into the shop for their own improvements.

The cockpit received controls and displays needed for the air launch and ferry missions. These included a sideslip indicator because the 747 had a tendency to yaw when mated with the orbiter. Other new equipment included telemetry at frequencies close to one gigahertz and transponders at frequencies five times higher. Under a separate contract, the 747's engines—JT-9Ds from Pratt and Whitney—were refitted to boost the maximum thrust from 43,500 to 46,950 pounds. They also were modified for water injection. This sprayed water into the engines' hot internal airflow during takeoff, cooling the air and making it denser so as to burn more fuel. Water injection promised additional takeoff thrust for use in ferry missions, when the 747 would carry both the orbiter and a full load of fuel.

The work was completed early in December 1976. The 747, well modified, flew on a short hop from Everett to Seattle's Boeing Field, for a flight-test program that ran through the Christmas holidays. In January 1977 Boeing turned the plane over to Rockwell, which flew it to Edwards Air Force Base for use with the Enterprise. This too was ready, again following several years of preparation.[30]

Construction, Roll-Out, Prelude to Flight

Structural assembly of this orbiter did not take place at Edwards, but at Air Force Plant 42 in nearby Palmdale. The work began as early as June 1974, but it took time for major components to arrive from Rockwell's subcontractors and from that firm's Space Division.[31] Meanwhile, it appeared appropriate to demonstrate through flight test that an aircraft resembling the orbiter could execute an unpowered approach and land with accuracy on a concrete runway. This had not been done; previous work with lifting bodies had had them land on marked airstrips within a dry lake bed, which offered far more room.

The new tests made use of another lifting body, the X-24B, which had been built for supersonic flight. It had a long, sharply pointed nose and a set of rocket engines. It lacked wings, but had a flat undersurface that produced lift. On a flight in August 1975 the NASA test pilot John Manke dropped from his B-52 mother ship, climbed to fifty-seven thousand feet, began his descent, and made his approach. Seven minutes after launch, he touched down on the planned landing spot, five thousand feet along Runway 04/22, the fifteen-thousand-foot main concrete runway at Edwards. Two weeks later NASA test pilot Mike Love duplicated this feat as he glided down from Mach 1.5 and seventy-one thousand feet. "We now know that concrete runway landings are operationally feasible," said Manke. "Touchdown accuracies of plus or minus 500 feet can be expected."[32] By then hardware for the orbiter was well along in assembly. The process had begun with structural design. Preliminary studies defined requirements and gave what amounted to an outline of the solutions, but these studies did not produce working blueprints that could go to the shop floor. These blueprints emerged over time through a well-defined process that relied on an interplay between design and analysis. Stress analysis was the key, to ensure that shuttle components could withstand their loads without buckling or bending. There were aerodynamic loads from dynamic pressures on the airframe; there also were propulsion loads from the thrust of rocket engines. Designers therefore had much to learn from specialists in those fields.

Stress analysis relied on well-defined mathematical methods. These now were implemented in computers through programs called NASTRAN, NASA Structural Analysis. Engineers, drawing on their experience, laid out patterns of ribs and spars for the wings and of frames and other supports for the fuselage. They did not do this merely once; there was a great deal of interplay. Fuselage people might argue that they could cut the overall weight if the wing would carry more load. Wing people would respond with design changes, studied using NASTRAN, and would answer this argument. Specialists in manufacturing kept watch, particularly when a blueprint called for titanium or a heat-resistant steel. Drawing on the experience of machinists, these experts could recommend changes that would enhance manufacturability. Interfaces drew continuing attention. Major components of the orbiter, such as the fuselage and wings, were being built by different companies; managers had to be sure that they fit together. Engineers learned new lessons as their designs and analyses matured, leading to further changes.[33]

It was of great importance to maintain, at all times, a complete definition of the shuttle's current design configuration, and to systematically understand the consequences of any pro-

Fig. 14. *X-24B lifting body, escorted by an F-104 chase plane. The nose-high attitude of these aircraft produced more lift. (NASA)*

posed change. Within the program management, assessment of these consequences was the responsibility of change control boards. Leroy Day, deputy program manager at Headquarters, describes such a board as "a combination of people at particular management levels who sit together to see if what's being proposed today will affect any other part of the system, as well as to examine the cost and schedule implications of something new that somebody wants to do." Level IV boards dealt with design issues in individual projects, such as the main engine. Boards at Levels III, II, and I addressed changes in requirements. The decision to delete the jet engines came at Level I.

For example, a manager working on the main engine might seek to replace one material with a different one. This change would have no impact on engine reliability or anything else. "Then," Day notes, "he had his own cost target and everything to stay within, and he could make that change himself. He didn't have to take that to the next level. On the other hand, if

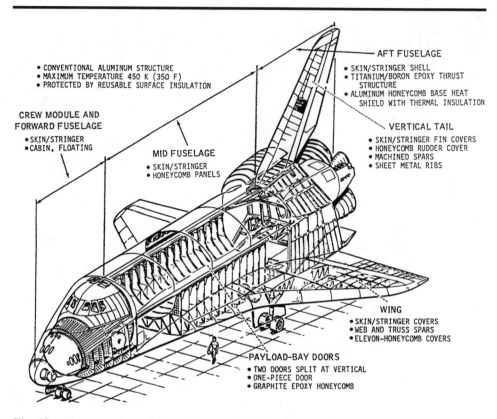

Fig. 15. *Cutaway view of the orbiter, emphasizing structural details. (Rockwell International)*

he wanted to change something that would interact with the way the engine was controlled from the Orbiter, then he had to bring this kind of a change to the Level II change board where it would be discussed with representatives from all other parts of the program. If there was a question on what the impact was, then it was studied in depth and brought back for a subsequent decision." Day describes such boards as "the fundamental building blocks of the organization, the thing that really kept the program on target."[34]

Through such means the design process converged, yielding lightweight configurations that appeared optimal. During 1975 production of major elements of the orbiter went forward as well. The forward fuselage came together at Rockwell's plant in Downey, California. The mid-fuselage, built at Convair in San Diego, reached Plant 42 and was joined to another Rockwell product, the aft fuselage, with three enormous holes for the engines. Wings, shipped from Grumman, were mated to each side. The forward fuselage went to Palmdale in early

November. By the end of that month, the orbiter was in the state depicted on that *Astronautics & Aeronautics* cover: wings in place, vertical fin from Fairchild Republic in place, fuselage open along its entire length and resembling a canoe.[35]

What was happening at that moment, late in 1975, within the overall shuttle program? Nearly four years had passed since Nixon had given his assent. Some forty thousand people were employed within the effort—a buildup that was continuing. Peak employment of forty-seven thousand workers was scheduled for 1977 in hundreds of commercial plants around the nation.

Rocketdyne, builder of the main engine, had fabricated and tested its principal components: thrust chamber, turbopumps, propellant injectors, and a nozzle. These had been exercised on company-owned stands at Santa Susana, California. Some one hundred component tests had been conducted. Moreover, a prototype of a complete engine was in place at NASA's main rocket center, the National Space Technology Laboratories in Mississippi. This engine had already fired more than twenty times; engineers were working toward its first burn at rated power level.

Subcontractors were proceeding with the manufacture, heat treatment, and machining of the prototype case for the solid rocket motor. Thiokol, the prime contractor, was fabricating a prototype of a flexible bearing that would allow the nozzle of an SRB to swivel in flight for thrust-vector control. Work on these solid boosters was progressing smoothly with no major problems. Smaller engines and power units were also advancing. In Sacramento, California, Aerojet was proceeding with rockets for the orbital-maneuvering system. Managers had tested candidate injectors and had selected a baseline design. The firm of Marquardt was testing complete prototypes of forty-four small thrusters for reaction control. The Auxiliary Power Unit (APU) was to power the hydraulic system, with hydrazine as its fuel. It too had undergone full-up operation at its manufacturer, Sundstrand.

Martin Marietta, builder of the external tank, had contracted for all the major weld tools. These were being installed at NASA's Michoud Assembly Facility. Welding of an initial set of panels for a test article had been completed in September. Fabrication of a complete prototype, to undergo structural testing, had begun in July. In addition, new construction was under way at Kennedy Space Center. In October concrete paving had been completed for a landing runway and towway. Workers were also beginning to build the Orbiter Processing Facility, which was to provide overall maintenance and preparation for reflight. Development of the Launch Processing System was also progressing, as an array of computers and sensors that would monitor the orbiters closely during preflight check-out.[36]

A year earlier, in January 1975, George Low had advised the OMB, "We would soon be taking the steps which would make it impossible to launch a backup Skylab or, for that matter, a Saturn V." Those steps included major reworking of existing Apollo facilities at the Cape. Launch Complex 39 was receiving substantial change, as was the interior of the immense Vehicle Assembly Building. On one of the Mobile Launch Platforms that had carried the Saturn V, the Launch Umbilical Tower, 380 feet tall, was being dismantled.[37]

Shuttle avionics and subsystems were also in development. IBM was to provide five on-board computers for each orbiter; these were in production. Inertial measurement units, built by Singer-Kearfott of Little Falls, New Jersey, were to provide information on attitude and velocity change. These had completed acceptance testing and had been delivered for installation. Other black boxes were under test at Rockwell's Avionics Development Laboratory in Downey. Some of these electronic systems would control the hydraulics, and complete flight-control systems were receiving verification. At the assembly plant in Palmdale, automatic check-out equipment was in use. Within the partially completed orbiter, high-voltage tests had verified the integrity of the wire harnesses.[38]

The completed orbiter, designated OV-101, served as the test vehicle for the Horizontal Ground Vibration Test, conducted in Palmdale. Earlier vibration testing had used an accurate structural model, at one-quarter scale, with water in its external tank to simulate liquid oxygen and air replacing the very lightweight liquid hydrogen. The new tests gave engineers their first opportunity to verify their mathematical models by taking data on the structural dynamics of an actual flight orbiter. In one set of tests, OV-101 was loosely supported, to simulate its structural behavior during reentry and landing. In a second series, this orbiter was held rigidly at its external-tank supports to represent the configuration during ascent. The tests vibrated this vehicle at frequencies from 0.5 to 50 hertz, determining natural or resonant frequencies and their damping. Other measurements determined frequency response at the locations of sensors used for guidance and control.[39]

These vibration tests took place during the summer of 1976, shortly before the roll-out. The roll-out of a new aircraft is somewhat like the entrance of a queen, and NASA was ready to celebrate with a brass band and plenty of red, white, and blue bunting. In keeping with the Bicentennial, the orbiter was to be christened Constitution. But one well-connected space buff, Richard Hoagland, had other thoughts.

To call Hoagland an enthusiast would be an understatement. He had turned up at NASA's Ames Research Center during the summer of 1975, where a group of academics had been studying concepts for space colonies. He gave a talk and spoke with high emotion, as if addressing a revival meeting, leading one participant to write, "Reverend Hoagland came and held a service for us." Years later he would actively assert that the "face of Mars," an array of natural topographic features having some resemblance to a human face, offered strong evidence that this planet had hosted an advanced civilization.

Hoagland, however, had worked as a science advisor within Walter Cronkite's organization at CBS and was intimately associated with the fans of the TV series *Star Trek.* He considered that OV-101 should be called *Enterprise,* after the starship, and persuaded his fellow Trekkers to bombard the White House with a hundred thousand letters that demanded this name. Here indeed was the voice of the people, a voice that President Gerald Ford could not ignore.[40] Within NASA, some officials disliked that name, as it suggested a link between the shuttle program and the television series. Others supported it, asserting that it would give the program ready recognition. The final choice, however, was Ford's. On 8 September he met with James Fletcher and gave his assent.[41]

The roll-out took place nine days later. Two thousand people showed up, some of them driving from Los Angeles along the Antelope Freeway. Morning mist was in the air along that route; mountains stepped into the distance in tiers, successively hazier and less distinct. Palmdale, however, was warm and sunny. Guests included actors Leonard Nimoy, George Takei, and De Forest Kelly. On *Star Trek,* they had played the roles of Mr. Spock, Mr. Sulu, and Dr. McCoy. The band, in fact, played the theme from that series. Then the orbiter, white and black with the name "Enterprise" prominent on its side, came into view as a tractor pulled it from behind a wall. Spectators were surprised at its size. With a length of 122 feet, it was about as large as an airliner. The seats for the crowd were close to it, and people were free to walk beneath it and touch it.[42]

It was not much of a starship. It lacked all propulsion; the main engines, orbit-maneuvering engines, and attitude-control thrusters were dummies. In place of its thermal protection, blocks of polyurethane foam covered the surface. The nose and wing leading edges, which would be of temperature-resistant carbon-carbon composite, instead were of glass fiber. Enterprise had fuel cells for on-board power, but these held their hydrogen and oxygen as gases in high-pressure tanks, not as liquids in cryogenic dewars. Hence they could not run for very long.

The flight deck lacked many instruments and displays that would be needed for ascent to orbit. The crew quarters needed such amenities as the galley, the waste-management system, and mid-deck lockers. The payload bay was not fitted out to accommodate its payloads, while the bay's doors had no hydraulics and no radiators to get rid of waste heat. In flight, the crew was to lower the landing gear by triggering explosive bolts and letting gravity do the rest. But this gear lacked a hydraulic system for retraction.[43]

By comparison with later operational orbiters, Enterprise was little more than an aluminum shell swathed in Styrofoam. Such critical elements as the propulsion and the thermal protection were present merely as mockups. Even so, it *looked* like a space shuttle, which raised a question: how could something so incapable of space flight appear so convincing? The answer lay in aspects of design and development. Within the overall shuttle program, Enterprise was the easy part. It was an aluminum airframe, a prototype that was simple to design and construct. Its design was a serious exercise, with engineers preparing blueprints for every one of a plethora of structural parts, while other blueprints presented subassemblies and larger structural assemblies, and with the wings and other major components undergoing stress analyses in computers. After this, it was easy to fabricate the parts and assemble them, for everyone had long experience in manufacturing aluminum airplanes.

Fig. 16. *Enterprise on public display following roll-out, September 1976. (NASA)*

Structural design was demanding but straightforward, lending itself to mathematical precision. Thus a reasonably mature design could be in hand during so early a year as 1976. The on-board systems of Enterprise were far less mature, but this was acceptable. They did not demand long life, for they were to operate only during the limited time of one of John Kiker's airborne test flights. Nor did these systems face anything like the demands of an orbital mission. The computers of Enterprise had to accommodate only a modest repertoire of commands. The need for electric power was brief. Such systems thus were prototypes in their own right, highly restricted in capability but nevertheless adequate to the needs of the moment.[44]

Yet if Enterprise was the easy part, an operational space shuttle stood as the hard part. The difference lay in development, a process that took years. The Space Shuttle Main Engine (SSME) was already enmeshed in this. As early as 1971, a prototype thrust chamber had fired successfully—for fractions of a second. But it took another decade to bring the complete engine to a level of development whereby astronauts could ride it—and trust it with their lives.

Like Enterprise, that first thrust chamber was a very early design that had been prepared with care and then fabricated. Yet while photos of its tests looked impressive, this prototype foreshadowed Enterprise by dodging most of the difficulties of a working rocket engine. The 1971 test model lacked turbopumps, which are very demanding, while its brief firing durations prevented it from becoming so hot that it would burn a hole in the side and explode.[45]

For the SSME as well as for other systems, the devil was in the details, and development meant attention to detail. It was not hard to change blueprints, but it often was hard to learn what change was needed. The lessons came from developmental tests, starting with brief durations and gentle operating conditions, progressing in time to prolonged durations and demanding conditions. In the course of any test, a failure might arise. It had to be diagnosed and pinned down, its source and nature discovered. Only then was it possible to change the blueprints and to make the needed design changes—amid general understanding that another problem might show up next month.

Complex systems, such as the SSME, could be certified for flight only after they showed through test that they could stand up to durations and conditions even more severe than those of an actual mission. In turn, test items had to do this not just once but repeatedly, for many problems did not show themselves until the system had undergone long and repeated operation. The process of development, of increasingly severe tests with problems found and fixed, then could take a fragile and untrustworthy prototype and turn it into a rugged, robust system that held a demonstrated reserve of strength and capability. But this process required a great deal of time.[46]

Enterprise looked like a shuttle, but it was not one. Yet it held more than outward show; plans called for it to be rebuilt in time to an orbital configuration, with thermal protection and SSMEs. For the moment, it was well-suited to its limited purpose. Mated to the 747, it was to help check out the joined pair for ferry operations, then fly from that carrier's back in the subsequent landing tests.

On the day after the roll-out, Enterprise left its hangar for a second time and went on display at an open house for Rockwell employees. It then returned to its hangar for checks of its on-board systems (yes, it did have them). Late in January 1977 it was mounted to a ninety-wheel trailer, again being supported at the external-tank fittings. Moving at three miles per hour, a diesel tractor towed it along back roads that spanned the thirty-six miles between Plant 42 and Edwards AFB. The trip called for more than clearing the route of traffic; power lines had been relocated, while street signs were mounted on hinges that allowed them to fold beneath the wings. When this wide load reached Edwards, its 747 was waiting.[47]

NASA-FRC had a new name: the Hugh L. Dryden Flight Research Center. This center had a large steel framework, the Mate-Demate Device, which soon surrounded the orbiter. Cables hoisted it high. The 747 rolled beneath it; workers secured it to that carrier's back. The mated combination was towed to an Air Force hangar for ground vibration tests and weight and balance calculations. These were preludes to further tests. In these the 747, Enterprise on its back, first was to taxi down a runway and then would lift into the air, simulating ferry.[48]

The Approach and Landing Tests were to follow, and the astronauts who would fly the orbiter were ready: Joseph Engle, Charles Fullerton, Fred Haise, and Richard Truly. Haise was a veteran of Apollo. He had faced life-threatening peril as a member of the Apollo 13 crew in 1970, when the explosion of an on-board tank, while en route to the moon, disabled the spacecraft by knocking out its electrical power. Haise helped save his crew through his close knowledge of the electrical system. He also was an expert on the Lunar Module, which served as a lifeboat and enabled the crew to return safely to Earth.[49] Selection of the other three men showed that the torch was being passed to a new generation, for none of them had flown Apollo. Nevertheless, all three were graduates of the Aerospace Research Pilots School at Edwards AFB, with Truly staying on as an instructor. Indeed, Truly was as able a man as ever earned an astronaut's wings, for he went on to pursue a career as a high-level administrator while rising to the rank of vice admiral.

Outstanding aviators seek hazardous duty. Serving with Fighter Squadron 33 in the early 1960s, flying combat jets from the carriers USS *Intrepid* and *Enterprise,* he specialized in night carrier landings, which took away normal visual cues. He graduated quickly from test pilot to astronaut, for in November 1965 the Air Force chose him as one of eight men who were to fly the Manned Orbiting Laboratory. Following that program's demise in 1969, he transferred to NASA. He left that agency in 1983 to head the Navy's Space Command with responsibility for that service's surveillance and communications satellites. But he had made three flights to orbit in the shuttle, two of them as its commander. He returned to NASA to take over the shuttle program in February 1986, only weeks after loss of the Challenger. As associate administrator for space flight, he directed the efforts that restored the shuttle to service. In 1989 President George Bush named him NASA administrator; he remained in this office until 1992.[50]

Like Truly, Engle had no experience as an Apollo crew member. But this did not stop him from qualifying as an astronaut, for he made sixteen flights in the X-15. He repeatedly topped Mach 5; he also flew three missions that reached above fifty miles in altitude, thus meeting an Air Force criterion that gave him this qualification. This experience was invaluable, for the X-15 was the last airplane before the shuttle; its speed and altitude records would not be bettered until the shuttle flew to orbit. As a high-performance aircraft that made unpowered landings, the X-15 also helped prepare Engle for the ALT series.[51]

Fullerton's career also resembled Truly's, for he too had been a Manned Orbiting Laboratory astronaut, transferring to NASA as well when that program died. He held bachelor's and master's degrees from California Institute of Technology in mechanical engineering. Significantly, he did not come up by flying hot jets or rocket planes. He served in the Air Force with bombers, flying the B-47 for the Strategic Air Command and then working as a

Fig. 17. *The Mate-Demate Facility at Edwards Air Force Base. (NASA, courtesy Dennis R. Jenkins)*

bomber test pilot at Wright-Patterson Air Force Base, a major developmental center. It was a long way from Wright-Patt to Edwards; when he qualified for MOL, this showed that his talents were exceptional indeed.[52]

To prepare for the ALT flights, these astronauts used two Grumman Gulfstream business jets that had been heavily modified for use as trainers. The flaps swung up as well as down; fuselage-mounted panels generated side forces, and the thrust reversers could be used in flight. With landing gear extended and speed brakes deployed, these Gulfstreams reproduced the high drag of Enterprise. Further simulation appeared on the flight deck. In the right-hand seat, an instructor pilot flew with standard instruments and controls. An astronaut, in the left-hand seat, had an orbiter stick controller, orbiter attitude and systems instruments, and a single cathode ray tube display identical to three that were mounted on the orbiter's cockpit panel. Within each Gulfstream, an on-board computer received inputs from the astronaut's controls and flew the airplane by giving it the flight characteristics of the shuttle.

At McDonnell Douglas in St. Louis, dual ground simulators—one for the 747, the second for the orbiter—evaluated separation maneuvers. These flight simulators were realistic indeed; they included complete cockpits for the two aircraft, with the orbiter simulator resting on movable supports that reproduced the accelerations expected during actual air launches. Again, computers received inputs from the aircraft controls and generated appropriate cockpit displays. With the 747 in high-drag configuration—engines at idle, landing gear and spoilers extended—the orbiter lifted well clear within two seconds, climbing to 120 feet above the 747 during the third second after launch.[53]

First Flights

Enterprise, both when towed from Plant 42 to Edwards and when mounted atop its 747, sported a large fairing or tail cone extending to the rear of the aft fuselage. This smoothed its aft airflow. Without this fairing, that flow would be highly turbulent, buffeting the 747's vertical fin and making it more difficult for the 747's flight crew to accomplish the precision maneuvers needed for air launch. This fairing also reduced drag from the orbiter, increasing the 747's ferry range. During ALT flights, this drag reduction also helped the orbiter to glide less steeply and to stay aloft longer.

But the shuttle certainly would not fly to orbit with this tail cone. The goal of the ALT program, the focus of its most demanding flights, was that Enterprise had to glide to safe landings with this cone off. Inevitably this meant steeper and more sudden descents, which demanded caution. "We will approach it incrementally," a Boeing manager told *Aviation Week* in 1976. "First, some high-speed taxi tests, then a low-speed flight. Then some simulated launches at launch altitude, and if everything is okay, a launch with the cone off."[54]

The mated 747 amounted to a new aircraft, a heavy one with an unusual shape that had never flown before. Thus, although the standard 747 had accumulated millions of flight hours in routine service with airlines, the mated 747 had to start again at the beginning, with taxi tests. The plane with Enterprise on its back would not take off and head for the wild blue yonder, at least not yet. Instead, it was to trundle down Runway 04/22, the main concrete runway at Edwards, under power from its engines. These taxi runs would evaluate the technique of setting thrust for takeoff. They would also assess directional stability and control, elevator effectiveness during rotation prior to takeoff, airplane response in pitch, thrust reverser effectiveness, use of the 747's brakes, and airframe buffet.

Three taxi tests took place within the program, all on the single day of 15 February 1977.[55] The Dryden research pilot Fitzhugh Fulton was at the 747's controls, and he did not pour on the power by immediately shoving the throttles forward. The mated pair was top-heavy; a sudden surge of thrust could have caused it to tip back. Fulton therefore accelerated under intermediate power. He used full takeoff power only after reaching 55 knots, beyond which the elevators had enough authority to hold the 747's nosewheel on the concrete.

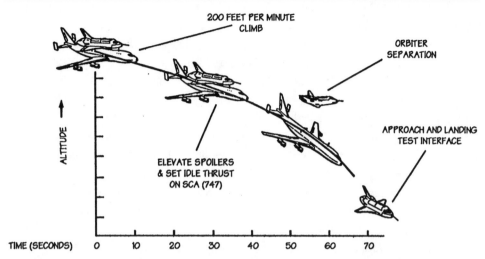

200 FEET PER MINUTE CLIMB

ORBITER SEPARATION

ALTITUDE

APPROACH AND LANDING TEST INTERFACE

ELEVATE SPOILERS & SET IDLE THRUST ON SCA (747)

TIME (SECONDS)   0   10   20   30   40   50   60   70

Fig. 18. *Approach and Landing Test flight profile. (Rockwell International, courtesy Dennis R. Jenkins)*

The first taxi test reached 76 knots, well below takeoff speed. Fulton then reversed thrust and slowed to 23 knots before applying the wheel brakes. Inspection showed no damage or overheating within the wheel assemblies, and Fulton received permission to turn his plane around and taxi in the opposite direction, at higher speed. This time he reached 122 knots. Fulton evaluated the elevator effectiveness during this run, raising the nosewheel momentarily between 95 and 100 knots. Again he reversed thrust and slowed, applying the brakes at 20 knots as his plane rumbled to a stop.

The third taxi test simulated an aborted takeoff, making good use of the fifteen thousand feet of runway length. Fulton accelerated to 137 knots, then cut the engines from takeoff power to idle. Pulling back on the controls, he applied elevator and raised the nose to a 5-degree pitch-up. The 747 rotated smoothly; Fulton held the nosewheel off the runway for some fifteen hundred feet before lowering it again. He then pushed the throttles forward and reversed thrust. This time he carried out a more demanding test of the brakes, braking between 49 and 40 knots. He indeed had come close to taking off; the plane would have lifted from the runway at around 145 knots and a pitch angle of 6.5 to 7 degrees.

The tests took place amid the coolness of early morning, for in the California high desert in February, temperatures can be well below freezing. Hence the temperatures of the brakes and wheels never topped 140 degrees Fahrenheit, which was acceptable. There was no aerodynamic buffeting or buildup of vibration. Fulton declared that his 747 responded so well that at times the flight crew could not tell that the orbiter was atop the fuselage, with a weight of 143,600 pounds. "As is oftentimes the case," he added, "the actual carrier aircraft-orbiter combination handled better than what we experienced in the simulator."[56]

The success of these taxi tests showed that the piggybacking pair was ready for takeoff. The next part of the program called for the 747 to fly, still mated to its orbiter. A series of six flights were to take place at increasing speeds, to evaluate handling qualities and airworthiness. Engineers wanted to learn the flight conditions that produced buffet, in which unsteady airflow around the orbiter, even with tail cone attached, would shake the 747's tail surfaces. They also were interested in flutter, which would vibrate a control surface such as an elevator or rudder. This was seat-of-the-pants flying; a pilot checked for flutter by moving a control and feeling any vibration that might result. "Once we're airborne," said Fulton, "the thing that will be of interest to all of us is any buffeting from the orbiter on the tail of the 747. If some unusual shaking or vibration levels are encountered during any of the flights, we'll back off and fly at a previously cleared airspeed."[57]

The prophecy of Leonardo da Vinci gained fulfillment on 18 February 1977 as the great bird made its first flight on the back of the great swan. Their appearance spoke of power, the orbiter white in its Styrofoam, the 747 looking silvery, the word "American" faded but visible clearly on its side. If this event did not fill all writings with renown, it at least merited two pages of text in *Aviation Week,* along with four pages of photos.

The 747 rotated smoothly to a seven-degree pitch and flew off the runway at 142 knots. Fulton kept the landing gear down until there was no chance of having to land immediately on the dry lake bed. He and his copilot, Tom McMurtry, soon found that stability and other handling characteristics were better than the simulator predictions. Followed closely by chase planes, they climbed to 16,000 feet and performed flutter and autopilot tests. Next came an airspeed calibration with a Cessna A-37, instrumented as a pace airplane, flying alongside.

Cruising at 250 knots, the flight crew evaluated stability and control, while load cells at the orbiter's supports measured its lift. They descended to 10,000 feet and slowed to 174 knots, for airspeed calibration at a variety of landing gear and flap positions. Fulton made a practice landing approach, finally touching down at 143 knots. He applied the brakes sparingly, allowing the plane to roll to the end of the runway.

The second such flight took place four days later, expanding the envelope to 265 knots at the top altitude, 22,600 feet, and 285 knots at 16,000. At that peak, and again at 16,000 feet, Fulton conducted checks of flutter, airspeed calibration, and stability. The top speed exceeded the target airspeed for air launch, 270 knots. The crew also used a high-power setting, 46,900 pounds of thrust for each engine. They had not planned to do this, but found that they needed the extra power to achieve 265 knots at maximum altitude.

This flight was the longest in the series, as they stayed in the air for more than three hours. The crew had a checklist with twenty-nine tasks to perform. They accomplished them

Fig. 19. *Enterprise and its Boeing 747 taxi to the runway in preparation for the first flight, on 18 February 1977. Note the tail cone mounted to the rear of the orbiter. (Dennis R. Jenkins)*

all, along with an item left over from the previous flight. This tested the rudder forces needed to maintain heading with three engines at full power and one at idle. The rudder demanded greater deflection at reduced airspeed. But even at 120 knots, close to stalling speed, the rudder maintained its effectiveness, preventing sideslip.[58]

Takeoff weight was 625,500 pounds, including the 143,600-pound Enterprise along with fuel for the hours of flight. But the standard commercial 747-100 series could take off at a gross weight as high as 735,000 pounds, including up to 350,000 pounds of fuel. "We are flying the airplane at much lower gross weights than a heavyweight 747 would be taking off from L.A. going to London," said Fulton. "So we aren't really taxing the airplane a lot." Carrying the orbiter "just puts the load in a little different place."

"Our overall impression, based on the two flights, is that the airplane is handling extremely well," he added. "We've seen a slight increase in the aerodynamic noise and buffeting as the speed had increased, but both conditions still are within the acceptable range. The most important thing we're trying to do on these first flights is to satisfy ourselves that the combination is aerodynamically stable. We have flown two missions and now need two more to completely clear the flight envelope we want."[59]

The third flight was on 25 February. Immediately after takeoff, Fulton simulated an engine-out maneuver by chopping the outer right engine to idle. The crew then made a successful three-engine climb-out to five thousand feet. "The carrier aircraft is very docile in the simulated engine-out tests," Fulton said. "We have more than enough rudder to control engine-out situations, which confirms the preflight predictions."

The flight crew concluded their extensive checks of flutter and stability, as they made shallow dives from altitudes up to twenty-six thousand feet. Maximum speed was 370 knots, well above that planned for separation. At 280 knots, the pilots noticed a considerable increase in buffeting. Fulton later said that "it seems like between 270 and 280 knots, the buffeting sort of takes a quantum jump in intensity." This appears to have resulted from resonance, with the frequency of eddies in the disturbed airflow, aft of the orbiter, matching a natural frequency of the 747's tail surfaces. There was distortion as well; a chase pilot reported seeing skin ripples on the orbiter's tail cone, again caused by disturbances in the flow. Fortunately, the planned separation speed of 270 knots was below the region of increased buffet and was not expected to produce problems during air launch.

The program had called for six such "captive-inert" flights with the orbiter unpowered and unpiloted. But these first three had been so successful that Deke Slayton, the ALT manager, canceled the last of them. The final two flights were to conduct the maneuvers of an air launch, though without such a launch, for Enterprise was to remain on its carrier's back.[60]

Flight four took place on the last day of February. Fulton climbed directly to twenty-five thousand feet and pushed over to initiate a series of shallow dives that simulated those of separation. As they dived to separation speed, the crew deployed the 747 wing spoilers, which increased the drag. They checked buffet levels at speeds up to 288 knots. "The buffet generated by the spoilers is very light and doesn't come through the regular buffet very strongly," Fulton declared. "The drag increase as the spoilers deploy also is not that noticeable." Following the final climb to 25,000 feet, Fulton used the engines' high-power setting to climb to 28,565 feet, higher than on the previous flights. He executed a shallow dive to 22,000 feet for more spoiler tests and checks of buffet levels. Next came an emergency descent as he chopped all engines to idle, extended the landing gear, and deployed spoilers anew. The plane dropped to 16,000 feet. Then, power restored, Fulton flew down the glideslope in a runway approach. He did not land; he executed a missed approach, holding his wheels 20 feet above the concrete and then using full power for a go-around. This too went smoothly. He ended the flight by demonstrating that the mated pair could land on short runways, such as the 7,500-foot strip at NASA-Marshall. Fulton aimed his plane carefully and touched down within the first thousand feet of the long Edwards runway. Braking moderately, he brought the 747 to a stop at the 5,800-foot mark.

The final flight, on 2 March, conducted complete simulations of two orbiter launch profiles. These called for the 747 to climb to maximum altitude and push over into a shallow dive to accelerate to the 270-knot separation speed. Using the high-power engine settings, Fulton started the first dive at 28,600 feet, the second at 30,100 feet. Descending at an angle of 5.7 degrees, the plane reached maximum speed of slightly above 280 knots. At this speed, Fulton cut the throttles to idle and deployed spoilers to the maximum air-brake position. This placed the 747 in a high-drag configuration while the orbiter, angled upward on its mount by 6 degrees, was generating high lift. This produced rapid vertical separation; in effect, the orbiter dropped the 747. Load measurements at the orbiter's supports showed a separating force approaching 0.8 g at conditions of release—as much as anyone wanted. Any increase was likely to cause Enterprise to pitch up, lose forward speed, move rearward, and perhaps strike the 747's vertical fin.

Fulton approached the runway and touched down in another short-field landing, duplicating that of Flight 4. "The landings certainly indicate we could go to heavier braking and get into even shorter fields if necessary," he told *Aviation Week.*[61]

What did these taxi tests and flights accomplish? They gave assurance of safety in the air launch, at least with the tail cone on the orbiter, and demonstrated acceptable levels of flutter and buffet. They gave flight data that could compare with the data from simulators, bench-marking the latter. They gave direct measurements of separation forces during maneuvers. The 747 qualified for use in ferry, as the mated combination demonstrated success in a host of procedures: engine-out takeoff, aborted takeoff, loss of power in an engine during flight, emergency descent, missed approach, go-around following a refused landing, short-runway landing.[62]

In all this, Enterprise had been mute. It had carried no crew, serving merely as an inert aerodynamic mass. Soon it would be time for astronauts to sit in its flight deck. Soon this orbiter would fly on its own.

The Shuttle Takes Wing

Taxi tests had taken only a day, in mid-February, while the captive-inert tests, qualifying the 747, had covered no more than the following two weeks. But the ALT plan now called for nearly three months to elapse before astronauts would board Enterprise for the next round of flights. This allowed engineers at NASA-Johnson to refine their computer programs and mathematical models, using data from those tests. Those codes modeled the separation and descent of the orbiter. With them, astronauts training in simulators could achieve greater realism as they practiced and rehearsed.

NASA also used the time to work on the orbiter. It needed additional equipment, further certification of subsystem performance. In particular, the elevon actuators had to go back to the manufacturer for final qualification tests; then they had to be reinstalled. The schedule called for resumption of flight on 26 May, but this date slipped. Balky APUs were part of the reason; their proof testing took longer than anticipated, for they had leaky seals. Meanwhile, astronauts Engle, Fullerton, Haise, and Truly continued their practice sessions in the ground simulators and the Gulfstream aircraft.[63]

Slayton set a date of 17 June, but that day brought three new problems: failure of an inertial measurement unit, trouble with two of the four primary flight-control computers, and a fault with ejection seats. These were fixed by the following day, allowing Haise and Fullerton to board the orbiter as it rested atop its carrier. Most of the orbiter's on-board systems were operating, including two of three APUs and ammonia boilers in an active thermal-control system. But as the 747 was being towed into position to start its engines, its air-conditioning system sucked in toxic fumes from those boilers' vent tubes. Fulton, still the 747 commander, shut the cabin air vents as he and his crewmates donned oxygen masks. The ammonia dissipated, and the mated pair soon was on the runway, ready for takeoff.

Fullerton later said that the height above ground of the orbiter cockpit gave a spectacular view. Neither he nor Haise could see any part of the 747, which made it feel as if they were flying alone. In fact, they would not do this until the next round of tests. The present series called for "captive-active" tests, with the orbiter piloted and powered up, while still remaining firmly attached to its carrier.

This flight made no attempt to push the envelope. It was airborne for less than an hour, flew below fifteen thousand feet, and did not exceed 180 knots. A DC-3 might have served as a chase plane. Indeed, the cruise speed was so slow that the 747 had to fly with flaps extended. But the crew was not out to break records; they wanted to see how Enterprise would perform in the air. Its working systems included APU, hydraulics, active thermal control, and electrical, the latter using this orbiter's fuel cells. All performed well.

The orbiter had a split rudder that served as a speed brake, with its two halves splitting open and angling to the sides of the vertical fin. The maximum deflection angle of each panel was forty-five degrees, listed as 100 percent speed brake. Use of this brake proved to have a pronounced effect on the big 747, as Haise and Fullerton conducted tests at 60, 80, and 100 percent. Fulton found that at the last setting, drag was so high that he had to increase the 747's engines from cruise to climb power in order to maintain altitude.

In Houston, the Mission Control Center at NASA-Johnson took primary responsibility for flight operations. The earlier captive-inert flights had been controlled on the scene, at NASA-Dryden, and this was the first time Houston had controlled a shuttle in flight. This called for good real-time communications links, which encountered some interference from a transmitter at Miramar Naval Air Station near San Diego, more than 150 miles away. Still,

Houston had communicated with astronauts as far away as the moon. California was much closer.[64]

The second captive-active flight took place ten days later, with astronauts Engle and Truly replacing their predecessors. Flutter tests were high on their agenda; the 747 had thoroughly mapped its own flutter, but the earlier captive-inert flights had left much to learn concerning that of the orbiter. Haise and Fullerton had carried out some initial flutter checks on the previous flight. The new ones involved the speeds and altitudes of air launch and free flight.

Fulton took off and flew at 230 knots. The astronauts moved their control surfaces and felt no flutter. Then Fulton moved his own controls, but this merely made Enterprise rock slightly. A speed-brake test followed; when the brake was retracted from 100 percent to zero, the effect on the 747 again was dramatic, for its rate of climb increased sharply. The two flight crews conducted additional flutter and speed-brake tests at 270 knots.

Next came pitch-over at 20,600 feet and a separation run. Fulton entered his shallow dive, cut his engines to idle, and deployed his spoilers. The astronauts did not merely go along for the ride; they set their elevons at predetermined positions, to determine the best setting for the release. Following this run, at 13,700 feet, Fulton flew through the microwave beam of an automated landing system. This would provide highly accurate position information to the orbiter during its landing approaches. The fly-through allowed the orbiter crew to make instrument readings of heading, distance, course deviation, and glideslope. Fulton then turned the 747 in a nearly complete circle, lined up with Runway 22, and landed.

Back on the ground, trouble appeared within an APU. It had had a minor leak, but the leak increased as a result of the second flight, with the unit losing both oil and hydrazine fuel. The highly toxic fuel needed removal, which delayed the third flight until late July. Technicians used this time to install a pair of hundred-gallon tanks for hydraulic fluid, change out two APUs, replace one of the five on-board computers, and put in new actuators for the main landing gear.[65]

This third flight was a complete dress rehearsal of a separation maneuver, including all elements other than actual launch. Once more, Haise and Fullerton were on the orbiter's flight deck; again, its APU, hydraulic, power, and cooling systems were powered up. Soon after takeoff a warning light went on, and Fullerton reported that the temperature of one APU "had gone off scale to the overheat side." He shut it down, not knowing that the problem was with the sensor rather than the unit. The APU later proved to have worked properly. Enterprise continued its flight with two rather than three working APUs.

Using his engines' maximum thrust, Fulton drove his 747 above 30,000 feet and pushed over, then cut his engine power and deployed spoilers. His speed increased to 272 knots as he approached the launch point at 25,620 feet, and he reported that he was "launch ready." He followed this with an approach to Runway 17, which had the microwave landing system, and aborted this approach to land as usual on Runway 22. As a final touch, Fullerton lowered the shuttle landing gear, with its main-gear trucks straddling the 747's fuselage. This was the first time these wheels had been deployed during an ALT test.[66]

That was it; there was no further need for captive flight tests. The next time Enterprise took to the air, on 12 August, it separated and came down on its own. Science writer Curtis Peebles was present, reporting for *Spaceflight:*

> *It starts with a long drive through the Southern California night. The hills and towns speed past merging into the high plains of the Mojave desert. Long trails of red tail-lights stretch across the desert to the parking lot.*
>
> *In the darkness, lights glow. Tiny figures move about. Slowly dawn banishes the stars. The clouds glow purple and red and so day comes to Edwards Air Force Base. You wait; the cars keep coming in a steady stream. Check the cameras and take a few pictures; focus the binoculars and wait.*
>
> *Slowly, the 747 and the Space Shuttle "Enterprise" back out of the support tower and move down the taxi way. Two T-38 chase planes wheel overhead; they, too, wait. Slowly, the 747/Enterprise moves to the runway.*
>
> *Crowds gather at every vantage point. They cover the two railroad lines; symbols of another transport system, another time, another frontier. As it sits on the runway, you begin to understand, to believe—it is going to happen. The wait is over now, 8 A.M.; a cry of thunder echoes across the base. Slowly, deliberately, the 747/Enterprise moves. Picking up speed, the ties with the Earth are broken. Trailed by five chase planes, it enters the sky. Binoculars follow their travels. For almost an hour, final checks are made. Suddenly, all is set. The 747/Enterprise comes around. The countdown begins: "10 minutes, 5 minutes, one minute." Numbers running backward to zero.*
>
> *Thousands of eyes search for a tiny black speck in the Sun's glare. Mission control wishes them a good flight. The 747/Enterprise begins a gentle descent. As the assembled thousands seek a glimpse, Fred Haise pushes a square white button which detonates the explosive bolts which join the two planes and begins a new age. The radio announces "separation" and applause breaks out. The 747 appears behind it, a thin contrail. At its peak, almost lost against an immense sky, a white wedge. The contrail disperses and you search again; then a black dot materializes, taking form, becoming larger.*
>
> *The "Enterprise" is flying free. You watch it remembering forever these brief moments. That familiar shape, the black nose, the high rudder, the shroud covering the rocket engines—so like an airplane yet so much more.*[67]

As many as seventy thousand visitors were there, including a thousand representatives of the news media.

Haise and Fullerton were in the orbiter, veterans of two captive flights, with Haise commanding. Fulton made a routine climb to thirty thousand feet and pushed over. He stabilized his 747 in a pitch-down attitude of seven degrees, calling "launch ready" as he approached twenty-four thousand feet. As noted by Peebles, Haise immediately fired the explosive bolts that bound the two craft. This brought separation, marked by a sudden thump and a brief sharp upward lurch. Quickly there was a master alarm. A warning light went on as an on-board computer blinked out.

Haise rolled his orbiter to the right to clear the 747, which entered a diving turn to the left. It quickly became clear that Enterprise was handling well, for it still had three primary computers. Haise lowered his nose and stabilized at nine degrees of downward pitch. Next came a practice landing flare to check the orbiter's handling characteristics at low speed. Gliding at 250 knots, he pulled up, raised the nose to eleven degrees of pitch, and executed a series of shallow-banked turns as his speed fell off to 185 knots.

During this flare, Mission Control at NASA-Houston made a mistake. Controllers believed the orbiter had remained steady in altitude during the flare and computed its lift-to-drag ratio accordingly. A Houston communicator told Haise that this ratio was well below the expectation, which meant that Enterprise would have to land very soon. Actually, the orbiter had plenty of lift, for it had not held level in altitude; it had climbed several hundred feet. "I climbed some since I pulled up faster," Haise told *Aviation Week.* "I could not tell any difference in the vehicle's response as the speed bled off. Of course, we were not doing any super maneuvers. I was making only small inputs in pitch and was limiting the bank angles to ten degrees. All the time I was doing this, we were slowing down and the vehicle's response characteristics looked the same."

Haise now dropped the nose again to pick up speed for the approach. Fullerton took the controls and made a ninety-degree turn; Haise made one as well, placing his craft on final approach. But because he had followed the advice of Houston, he expected to drop relatively rapidly while accelerating somewhat slowly. Having more lift-to-drag ratio than Houston expected, his orbiter did no such thing; it stayed high and built speed quickly. Haise now realized that he was too high and too fast. He used his speed brake and continued to aim for the runway, but he realized he would overshoot. There was nothing he could do.

His only choice was to execute the overshoot, staying in the air until his speed could fall off. He flared and touched down, two thousand feet beyond his aim point. Fortunately, he was using Runway 17, seven miles long as marked on the lake bed, with the lake bed itself providing a further five miles of overrun. Enterprise landed with a very low sink rate of under one foot per second. The speed at touchdown was as planned: 185 knots, or 213 miles per hour. The orbiter rolled for over two miles, with 100 percent speed brake and minimal use of the wheel brakes. "I was hot and long," Haise admitted, "but that was not a big problem. For the first flight we were conservative. We wanted to make sure we got it on the ground and weren't going to worry about the aim point." He had used his high speed to hold his craft just a few feet above the runway, settling down softly. The orbiter then threw up a large dust cloud from which the vertical fin continued to protrude. Spectators applauded; Enterprise rolled to a stop and the 747 flew overhead, in what Peebles called "an eagle saluting the success of its fledgling."[68]

A second free flight now lay immediately ahead, with one objective being a test of the microwave landing system on Runway 17. But a few days after the first flight a tropical storm swept across the high desert, dumping several inches of unseasonable rain on Edwards. Runoff water pooled on that runway, flooding some two miles of its length. It would take up to three weeks to dry, so program officials decided to switch to an alternative, Runway 15, another lake bed strip.

"It's been our backup all along," said Slayton. "We said at the beginning of the program that if we got rained out on 17 we would go to 15, which is higher and on the west side, so it tends to dry faster due to the prevailing winds." Its length was 3.7 miles, short by Edwards standards, but it was extended quickly to 5.5 miles by the simple method of lengthening its markings on Rogers Dry Lake. After all, when that dry lake truly was dry, much of its surface formed a natural runway. There would be no microwave-landing test on this upcoming flight, but opportunities for this test would soon recur.

Meanwhile, it was important to draw lessons from the first free flight. The orbiter's true lift-to-drag ratio now had been measured in flight test, not merely estimated in wind tunnels; its value would go into the astronauts' flight simulators. In addition, the only real problem had been the computer glitch at the moment of separation, and it was highly important to learn and correct its cause. If that computer had shut down prior to separation, the flight crew would have aborted the air launch and returned to base, with Enterprise still on the back of its carrier.[69]

The orbiter's primary flight control system had four general-purpose computers, each with an input-output processor. A backup flight control system used a fifth computer. They all came from IBM, with each performing up to 480,000 operations per second. Though the four primary computers worked independently, they were expected to agree; if any of them did not, the other three would overrule it. On the 12 August flight, the failed computer had lost synchronization with its fellows, which had voted it out of the system.

What had produced this loss of synchronization? Avionics specialists spent two weeks operating the computer system under simulated conditions at separation—and traced the problem to a printed circuit board in the faulty machine's input-output processor. That board had been soldered in a way that proved to be faulty and had transmitted commands erratically. This brought the synchronization problem; once this single board had been isolated, the avionics group succeeded in duplicating the sequence of events that had brought on the problem. One other circuit board had been soldered using the same methods. It was marked for replacement prior to the next free flight.[70]

That flight took place on 13 September, with Engle and Truly at the controls. They were to execute a tight turn, a maneuver that would dissipate excess energy during return from orbit, to reach the runway accurately. They also were to evaluate the orbiter's control and handling characteristics by carrying out specific motions in pitch, roll, and yaw and at higher speeds than in the first flight.

Engle and Truly were test pilots, possessors of the Right Stuff, ready to blast into the sky with rockets roaring. But they were well aware that much work in flight test calls for nothing more dramatic than to move the stick and rudder pedals in predetermined ways and observe the resulting vehicle motion. They did this during the second flight, moving the controls by computer as well as by hand.

They separated from the 747 and Engle entered a dive, reaching 300 knots. Truly carried through a set of computer-driven control tests. Engle made similar tests manually, knowing that the schedule was tight; they would be on the ground five minutes after air launch. "We got all of the test points in and were ready to make the turn," Engle said. "I had just put in the last of the roll inputs and damped it out when we got the ten-seconds-to-turn call. I went ahead and turned there rather than delaying any longer."

The turn indeed was tight, for they banked at fifty-five degrees. "I went right to 1.8 g in the turn and held that all the way down to below the 200-knot minimum target speed," Engle continued. "The airplane felt very solid all the way through." He pitched up as he turned; the airspeed fell off to 188 knots. He then stabilized his craft at 195 knots and carried through a second series of manual control tests. Now Truly took over, conducting additional control tests using the computer.

"I was managing the energy so that I could give the aircraft back to Joe at the right airspeed, right altitude and the right place in the sky," said Truly. "We had planned the mission so that I would fly the orbiter down to 2000 feet, which we felt would be a good place to give Joe plenty of time to get the landing set up." Approaching Runway 15, Engle aimed at a patch of green grass near the edge of the lake bed, a landmark that was easy to spot. The orbiter settled smoothly onto the runway at a measured distance of 680 feet from that target point. Again the orbiter made good use of the length of its landing strip, for it took over a minute to come to a stop, in a roll-out that once more exceeded 10,000 feet.[71]

Ten days later Runway 17 was dry, which meant that the third free flight could use the microwave landing system. This system, built by Cutler-Hammer, had a transmitter that produced a beam in the shape of a triangle or pie wedge, oriented vertically. This thin wedge swept from side to side, covering a range of approach azimuths, while its angle defined glideslopes as steep as thirty degrees. This was ten times steeper than the conventional powered approach of an airliner. This microwave system was a key element of the shuttle program, for it was to guide orbiters to a runway during their returns from space.[72]

Haise and Fullerton again were on the flight deck. They separated from the 747, then followed the flight plan by executing another tight turn, followed by more tests of response to the aerodynamic controls. "When we completed that," said Fullerton, "we were very close to the planned eleven-degree glideslope as defined by the microwave landing system, so I pushed over and centered the guidance steering needles."

The system placed Enterprise under automatic control, with this orbiter rolling suddenly to acquire the exact runway heading. As it lurched, Fullerton inadvertently touched his control stick, causing the system to revert to manual operation. He then reengaged the microwave system, resuming automatic flight. "It was absolutely smooth and was headed right for the advertised aim point on the lakebed," he told *Aviation Week.* The automatic system flew the orbiter as it descended from 6,500 to 3,000 feet. Haise then took over, made a flare, and touched down at 187 knots.[73]

Like the captive flights, the first three free flights had kept the tail cone on the orbiter. Now it was time to remove that fairing and to land the shuttle in the configuration of a return from orbit. Everyone knew that when mated, and without this cone, the orbiter was likely to produce more buffet on the 747. There was plenty to start with; during an early captive flight, with the cone on, Fulton had remarked that his 747 "reminds me of the old B-36 days. The B-36 used to shake like this in turbulence, but this old bird shakes like this all the time."[74]

Engineers hoped to reduce the effects of buffet by installing a yaw damper, a thousand-pound weight mounted on springs in the 747's nose. With this big plane rocking in the turbulence of the orbiter's slipstream, this weight tended to oscillate strongly and to help keep the plane's motions more steady. Slayton also anticipated a cautious procedure in the flights. He considered ordering a new captive flight, with tail cone off, specifically to examine levels of buffet before anyone could determine that the 747 indeed could safely release the orbiter. That would happen on the next flight.

But he wanted to reduce the number of ALT flights, to save time in program development, and he found that he could combine both flights into one. Fulton was to fly his 747 at progressively increasing speeds, from 180 knots up to 250, with the latter being faster than the speed for air launch. He also would make a simulated separation, and specialists on the ground would examine the data while the 747 was still in the air. "We'll have about twenty minutes for the ground team to look at the data and convince themselves that it is good," Slayton said. "If they aren't happy about it, we'll run another simulated launch and land. But if things are looking good, we'll go ahead and launch the orbiter."

With tail cone off and in free flight, Enterprise faced its own demands. Its lift-to-drag ratio fell from 8.5, with cone on, to 4.5. The glideslope angle increased accordingly, from eleven to twenty-two degrees, forcing the orbiter to dive sharply at the runway. The descent, from separation to touchdown, would not take five minutes and more, as in the cone-on free flights. The time would be two and a half minutes, at most.[75]

Engle and Truly were the shuttle pilots on 12 October, Columbus Day, as they and Fulton made the attempt. Buffeting indeed was more severe than on the cone-on missions, but Fulton found that it did not bring safety concerns. Though quite noticeable, the buffeting was less of a problem than people had feared.

Separation occurred at 245 knots, slower than during previous flights. Though the free flight was brief, there was time for tests of the orbiter's response to its controls. This information, gleaned in flight test, was important for accurate simulations, for the orbiter now had the shape of a shuttle returning from orbit. Engle and Truly took turns with the controls, diving at more than twenty-five degrees while on final approach. A project report summarized what they learned: "The tail-cone-off configuration showed no noticeable differences in handling qualities from the tail-cone-on configuration. Any increase in airframe vibration due to buffet at the aft fuselage was not noticed by the crew. The difference in tail-cone-off performance, however, was spectacular. Lift/drag modulation using both airspeed and speed brakes was much more apparent in the tail-cone-off configuration."

They flared and landed at 189 knots, which was expected, and with a sink rate of 3.5 feet per second, which was rather high. Previous free flights had used light braking and long runway roll-outs, but this time they braked heavily, testing the orbiter's ability to stop more quickly. This cut some 4,000 feet from previous roll-outs, as the orbiter came to a stop a mile after the nosewheel touched down.[76]

This braking exercise was a preparation for the fifth and final free flight, which was to address one more task: to land on a ten-thousand-foot concrete runway. Such a landing strip, with length of fifteen thousand feet, now existed at Kennedy Space Center. To land on it, when returning from space, a shuttle needed accuracy; its pilot would not have the whole of Rogers Dry Lake to accommodate mistakes, while the unpowered landing gave no second chance. In addition, the returning orbiter had to stop before running out of pavement.

Haise and Fullerton flew this mission on 26 October, and the flight plan was simple. As Slayton put it, "We're going to do absolutely nothing except separate, come in and touch down at five thousand feet down the runway." They were to land on Edwards AFB's Runway 04, fifteen thousand feet in length. The aim point, one-third of the way along its length, left ten thousand feet for the landing and roll-out. The landing went awkwardly, for as they crossed the runway threshold, they were twenty knots too fast. Haise deployed his speed brake and used his elevons, trying to pitch down in an effort to force his craft onto the pavement. Instead, his elevons acted somewhat like flaps and actually increased his lift. Enterprise approached within inches of the hard surface, close to the planned five-thousand-foot mark, and suddenly rose again as if bouncing. It also rolled sharply to left and right as Haise struggled with his controls.

Fullerton saw that they were in a "pilot-induced oscillation," in which Haise's own control movements were making things worse. He knew the orbiter could stabilize on its own and

119

Fig. 20. *Enterprise separates from its carrier aircraft for a free flight with tail cone off. (NASA, courtesy Dennis R. Jenkins)*

told Haise to loosen his grip on the stick. Haise described this as "the normal thing you have to do to stop that sort of thing. So I let go of the stick and it stopped." They touched down a thousand feet past the planned mark, bounced, stayed in the air for another two thousand feet, and then came down for good. But although they had landed hot, with excessive speed, hard braking brought them to a stop with three thousand feet of runway still in front of them.[77]

Was there a problem, perhaps one that would call for another test flight and landing? At NASA Headquarters in Washington, the administrator Robert Frosch, successor to Fletcher, examined this issue closely. The high landing speed appeared to have resulted from the orbiter's drag being less than expected. Nor was this new; Myron Malkin, the program manager, wrote that "all four of the previous flights had landed long. In retrospect, somewhat more significance should have been attributed to these longer-than-planned landings." The significance, again, was that the drag had been low.

Fig. 21. *Enterprise, with tail cone off, dives steeply toward Runway 22 at Edwards Air Force Base. (NASA)*

That flight had shown no real surprises; its technical aspects were well understood. Astronauts could draw on its lessons when practicing landings in flight simulators, which now would incorporate the reduced drag. Malkin wrote, "An actual flight demonstration would be of little value." Frosch concurred: there would be no further flights.[78]

What did the ALT program accomplish? Through operational experience, it gave benchmarking data for the flight simulations that were the working tools of day-to-day astronaut training. The data refined the simulators' mathematical models, as when the fifth free flight dramatically showed low drag. The test flights demonstrated tight turns for energy management in a landing approach, automated approach using the microwave landing system, and unpowered landing on a standard runway.

The ALT series now was over. The next time a shuttle orbiter executed a piloted approach and landing, it was returning from space following its initial flight to orbit in April 1981.

The Odyssey Continues

Following completion of the ALT flights, additional test flights verified that the 747 could carry Enterprise on long ferry missions. The first such mission carried Enterprise to NASA-Marshall for extensive structural tests. For this purpose, NASA for the first time assembled a complete space shuttle—orbiter, external tank, solid boosters—with all elements of proper size. The ET held water rather than liquid oxygen; the propellant in the SRBs used salt in place of oxidizer, rendering it inert. Still, like the roll-out at Palmdale, this fully assembled vehicle also offered a glimpse of what would come.

Studies of structural loads were part of this effort; another was to determine the proper placement of flight-control sensors. The shuttle's elements would bend and flex in flight, under their loads, and sensors in the wrong places would respond to these structural motions, corrupting their measurements of motion along the trajectory. This would lead to false readings, perhaps causing the mission to abort. The solution lay in finding "nodes," locations where bending was at a minimum. But these could be found only through careful tests on a full-sized vehicle, fully representative of the operational shuttle in all dimensions and particulars of structural design.[79]

Initial work, during the spring of 1978, mounted Enterprise to the ET only, with no SRBs. This simulated the configuration following SRB separation. By varying the amount of water in the tank, engineers simulated different propellant levels. They studied vehicle characteristics immediately after separation, midway through flight to orbit, and just before reaching orbit. Tests including the SRBs followed, from September 1978 to February 1979. In one series these boosters were full of inert propellant, simulating liftoff. Then, after replacing these filled boosters with empty casings, a final series represented conditions at SRB burnout, immediately prior to separation.[80]

Following this activity, with the tests completed, Enterprise had been slated to return to Palmdale for refitting as a flight orbiter. But lessons from the ALT tests changed the design in significant respects. The wings and mid-fuselage were to be noticeably stronger. In addition, some aluminum castings were to be made of titanium to save weight. Hence a refit of Enterprise would be costly and would take time. Fortunately, there was an alternative: a structural test article, STA-099, which was also under construction in Palmdale. Its design incorporated many of the changes and could be more readily rebuilt for operational service. The space program had not previously used a specific prototype both as a structural test article and as a flight vehicle, but careful study showed that this was feasible. Accordingly STA-099 received a new designation, OV-099. The world would know it as Challenger.

Nevertheless, Enterprise still had roles to play. In April 1979 it went to the Kennedy Space Center, where workers mated it anew to an ET and a pair of inert solid boosters. Once

Table 3.1. Flight Tests of Enterprise, 1977

Test Series	Date	Crew	Maximum Duration	Maximum Speed (knots)	Maximum Altitude (ft.)
Captive-inert,	18 Feb.	—	2:05:00	250	16,000
no crew	22 Feb.	—	3:13:00	285	22,600
	25 Feb.	—	2:28:00	370	26,600
	28 Feb.	—	2:11:00	370	28,565
	2 Mar.	—	1:39:00	412	30,000
Captive-active,	18 June	Haise, Fullerton	0:55:46	181	14,970
with crew	28 June	Engle, Truly	1:02:00	270	22,030
	26 July	Haise, Fullerton	0:59:53	271	27,992
Free flight	12 Aug.	Haise, Fullerton	0:05:22	270	28,000
with tail cone	13 Sept.	Engle, Truly	0:05:31	300	24,000
	23 Sept.	Haise, Fullerton	0:05:34	250	21,400
Free flight,	12 Oct.	Engle, Truly	0:02:34	241	20,534
no tail cone	26 Oct.	Haise, Fullerton	0:02:02	245	19,900

Note: For free-flight tests, "Maximum Duration" is time of actual free flight, from separation to touchdown; "Maximum Altitude" is that of separation.
Source: NASA SP-4012, vol. 3, p. 118.

again, this assemblage looked like a complete shuttle. It rode atop a Mobile Launch Platform to Pad 39-A early in May, spending three months there as technicians practiced maintenance activities and crew-escape procedures.[81]

It then returned by ferry flight to the West Coast, where it spent the next several years at Edwards AFB and Palmdale. In 1983 it launched a new career, one of public display, for with the shuttle now making flights into space, interest in it was high indeed. Enterprise was at the Paris Air Show during May and June of that year, and it visited West Germany, Italy, England, and Canada. In April 1984 it showed up at a world's fair in New Orleans, garnering further acclaim.

Enterprise had one more task to perform within the space program. Later in 1984 it was ferried to Vandenberg AFB, where the Air Force was finishing the construction of a set of launch and maintenance facilities to rival those at Kennedy Space Center. Again this orbiter received an ET and a pair of boosters; once more it served to verity the facilities and procedures. The Vandenberg site reached completion in October 1985—just in time for the Air Force to turn decisively away from the shuttle following the loss of Challenger in 1986. Still, Enterprise had done its part for national defense.

In May 1985, following completion of its use at Vandenberg, it returned to NASA-Dryden. Later that year it put in an appearance once more at Kennedy Space Center, awaiting a long-term storage location at Dulles Airport near Washington, D.C. In November it was ferried to Dulles and its ownership transferred to the National Air and Space Museum.

This vehicle continued to find uses. There was concern that the brakes of an operational orbiter might fail following touchdown; if that happened, an arresting barrier might prevent the vehicle from overrunning its runway. The Air Force and Navy use such barriers; NASA built one at Dulles, and in June 1987, Enterprise was slowly winched into it to show that this orbiter could avoid being damaged.

A somewhat more impromptu test in 1990 involved an antenna for the Shuttle Amateur Radio Experiment. No operational orbiter was available as a test bed for the new antenna, while windows of shuttle mockups lacked the right type of glass. Enterprise, still at Dulles, filled the need. Experimenters mounted their antenna within its window, which indeed was of the proper material. An antenna based on this design subsequently flew aboard the orbiter Discovery, early in 1995, and communicated with the Russian space station Mir.

For Enterprise, however, such moments in the sun proved few and far between. Engineering studies considered its suitability for refurbishment as an unpiloted orbiter, within the operational fleet, but these studies brought no follow-up. Indeed, NASA officials have taken to cannibalizing this vehicle, removing some parts to support structural tests at the Johnson Space Center. Enterprise remains in a hangar at Dulles to this day, still in storage, pending construction of a major new facility that can place so large an aircraft on open display. Like the "Fir Tree" of Hans Christian Andersen, this craft awaits the moment when it can come out of the nation's aeronautical attic, to stand once more in the open as on the day of its roll-out in 1976, and to find itself surrounded again by throngs of people.[82]

CHAPTER FOUR

Propulsion I:
The Space Shuttle Main Engine

How does NASA design a new type of rocket engine? "You start by making a parts list," recalled Sam Hoffman, president of Rocketdyne during the Apollo era:

> That'll be your plan. Now what are the major components? Well, it's gotta have a thrust chamber assembly. It needs a turbopump assembly. It needs an injector. Also a throat and nozzle. I write down these words, leave a lot of blank space under each one, and knowing rockets I can list other subassemblies that make up each main component. Then I can further subdivide, down to the level of individual parts.
>
> Each part and subassembly has its own design, its own engineering drawing. Eventually I'll have five hundred or a thousand drawings listed in my plan. Then, knowing how many engineers I have, and how rapidly they can get the work out, I can estimate how long it'll take to prepare the design. If I need a special steel, I can order it as a long-lead item, so it won't be a bottleneck. If I need to get the design in earlier, I can bring in more men. That's how my plan develops.[1]

This procedure, oversimplified even as Hoffman described it, was only the first step on a long journey toward a rugged and reliable operational engine. That initial step gave a preliminary design, capable of being fabricated in the shops and of being tested, but that first try at a design held hidden flaws, which could come to light only through a meticulous process of extensive test and development. This process required much of the 1960s for the main engines of Apollo and much of the 1970s for the Space Shuttle Main Engine.

Design flaws made themselves known by producing malfunctions in engines under test. At times, such malfunctions led to engine fires on the test stand. These blazing infernos caused their own damage, which made it all the more difficult to learn the source of each problem. These were among the standard difficulties of rocket engineering, but they imposed particular demands on the SSME. It was to be reusable, which called for long life. It also was to be man-rated, which meant that this engine needed particularly high levels of reliability.

As early as June 1971, preceding the contract award to Rocketdyne, a lengthy summary document gave detailed descriptions of the SSME's particulars.[2] This engine was of a new and very demanding type, sustaining very high internal pressure to achieve the highest possible exhaust velocity. To raise this velocity to the maximum, the SSME employed a "staged combustion cycle" to drive its turbopumps. Liquid-rocket engines drive their turbopumps using flows of hot gas. The standard approach, used on the main engines of Apollo, featured the "gas generator" cycle. This tapped a small flow of propellants that burned within an auxiliary chamber, producing a powerful stream of high-pressure gas that drove the pumps' turbines. But after exhausting from the turbines, this gas simply flowed overboard. It went to waste, contributing almost nothing to the thrust.

This was acceptable in Apollo-era engines, in which this flow was not large. In the H-1 engine that powered the Saturn I first stage, for example, it took only 2 percent of the total propellant flow to run the pumps. In the J-2, used in the upper stages of the Saturn V, it was 1.3 percent. But the SSME needed more, for it was to achieve an exhaust velocity of 14,640 ft/sec, for a 7 percent improvement on the 13,700 ft/sec of the J-2. Though modest, this increase was crucial. It would prevent the shuttle from growing fat with fuel and excessively large.[3]

To win this improvement, the engine's chamber pressure was not to increase merely by its own 7 percent. It was to quadruple, from 780 psi in the J-2 to 3,000 psi. The power for the pumps would quadruple as well, along with the demand for propellant in a gas generator. Studies showed that this threatened to defeat the SSME's purpose by bringing diminishing returns, with most of the performance increase being lost within the outflowing turbine exhaust. The solution called for feeding this exhaust back into the combustion chamber, where it could contribute to the engine's thrust and would not go to waste.

To recycle this turbine exhaust, Rocketdyne and NASA turned to the staged-combustion cycle. In effect, it burned the hydrogen fuel twice: first in preburners and then in the main combustion chamber. The preburners burned this fuel with a modest stream of oxygen, producing flows of hydrogen-rich gas at high pressure and limited temperature. This gas drove the turbines—and then exhausted into the main chamber, to burn with the rest of the oxygen. The exhaust from this chamber then expanded through the nozzle, yielding the engine's thrust.[4]

The concept of staged combustion had long been known. It attracted engineers because it promised the highest performance. Still, it was highly demanding. It called for extreme internal pressures, as noted, which were far higher than standard practice could accommodate, at least during the 1960s. In addition, it was extraordinarily sensitive, which made it difficult to design.

"I would guess that sometime in the late '40s, early '50s, people started looking at the staged combustion cycles," recalls Henry Pohl, chief of the Power and Propulsion Division at NASA-Johnson. "As far as an idea, the staged combustion cycles were being studied rather extensively when I came into this business in the mid-'50s, late '50s." However, it was not practical to pursue: "On a staged combustion cycle everything has to operate in sequence, and if you miss *one,* just a little bit, it throws everything else off. For example, if the efficiency of the turbine is not quite as high as you calculated, you have to have more fuel put through it; you have to change the whole design. Before the days of the computer it was not really possible to iterate those cycles that well."[5]

Pratt and Whitney encountered similar difficulties during the 1950s as it pursued the RL-10 that used the "expander" cycle. Hydrogen fuel cooled the thrust chamber; then, being hot and under pressure, it expanded through a turbine. After exiting from the turbine, the hydrogen flowed to the combustion chamber, to burn and produce thrust. This arrangement foreshadowed staged combustion by using the hydrogen to drive turbopumps, while eliminating the waste of a gas generator cycle. It was suitable for small high-performance engines such as the RL-10, which developed 15,000 pounds of thrust. But the RL-10 also proved very difficult to design.

"Pratt and Whitney had that contract, I guess, in the '56, '57 time frame," Pohl continues. "They went through many, many iterations coming up with a satisfactory engine. Of course, every time they missed one—if the point was not quite as efficient as they had planned—they'd have to go redesign the whole cycle. And it was not until we got the computer that you could run through many, many iterations to optimize the system all the way up and down."[6]

The tightly coupled integration of the expander cycle was new. Pohl adds that in early engines, as on the V-2, "the designers made use of a modular concept, where they could essentially treat the power system that drove the turbopump as a separate module and design it independently." The turbopump systems of the V-2, Viking, and X-15 indeed were independent; they even had their own propellant, hydrogen peroxide, which generated hot steam to drive the turbines. Even in gas-generator engines—on Atlas, Titan, and Thor, as well as Apollo—the coupling between the turbopumps and the engine as a whole was loose enough for the design problem to remain tractable. The RL-10 thus foreshadowed a day when improving tools of analysis—computers—would permit performance improvements that previously had stood out of reach.[7]

During the 1960s, with support from the Air Force, Pratt went on to take the lead in early work on a staged-combustion engine, the XLR-129. Designs called for 250,000 pounds of thrust. Computers entered general use in the course of that decade; with this, and with Pratt's work showing promise, the staged combustion rocket engine now had the opportunity to come into its own. In doing so, it illustrated anew the interplay between high-level management and contractor design work that defined the program management of the space shuttle.

The basic decisions came from NASA Headquarters in October 1969; there would be an SSME rather than a new version of an earlier engine, and it would use staged combustion.

However, this decision left unanswered many technical questions, which again could be resolved only through design studies. There were a number of ways to craft such an engine, an important issue being whether there should be a single preburner or dual preburners.[8]

A single-preburner design promised simplicity, but two preburners offered more accurate control. The only way to choose was to have engineers treat the alternatives by designing both types of engine and have them compared. Pratt went for a single preburner; Rocketdyne preferred two. Willy Wilhelm, a senior Rocketdyne engineer, notes that Pratt's configuration did not lend itself to fine control: "You have to do it by throttling the main propellant lines; that's all you've got." Dual preburners permitted adjustment of the power from each of them, which proved to be considerably easier.[9]

Rocketdyne won the contract. Its SSME indeed used two preburners, each driving its own turbopump. One pumped liquid oxygen; the other handled liquid hydrogen. This arrangement gave tight control over both the thrust level and the mixture ratio, or ratio of oxygen to fuel. The latter required attention to ensure that the shuttle would not run out of one before the other.

An engine-mounted computer exercised this control. Sensors within the engine measured pressure inside the main chamber, which corresponded to thrust, as well as fuel flow rate. Knowing the thrust, the computer calculated the oxygen flow rate. It then gave commands to the oxygen valve on the oxygen preburner, thus governing the flow of oxygen to this chamber. That determined the thrust level, for more oxygen meant more thrust. The computer also sent commands to the oxygen valve on the fuel preburner to control the flow of liquid hydrogen to the combustion chamber, which adjusted the mixture ratio.[10] This design also made it possible to throttle the SSME, changing the thrust level while preserving a desired mixture ratio. Throttling these engines reduced dynamic pressure on the shuttle as it rose within the atmosphere, limited the g-forces on the flight crew, and kept the forces from becoming too severe.[11]

The X-15 had throttled its own engine for similar reasons. Its fuel and oxygen pumps were mounted on a common shaft and were driven by the same gas generator and turbine. This turbine had a speed control; when the pilot made it slow down, it pumped less propellant and the thrust dropped accordingly.[12] The SSME was more sophisticated. Its computer executed throttle commands; it did not merely rely on a control system. The computer did this by associating the commanded levels of thrust and mixture ratio with corresponding values of combustion chamber pressure and fuel flow rate. It then adjusted the oxygen valves on both the fuel and oxygen preburners, to drive measured values of the pressure and flow rate to equal the desired ones.

The SSME had major components that operated in a sequence of increasing pressure. At the rated power level of 470,000 pounds of thrust, one could proceed upstream:

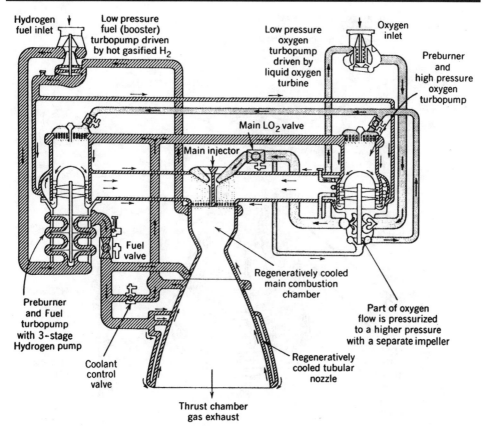

Fig. 22. Flow diagram for the Space Shuttle Main Engine. (Rocketdyne)

Main combustion chamber		3,010 psi
Oxygen turbopump:	turbine exhaust	3,318 psi
	turbine inlet	5,020 psi
Oxygen preburner		5,039 psi
Oxygen pump discharge		7,211 psi

Such internal pressures were high indeed; the oxygen pump discharge pressure sufficed to push a column of water to a height of three miles. The turbopumps demanded correspondingly high power levels. The oxygen turbopump was rated at 23,068 horsepower. The unit for liquid hydrogen, which is bulky and voluminous, was to achieve 61,420 horse.[13] This was

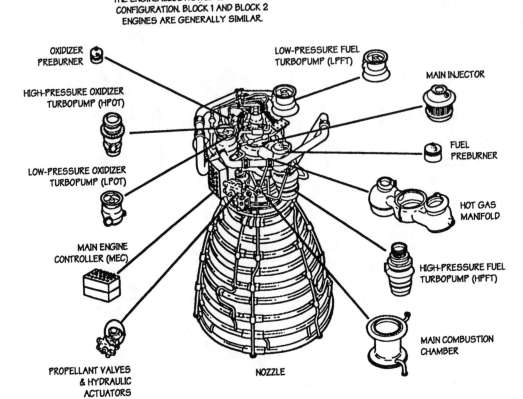

THE ENGINE ILLUSTRATED IS A PHASE II
CONFIGURATION. BLOCK 1 AND BLOCK 2
ENGINES ARE GENERALLY SIMILAR.

OXIDIZER
PREBURNER

HIGH-PRESSURE OXIDIZER
TURBOPUMP (HPOT)

LOW-PRESSURE OXIDIZER
TURBOPUMP (LPOT)

MAIN ENGINE
CONTROLLER (MEC)

PROPELLANT VALVES
& HYDRAULIC
ACTUATORS

LOW-PRESSURE FUEL
TURBOPUMP (LPFT)

MAIN INJECTOR

FUEL
PREBURNER

HOT GAS
MANIFOLD

HIGH-PRESSURE FUEL
TURBOPUMP (HPFT)

MAIN COMBUSTION
CHAMBER

NOZZLE

Fig. 23. *Space Shuttle Main Engine components. (Rocketdyne, courtesy Dennis R. Jenkins)*

similar to the 55,000 horsepower that drove the *Titanic* in 1912, in an era when the engine and boiler rooms of large steamships covered an acre of area below decks. For the SSME, the counterpart of that vessel's boilers, engines, and screws was the preburner and turbopump. This installation could fit on your kitchen table; it was two feet across and four feet long, and it both generated and used this power within that compact space. This compared with the 8,587 horsepower of the J-2 fuel turbopump—a sevenfold increase. Needless to say, it took more than engineering drawings from Sam Hoffman's technical staff to turn such concepts into working hardware that astronauts could rely on. Still, those drawings were an important start.[14]

Structural analysis again was important. The "combustion devices," the main combustion chamber and preburners, had to withstand their internal pressures without blowing up. They were to do this in the face of the intense vibration and severe noise that produced the

130

engine's roar. They had to stand up to repeated cycles of being heated. They also had to re-sist attack by hydrogen, which makes many metals brittle. Valves, turbopumps, ducts, and other highly stressed components had to meet similar requirements.

Calculations of heat transfer played a vital role, for these established the technical de-tails of methods for cooling. The nozzle, channeling the hot exhaust, was to be built with an inner wall consisting entirely of thin tubes carrying flows of hydrogen. Heat-transfer work defined the size and thickness of the tubes. Turbopump bearings were to be cooled using flows of liquid hydrogen. Calculating the design requirements for the channels that carried these flows was an art in itself, which called for a close understanding of fluid mechanics.

Metallurgy was essential. Metallurgists knew how to avoid hydrogen embrittlement through careful selection of materials. The design of turbines and their blades also drew on work in this field. The turbines were to be as small as possible, which meant driving them with flows of gas that were as hot as possible. How hot? The limit was set by "creep," wherein blades at high temperature slowly stretched under the inhuman stress of centrifugal force from their rapid rotation. Turbine designers kept close watch on new alloys that could resist creep—while making sure that their properties were well demonstrated and well characterized.[15]

These were some of the issues that guided Rocketdyne in preparing blueprints for the first engines. Those engines launched the process of test and development, which was far from easy. Development of the SSME spanned the full decade of the 1970s and even then gave only a bare-bones model, lacking long life and a reserve of additional thrust. It had been intended to operate at 109 percent of rated power, each engine serving for fifty-five flights. The first flight units reached only to 100 percent and were replaced after six flights.[16]

Starting the Project

Prototype hardware for an SSME took shape as early as 1970 in test versions of the pre-burners, injector, and main combustion chamber. This main chamber had to withstand a heat flux of 80 BTUs per square inch per second, which equaled that of a reentering ICBM nose cone. This heat flux came from hot combustion gases at temperatures up to 6,000 degrees Fahrenheit. The chamber also had to contain the internal pressure of 3,000 psi. Hence it de-manded a material with high thermal conductivity, high strength, and high ductility, the lat-ter property permitting this reusable engine to accept repeated cycles of use. Rocketdyne built the chamber of NARloy-Z, a proprietary copper alloy that contained silver.[17]

Workers began by forging a solid piece of this metal, to form the liner or hot gas inner wall of the chamber. They machined the inside diameter to the specified contour and then ma-chined the outside diameter, to bring the thickness to the required values for cutting of slots for cooling passages. A lathe carried devices to space these channels, 280 in number, around the periphery. Templates guided the cutting tool, controlling the depth of cut of each slot. An ultrasonic micrometer, accurate to 0.0003 inches, used high-frequency sound waves as a form of radar and measured the wall thickness.

Now it was time to close out the coolant channels. Technicians filled the slots with wax, then cleaned the outer surface and covered it with an electrically conductive coating. Then the liner went into an electroforming tank, receiving a thin protective copper layer. Deposition of nickel followed, as the liner rotated within the tank to assure a uniform coating. Heating of the liner melted the wax, freeing the channels; machining the electroformed nickel layer gave a desired outside contour, which was shaped to mate with a support structure, welded in place over the nickel. Inlet and exit manifolds, welded to the unit, completed this combustion chamber as a subassembly.[18]

Rocketdyne exercised similar care in fabricating other combustion devices: dual preburners, hot-gas manifold, injector. Together they formed a combustion system assembly, a prototype of a partial engine that lacked nozzle and turbopumps, receiving propellants from pressurized tanks. Tested in Nevada in early 1971, this assembly demonstrated ignition, operation above the design pressure of 3,000 psi, cooling with less than the design coolant flow, and stable thrust. The success of these tests gave a critical margin to Rocketdyne, as it nosed out a highly qualified competitor, Pratt and Whitney, to win the SSME contract.[19]

Nevertheless, these combustion devices were the easy part of the problem. The hard part called for development of the turbopumps and of a complete SSME, integral with those pumps. Following contract award in July 1971 it took several years before Rocketdyne could begin. At the outset, Pratt and Whitney appealed NASA's choice of Rocketdyne by lodging an appeal with the General Accounting Office; this took eight months to resolve.[20] In May 1972, the contract award confirmed, NASA and Rocketdyne turned to the first item of business: final determination of the design requirements.

An important specification was already in place, for a letter contract dated 5 April 1972 set the thrust of the SSME at 470,000 pounds in vacuum. However, project managers needed much more. Rocketdyne's proposal had been highly specific, including an executive summary, seven volumes of technical discussion, five on management, and 81 with supporting material. Discussions between NASA and Rocketdyne went into similar detail, leading to the release of two documents that defined what NASA wanted and what Rocketdyne was prepared to build.

The Interface Control Document (ICD), in February 1973, described the engine as an element of the complete space shuttle: SSME dimensions, weight, and center of gravity; dimensions, tolerances, and structural capabilities of all physical interfaces; electric power and frequency requirements; computer formats for data and command, and failure responses; engine environment; SSME performance requirements. The Contract End Item Specification (CEI) came out in May. It defined the detailed requirements for engine check-out, start, operation and shutdown; engine service life and overhaul; design criteria for thermal, vibration, shock, acoustic, and aerodynamic loads; material properties, traceability, and control of fabrication processes; control system redundancy requirements, and required safety factors.[21]

Each of these items demanded close attention. Traceability, for instance, began literally at the level of nuts and bolts. Charles Feltz, a senior engineering manager at North American, described these fasteners to the House space committee during the Apollo program. He said that their iron ore came from a particular section in a specific open-pit mine in the Mesabi Range, near Duluth, Minnesota. The bolts then took eleven steps to manufacture, with the product being certified at every step through meticulous tests. This certification applied to the ingot smelted from the ore, the billets forged from the ingot, the steel rod extracted from the billet, and the bolts themselves that were milled from the rod. The fasteners that resulted were some fifty times more costly than the ones people buy in hardware stores, but that was what it took to send astronauts to the moon.[22]

For the SSME, the ICD and CEI contributed to development of detailed Design Verification Specifications, for the engine as a whole as well as for turbopumps and other components. These specifications also drew on the contract Statement of Work as well as on Rocketdyne design standards. Each detailed requirement was identified by source, specifying methods for proving that the design met the requirement and that the requirement was valid.

To establish these proofs, engineers expected to rely on analysis, hardware inspection, laboratory or bench tests, and hot-fire tests first of subsystems and then of complete engines. For instance, turbopump bearings and seals were tested within their liquefied-gas propellants, at operating speeds, prior to the first turbopump test. For the SSME as a whole, these specified requirements formed the basis for the entire development program, which extended into the early flights. This development program included a total of 4,566 laboratory tests and 1,418 subsystem hot-fire tests.

Even with these program documents in hand, it still was not possible to build and test prototype turbopumps. Rocketdyne had to carry through major modifications of existing test facilities at its Santa Susana Field Laboratory, to accommodate turbopumps, combustion devices, and combinations of components. At a site called Coca a turbopump test stand had some two thousand valves, most of which were set by hand but two dozen of which were operated remotely by servocontrollers. Propellants for preburners came from a system rated at 14,000 psi with valves as big as a closet and weighing as much as five tons.[23]

Another long-lead item, the SSME control system, went into development at Honeywell, a subcontractor. It held two programmable computers for redundancy, each with sixteen thousand words of memory. This controller was to be mounted to the engine itself, operating amid its vibration and extreme levels of noise. In operational service, it was to perform an automatic check-out of its engine, isolating faults to the level of replaceable units. It would regulate and monitor the engine and its components during start, providing automatic shutdown if it detected a problem. The controller also would operate its engine during powered thrust, varying the thrust level and propellant mixture ratio upon demand while continuing to monitor systems and performance parameters. Finally, it was to shut down the engine safely, either by noting an emergency or in response to a command.[24]

During the winter and spring of 1974 it became apparent that both efforts were in trouble. The activity at Coca encountered delays, and NASA's George Low wrote that "facility construction at Santa Susana appears to be entirely out of control. None of the people at Rocketdyne or at Marshall seems to have recognized that an overrun on a facility requires congressional reprogramming. The total overrun appears to be of the order of $4 million over and above the $18.7 originally allowed." These problems slipped the schedule by five months, with the start of engine testing being put off from December 1974 to May 1975.[25]

There also was bad news concerning the engine's computer, with Low noting that Rocketdyne "has done a very poor job so far on controlling costs and on controlling its main subcontractor, Honeywell." Furthermore, Honeywell "has done a lousy job on this, just as they have on the Viking computer." That company's project "is in major cost, schedule, and weight difficulty. Honeywell is just not performing." Low added, "The Rocketdyne overrun problem is estimated to be between $50 and $100 million."[26] This reflected more than the direct expense at Coca and Honeywell; it also accounted for added costs as the project staff continued to draw salaries while waiting for testing to begin. The Rocketdyne project manager, Paul Castenholz, was the man on the spot, and he remembers some rather unpleasant exchanges with Rocco Petrone, the new director of NASA-Marshall.

Petrone "didn't want to hear what I had to say, did not want to hear that we would need more money. He did not want to hear my analysis of why. He ranted and raved; he talked about how when you're in combat, you do what you've got to do. He didn't realize that I was probably one of the few people in the room who'd actually been in combat," having served for three years in the Pacific during World War II. "I left feeling that this guy's gonna be murder. I think he was a brute-force bully, and that doesn't get the job done."[27]

Castenholz certainly had experience; he had managed the J-2 program for Apollo, while his aggressive leadership had snatched the SSME contract when Pratt and Whitney thought the award was in the bag. Indeed, he had been a little too aggressive, for in directing the SSME program, he had tried to rely on Rocketdyne's rather limited background in electronics and computers. Low described this capability as "very, very weak, and they are only now beginning to bring in people from other divisions." Castenholz could have survived this, but the overrun meant he had to go. He was replaced during June.

In Low's words, "My own experience is that these things never get straightened out until you write a firm letter to the top man in the company." He and Fletcher wrote to Robert Anderson, president of Rockwell. This was far above the pay grade of Castenholz, who was a mere division vice president. Anderson sent Joseph McNamara, president of the Space Divi-

Fig. 24. *Rocket engine test facilities in the Santa Susana Mountains near Los Angeles.* *(Rocketdyne)*

sion, to conduct a review. McNamara told Low, "Castenholz had 'broken his pick' on the job and therefore had lost the confidence both of his people and of management." Castenholz was not dismayed, as he understood that project managers were expendable. He took a directorship at Rockwell's science center and then went on to the corporate offices, but his twenty-seven-year career as a rocket man was over.[28]

Castenholz's replacement, Norman Reuel, did not stay long as SSME program manager, for he had heart trouble. Reuel's replacement, Dominick Sanchini, was a veteran who had directed preparation of the SSME proposal in 1971. "He was tough," recalls J. R. Thompson, Sanchini's counterpart at NASA-Marshall. "There were a lot of days that lesser men would have caved. Given up—not push as hard, not try to challenge the team, not trying to make the schedule. It took a tough guy to do that."[29]

In his day-to-day routine, Sanchini presided over an important management activity: the five o'clock meeting. Rocketdyne's Bob Biggs describes this as "a daily ritual set aside to

135

recap that day's activity and progress (or lack thereof) on the most significant current problem." Sanchini conducted the meetings in his office, along with key members of his staff. Regular attendees included Willy Wilhelm or Paul Fuller, the chief program engineer; Jerry Johnson, an associate program manager, and Biggs, who directed the engine test program. Other senior managers, knowledgeable in turbomachinery or combustion devices, attended meetings that dealt with these components. J. R. Thompson, the SSME program manager at NASA-Marshall, was another frequent attendee. He lived in Huntsville, Alabama, but spent much time at the Rocketdyne plant near Los Angeles and maintained a permanent office next door to that of Sanchini.

At these meetings, people sitting at the table also included the leaders of "tiger teams." These were specialized groups set up to address specific problems, often drawing their membership from Marshall as well as from Rocketdyne. Members of a tiger team worked full time on the assigned problem, which often showed its presence in a fire or other accident on a test stand. These people, who typically were technical managers, held responsibility for identifying the cause. They had to bring the problem under control so as to permit safe resumption of testing, implement a design change to prevent recurrence, and provide proof to both corporate and NASA management that the problem had been eliminated or controlled. Biggs writes, "Of the first 20 special teams formed, Ed Larson, director, Design Technology, was assigned as team leader for half of them. It is likely that he would have been assigned to others except for the fact that he had not yet concluded an investigation of a previous problem."[30]

The staff now turned to preparation for test, working with a prototype engine, the Integrated Subsystem Test Bed (ISTB). Being throttleable, the SSME was to vary its thrust over a wide range. The ISTB was to operate at 50 percent of rated power, the minimum level planned for flight. It included turbopumps, combustion devices, controls, and a shortened nozzle designed to match its limited thrust. As hardware began to take form, Reuel and Thompson agreed that the first article of each major component would be allocated to the ISTB. The second article would serve for component testing, at Santa Susana.

The ISTB was to be tested at the National Space Technology Laboratories, the Mississippi Test Facility of Apollo days.[31] It stood amid thick pine forest; within the offices and test areas, dense foliage was seldom far away. The sun sank at night with dramatic suddenness, leaving the center in darkness.[32] But the return of rocket-engine testing meant that once again there were lights amid the dark, the stark and brilliant illuminations of test installations. When an SSME was in place for a night firing, its hot exhaust struck sprays of cooling water that flashed into steam, raising a thick plume of vapor that reflected the glows from its test stand. During the brief minutes of its firing, it would hold back the night. And within that state, one could cherish the hope that somehow there would be other lights, brighter and stronger, to drive shadows from the hearts of men.

Fig. 25. *Space Shuttle Main Engine with short nozzle. (Rocketdyne)*

Starting the Engine

Testing of components, at Santa Susana, preceded tests at the same level using the ISTB. The Santa Susana facility designated Coca-1 had separate test stands for oxygen and hydrogen turbopumps. Coca-4 was a single stand with two test positions. One served for igniters and preburners; the other conducted evaluations of the hot gas manifold, main injector, and main combustion chamber. Only a month elapsed between a component test at Santa Susana and the counterpart ISTB run in Mississippi. Managers would have preferred three months, but some SSME components proved difficult to assemble, which led to delays. Previous engines had been bolted together, but for the SSME with its high pressures, this would have necessitated heavy flanges. To save up to a ton of weight, the engine was welded, requiring more than four thousand welds. Many parts had complex shapes and demanded time and care in preparing welds that could pass Rocketdyne's meticulous inspections.[33]

In Mississippi, effort focused on two stands that had accommodated complete S-II stages during Apollo. These now had new tanks, holding 40,000 gallons of liquid oxygen and 110,000 of liquid hydrogen. Many of the staffers were being recycled as well, for these people were veterans of Saturn development who had stayed in the area. They had lost their NASA jobs as Apollo wound to a close and had found themselves seeking work amid a major nationwide downturn in aerospace. Nor could they sell their homes; the local area was one of small towns where the real-estate market was close to nonexistent. Mack Herring, the center historian, recalls that one man found work as a barber. Another got a used-car lot; still another had a bar in the nearby town of Bay St. Louis, while one fellow opened a garage to fix foreign cars. Still, in Herring's words, "Our guys were personal friends. We'd go to a watering hole and have a beer."[34]

As the ISTB took form, plans called for it to serve initially in learning how to start the engine. This was no simple matter of turning a key in the ignition; it required close attention to detail. The start sequence had to maintain a hydrogen-rich mixture in all parts of the engine at all times. If the propellant mix ever became oxygen-rich, the burn within the engine would become "component-rich," with portions of the SSME physically burning within this hot, high-pressure gas.

Overspeeding of the turbopumps was another danger. The preburners, which drove their turbines, were substantial rocket engines in their own right. The turbines, which received their exhaust, were about the size of a layer cake from your kitchen. They had very little inertia, due to their small size and light weight. They could accelerate from a dead stop to a destructive overspeed in less than tenth of a second, at which they would tear themselves apart due to centrifugal force. This meant that starting the turbines, within the blast from the preburners, amounted to using a sledgehammer to drive a tack into a wall—without breaking the wall. To prevent overspeeding, the turbines' pumps had to be ready to use their power by having a good hold on their propellants. It also was necessary to bring up thrust within the main combustion chamber, which would provide back pressure on the turbine exhausts and further limit their speed. Additional hazard came from temperature spikes due to improper flow of hot gas from this main chamber, for these could damage turbine blades.

Five valves governed the start sequence. "We could control all five, make them go precisely where we wanted," recalls Bob Biggs, Rocketdyne's manager of SSME system development and chief project engineer. "The difficulty was in learning where we wanted them to go." An error in valve position of 2 percent, or a timing error of a tenth of a second, could damage an engine severely.[35] Computer control was essential. The Honeywell controller was not ready for use, so the ISTB used a rack-mounted laboratory computer, located remotely. However, the control system included all required valves and actuators. Officials at the Office of Management and Budget kept track of the preparations for test, for the first engine ignition represented a milestone in the development program. Rocketdyne reached this milestone on schedule, conducting an initial countdown on the ISTB in May 1975 and achieving full thrust chamber ignition a month later. With this, the engine development test program was underway.[36]

The work proceeded cautiously, expanding the start sequence in small time increments. It took nineteen tests, twenty-three weeks, and eight turbopump replacements to reach two seconds into a five-second start sequence. It took an additional eighteen tests, twelve weeks, and five turbopump replacements before momentarily touching the minimum power level, 50 percent of rated thrust. This happened in January 1976 during a test that lasted 3.36 seconds. The engine needed another year before it was ready to run at rated power even briefly, and after that the start sequence continued to call for fine adjustments. The final protocol was not in hand until the end of 1978.[37]

The sequence began with a start preparation phase. Dry nitrogen flowed through the oxygen passages to eliminate moisture. Dry helium purged the hydrogen side, eliminating air as well as moisture. Helium was used because liquid hydrogen was cold enough to freeze nitrogen, and air, into forms of ice. With the engine well purged, cryogenic propellants flowed into the fuel and oxygen lines and pumps, to be held back by valves. Flows of cold gas, maintained for more than an hour, chilled the turbines to cryogenic temperatures. The main combustion chamber and nozzle were left at ambient temperatures. During engine operation, they would be cooled by hydrogen flowing in thin tubes or ducts.

At the start command, the main fuel valve opened. Liquid hydrogen, pushed by modest pressure in its tank, flowed downstream and entered the cooling passages of the thrust chamber. Absorbing heat from the warm metal, this fuel flashed into gas, producing a surge of pressure to begin driving the turbines. Hydrogen also flowed into the preburners and main combustion chamber. The pressure in this surge, however, was far from steady; it oscillated markedly, with pressure peaks and reductions occurring repeatedly during the next 1.5 seconds. The opening of other main valves had to be timed to coincide with these peaks. To accomplish ignition, spark plugs stood ready.

The key events that followed were the priming of the three combustors, filling their injectors with liquid oxygen as an immediate prelude to ignition. Priming was accomplished by opening oxygen valves. The valve on the fuel preburner opened first, priming that preburner 1.4 seconds following start. In came the oxygen, amid plenty of hydrogen; a spark plug ignited the mix. This brought a rapid acceleration of the fuel turbopump, driven by that preburner, which pumped more hydrogen into the system.

To prevent this turbopump from overspeeding, it was essential to produce a back pressure by lighting the main combustion chamber. Its main injector primed and ignited at 1.5 seconds. Engine operation still was hydrogen-rich, for there was pump pressure on the fuel but only tank pressure on the oxygen. But at 1.6 seconds, the oxygen preburner also primed and ignited. Now the oxygen turbopump went into action, with oxygen as well as hydrogen flowing freely into the engine. During the next 2 seconds, the turbine speeds and the pressure in the main chamber rose to their full values. Five seconds after start, the engine was in stable operation at rated power level.

The priming events occurred in response to valve commands during the first two-tenths of a second following start command. These events thus were delayed responses, for it took time—fractions of a second—for valves to physically open and for propellants to flow and fill the injectors. Timing was critical, as these events were keyed to the early pressure oscillations. It was essential for the primings to take place during a pressure peak. If any of them occurred during a pressure drop, a quarter-second later, hot gas from that combustor would produce a damaging temperature spike, flowing backward into the injector or turbine.

If the fuel preburner ignited late, the fuel turbopump would lack time to come up to speed. If the main combustion chamber lit up early, it would produce excessive back pressure that would prevent proper acceleration of this turbopump. In either case the engine would receive inadequate fuel flow and would run oxygen-rich, resulting in extensive burning of its hardware. If the oxygen preburner ignited early or the main chamber fired late, the oxygen turbine would fail to receive back pressure and could overspeed to its destruction. Traces of positions of the five valves, during the start sequence, showed intricate patterns of dips and advances. In Biggs's words, "Every one of them was the result of a problem," with each detail in an individual trace addressing a particular difficulty.

Even when a reliable start sequence was in hand, early in 1976, engine development was just beginning. This sequence merely allowed the engine to run at minimum power level. Engineers hoped to advance quickly toward rated power but they encountered roadblocks; it took another year to reach this level. The reason was that with the engine now started, there was plenty of opportunity for new and unanticipated problems to come to the forefront. These emerged during 1976 and 1977 in both the fuel and oxygen turbopumps.[38]

The Fuel Turbopump

Early tests of the fuel turbopump, both as a component and as part of the ISTB, showed considerable vibration, which is a well-known scourge of engineering designs. However, it took time to appreciate that this signaled a real problem. Many of the tests were short, as part of start-sequence development. In addition, the vibration appeared to stem from inadequate performance of a balance piston that damped out side forces within the pump. A redesign led to a piston that proved satisfactory. Then, with new engine tests reaching for higher thrust and longer duration, vibration in the fuel turbopump returned with a vengeance.[39]

On 12 March 1976 the ISTB was scheduled for a run of sixty-five seconds at 50 percent power, with one second at 65 percent. The test reached this thrust level, the highest achieved to that time, but the engine shut down twenty seconds too soon due to failure of the fuel turbopump. Afterward, examination of the pump showed that it had seized due to failure of bearings that supported its shaft. Data from the test revealed two major departures from normal performance.

The turbine gas temperature had increased by almost 200 degrees Fahrenheit during the run. This meant that the turbine had experienced a significant loss of efficiency. In addition, vibration measurements showed a powerful oscillation of large amplitude, at a frequency of close to half the turbopump's rotational speed. Specialists immediately recognized this as an instability in the rotor dynamics known as "subsynchronous whirl."[40]

In normal operation, the rotor was to spin freely while being held tightly within its bearings. Subsynchronous whirl caused the rotor to roll while spinning, like a pipe inside a barrel. The rotor was rather flexible and would bend, allowing turbine blades and pump impellers to scrape against the inside of their casing. As the blades abraded, the turbine lost its tight fit within its housing, permitting hot gas to leak past. That brought the temperature rise and the loss of efficiency. Meanwhile, the whirling rotor placed considerable stress on its bearings, causing them to fail as well.

Subsynchronous whirl was well known within the field of rotordynamics and thus presented NASA and Rocketdyne with a well-posed problem that could yield to a concentrated attack. The leaders of this effort were Matthew Ek, Rocketdyne's chief engineer, and Otto Goetz, the leading expert in turbomachinery at NASA-Marshall. As the work expanded, they brought in consultants from industry, government, and universities in the United States and even from Great Britain.

The activity began with a review of prior experience on eight previous federal programs that had built liquid-hydrogen turbomachinery. Four had shown subsynchronous whirl: the J-2, two projects within an Atomic Energy Commission program that had sought to build nuclear-powered rocket engines, and a Pratt and Whitney turbopump for an engine with 350,000 pounds of thrust. Investigators searched the published literature, while researchers constructed five separate mathematical models. The work led to identification of twenty-two possible contributors to this instability. Two of these stood out: inadequate rotor stiffness and hydrodynamic disturbances produced by internal seals.

Ek and Goetz did not eliminate subsynchronous whirl within the fuel turbopump. But over a ten-month period they raised the speed of onset from 18,000 rpm, below the minimum power level, to more than 37,000 rpm, which was higher than the maximum operating speed.

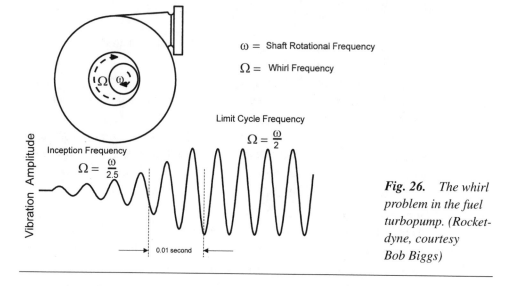

ω = Shaft Rotational Frequency

Ω = Whirl Frequency

Limit Cycle Frequency

$$\Omega = \frac{\omega}{2}$$

Inception Frequency

$$\Omega = \frac{\omega}{2.5}$$

Vibration Amplitude

0.01 second

Fig. 26. The whirl problem in the fuel turbopump. (Rocketdyne, courtesy Bob Biggs)

Engineers stiffened the shaft and bearing supports, while installing new seals that promoted stable rotation. Testing continued, reaching cautiously toward more rapid rotation.[41]

As these speeds increased, it became clear that turbine bearings were continuing to overheat and fail, for reasons unrelated to whirl. These bearings were immersed in the preburner's hot exhaust. They needed cooling, and designers expected to do this by flowing liquid hydrogen down the middle of the shaft. Rocketdyne's Joe Stangeland, director of turbomachinery, compares this to "putting a refrigerator in the oven." Test turbopumps carried instruments that had been added to learn about whirl; their data showed that the bearing was not obtaining enough coolant.

Detailed study showed that the flow of hydrogen was forming a vortex at the base of the pump shaft. This brought a substantial pressure drop that partially blocked the flow. The cure was simple: a baffle, about the size of a quarter. This spoiled the vortex and removed the blockage. Ek wrote that as a result of these design changes, "the whirl phenomenon disappeared." The solution of the whirl problem meant that for the first time, it became possible to operate the SSME for extended durations.[42]

The solution to the whirl problem unfolded amid a mixed set of developments during 1976. A notable event that took place in September, the Critical Design Review, certified that the SSME complied with requirements. In preparation for this review, a second test engine was installed in Mississippi with its own short nozzle. It also had the Honeywell controller. The ISTB remained in use, each engine firing about every two days.[43] Nevertheless, the whirl problem prevented the program from meeting some goals. NASA and Rocketdyne had wanted 8,000 seconds of accumulated running time prior to that review. They cut this to 2,500 seconds and had to settle for 2,104.

Even so, the firings broke new ground. In September, at the time of the review, a third test engine logged 1,100 seconds of total operation in three consecutive runs of 150, 300, and 650 seconds. Prior to these runs, no test had lasted more than a hundred seconds. "We tore the engine down following the three long runs," Dom Sanchini, the Rocketdyne program manager, told *Aviation Week.* "It was remarkably clean, with very few signs of wear and tear and very few indications that we'll have any long-life engine problems."

These tests took place at 50 percent of rated power,[44] and there was much interest in running an engine at 100 percent for a full sixty seconds. The program achieved this in March 1977 during a run of more than eighty seconds that spent sixty-one seconds at this level. The work now featured the full-sized flight nozzle, and an engine with this nozzle demonstrated successful throttling from 50 to 100 percent of rated power. This represented a prelude to flight, for operational engines were to vary their thrust during launch.[45]

The SSME was reusable, having an original life requirement of a hundred missions. The shuttle concepts of the day were two-stage; the SSME then needed a life of twenty-seven thousand seconds. This included six flights at an emergency thrust level, 109 percent of rated power, or 512,000 pounds of thrust. This emergency level was intended for use during abort, but NASA officials decided that they wanted the option of using it on every flight to increase the payload. They renamed it "full power."

Shuttle configurations after 1971 called for an SSME burn time of some five hundred seconds; the twenty-seven-thousand-second life then was equivalent to fifty-five flights. A fatigue analysis found that by cutting the requirement to fifty-five missions, the SSME could operate routinely at 109 percent without blowing up. Hence the shuttle indeed could use full power freely—if the SSME could be developed to provide it.[46]

Not for years would a single engine demonstrate twenty-seven thousand seconds in repeated tests on a stand. Yet it was possible to achieve fifty-five starts, as a highly demanding element of this life requirement. The ISTB did this first, undergoing sixty-seven tests by mid-1977, with Engine Two following closely at sixty-two. This showed the quality of the early design.

At Santa Susana, in a prelude to full-scale engine testing at 109 percent of rated power, component tests successfully operated the preburners, main injector, main combustion chamber, and short nozzle at this thrust level.[47] Here indeed was an expression of optimism, for while the SSME flew at 104 percent after 1982 and underwent considerable test-stand operation at 109 percent, it would not be certified or flown at that level. In this respect, it fell short of its specification. Nevertheless, the reach toward 109 percent, as early as 1977, showed that the spirit of adventure was alive and well. The program needed this spirit, for it was about to encounter technical problems that were serious indeed.[48]

The Oxygen Turbopump and Fuel Turbine Blades

On the Coca-1A stand at Santa Susana, one day in February 1976, the oxygen preburner and its turbopump were under test. Nineteen seconds after initiation, a flowmeter within the test facility failed, releasing instrument parts into the stream of liquid oxygen. These struck sparks when they hit a throttle valve, touching off a fire. The valve burned, reducing its resistance and allowing the oxygen to flow more freely. This overtaxed the turbopump; the pump's impeller now began to scrape against the interior of its housing. Like a Boy Scout rubbing sticks together, this scraping produced friction, and with liquid oxygen close at hand, this friction touched off another blaze.[49] The test stand went up in flames, and the consequences reached across the country. In Washington, Fletcher signed a statement determining that this stand "has been made inoperative by explosion and subsequent fire." In Mississippi, SSME testing was held up for most of that month, as engineers verified that their facilities would not face the risk of a similar fire. Tight funding prevented rebuilding of Coca-1A, and it would be missed.[50] A year later, another test-stand fire launched a major struggle with the oxygen turbopump, a struggle that continued for months. This unit contained a turbine and a pump, no more than two feet apart. The turbine ran on hot, dense, high-pressure hydrogen-rich gas from its preburner. The pump, inches away, drove a copious flow of liquid oxygen. This was like spewing gasoline from a fire hose within two feet of a large flame of a blowtorch, except that the SSME's combination was considerably more explosive.

Hence it was absolutely necessary to maintain a total separation between the turbine and the pump. Rocketdyne would willingly have packaged them as separate components, connected by a shaft. But that shaft would have had to penetrate their housings, and at the high pressures involved, there would have been no way to prevent leaks. Within their common housing, this separation called for more than good seals, which also could leak. The design used such seals to establish a zone, midway along the connecting shaft, that was continually filled with high-pressure helium. The pressure was high enough to create a true barrier, which neither the oxygen nor the hot hydrogen could pass. Leakage through the seals was outward, not inward, as helium slowly flowed on both sides of the barrier, exiting safely from the shaft.[51]

If the turbopump ran into trouble during a test, however, this barrier could readily break down, with explosive consequences. Bob Biggs writes that such explosions "are nightmarish events in rocket engine development programs," noting "the fiendish nature of the failure. Once a fire has been ignited in the high pressure LOX environment, it readily consumes the metals and other materials that make up the hardware. In most cases, the part that originated the failure is totally destroyed, leaving no physical evidence as to the failure cause. Program management is often left in a quandary as to what to do to prevent further occurrence of the failure."[52]

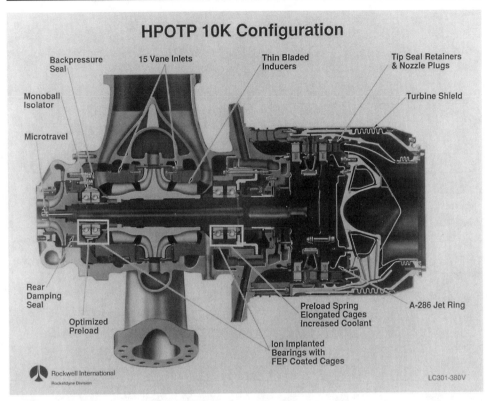

Fig. 27. *High-pressure oxygen turbopump. (Rocketdyne)*

Fuel turbopump accidents were bad enough, but they were mild by comparison. The computer sensed the fault and shut down the engine safely. This allowed the project staff to disassemble the bad unit, studying it in detail to learn the nature of the problem. By the time the computer detected a malfunction in the oxygen turbopump, however, it often was already too late for safe shutdown. The SSME was built of copper, nickel, and steel, which we do not regard as fire hazards. But in Biggs's words, "You can't burn an iron bolt with a match, but you sure can in liquid oxygen." Fires in oxygen turbopumps generally burned so much that it was difficult to discover what parts had failed or in what sequence the failure had spread.

Such fires were not new; they had occurred in other tests while developing oxygen turbopumps for earlier liquid-fueled engines. This experience helped to shape the SSME design, which had only four such fires, a modest number. Still, Biggs notes that following such an accident, "the program comes to a screeching halt until we find a reason for it, and prove that

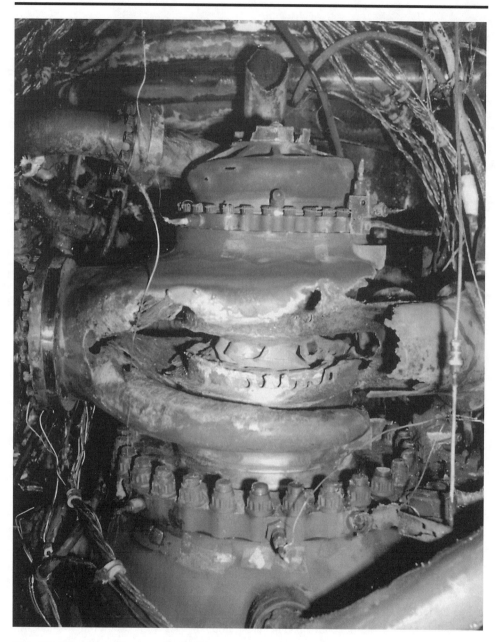

Fig. 28. *Remains of an oxygen turbopump, partially burned and severely damaged following a test-stand fire. (Rocketdyne)*

it's okay to start again. There's no other thought but, 'We've gotta find out what did this.' A lot of times this required working seven days a week. You cancel your vacation, which I've done. Or you get called back from vacation, which I've done."[53]

The first such fire occurred in Mississippi in March 1977. The test was to run at 75 percent of rated power for 535 seconds, which was as long as the available propellant supply would permit. The computer cut off the engine at 74.07 seconds, as it detected a falloff in the shaft speed of the oxygen turbopump. By then, however, an intense blaze was already burning within that component; an observer wrote that "the engine suddenly became engulfed in flame." A water deluge system sprang into action, saving the test stand, though it took nearly a full minute to douse the conflagration. The engine and turbopump were badly damaged.

As with the fuel turbopump failures, this accident brought a careful investigation. The formal report cited 33 possible contributors, each with substantiating or refuting evidence. Seven of them stood out. The test engine had been well instrumented, and the data showed that the fire started near an important liquid-oxygen seal. This seal had two rings, set side by side like engagement and wedding bands, with the shaft as the finger. One ring was stationary, fixed to the shaft housing. The other ring rotated, being mounted to the spinning shaft. A bellows, acting as a spring, pushed them together. But when rotating, the shaft-mounted ring generated hydrodynamic forces that pushed it slightly away from its mate, eliminating friction and wear while permitting a controlled leakage through the resulting gap.[54]

One did not have to be a rocket scientist to see how this arrangement could cause problems. The leakage rate might have been excessive. The hydrodynamic lift might have been inadequate, allowing the rings to rub together and create friction. The bellows might have failed, allowing large-scale leakage. This seal certainly needed redesigning, and Rocketdyne took other measures as well. Its engineers added instruments to other test engines. They reprogrammed the engine computers with new "redlines," measured limits that would bring shutdown if exceeded. They also strengthened the shaft's helium barrier by increasing the flow rate of this gas several times over.

Testing resumed in Mississippi in April, a month after the fire. The new redlines protected the engines, which continued to run with the existing seal. This seal indeed proved faulty; inspection showed both bellows failure and rubbing together of the rings. The replacement seal entered test in July. It was a "labyrinth," looking like a comb in cross section, mounted to the shaft housing while snugly encircling the shaft. Within this seal, numerous partitions created a succession of chambers, and leaking fluid had to make its way through all of them to escape. This new seal was intended for use as an interim measure, but it proved so successful that it remained as the permanent choice.[55]

Testing stepped up during the summer. The Number Four engine ran repeatedly for up to 425 seconds, at 70 percent of rated power. Late in August, as an important prelude to full-duration tests, Number Two operated at 100 percent for 301 seconds, close to the limit of the

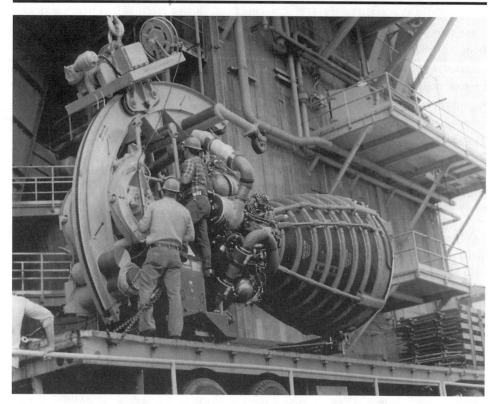

Fig. 29. *A Space Shuttle Main Engine arrives on a flatbed truck to be hoisted into position for testing. (NASA)*

test stand propellant supply.[56] Then in September, a similar test of Number Four ended in another oxygen-turbopump fire. Here was no sudden failure; the events that led to this accident unfolded over nearly three minutes.

The planned duration was 320 seconds, beginning with two minutes at rated power. At 133 seconds, the engine throttled back to 90 percent. Instruments detected a slight increase in vibration of the oxygen turbopump, with accelerometers on the turbine end showing a gradual rise in its intensity. Subsequent investigation interpreted this as a degradation of the turbine-end shaft bearings. At 185 seconds, the turbine-end vibration stopped increasing and began to decrease, while that of the pump end started to increase. The computer did not halt the test, for measured values were still within the redline limits. Nevertheless, this change in the vibration pattern showed that internal loads were being transferred from the turbine bearings to the pump bearings. Other data pointed to an increase in internal friction.

Fig. 30. *Test of a Space Shuttle Main Engine in Mississippi. (NASA)*

At 193 seconds, an increase in the liquid-oxygen temperature at the pump discharge indicated internal heat generation. At 200 seconds, the flow rate of liquid oxygen from the facility tanks began to deviate from the engine flow rate. This appeared to result from wear in an engine seal, which resulted in leakage. Meanwhile, the turbine power requirement rose slowly, indicating a continuing increase in internal friction.

At 275 seconds, the system began to degrade rapidly. Wear in another seal became evident, pump vibration further increased, turbopump performance fell off, and friction brought a further rise in the liquid-oxygen temperature. Hot gas leaked past the turbine seals. At 300 seconds, rapid malfunction of the turbopump brought an immediate halt to the flow of liquid oxygen. This created a water-hammer effect, as if a valve had slammed shut, and blew out a duct. Fire followed, burning control system wiring. Only then did the computer command engine cutoff—while it was already shutting down on its own.[57]

149

This accident threatened the overall program schedule for the entire space shuttle. The budget cuts of 1972-74 had slipped the planned date of first launch from March 1978 to June 1979, but NASA had won back some of this loss and now hoped to fly to orbit in March 1979.[58] This depended on having a healthy SSME, and that launch date would be jeopardized if Rocketdyne failed to fix this latest problem quickly.

Whereas the earlier oxygen turbopump failure had stemmed from a faulty seal, it was clear that this one had resulted from a gradual deterioration of the bearings on both ends of the turbopump shaft. The data from instruments was inadequate to trace the specifics, while the fire had burned up the evidence. Still, three issues stood out: unbalance in the turbine rotor, inadequate cooling of the turbine bearings, and poor load distribution and load-carrying capacity of both sets of bearings.

Eighteen days after the fire, Engine Two returned to test with additional instrumentation and new bearings. These carried modifications that increased the coolant flow and improved the rotor balance. The changes brought good results almost immediately, with drops in vibration levels and in maximum bearing loads. In subsequent work, investigators developed a mathematical model of rotor dynamics that helped to pin down the continuing sources of vibration. This brought the introduction of heavy-duty bearings and new bearing supports.[59]

There would be other oxygen turbopump fires, but they resulted from causes other than faults in the basic design. To that extent, at least, the modifications solved the problem and permitted development of a reliable unit. "The clue—and the solution—was always found in the data," recalls J. R. Thompson, the SSME program manager at NASA-Marshall. "We never learned as much as we did during an investigation coming out of a failure. That's the time you're really on the steep part of the learning curve." Joe Stangeland agrees: "On the shuttle, we pinpointed precisely the root cause of every single failure," using data from instruments.[60]

Even so, no one was ready to declare victory during 1977. In addition to the turbopump fires, testing during the second half of that year disclosed two new problems, unrelated both to each other and to the difficulties with the fuel and oxygen turbopumps. The first of them appeared in late August, when a test observer saw the engine catch fire and shut it down. Examination of the hardware showed that a hole had burned completely through the fuel preburner.

This preburner amounted to a combustion chamber in its own right. Its internal pressure of 5,572 psi was among the highest in the SSME system. It was compact, about four inches long by ten inches across. It had an injector, a plate studded with numerous short pipes that injected the propellants. Each pipe had a central protruding tube, called a post, that delivered liquid oxygen. Each tube or post was surrounded by a short cylindrical duct that carried hydrogen. This arrangement, with each post being surrounded by a duct, helped their gases to mix. Small recessed compartments on the preburner wall, known as acoustic cavities, promoted stable combustion by suppressing disturbances. The fire started with a flow of oxygen from a post at the periphery of the injector. This gas flowed as a localized recirculation. Metal in the

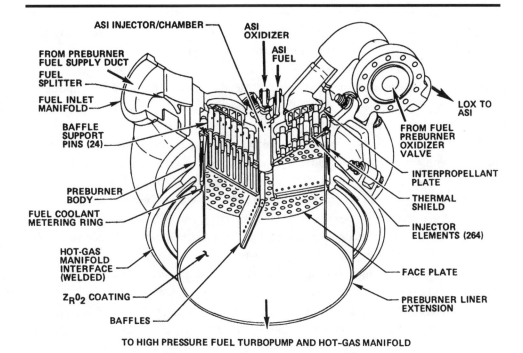

ASI INJECTOR/CHAMBER

ASI OXIDIZER

FROM PREBURNER FUEL SUPPLY DUCT

ASI FUEL

FUEL SPLITTER

FUEL INLET MANIFOLD

LOX TO ASI

BAFFLE SUPPORT PINS (24)

FROM FUEL PREBURNER OXIDIZER VALVE

INTERPROPELLANT PLATE

PREBURNER BODY

THERMAL SHIELD

FUEL COOLANT METERING RING

INJECTOR ELEMENTS (264)

HOT-GAS MANIFOLD INTERFACE (WELDED)

FACE PLATE

$Z_R O_2$ COATING

PREBURNER LINER EXTENSION

BAFFLES

TO HIGH PRESSURE FUEL TURBOPUMP AND HOT-GAS MANIFOLD

FUEL PREBURNER

Fig. 31. *Space Shuttle Main Engine fuel preburner. (Rocketdyne)*

acoustic cavity began to burn and acted as fuel, which fed this internal fire. It burned its way through the outer wall, and the preburner's high pressure turned this flame into a blowtorch.

This time, the needed design changes were not so difficult. By then it was clear that the preburner could maintain stable combustion without the acoustic cavities; these were simply eliminated. Engineers also deactivated the liquid-oxygen posts at corresponding locations, by plugging them up. The hole in the preburner was repaired by welding, and its engine returned to test after only five days. These modifications cured the problem of preburner burn-through, at least for that year.[61]

Late in 1977 it became clear that turbine blades in the fuel turbopump had a tendency to crack. This turbine had two stages or turbine wheels, mounting a total of 122 blades. Each was an inch long by half an inch wide, the size of a large postage stamp. At full power, each blade was to generate 600 horsepower, enough for a racing car. They were heavily loaded and under extreme centrifugal force; they also were uncooled, while operating at temperatures above 1,500 degrees Fahrenheit. Each blade thus faced a stress of 50,000 psi, ten times the pressure in the fuel preburner.

The blades were cast from a nickel-based superalloy developed by Martin Metals, an alloy that exhibited high strength and temperature resistance.[62] Conventional casting from the melt would have formed every blade as an interlaced array of tiny metal crystals, but they gained additional strength through "directional solidification." The castings solidified in molds, and the molds were cooled at one end. Metal crystals formed at that end and grew down the length of the blade as the cooling progressed. Within the metal, carbides migrated to grain boundaries and made the blades stronger. Ironically, this process had been invented at Pratt and Whitney, Rocketdyne's great rival; the metallurgist Francis VerSnyder had introduced it to improve the turbine blades of jet aircraft. Yet Rocketdyne was prepared to learn from its competitors and to purchase patent rights through licensing agreements.[63]

The first blade failure occurred in mid-November. It broke off and ripped through both turbine wheels, producing a cascade of debris and causing a great deal of damage. The turbopump shook violently; the computer detected the intense vibration and shut down the test. The engine had been running at 70 percent of rated power, well below levels sustained during earlier runs, and a temperature spike at more than 2,100 degrees appeared as the cause. Testing continued.

Two weeks later, another engine was running above 80 percent when the same thing happened. This time there was outright destruction, as debris from turbine blades caused the rotor to seize up, halting the flow of hydrogen. The engine continued to run briefly and oxygen still flowed, producing extensive burning and melting throughout the hot-gas system. Fortunately, although the damage was widespread, the engine contained it; there was no external burn-through.[64]

These two accidents brought a large and high-priority effort that sought solutions. Evidence from the failed hardware and from other blade samples showed that when a turbine had run repeatedly for a number of times, its blades tended to crack from fatigue, close to the root. Again this led to consultation with experts in government, industry, and academia, along with a comprehensive test program not only at Rocketdyne but also at General Electric, TRW, and Garrett AiResearch. Rocketdyne used an electrical turbine driver called the Whirligig, which could spin a turbine wheel to 38,000 rpm while measuring stress on the blades by using strain gauges.

This work did not yield a definitive solution, as turbine blades continued to require close attention through subsequent years, but the effort at least eliminated the causes of failure that brought the turbine-blade accidents of 1977. Those blades were firmly fixed in place but nevertheless had to be free to flex and vibrate. They had locked up and lost this freedom, imposing further stress that brought the fatigue cracks.

The turbine wheels received design changes that precluded blade lockup. Damping devices, attached to each blade, reduced stress due to engine vibration that contributed to blade failure. Inlet guide vanes, directing hot gas onto the blades, were reshaped to lessen the stress imposed by that flow. The blades rotated within internal seals that lined the interior of the housing. These seals were modified for closer tolerances, to prevent blade tips from rubbing against them. New checks during turbine assembly verified looseness of the blades.

Still, these changes took time to implement; Rocketdyne's Whirligig, for instance, did not enter service until the following summer.[65] And SSME development was already taking far too much time. The initial shuttle program schedule, in 1972, had called for the first flight to orbit on 1 March 1978.[66] That launch date had slipped due to budget cuts. Even so, it was disconcerting to note that as this calendar date approached, the engines were in Mississippi and were in no condition to fly anywhere.

During a mission, three such engines were to thrust at rated power for 520 seconds. No such triplet had yet fired, and although single engines had run at rated power, none had done this for that mission duration. The 301-second rated-power test of August 1977 had not been repeated; indeed, less than 5 percent of the total accumulated run time had been at that power level. No test had yet achieved 109 percent, not even briefly.[67]

A test series in February 1978 brought further disheartening results. It used Engine Two, a workhorse dating to 1976, now equipped with a fuel turbopump that had been modified to protect the turbine blades against cracking. A run on 2 February was scheduled for 100 seconds at 70 percent and was successful. The next six tests, during the following three weeks, brought increasing disappointment.

On 10 February the test was scheduled for 310 seconds at 90 percent. It shut down at 96.33 seconds due to failure of a turbopump speed sensor. This was essential; that instrument was needed to detect any overspeed or loss of power, and it had to work properly for the test to proceed.

On 12 February the run was planned for 320 seconds. It cut off at 4.04 seconds and 86 percent, due to excessive vibration of the oxygen turbopump. On 14 February the engine was slated for 300 seconds at 100 percent. It stopped at 4.2 seconds and 99 percent, due to vibration in a pump. The source was "cavitation," a rapid growth and violent collapse of bubbles due to low instantaneous pressure, as fluid flowed over impeller blades. Cavitation was unacceptable; it could prevent the impeller from pumping effectively, leading to a destructive overspeed. On 15 February the plan called for 300 seconds at 100 percent. The test halted at 11.3 seconds due to pump vibration, again produced by cavitation. On 17 February the run was scheduled for 310 seconds at 100 percent. It shut down at 3.57 seconds and 89 percent due to vibration. This stemmed from a condition of resonance, which amplified the intensity. On 21 February the test was slated for 300 seconds at 95 percent. It cut off at 6.08 seconds and 91 percent, too late to avoid a serious accident.

The engine had a new oxygen valve, which allowed too much flow early in the start sequence. This brought a temperature spike in the oxygen preburner, damaging a turbine tip seal and reducing the turbine efficiency. The engine computer tried to compensate for this by

153

further opening the valve, to let in more oxygen. But with its damaged turbine, the engine could reach only 91 percent of rated power. The controller detected a preburner temperature that was too high, violating a redline; recognizing this, it commanded a shutdown. Still, it took time for the valves to close in response, and they closed in a way that boosted the oxygen pressure substantially. The preburners, built to run fuel-rich, now received a surge of oxygen and burned far hotter than intended. Their hot gases melted parts in both turbines. Again there was no burn-through, but the engine was out of service for a month as Rocketdyne replaced both turbopumps.[68]

These events followed 1977, during which overall testing had also fallen short. The Mississippi test stands had certainly been active, running an engine about once every three days. Yet though nearly all runs were at reduced power, their durations averaged less than a hundred seconds, only one-fifth of the mission requirement. This resulted from numerous premature shutdowns. The original test plan of 1973 had called for cumulative run time to reach 38,000 seconds by the end of 1977. The actual total, 13,507 seconds, was barely one-third of this mark. As noted, little of it was at rated power.[69]

Were there basic flaws, perhaps not only in the engine itself but in the project management? By late 1977 the SSME was ready for a review that would indeed be critical (and was not the Critical Design Review). It was a program assessment conducted by a high-level technical commission, sponsored by members of the Senate. This panel did not have the ulterior motive of cutting the space budget to free up funds for social programs. Its members supported the shuttle.

The Need for Testing

In mid-December of 1977 Senators Adlai Stevenson and Harrison Schmitt wrote a letter to NASA administrator Robert Frosch. Stevenson chaired the Senate's space subcommittee; Schmitt had been a scientist-astronaut and had walked on the moon during Apollo 17. Together they called on Frosch to have the National Research Council conduct an independent assessment of the SSME. The NRC was an arm of the National Academy of Sciences, which dated to the Civil War. They were not formally part of the government; the National Academy was a private corporation. Membership was a high honor and distinction, sometimes counting as a prelude to a Nobel Prize. These organizations were on call to provide expert counsel on technical issues, both to Congress and to federal agencies such as NASA.

Frosch responded to this letter by awarding a contract to the NRC, which put together the review committee. Its chairman, Eugene Covert, was a professor at MIT. Its members included a leading metallurgist, two vice presidents from rocket-building companies, and two Air Force directors of rocket-engine development and test. Allen Donovan, another member,

154

had been part of a small group at TRW that directed development of the Air Force's big missiles during the 1950s. Another panelist, Boeing's Maynard Pennell, had shaped the design of that company's airliners from the 707 to the supersonic transport. A special advisor, Richard Mulready, had managed Pratt and Whitney's work on engines closely resembling the SSME, and had narrowly lost to Rocketdyne in competing for its contract.

Members of this commission met for six days during January and February 1978 and visited Rocketdyne as well. Their work thus was cursory indeed; in no way would it compare to the depth and intensity of the accident investigations that had followed fires on the test stands. Yet these men had broad experience and were prepared to offer useful comments. Testing of the SSME drew their particular concern.[70]

At the outset, NASA and Rocketdyne had expected to conduct a major program of tests of components, such as turbopumps, in parallel with firings of complete engines. Some components indeed were tested separately; these included the preburners and main combustion chamber, valves, nozzle, and computer controller. Turbopumps were also on the agenda, and Rocketdyne's Coca-1 facility served for their evaluation. However, the destructive fire of February 1976 prevented Rocketdyne from continuing with component testing of the oxygen turbopump.

That test stand could have been rebuilt, and NASA's Fletcher declared that "repairs to that facility are of greater urgency than the construction of new facilities." Still, no one wanted to hold up testing in Mississippi pending restoration of Coca-1 and resumption of its turbopump work, and program officials decided that they could dispense with the separate turbopump development activity at Santa Susana. Instead they chose to conduct turbopump development by running these units as part of complete engines. For these components, the engine itself was to serve as the test facility.[71]

This was not a decision made casually. The extreme pressures of the SSME had required heavy hardware at Coca, and efforts to use it proved to be difficult, costly, and not very fruitful in acquiring useful data on the turbomachinery. This brought a conclusion that turbopump testing at Santa Susana, even with a repaired Coca-1, could not be conducted at a pace that would help support the overall development program. Moreover, even when components were successfully operated as separate units, there was reason to question the results. On the component test stand, during start-up, thermal transients differed from those of a complete engine. Starting transients also were different. The oxygen turbopump, a prime focus of concern, took 8 to 9 seconds to come up to full speed when operating with its preburner on the component stand. On the engine, it needed only 4.5 seconds, which was more severe.[72]

Rocketdyne discontinued testing of oxygen turbopumps following the Coca-1 fire. Start-up tests had indeed featured prominently, the runs averaging less than seven seconds. Work with fuel turbopumps continued on the undamaged Coca-1B installation, but at a desultory pace. In September 1977 it too was shut down. The Covert committee report summarized the work (table 4.1).

Table 4.1. Component Tests at Coca-1

	Assemblies Tested	Number of Tests	Cumulative Time (in Seconds)
Coca-1A, oxygen turbopump	3	24	161
Coca-1B, fuel turbopump	6	27	111

Source: Covert, Technical Status, 28.

These cumulative times covered twenty-seven months of operation. Lee Webster, a panelist who managed rocket-engine testing at the Air Force's Arnold Engineering Development Center, offered a trenchant comment: "Normally, such testing for past rocket developments would have accumulated at least ten times more testing." The modest totals at Santa Susana compared with close to ten thousand seconds of complete engine operation in Mississippi as of the end of that September.[73]

The committee urged NASA and Rocketdyne to "explore means of acquiring and operating a component-development test rig for the rotating machinery." Pratt and Whitney's Mulready added helpfully that his company had built just such an installation and had used it for its own turbopump development tests. The report cited recent problems with the oxygen turbopump, and William Rostocker, the commission's metallurgist, reviewed the problems with turbine blades. The panel wrote,

> The lack of an adequate test rig to conduct component tests expeditiously outside the engine appears to be the primary reason that the turbopump problems were not discovered before last summer, during the initial engine tests at the rated power level.
>
> NASA and Rocketdyne hold that the engine tests at rated power level have provided the data necessary to identify any deficiencies in the engine and the high-pressure turbopumps, and the data necessary to formulate corrective actions. Insofar as the committee is aware, however, no rocket engine using turbopumps has ever been produced without extensive development testing of the turbopumps first as components.[74]

NASA would have none of this. Its officials were committed to the continuing use of complete engines as turbopump test facilities. Within days after the release of the report, Frosch presented testimony to Adlai Stevenson's Senate subcommittee: "We have found that the best and truest test bed for all major components, and especially turbopumps, is the engine itself."[75]

Nevertheless, program managers proved receptive to another recommendation of Covert's group, that "Rocketdyne acquire an additional engine and critical parts to accomplish the required testing." The company responded by reactivating a test stand at Santa Susana and installing the old ISTB, refurbished and ready now for further use. Complementing the continuing work in Mississippi, the new test series at Santa Susana got under way in November 1978. This effort took the lead in conducting initial trials of new modifications, which then could become part of mainstream SSME development.[76]

During 1978 activity in Mississippi hit its stride. That center's engines incorporated modifications intended to overcome the oxygen-turbopump and turbine-blade problems of 1977, and engineers hoped that these would be the last major design changes. The turbine-blade failures had occurred after about thirty-four hundred seconds under test, and the new engine runs had to accumulate more than that total time before anyone could have confidence that the blades could achieve long life.

Engine Two returned to service in March, following replacement of its turbopumps in the wake of the mishaps of February. It did not remain in test for long; it displayed a new problem, in the main injector. Some of its oxygen posts cracked, releasing oxygen and producing localized fires that damaged adjacent injector elements. It was back to Rocketdyne for this engine, where it was slated for an extensive rebuilding. This faithful old horse had accumulated 122 tests, far exceeding the 55 starts specified as a requirement. A new SSME, Engine Five, took its place in Mississippi. It quickly achieved a milestone, for on 10 May it ran at rated power for full flight duration, 520 seconds. It was the first engine to do this, improving substantially over the 301-second rated-power test of the previous August. This 520-second run needed more propellant than the test-stand tanks could hold. The test area adjoined the Pearl River, the eastern border of Louisiana, and was accessible by water. Barges held additional propellants, which were pumped into the run tanks during the test. This practice became standard for long-duration firings.[77]

After only a few weeks, Engine Five developed the same problem with oxygen posts that had appeared in Engine Two. This was worrisome, for Five had entered service only recently, whereas Two had been in test for two years. However, most tests of Two had run at reduced power. They both had about a thousand seconds at or near rated power, and this higher level had brought the cracks. The posts needed strengthening, and they received steel supports, bolted in place. The problem disappeared.[78]

In Mississippi, a parallel effort tested a three-engine cluster. This work used the center's main stand, which towered to a height of more than four hundred feet. During Apollo it had fired complete Saturn V first stages, with 7.5 million pounds of thrust. The shuttle's three engines developed 1.1 million pounds at sea level, at 375,000 pounds per engine—and the site was open to visitors, who could watch from half a mile away.[79] This triplet was part of the

Fig. 32. *Space Shuttle Main Engine under test at Rocketdyne facilities near Los Angeles. (Rocketdyne)*

Main Propulsion Test Article (MPTA), which amounted to a complement of Enterprise. That orbiter looked like a shuttle, duplicating its approach and landing characteristics while reproducing its flight deck. However, it was completely innocent of propulsion. The MPTA looked nothing at all like a shuttle, but it had what Enterprise lacked. It included an aft fuselage, complete with the main engines, and carried an external tank that held its propellants. A truss replaced the rest of the orbiter, providing mounting points for that tank. Tank, truss, and aft fuselage then made up the MPTA.

Rockwell's Space Division, rather than Rocketdyne, built it. As with Enterprise, this gave that division experience in constructing and exercising hardware that included major elements of an operational shuttle. Significantly, the MPTA tests were not part of the formal SSME development program, even though they produced data that were useful for that purpose. The MPTA was a Rockwell project, which had the purpose of exercising the complete main propulsion system, including the external tank with its pressurization, feed lines, and valves. NASA ordered the three engines of MPTA as production models built by Rocketdyne and then supplied these engines to Rockwell as government-furnished equipment. Rocketdyne specialists were on hand in Mississippi to provide support in the field, but the people who actually ran the hot-fire MPTA tests were from Rockwell. Nor did these hot-firings count as parts of the testing for SSME qualification, which was entirely in the hands of Rocketdyne. Instead, the work with the MPTA, as with Enterprise, was treated as tests of the orbiter.

Workers in Palmdale assembled the MPTA, sans engines and external tank, and shipped it to Mississippi in June 1977.[80] The big tank arrived from Michoud in September. The three main engines came from NASA; they underwent acceptance testing and then were installed within the MPTA. The first firing occurred in April 1978—for 1.2 seconds. Three other firings took place from May to July, reaching 100 seconds and 90 percent of rated power. "It was just fantastic," the test manager said after one such run. "After it was over, we stood around for about an hour and marveled at the way it went off." The engines then were removed for further use as individual units while the MPTA awaited its next series.[81]

On another stand, an engine carried extensive instrumentation to study the internal environment of the oxygen turbopump. The instruments included pressure and temperature gauges, strain gauges, accelerometers to measure vibration, and an electrical device to detect any displacement of a shaft or bearing. This test series did not go well; of fourteen runs, only three reached the scheduled durations, as the computer commanded shutdown during the other attempts. Then, on 18 July, a planned 300-second run ended at 41.81 seconds with excessive vibration—and a major fire in the oxygen pump. Despite the safeguards, here again was a destructive accident. This time the array of instruments was sufficiently complete to permit rapid determination of the cause. The bearing-displacement sensor had failed structurally; the rotor had rubbed against it and ignited it. One could say that this sensor had

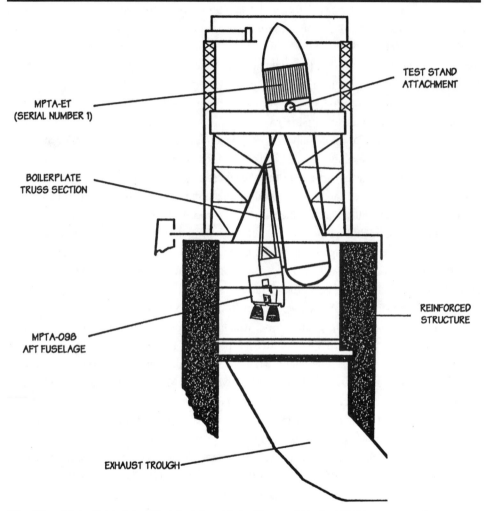

MPTA-ET (SERIAL NUMBER 1)

TEST STAND ATTACHMENT

BOILERPLATE TRUSS SECTION

REINFORCED STRUCTURE

MPTA-098 AFT FUSELAGE

EXHAUST TROUGH

Fig. 33. *Main Propulsion Test Article with its External Tank. (NASA, courtesy Dennis R. Jenkins)*

caused the accident it was intended to forestall. Significantly, the turbopump itself, with its design modifications, was not at fault.

Meanwhile, Engine Five was receiving modifications to all three injectors. A technician left a handkerchief-size rag in a preburner fuel manifold, which was akin to a surgeon leaving something inside a patient. Rather than disassemble the preburner, people applied heat to it while running oxygen through it, causing the cloth to burn away. Following this "rag roast," this engine qualified for further testing.[82] Nevertheless, the delays in Mississippi carried over

Fig. 34. *The Main Propulsion Test Article External Tank in position in Test Stand B at NASA's National Space Technology Laboratories in Mississippi. Note that the test facility is accessible by water. (NASA, courtesy Dennis R. Jenkins)*

to the entire shuttle program. The Covert report, in March, had looked ahead to a first flight to orbit in March 1979. This launch date slipped to June 1979 and then into the fall.[83]

Testing resumed on 12 August 1978. During the next six weeks the SSME development effort struck gold. Engine Five ran off five successful firings in a row at rated power and 520-second duration. With other tests, this engine accumulated more than 5,000 seconds of trouble-free operation, more than 4,500 seconds of that being at or above 100 percent of rated power. Indeed, during one run it spent three hundred seconds at 102 percent. The SSME thus took its first significant step beyond its rated thrust level.[84]

Subsequent work addressed issues of abort. Mission rules mandated that the shuttle had to be able to return safely if a significant problem developed at any time during ascent; ditching in the ocean was not an option. During the first 250 seconds following liftoff, the orbiter was to abort by returning to its launch site. It was to do this by turning around to point its SSMEs in the direction of flight, to use them as big retro-rockets. The orbiter would literally come to a stop, then reaccelerate backward to gain enough speed for the return. But if the shuttle was too far along in ascent to do this, it would abort to orbit, continuing into a low orbit from which it was to reenter as soon as it was within range of an emergency landing site.

The most likely causes of abort involved loss of thrust in an SSME. The remaining propellant supply, within the external tank, then would allow the surviving engines to burn longer. Abort to orbit required up to 665 seconds of engine operation, and Engine Five demonstrated this capability by firing for this duration on its test stand. Return to launch site was more demanding, calling for as much as 823 seconds. This engine achieved that duration as well, at 97 percent of rated power.[85]

On 25 September John Yardley, NASA's associate administrator for space transportation systems, addressed a subcommittee in the House of Representatives and looked ahead to a first flight that might be as little as a year away: "A detailed Shuttle program review has been completed to permit an accurate, updated assessment of cost, schedule and performance. The review showed that substantial progress has been made in the program this year. It also showed that all program elements could be ready for a September 1979 first manned orbital flight (SS-1) if all planned tests were successful. . . . September 28, 1979, therefore, has been set for SS-1 as an internal target working schedule." By the end of 1978 the cumulative SSME test time came to 34,810 seconds, for an increase of more than two and a half times the total of a year earlier. Nearly half the testing of 1978 was at rated power or higher.[86]

What did NASA gain from this? The work validated the design changes within the oxygen turbopump, demonstrating that this unit could operate reliably. This complemented the earlier resolution of the subsynchronous-whirl problem in the fuel turbopump.[87] The extensive tests also confirmed the value of the improvements to the turbine blades, for no complete blade failures occurred in the fuel turbopump after March 1978. Nevertheless, these blades continued to demand close attention, for some of them continued to develop small cracks in their surfaces. These did not propagate into the solid metal, and experience showed that this turbopump could operate for at least fifteen start-stop cycles with no blade failures. This brought a new practice: disassembling the turbine after fifteen cycles and removing the blades for a close look.

This included a metallurgical examination, with the blades being cut, etched, and scrutinized under a microscope. The data from these studies went into a Rocketdyne project that sought to define procedures for scheduled inspections and blade replacements in operational engines. At NASA-Marshall, a related effort sought to learn the rate at which the cracks would grow, again with an eye toward determining how often the blades would need examination. Significantly, this work did not seek to prevent the cracks from forming. The emphasis was on finding ways to live with them by keeping close watch. In the words of the Covert

Committee, "The present planning allows the use of blades with small cracks in their plat-forms for engines in the first manned orbital flight. Such a procedure is contrary to current, conventional practice. . . . Every effort must be made to establish a technical and statistical basis justifying the use of blades with platform cracks in early flights."[88]

The engine itself remained extraordinarily sensitive, experiencing a serious problem in starting as late as October 1978. That test began with two unrelated events: torque in the fuel turbopump was slightly higher than normal, while a misaligned actuator meant that the main oxygen valve was 2 percent further open than indicated by its position measurement. In com-bination, these events brought a modest reduction in the speed of the fuel turbopump—and an early priming of the main combustion chamber. This chamber generated back pressure that prevented the fuel turbine from building up power in timely fashion. Fuel flow was in-adequate, while the open valve brought excessive flow of liquid oxygen. The preburners ran oxygen-rich, which brought a severe burnout of the turbines and hot gas system.

Engine Five retired with honor in November, marked for detailed disassembly and close inspection of its parts. It had racked up more than 12,000 seconds, the most time accumu-lated on a single engine to that date.[89] Yet as testing continued with other engines, it still was far too soon to claim victory. Following start-up of a test on 5 December, and only 3.5 sec-onds into the run, an explosion occurred within a heat exchanger. This component drew on engine heat to produce a flow of gaseous oxygen that pressurized the liquid-oxygen tank. This pressure was necessary to deliver this fluid to its pumps.

The Covert report had warned of this: "The heat exchanger . . . is an extremely compli-cated welded assembly. It is located in a position where inspection is difficult. While no re-cent incidents have been attributed to heat exchanger failures, the committee is concerned that a failure of the heat exchanger could be catastrophic." The report added, "The heat ex-changer system poses a potential threat to the total shuttle system." The 5 December accident indeed was serious, for it also damaged the oxygen turbopump. But it did not halt the test program, which continued through the month.[90]

Two days after Christmas, the program received an unwelcome gift. An engine appeared to be running normally, halfway through a 520-second run. Suddenly, in the words of an ob-server, "the engine burst into yellow and orange flames. These flames included flashes like a roman candle and appeared more intense and voluminous than any previous failure." An-other observer wrote of "a violent fire. The engine area of the test stand was completely en-gulfed in smoke and flame and I couldn't see anything else." Test conductors immediately turned on the "firex," the facility's built-in water deluge system, but it took repeated use of their sprays to douse the flames completely.[91]

This was damaging indeed. Managers held up further testing while they sought the causes of the two nearly simultaneous failures. The target launch date of 28 September 1979 went out the window as it slipped into November, with officials appreciating that a further

163

delay into 1980 was only too likely. The 27 December fire destroyed that engine, which had been slated for use in the MPTA. A replacement was available, but it needed qualification testing, which meant more time lost. Myron Malkin, NASA's shuttle program director, placed the cost of the accident at $15 to $20 million, adding that any similar failure would severely affect the shuttle program's budget and schedule.[92]

The accident investigation made use of a liquid-oxygen feed system at NASA-Marshall, which allowed study of the flow under controlled laboratory conditions. Engineers traced the fault to the main oxygen valve—and found that this far-reaching failure had started with nothing more than a loose screw. Turbulence in the high-speed oxygen flow through the valve had created vibration that loosened this fastener and caused a metal part to shake excessively. It rubbed against some adjacent fittings, creating friction that ignited them. The fire brought an increase in oxygen flow. Then within one-tenth of a second the fuel turbine overheated, the oxygen turbopump failed, and the main oxygen line blew out.

Flaws in design had brought this destruction, but the solutions were simple, including the use of new screws that locked in place. An intensive effort carried through the redesign and fabricated new parts, while the oxygen valves on the two preburners were similarly modified. New instruments were mounted to the valves of a test engine, to better define their operating environment. Testing resumed at the end of January 1979, initially at Santa Susana, where Rocketdyne's new facility now showed its value. The company brought in extra people and ran the test stand around the clock, with a two-shift firing crew and a third shift for maintenance.[93]

The heat-exchanger problem proved more recalcitrant. "The failure of the heat exchanger in December remains unexplained," a knowledgeable specialist told *Astronautics & Aeronautics,* "and it gives you a very soggy feeling. These incidents occurring so late in the test program just do not inspire confidence." "We haven't been able to find anything wrong from a design standpoint," Rocketdyne's Dom Sanchini told *Aviation Week.* "We have to suspect and generally believe that somehow the problem existed in the fabrication process or in some retrofits we did to that area of the engine that somehow damaged the heat exchanger."[94]

Here indeed was the explanation. A worker had used arc welding to repair brackets on that exchanger's coil while the component itself remained within the engine. This welding job weakened part of the tubing in the coil, which sprang a leak. The lost strength went undetected because existing procedures did not call for inspection or proof-testing of the reworked part. Rocketdyne responded by adding pressurization tests and leak checks to the fabrication process, and it continued to use the existing heat-exchanger design.[95]

Meanwhile, the renewed engine test program turned in good results. Following resumption of testing on 30 January, engines with modified valves ran up some 5,000 seconds of firing time and showed no sign of similar problems. The program carried out twenty-two consecutive runs without a premature engine cutoff, followed by a similar streak of fourteen more runs. All reached their planned durations; several went for 520 seconds, and another went for the abort time of 823 seconds.[96]

One still could not describe the SSME as reliable. Engine Five may have operated for a total of twelve thousand seconds during 1978, but it did not do this with a single assembly of hardware. It received numerous changes of components; in this respect, it resembled the farmer's axe that got five new blades and six handles but was still the same old axe. Indeed, during early 1979, any engine that ran for four hundred seconds required at least one component changeout, on average. Nevertheless, by then the turbopumps were beginning to show a semblance of long life. The mean time between premature cutoffs, due to malfunction of a turbopump, exceeded ten thousand seconds. This was barely enough for six shuttle flights, each with three engines, but it suggested that these units might soon merit full confidence.[97]

Preludes to Flight

At the start of the test, a bright orange glow appeared at the top of the flame bucket, the enormous curving duct built to receive the exhaust. Sprays of water filled this bucket with a grayish mist, and when the exhaust struck these sprays, the water flashed into thick white steam. This vapor erupted with great force, rushing from the flame bucket like an explosion of gas from a volcano. A bellowing roar assaulted the ears as the steam formed a plume that drifted with the wind, rising into the air after some hundreds of yards.

The roar and the flow of vapor continued without letup, minute after minute, with the steam continuing to gush forth as if sprayed from an immense hose. A cluster of three SSMEs was in action, booming forth with the thunder of well over a million pounds of thrust, running for full duration of 520 seconds. Visitors could readily find themselves caught up in exaltation, carried away emotionally by the power they were witnessing.

Within the dense plume of vapor, rain condensed and began to fall. As it fell more quickly, the sun formed a rainbow between the plume and the ground. The white steam rose higher; the rainbow brightened and grew larger. Now people's ears, adapted somewhat to the cacophony, discerned a thin high whining noise. This came from the turbopumps. As the test ended and the engines prepared to shut down, this whining seemed to grow louder, giving way quickly to the sound of a squeaky door when the pumps stopped. The tail of the plume drifted downward; its whiteness dissolved into the blue of the sky as it gave up the last of its rain. The rainbow glowed brilliantly for a few more seconds before fading, as if to salute this success.[98]

There indeed was much to salute, for during 1979 and 1980 the program reached beyond development testing to enter a phase of certification for flight. The goal now was not to uncover flaws in design; rather, it was to prove that the engine could be trusted. John Yardley, NASA's top space-shuttle manager, set down firm requirements. "Making sure we could run

Fig. 35. *Space Shuttle Main Engine firing on Test Stand B in Mississippi. (NASA)*

the whole mission duration," recalls J. R. Thompson, SSME program manager at NASA-Marshall. "Making sure we had plenty of margin. Making sure we could run the abort missions." Yardley froze the engine design, ensuring that certification would proceed on a stable SSME configuration.[99]

Flight certification used single engines, not the MPTA's cluster, and the units that served for certification were not the ones that were to fly. Flight engines ran three times, for 520 seconds as well as for shorter durations, and then went to Kennedy Space Center. Certification engines were tested far more extensively, which used up much of their life. However, both sets conformed to a common engine design.

The work during 1978 with Engine Five, succeeding at both abort and mission durations, amounted to practice for flight certification. NASA defined its certification requirements in terms of a set of thirteen tests, called a cycle. They included evaluation of the start sequence, followed by calibration tests to verify compatibility between hardware and software. Then came firings at rated power: four of mission duration, 520 seconds, with abort simulations at 665 and 823 seconds. The series also included a 425-second run at 102 percent. Two engines served initially for certification; each went through two cycles and the testing required complete success. If any engine shut down prematurely, that cycle did not count and had to start over from the beginning. Sanchini described this as "the most stringent criterion ever imposed on a rocket engine prior to its first manned flight."

A production engine, number 2004, took center stage in the certification program. It went through its first cycle between March and June 1979. Following disassembly and close inspection, it went back for the second cycle, completing it in February 1980.[100] It then returned to development testing—and became the first engine to run at 109 percent. It did this during four runs in March and April, racking up 1,090 seconds at that power level. A second engine then entered the certification process, completing its two cycles in August and December. With this, the SSME qualified for the first ten shuttle missions.[101]

These tests, like those of the MPTA, were open to view; one test area even had a grandstand for visitors. At 109 percent, bursts of steam flew from the flame bucket as if shot from cannon. A shift in the wind could blow the plume directly at the spectators. A solid gray fog then engulfed them, chilly and wet even on a warm day, as rain condensed within the cloud and blew in their faces. They would be lost within the thick clamminess, while still the engine roared. Following the test, they found their cars splotched with dirt from the rain. Its water came directly from the nearby Pearl River and was fed into the flame-bucket sprays without filtering.[102]

Operation at 109 percent was doubly demanding, for it required commensurate increases in both the temperature and pressure of engine components. The propellant flow rates also increased, which meant that the turbines had to develop far more power. The fuel turbopump went from 61,402 horsepower at 100 percent to 74,928 at 109 percent, an increase of more than one-fifth. This happened because this unit not only had to pump more hydrogen but had to raise it to a higher pressure. Still, the design of the turbopump allowed for this. The work at 109 percent achieved success, at least in those early tests.[103]

Cumulative firing time was another goal. During 1978, John Yardley set a requirement of 65,000 seconds of single-engine test operation prior to the first flight to orbit. The program reached this mark in March 1980, and it is appropriate that Engine 2004 was the one that did it.[104] A year later, on the eve of first flight, the tally was 99,379 seconds in the course of 669 firings. The triplets of MPTA brought the total to 110,153 seconds, in 726 tests.

How did the program achieve this? It did not happen by increasing the testing rate, which remained around 140 per year after 1977. Rather, the average test duration increased smartly, as programmed run times rose dramatically while premature cutoffs diminished.[105] It took until the fall of 1977 to reach ten thousand seconds within the entire program. In 1980 this was no more than a single engine was expected to achieve in certification testing.[106]

During 1980, amid growing engine maturity, the mean time between premature cutoffs reached 15,000 seconds. The mean time between component removals rose to a thousand seconds. This increasing maturity even reached to Santa Susana, where new changes first went under test. Such changes brought ample opportunity for early cutoffs. But from December 1979 to August 1980 this facility ran off thirty-five firings in a row, all at scheduled duration.[107]

The MPTA gained its own successes, but less rapidly; with three engines, there were three times as many things that might go wrong. Following the test series of 1978, it returned to action in May 1979 with a 1.5-second ignition demonstration. The next attempt, in mid-June, reached rated power for the first time. But while it had been programmed for 520 seconds, it cut off after less than a minute. This resulted from a faulty sensor that erroneously reported high vibration.[108] Early July brought another try for 520 seconds. Eighteen seconds into the test, one engine's main fuel valve cracked its housing. This brought a substantial hydrogen leak. The flow of fuel fell off; the preburners began to run oxygen-rich, and the turbine temperatures increased sharply. The computer caught it in time, before that engine could burn up, and all three engines shut down. The hydrogen leak was so severe that it raised the pressure in the aft compartment, causing significant structural damage by blowing off parts of the engine housings.

This accident showed how a problem could lurk undetected until disclosed through long-continued testing, for that valve had served in fifty-five previous firings, accumulating more than eight thousand seconds of operation. Hence testing of other engines could continue, with valves that had seen less use, while the investigation went forward. The crack occurred at a corner within the valve; it had been rounded to relieve stress, but not enough, so engineers rounded it further. They also redesigned the housing for an overall reduction of stress.[109]

Four months passed, as managers modified the MPTA with unrelated changes. In November 1979 they tried again for 520 seconds. It cut off at 9.7 seconds when a sensor detected excessive pressure near a turbopump seal. Then as all three of the SSMEs were shutting down, a hydrogen pipe broke within the nozzle of a different engine. Lacking fuel, it ran oxygen-rich as the shutdown proceeded, sustaining damage both internally and to its instrumentation. The two failures had different causes; it was disheartening that a malfunction in one engine, even when caught, could trigger an unrelated problem in a second SSME.[110] It was all the more disheartening because a similar nozzle failure had occurred during the previous May, in a test at Santa Susana.

Twice in each run, during start-up and shutdown, a nozzle took structural loads due to rapidly changing pressures from the exhaust. These loads caused the nozzle to flex, which in time produced fatigue. Nozzles were constructed to resist fatigue, but the one tested in May had already seen forty-five previous firings, and a section of tubing broke off as that test was shutting down. That tubing carried hydrogen; again, its lack caused the engine to run oxygen-rich. It did this only momentarily, as the engine came to a stop, but that was enough to burn the nozzle, main combustion chamber, main injector, both turbines, and both preburners.

As with the fuel valve, testing continued with nozzles that had seen less use, while investigation proceeded through much of 1979. It found stresses that were brief but unexpectedly strong, and that indeed had caused the pipe to break through fatigue. Engineers dealt with this by redesigning those hydrogen lines to reduce stress and imposing limits on life in service prior to replacement. Against this background the November failure was puzzling. That nozzle's tubing had given way after only eight tests and at measured strains that were not high enough to produce breakage through fatigue.[111]

That hydrogen duct had failed at welded joints. Upon examination, the welds proved to be soft. They should have been made with Inconel 718 welding wire, but studies with an electron microprobe X-ray analyzer showed that the wire had been Inconel 62, with only half as much strength. Rocketdyne responded by testing its supplies of welding wire—and found two lots with the wrong material, both from the same vendor. A chemical-etch process allowed instant recognition of Inconel 62 in other welds, several thousand of which were found to be faulty. Many other nozzles also had soft welds. All of them were reinforced with electrodeposited nickel plating.

That November accident had started with a problem in a turbopump seal, which had leaked from erosion as that seal rubbed against its shaft. Here the design changes were more easily identified; they included use of carbon as a seal material, which eroded less readily. With these changes in hand, the MPTA was scheduled for another try.[112] This time it worked, and it did more. Between December 1979 and January 1981 the MPTA ran off a string of six successful tests, all at planned duration and all reaching beyond 520 seconds. The firings demonstrated a good deal of flexibility, gimbaling engines on their mounts, throttling them through a range of thrust levels, shutting some down early while others continued under power. Throttling would reduce dynamic pressures and g-loads during ascent; the shutdowns represented engine-out operation. Other exercises simulated the inability of an engine to throttle when commanded. The cluster ran at 102 percent during two tests, while the last in the series reproduced an abort. This one shut down an engine at 239 seconds, as if it had failed in flight. The two other SSMEs continued at rated power for another 390 seconds, enough to allow the orbiter to return to its launch site.

Not everything went as planned during that year. An MPTA run in February 1980 went for only 4.6 seconds due to a faulty start. Another firing, in November, terminated at 21.74 seconds as a computer detected high temperature in its fuel turbopump. That engine had begun to run oxygen-rich because it had . . . a hole in its nozzle, which burned through coolant tubes that carried hydrogen, reducing its flow. This was no emergency; the nozzle was old and had seen much use, while standard types already included better construction.[113]

169

A potentially more serious accident occurred on 12 July, as fire broke out and the engines shut down some 106 seconds into a 541-second run. One engine had a hole in its fuel preburner, similar to that of a test attempt in August 1977. As with that earlier failure, oxygen had flowed from a post in the injector and burned the metal. This time, however, it happened for a different reason. The injector face plate, which carried the posts, had deformed and bowed outward. This restricted the flow of hydrogen through a tube that surrounded that post and created a localized oxygen-rich condition. No other preburner showed a bowed face plate, while the failed unit proved to have shown more instances of overheating or minor wear than all others combined. Even so, this problem merited close attention, for Engine 2004 had experienced something similar: a burn-through of its main injector. That component had accumulated 14,210 seconds of operation, but officials were reluctant to dismiss the issue by saying that this injector had merely exceeded its life limit.

Investigators studied the flow of gases in preburners and found flow paths that could contribute to burning the metal. Engineers introduced design changes that blocked these paths. Preburners were given an internal shield to resist burning; other changes ensured that hydrogen had enough pressure to flow into any hole in the component's inner wall and cool it. An engine ran on its stand with deliberately damaged injector posts. No burn-through occurred; temperatures remained within safe limits, showing that the SSME now could run properly even with a malfunctioning injector.[114]

For the MPTA, the failures of July and November 1980 led to delays. Following a successful test around Memorial Day, the next fully satisfactory firing did not take place until after Thanksgiving.[115] Still, this work helped to qualify the orbiter for flight, as the propulsion system achieved goals in development. The engine's growing maturity stood out as well. As recently as December 1978 the failures of the heat exchanger and oxygen valve had threatened to bring the program to its knees. In July 1980 the preburner fire was followed by an oxygen turbopump explosion. These accidents were serious, unrelated, and nearly simultaneous. Even so, the effort took them in stride.

The turbopump accident began with an electrical surge that knocked out the power supply in a data channel. A backup sensor was in use, but in an unrelated event, it dislodged in position. Together, these malfunctions meant that the computer received a highly inaccurate measurement of the pressure in the main combustion chamber. It responded by adjusting valves in a manner that pumped a mere trickle of liquid oxygen into a large flow of very cold fuel, much of which was liquid hydrogen straight from its tank. This chilled the hot combustion products; within the oxygen turbopump, steam produced by combustion actually froze, leading to a buildup of ice. The ice unbalanced the shaft; its pump rubbed within its housing and produced friction; liquid oxygen was close at hand. The resulting fire blew out the oxygen duct. This was unfortunate, but investigators concluded that nothing was wrong with the design of the oxygen turbopump itself. In that sense, the unit survived a trial by fire.

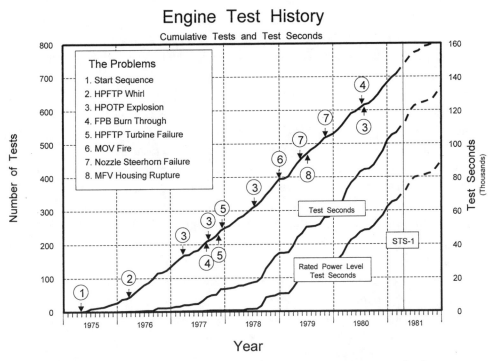

Engine Test History
Cumulative Tests and Test Seconds

Fig. 36. *Space Shuttle Main Engine test history during development. (Rocketdyne, courtesy Bob Biggs)*

The sensor installation was modified. The computer program received a new redline to guard against, of all things, freezing temperatures within the turbine.[116]

This was the last important failure within the test and development program, and by early 1981 the SSME qualified for flight. Its record of achievement included the MPTA firings, which exercised engine triplets in a flight configuration. It included certification testing, with the program completing eight certification cycles prior to the first mission. The record also showed 99,379 seconds of total test operation of single engines, far exceeding the requirement of 65,000.[117]

John Yardley had set that requirement, drawing on his experience with fighter aircraft at his home firm, McDonnell Douglas. A standard ground rule called for a new fighter to conduct forty missions during flight test before it could be released for operational use. The SSME followed this; during each flight of a shuttle, its three engines were to burn for 520 seconds, so that forty flights corresponded to 62,400 seconds. On this basis, the SSME was at least as safe as a new Air Force or Navy fighter—and the shuttle's astronauts were test pilots, who were accustomed to flying hot jets even in their earliest and most dangerous stages of flight test.

For the SSME, further safety came from the process of development. At the outset, project managers set forth its Design Verification Specifications, which formed a basis for the test program. The goal of the testing had been to prove that the design met the specified requirements, and this proof demanded the successful completion of a large number of individual tasks. Prior to the first flight to orbit, Rocketdyne had to complete 991 such tasks, which the company did.

Failure modes analysis, carried to considerable depth, gave further guarantees. As Bob Biggs describes this process, "It goes through the engine design and analyzes every failure that could happen, and puts in writing what design features keep that failure from happening and what inspections and test requirements would detect an unsafe condition relative to that failure mode. And it references all documents associated: specifications, drawings."[118]

Astronaut safety gained additional assurance because the on-board engines mounted fault detection systems that could shut them down in flight, if that became necessary. These systems closely resembled those that protected SSME engines during test-stand firings. Sensors checked critical parameters: gas temperatures within the turbines of the turbopumps, which could not be too hot; pressure in the main combustion chamber, which could not fall too low; and pressure in the helium purge along the main shaft of the oxygen turbopump, which provided a barrier that separated the oxygen from the hot and hydrogen-rich gas of its turbine. If any measured values fell outside permitted redline limits, that engine would stop.

The SSME was "man-rated," which imposed special requirements. The capacity for safe in-flight shutdown was at the top and gave an advance over prior experience. If a shuttle lost thrust from one engine, it could continue onward with the power of the two survivors. The Atlas booster also had three engines, but it lacked this capability. If one of them failed, the two others could not simply gimbal to point through the center of mass and fly on. Instead, the vehicle splashed into the Atlantic.

Design redundancy supported the shuttle's engine-out arrangements. The Honeywell engine controller had dual circuits; if one went out, the other could take over. If both went out, that engine still could shut down safely. The failure modes analysis gave a third element of man-rating. Such analysis was standard in the design of any rocket engine, including those for unpiloted launch vehicles, but the SSME project carried it to unusual levels of understanding.[119]

When astronauts John Young and Robert Crippen flew the shuttle Columbia to orbit on its first ascent, in April 1981, they had the pleasure of knowing that as recently as the previous July, the SSME had twice caught fire on the test stand. But they also could appreciate the meticulous care that had winnowed out the causes of failures and verified the engineering solutions. When Columbia rose toward space on its five pillars of flame, three of those pillars came from the Space Shuttle Main Engine.

CHAPTER  FIVE

Propulsion II:
SRB, ET, OMS, RCS, APU

The topic of space shuttle propulsion includes far more than the main engine. At liftoff and during the first two minutes of flight, most of the thrust was to come from a pair of solid rocket boosters. The three Space Shuttle Main Engines drew their propellants from the External Tank (ET). These engines shut down just below orbital velocity; the orbiter then was to complete its ascent using its Orbital Maneuvering System (OMS), with a pair of auxiliary engines in pods mounted at the tail. Small rocket thrusters, within the OMS pods and close to the nose, were to provide attitude control; they were part of the Reaction Control System (RCS).

In addition, both the orbiter and the SRBs relied extensively on hydraulic systems. These steered the vehicle by gimbaling the SSMEs and the SRBs' exhaust nozzles, moved the elevons and rudder, deployed the landing gear, and operated the brakes and SSME valves. The hydraulics obtained their pressure from Auxiliary Power Units. These were propulsive systems in their own right, for the shuttle APU relied on a turbine driven by hydrazine, itself a rocket propellant.[1]

The Solid Rocket Booster

Where the SSME pushed the frontiers of engineering, the Solid Rocket Booster was planned from the start as a low-risk venture that would make maximum use of existing technology. This cut costs; it also reflected the fact that large solid rockets were already approaching a definitive form, in the 120-inch solid boosters of the Titan III.[2] In turn, this conservatism allowed NASA's Marshall Space Flight Center to take a leading role in SRB design and development. Marshall had become famous for its expertise in liquid rocket engines, particularly those of Apollo. Its staff had no significant background in solid rockets. Nevertheless, NASA-Marshall boldly took on the role of SRB prime contractor, with Thiokol Chemical in a supporting role.

In May 1973 Fletcher issued a ground rule: "SRB is to be designed in-house with the exception of the SRM." The two were not the same, and the differences were far from trivial. In taking on the SRB, Marshall accepted a role somewhat analogous to that of Lockheed when that firm became prime contractor for the Navy's Polaris missile. Polaris had a solid rocket motor built by Aerojet, but the missile as a whole also had a guidance system, a nose cone for reentry, and provisions for a nuclear warhead. Similarly, the SRB had structural elements of substantial size, to mate it to the External Tank. It had a recovery system with parachutes, along with an electrical system and rate gyros as well as other instruments.[3]

"We were the prime contractor for the solid rocket booster during the development phase," recalls George Hardy, the SRB project manager at Marshall from 1974 through 1982. "We became the systems managers and the systems integrators; we designed a number of the subsystems. We functioned and performed much the same way a prime contractor would. We made certain make-or-buy decisions; we made decisions on what we would design in-house versus what we would contract to be designed out-of-house. We did design a number of the systems in-house, such as the structural components. And then we contracted out of house to have those units fabricated." Hardy's colleague, W. P. Horton, the SRB chief engineer, adds that "Marshall didn't do the detailed development of the SRM. Thiokol had a—I guess we didn't call it 'prime contract'—but they had the development, fabrication, and qualification of the motor contract. We did not do that in-house, but we integrated it into the rest of the system like the structures, the thrust vector control, the electronic black boxes, and the parachutes. All that good stuff."[4]

From NASA's perspective, this arrangement brought several advantages. At a time when Marshall needed work, Fletcher's decision gave more to this center, for rather than merely supervising a contractor, it actually would *be* a contractor. Marshall's new role drew on its deep experience in designing and building rockets—which, again, needed far more than merely their engines. Horton asserts that the decision saved money by continuing to use existing NASA resources at Marshall. Hardy adds that it also promoted flexibility in making design changes, early in the program.[5]

Close to NASA-Marshall, in Huntsville, Alabama, was a small company called United Space Boosters (USBI). It had been on hand to help that center in its activities. As the shuttle's first flight approached, Marshall gave it the role of booster assembly contractor at Kennedy Space Center. In Horton's words, USBI "took facilities that we'd furnished them, and hardware that we'd furnished them, and instructions that we furnished them, and they assembled the SRBs and checked them out at the Cape." Marshall continued to hold its role as prime contractor for the SRB, as the shuttle proceeded through its first six flights. Beginning with Flight 7, however, USBI took over as prime contractor; as Horton puts it, "The equivalent of Rocketdyne on the engine and Rockwell on the orbiter." This state of affairs persisted until 1999, when USBI became part of United Space Alliance, a commercial firm that was responsible for shuttle launch preparations.[6]

In designing the SRB, Marshall worked with what was already known. In Hardy's words, "The objective—if I could kind of overstate—was to avoid inventing anything new, stay within the state-of-the-art at all times, if possible, and even beyond that, stay within the state of the experience. And I make a distinction between the two because state-of-the-experience means to me that it's well-characterized, it's been demonstrated, it's not new."[7] Still, the SRB was not merely a scale-up of the Titan III's 120-inch standard. It broke new ground in its method for "thrust vector control" to steer the vehicle by changing the direction of the SRB exhaust. The technique that had served on the smaller Titan III was inadequate for the shuttle. Further, in Hardy's words, the SRB was "the first to be developed and put into a program man-rated, and also, the first to be designed for recovery, refurbishment, and reuse." (According to UTC, both five- and seven-segment Titan III solid boosters had been man-rated as well. However, this happened well after development of the basic Titan III solid motor.)[8]

"Man-rating," certifying the SRB to carry astronauts, was particularly demanding. No one had ever flown atop a solid. To qualify for this role, the SRB required higher safety factors, which gave greater margin against failure. These added weight, which was a shuttle designer's enemy. Fortunately, the SRB had a great deal of leeway. As Hardy put it, "For twelve pounds of weight in the boosters you penalize yourself one pound of weight for payload. This is because the boosters only fly the first two minutes of flight and are separated from the shuttle vehicle. It was not a severe weight penalty or a severe tradeoff." This contrasted with the tradeoff for the SSME and ET, wherein twelve pounds of extra weight incurred a penalty of twelve pounds of payload, for a ratio of one to one rather than twelve to one.[9]

In contrast to the SSME, where the components were very tightly integrated, one can view the development of the SRB in terms of distinct elements: propellant, casing, nozzle, thrust vector control, ignition, parachutes and recovery equipment. With the exception of the parachutes, which were new, these drew on experience with the Minuteman ICBM, the Titan III, and with an experimental program of the 1960s that had built and fired several 156- and 260-inch solid motors. The SRB was of 146-inch diameter; hence the lessons of that program were highly relevant, particularly in thrust vector control.[10]

Propellant: The SRB used a standard mix: a rubbery binder, PBAN (polybutadiene-acrylic acid-acrylonitrile); aluminum powder for fuel; and ammonium perchlorate, $NH_4 ClO_4$, as the oxidizer. The formula called for 12 percent PBAN, 16 percent aluminum, 70 percent perchlorate, and 2 percent epoxy curing agent.[11] Thiokol had pioneered its large-scale use, having processed more than 150 million pounds for its previous rocket programs. "This is the best example we have of using available technology," said John Thirkill, Thiokol's deputy di-

175

rector for the SRB. "Over the last fifteen years, we've loaded more than 2,500 first stage Minuteman motors and around 500 Poseidon motors with this propellant."

In addition to using an existing propellant, the firm planned to use existing facilities. "No new buildings will be needed at the Wasatch Division for production of space shuttle motors," declared John Wells, a senior manufacturing engineer. "Most of the existing facilities, tooling and processes in use for Minuteman and Poseidon are also completely applicable for the shuttle." In particular, the firm expected to use its standard methods for manufacturing large solid motors.[12]

A segment of steel casing would arrive, cylindrical in shape. Workers sprayed a primer on its interior surface and then rolled rubber insulation within it, as a lining, with this insulation being up to three inches thick. A complete SRM used twelve tons of such insulation. With this rubber being bonded and cured, the segment received a coating of polymer to create a tight bond between propellant and insulation. The filling followed. Each SRB required 1.1 million pounds of propellant, which was mixed in 7,000-pound batches within six-hundred-gallon vats. A gantry crane lifted the segment and placed it within one of twelve casting pits. Workers fitted a mandrel within the segment, to mold a central perforation, and lowered a vacuum bell to enclose the segment. This bell prevented bubbles of air from forming within the propellant fill, for such bubbles could prevent the motor from burning as programmed. Propellant flowed into the case from three separate hoppers. With casting completed, the vacuum bell was removed while the filled segment remained within its pit, to cure for four days at 135 degrees Fahrenheit. Workers removed the mandrel and took the filled segment to an X-ray inspection facility. After passing this inspection, it was ready for use.[13]

The SRB increased its design thrust as the effort proceeded. The planned thrust rose from 2.5 million pounds in 1973 to 2.94 million during the early flights.[14] This was the average over an entire launch; it varied as the burn proceeded. The forward portion of each propellant charge was cast with a central perforation in the form of an eleven-point star, with the star's points being long and narrow. This allowed the thrust to vary as planned. The star initially exposed a large burning surface area, for peak thrust after liftoff. The burn spread from the points, widening them and reducing this burning surface. This lessened the thrust and hence the buildup in flight velocity, to prevent overstressing the vehicle through excessive aerodynamic pressure. The SSMEs also helped by throttling back to 65 percent of rated power, to further diminish this pressure. Then, ascending amid thinner atmosphere, the shuttle was free to accelerate anew. The star was completely consumed at fifty-two seconds, leaving a cylindrical perforation that would widen, for a useful increase in thrust.[15]

Some 20 percent of an SRB's exhaust, by weight, was hydrogen chloride. In addition to being toxic and corrosive, it also posed a threat to the earth's ozone layer. The shuttle was planned to fly up to sixty times per year; hence it would pump plenty of exhaust into the upper atmosphere. NASA responded by considering whether to launch with the standard

SRB solid propellant during only the first few years of shuttle operations and then to switch over to a modified SRB. The concept called for a layered propellant charge. The new SRB was to use the standard type below sixty-five thousand feet, an altitude where ozone is dense. Above that altitude, with the standard mix all burned, the SRB might use a formulation that replaced most of the ammonium perchlorate with a different oxidizer, ammonium nitrate. This would greatly reduce the production of hydrogen chloride—while also threatening to reduce the boosters' performance.

With standard propellant, the exhaust velocity was 8,436 ft/sec, hardly more than half of the SSME. This fell to as little as 6,100 ft/sec when using ammonium nitrate, which was unacceptable; it meant a loss of up to seven thousand pounds in payload. It appeared possible to regain some of this lost performance by mixing in some HMX, an energetic nitrogen compound that could serve as an oxidizer.[16] Even so, the modified SRB would add $120 million to the shuttle program as its cost of development. It increased the cost per flight by $1.6 million and still cut the payload by two thousand to five thousand pounds.

Fortunately, this proved unnecessary. When NASA initiated its studies in August 1974 there was reason to believe that a full-up program of sixty flights per year could deplete the ozone by as much as 2 to 4 percent. Less than two years later, new research in atmospheric chemistry showed that the depletion would not exceed 0.6 percent and might readily be much less. This appeared acceptable, allowing NASA to proceed with the use of standard propellant. In May 1976 John Yardley, the associate administrator for space flight, recommended to Fletcher that NASA place the alternative-propellant effort on hold. Subsequent environmental studies supported the findings of 1976, and the operational SRB used the standard propellant in full.[17]

Casing: Large solid rocket motors typically had casings of steel. For the SRB, this alloy was D6AC, a standard type that had previously seen extensive use for both Minuteman and Titan III.[18] To avoid the need for X-ray inspections, the casing segments were free of welds. The subcontractor, Ladish Company in Cudahy, Wisconsin, used hot operations to punch a hole in a steel block or billet and then to expand the hole, shaping the billet into a rolled ring forging that had the SRB diameter, 146 inches. Additional work then squeezed this ring into a cylinder with a length of 164 inches. This activity took about sixty days. The segment then went by rail to the firm of Cal Doran near Los Angeles, to undergo heat treatment for strength and toughness. This required two weeks. It then was shipped to Rohr Industries in Chula Vista, near San Diego, for a month of final machining. Following this, the segment was ready for transport to Thiokol, again by rail, and to be loaded with propellant.

At Rohr, a critical set of operations prepared the segments for assembly into a complete booster. Here too the joining was to use no welds, relying instead on mechanical joints of the "tang-and-clevis" type. The tang was the bottom few inches of a cylindrical casing. It fitted into the clevis, a deep groove surrounding the top of the adjacent segment. Pins milled at

Rohr, 180 for each joint, fitted through holes drilled in both tang and clevis, holding adjacent segments as if they had been riveted while permitting easy disassembly.[19]

Tang-and-clevis joints had proved their merits in operational service with the Titan III. For the SRB, they initially demonstrated their mechanical strength in tests that stressed these joints until they burst. The tests included such anomalies as oversize holes, flawed or missing pins, and specimens that were highly corroded or showed low toughness. The required safety factor was 1.4; that is, the joint had to withstand a stress 1.4 times greater than it could expect to encounter in service. In the individual tests, the demonstrated safety factors ranged from 1.72 to 2.27, well above the requirement. Clearly, these pinned joints were strong indeed.[20]

By itself, a tang-and-clevis joint could not withstand the hot gases from burning propellant. These gases would flow through the joint as if it was a channel, eroding the pins and the steel to produce a catastrophic burn-through. The Titan III guarded against this by filling the joint with putty. An O-ring, a thick rubber band, surrounded the tang and acted as a seal to hold the putty against internal pressure. For the SRB, Thiokol and NASA did more. In the words of NASA's William Horton, a year and a half before the Challenger disaster, "I don't know that Titan has ever had a failure because the single O-ring at the stack joint failed. I don't think they've ever had a failure. But, since we are putting a man on it, we put in a second O-ring, the redundant O-ring in order to increase the probability of success appropriate to a man-rated vehicle."[21] As events would show, the combination of demonstrated flight experience, large safety factors in strength, and use of an extra O-ring was not enough. Indeed, together they bred a fatal complacency.

Nozzle: The flame within a solid motor burned at 5,700 degrees Fahrenheit, which was hot enough to boil iron.[22] What was to prevent it from destroying the booster? A liquid-fuel engine relied on regenerative cooling, as it circulated hydrogen through numerous small tubes or channels, but this was out of the question with solid propellants. Insulation helped; it protected the casing as the flame front approached the wall. The nozzles of large solid motors relied on a third approach, for they were lined with thick slabs of ablative material. Like a reentering nose cone, this ablative layer could slowly decompose, vaporize, and erode as the burning proceeded.[23]

For the SRB nozzle, the basic ablative material was carbon cloth phenolic, a cloth woven of carbon fiber and strongly impregnated with phenolic resin. It cost thirty dollars per pound, and each SRB used it by the ton. Layers of this substance protected the throat as well as other regions that faced the full severity of the hot gas flow. Silica cloth phenolic, woven from silica fibers, protected parts where the thermal environment was less demanding. Glass cloth phenolic served as insulation.

These materials came from vendors in the form of tape, with widths from three-fourths of an inch to thirteen inches. Rolls of tape fed a wrapping machine that laid the tape in plies on a rotating mandrel. A blast of hot air, at temperatures up to 700 degrees Fahrenheit, soft-

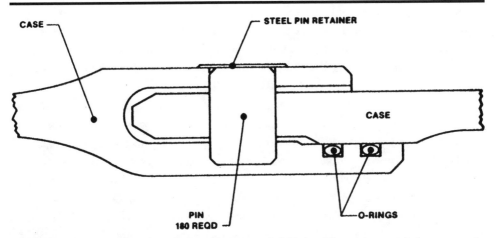

Fig. 37. *Cross-section of a Solid Rocket Motor field joint. The casings and pins are steel; interior of the SRM is at the bottom. Note the use of two O-rings to seal the joint. (NASA)*

ened the resin. A roller pressed the tape against the substrate, with a force of up to three hundred pounds for each inch of tape width. After rotating past the roller, the tape was exposed to a flow of carbon dioxide at -60 degrees. This prevented the resin from curing and produced a hard, solid surface as a substrate for the next ply.

Each nozzle used five tons of carbon cloth phenolic, two tons of glass phenolic, and one ton of silica phenolic, all tape-wrapped in this fashion. Finished carbon lay-ups were cured in a hydroclave, which used water to apply heat and pressure. Other lay-ups went into an autoclave, which used carbon dioxide. Cured components were machined using diamond cutting tools, achieving tolerances as close as 0.0025 inches.[24]

To achieve thrust vector control, the nozzle was to swivel by up to 7.1 degrees in pitch and yaw. Designers avoided the use of sliding surfaces, which could prove difficult to seal against leaks of hot gas. Instead they used a flexible support or bearing, built from ten steel plates interleaved with eleven layers of rubber. Similar flexible bearings had flown previously, but this was the largest ever built. Within the hot gas flow path, the bearing lay in what amounted to a backwater, removed from the full force of this exhaust. Nevertheless, some gas would reach it, which meant that this flexible support needed flexible thermal protection. It obtained this from a "boot," a barrier of laminated rubber that eroded or burned away at a calculated rate and that was thick enough to hold out until the motor expended all of its propellant.[25]

Thrust vector control: The Titan III lifted off using only the thrust of its solid boosters; the liquid-fueled engines in the first stage ignited at altitude. Those boosters used fixed nozzles, canted outward to cause their thrust directions to pass through the center of gravity of the complete vehicle. For thrust vector control, each booster carried a long tank that held a pressurized liquid, which could be squirted into the exhaust from within the nozzle. This injectant flashed into gas, creating a localized shock wave within the supersonic flow of exhaust, and this shock deflected the direction of the flow. This technique, "liquid injection thrust vector control," was well understood and well proven.[26]

At the outset of the space shuttle program, there was interest in using this method anew. The orbiter was to steer by gimbaling its three SSMEs following separation of the SRBs. With those SSMEs going the Titan III one better by igniting at liftoff, it appeared natural to rely on them for steering throughout the flight. This could permit the use of a fixed-nozzle SRB that might also rely on liquid injection.

For several reasons, this approach proved infeasible. With the SRBs being widely separated from the shuttle centerline, their nozzles required cant angles of eleven degrees. This large cant angle caused a loss of thrust, requiring each SRB unit to hold as much as 52,000 pounds of additional propellant. The angle also produced a side load of 790,000 pounds, which was to be taken up by the structure of the External Tank that supported the SRBs. During abort, these boosters would need quick separation—and would demand separate rocket motors, to overcome this side force. All these features added to the cost of the SRB program, while imposing weight penalties.

Twin canted SRBs demanded close alignment in their thrust vectors, driving up the number of static tests and their cost. In addition, the high cant angle meant that the internal environment of a burning SRM was not adequately predictable. Hence the program would need even more static tests, along with large safety factors and increased component test and analysis.[27] For all these reasons, canted SRB nozzles were out; the nozzle had to be straight. Liquid injection also was out; the large size and high thrust of the SRB resulted in excessive requirements for fluid. The SRBs needed enough deflection of the thrust vector to cope with shutdown of an SSME amid high wind. NASA's William Horton notes that "with the thrust that we're talking about, maybe we could get up to two or three degrees" with liquid injection, which was not enough.[28]

The alternative was a movable nozzle, pushed by hydraulic actuators. The SRB needed 5 degrees of swivel in the plane of each actuator, with these units being set at 90-degree angles; two of them worked together to produce 7.1 degrees of deflection in yaw and pitch. As early as 1964, Thiokol had demonstrated a fully gimbaled nozzle within a 156-inch motor that developed 1.5 million pounds of thrust. In a parallel effort, Lockheed introduced its own gimbaled nozzle that incorporated a flexible bearing called LockSeal. Thiokol adopted this Lockheed bearing and used it in the SRB. With this background, development of the steerable nozzle went forward in straightforward fashion.[29]

For operational service with the shuttle, thrust vector control brought options that went beyond safe abort. The SRB thrust greatly exceeded that of the SSMEs, while these boosters were set much farther from the centerline. This multiplied their control authority, making it easy to conduct an important maneuver at liftoff, wherein the shuttle rolled to align itself with the launch azimuth. As Horton stated in 1984, "One of the very first things we do, coming off the pad, is to roll downrange; and that's a pretty healthy maneuver. Gimbaling only the SSME actuators just couldn't accomplish that. The required roll maneuver couldn't be made with a non-gimbaled SRB. And then, as we fly the trajectory, we vary the gimbal angle. With a fixed nozzle, I'm not sure how you'd avoid a severe payload penalty."[30]

Ignition: In starting the SSME, many things could go wrong and produce fire or other serious damage. Large solid motors had their own dangers. Their igniters had to spread flame over the entire burning surface area, causing the internal pressure to increase with no over-pressures, combustion instabilities, combustion oscillations, damaging shock waves, ignition delays, or extinguishment. During ignition, the SRB had specific requirements. Its thrust was to reach 1.9 million pounds within 0.15 to 0.45 seconds. This thrust was to increase by no more than 150,000 pounds during any period of ten milliseconds. There could be no ignition pressure overshoot, and the ignition characteristics of separate SRBs were to be repro-ducible; during the ignition transient, the maximum thrust imbalance between twin SRBs was not to exceed 300,000 pounds. The SRB igniter had to meet all these requirements.[31]

During the mid-1960s, Aerojet had successfully tested 260-inch solid motors by using a solid rocket of 250,000 pounds of thrust as an igniter, shooting its flame through the nozzle of the big motor. The Titan III boosters followed a more practical approach by using a head-end igniter, and the SRB did this as well. Its igniter was a solid-propellant rocket in its own right, a small one about the size of a trash can, with forty bolts securing it against internal pressure. The SRB built its fire in several stages. The core of the igniter was a safety and arm-ing device from Consolidated Controls, a subcontractor to Thiokol. This device held two NASA Standard Initiators, electric components that fired on command.

When armed and fired, each initiator burst through a thin aluminum barrier and ignited a 1.4-gram charge of pyrotechnic, a granulated mix of boron and potassium nitrate. This booster charge ignited the main charge, with eighteen grams of the same pyrotechnic. Its flame quickly spread to the initiator in the main igniter, with 1.5 pounds of solid propellant. This ini-tiator lit the main igniter, with 94 pounds of that same propellant, and the flame from this ig-niter, spraying through the interior of the SRB, was what started the burning of its 1.1 million pounds of rocket fuel.

For the main igniter, the initial choice of propellant drew on Thiokol's experience with the first-stage motors of Minuteman and Poseidon. This led to specification of a mix that con-tained only 2 percent aluminum powder. When tested, the igniters showed combustion in-stability. This appeared within the igniter, not in the SRB itself, but still it had to be sup-pressed. Specialists at Thiokol consulted with colleagues at Edwards Air Force Base and at the Naval Weapons Center in China Lake, California. They decided that the best solution was to change the igniter's propellant. The new formula had 10 percent aluminum. It reduced the instability (measured as a pressure oscillation amplitude) by 77 percent, resulting in an igniter suitable for use in full-scale firings of SRB prototypes.[32]

Parachutes and recovery equipment: The SRB was designed for reuse, with the casing and other major components qualifying for twenty flights. To get the most out of each booster, NASA was willing to go to considerable lengths. Thus, within the nozzle, the flexible bearing was built for ten uses before replacing its pads of "elastomer," or rubber. B. C. Brinton, Thiokol's nozzle manager, wrote that after each flight this bearing was to be removed and tested. He added helpfully that if the tests showed damage, "the bearing metal parts can be refurbished and reused by heating the bearing in an oven until the elastomer is reduced to ash. The parts then will be cleaned and the bearing remolded with new elastomer pads."[33]

Reuse of the complete SRB called for more. It was to separate from the shuttle at 150,000 feet and 46,000 ft/sec, coast to a peak altitude of 220,000 feet, then reenter the atmosphere at Mach 5. Tumbling during entry, it would slow to one-tenth this speed to drop into the ocean some 160 miles downrange from the launch site. The steel of the casing was half an inch thick and certainly could withstand stress, but it was out of the question for an SRB to simply fall into the water with no further deceleration. It needed parachutes, and they would be enormous, for an empty SRB weighed 170,000 pounds. It was the largest and heaviest object to be retrieved using chutes.

NASA conducted drop tests of Titan III motor cases and nozzles during 1973 and showed that these booster components could survive water impact when falling at speeds up to 100 ft/sec. Next came drop tests of scale models of the SRB during 1973 and 1974, at NASA-Marshall and the Naval Surface Warfare Center in White Oak, Maryland. These led to a conclusion that an optimum speed at water impact was 85 ft/sec, equivalent to a free fall from 112 feet. More rapid impacts carried risk of damage; slower ones demanded a more costly recovery system. This optimum speed was available using parachutes only, with no need for a terminal retro-rocket.

Research in wind tunnels followed, using scale models. This work examined the aerodynamics of a falling SRB, particularly at the subsonic speeds where it would begin to deploy its chutes. Models of the chutes themselves, including drogue and drogue pilot as well as main, also went into the wind tunnels. This effort examined deployment methods for the drogue parachute, using models at one-eighth scale. The studies led to selection of a recommended method, which was later adopted and proved successful. Subsequent work demonstrated deployment of the drogue pilot at full scale and true airspeed. This took place at the Rocket Sled Test Facility of Sandia Laboratories in Albuquerque, New Mexico. The first test used sixteen solid-fuel rockets with total thrust of 100,000 pounds and accelerated the sled down a 5,000-foot track to a peak velocity of 465 ft/sec. The second run used twenty-one such rockets and it also went well. These tests showed successful deployment with the SRB in unfavorable flight attitudes and at excessive speed.

A final set of exercises featured air drops at the National Parachute Test Range near El Centro, California. A test vehicle reproduced the nose of an SRB, with its drogue and main chutes; lead ballast gave this vehicle a weight of forty-eight thousand pounds. A B-52 from NASA-Dryden made six drops, with test configurations being drogue chute only, drogue with a single main chute, and drogue with a cluster. The single-main tests simulated the loads on each of the three main chutes of an operational SRB; the cluster reproduced the operational array. There could be no preliminary demonstration of parachute recovery of a full-sized SRB; this took place only during actual flight of a shuttle. But the B-52 drop tests gave data that supported the predictions made in wind tunnels. They also refined procedures for reefing, which protected the main chutes by initially allowing them to open only partially.[34]

In operation, the deployment of the chutes began near 16,000 feet, with the SRB falling at about 500 ft/sec. A barometric switch initiated deployment of the pilot parachute, with diameter of 11.5 feet. It pulled out the drogue, 54 feet across, which opened with two reefing stages. This slowed the descending SRB while halting its tumbling motion, to make it fall tail-down for easy deployment of the main chutes. At 9,000 feet, a second baroswitch triggered explosive charges that cut away the main chutes' container, allowing these three parachutes to deploy in turn. Each of them was 115 feet across and also opened in two reefing stages.[35]

Other explosive charges cut away the lower portions of the nozzles, to reduce their impact forces when striking the water. Each nozzle also had a bumper to prevent the impact from pushing it inward and tearing the flexible bearing apart. The SRB struck the ocean tail-first, taking on some water and floating more or less vertically, like a spar buoy. A light flashed and a radio beacon began to transmit, enabling a recovery ship to find it.[36] Crew members aboard ship lowered a nozzle plug, remotely controlled and guided by underwater television, and steered it into the SRB. It pumped out the water, enabling the recovered booster to float like a log. It now was ready to be towed back to port. With this, the recovery was complete.[37]

Development and Test: The View from Thiokol

Like the SSME, the SRM was designed in detail long before it ever saw a firing range. However, the fundamental simplicity of solid motors showed within the test program. The SSME required more than seven hundred hot-firings prior to launch of STS-1, accumulating 110,000 seconds of run time. The SRB fired only seven times, four for development and three as qualification for flight. Its cumulative run time was under a thousand seconds, less than 1 percent that of the SSME.[38]

Fig. 38. *Drop tests verified the design and deployment sequence of large parachutes used with the Solid Rocket Boosters. Upper right: a small chute supports a test component. (NASA, courtesy Dennis R. Jenkins)*

At Thiokol, major activity began in mid-1974, as NASA confirmed its contract award in the wake of Lockheed's unsuccessful protest. NASA issued a $5.5 million letter contract on 27 June, good for six months, to tide everyone over while negotiating the definitive contract for design, development, test and engineering. The parties reached agreement on 27 November, with the contract specifying total costs of $136,546,674, along with a fixed fee or profit of 3 percent and an award fee of 5 percent. Also in November, Thiokol awarded the contract for the casing to Rohr Corporation, with Ladish as the principal subcontractor.[39]

During 1975, the first full year of effort, activity focused on procurement of long-lead-time items and on fabrication of tooling. By the end of the year, tool design was 85 percent complete. Thiokol gave early attention to the nozzle's flexible bearing, working with test versions of full size as well as at one-quarter scale. This work emphasized the production of thick rubber pads with enough fatigue resistance to serve for the required ten flights. Some of the pads proved to lack this resistance, which led to redesign of their molds. This solved the problem, assuring that test bearings would show the desired life.[40]

Work on components continued during 1976. Thiokol built and tested three igniters, which showed the combustion instability that led to the change in propellant. The firm also built and tested three prototypes of flexible bearings. Two performed properly; the third disclosed a problem in bonding the rubber to the steel. A fourth prototype tested an improved bond and did this successfully. Also during 1976, Thiokol prepared for construction of the first full-scale demonstrator motor, DM-1. Ladish supplied eleven case segments for each complete SRM motor casing, nine of them cylindrical along with two others placed fore and aft, with all to be pinned together using tang-and-clevis joints. Thiokol assembled these case segments into four longer sections known as casting segments, which were to be filled with propellant. The first case segment arrived at Thiokol in September. Two months later, the firm began to fabricate the first nozzle, a task that took six months. This work included the tape-wrapping of its thermal-protection components, which were cured in the hydroclave and autoclave. The latter was substantially larger than any autoclave previously used within the plant.[41]

DM-1 stood as the focus of effort during the first half of 1977. Its four casting segments were filled one by one with loads of propellant. The first of them was a center segment, a simple cylinder, well removed from the ends. It took forty mixes of seven thousand pounds each, which were poured into the casing in an around-the-clock effort that lasted forty hours. The work was completed in March; the second center segment was finished a month later. Here too the effort went well, though one of its batches had to be scrapped when it began to cure and harden prematurely.

The other two casting segments brought minor problems. In filling the forward segment, with the igniter and star-shaped perforation, two mixes were lost due to inadvertent activation of a water deluge for fighting fires. The aft segment saw structural failure of a mandrel used in shaping its perforation, causing a delay of four days. Workers removed the damaged mandrel and proceeded with the casting. All four casting segments passed their inspections and were assembled in June within the test area to form the first complete prototype of a solid motor for the SRB.[42] This too was not easy, for this motor was assembled while lying horizontally. Vertical assembly was straightforward, as in joining a pair of case segments to form a casting segment. A crane lowered the top one onto its mate; the heavy weight of the former engaged the tang and clevis readily, allowing insertion of the pins. But in fitting together the casting segments, Thiokol needed more.

Those filled sections of SRM weighed more than three hundred thousand pounds and were too heavy for wheeled vehicles when moving them within the plant. Thiokol relied instead on air-supported dollies. Floors within the plant were kept very smooth, and these dollies rested on air bearings, pads that received a flow of compressed air and allowed it to leak. This formed a film of air no more than two- or three-thousandths of an inch thick, supporting the dolly and its heavy load, which now could move freely. Operating with pressures of only 20 psi, these units did the work of a fifty-eight-wheel tractor-trailer rig that subsequently carried SRB segments to a nearby railhead for transport to the Cape.

When lying on their sides, filled casting segments were substantially out-of-round. Heavy rings, mounted to the segments' mating ends, restored their shape. These segments rested on chocks that had their own air bearings, allowing a casting segment to rotate freely and align with its mate. The chocks were mounted to another air-support dolly. Three hydraulic rams pushed the segments together, engaging tang and clevis. Only then was it possible to insert the pins, often with taps from a mallet, to produce a trustworthy joint.[43]

DM-1 fired on 18 July, roaring dramatically and shooting a four-hundred-foot flame across the test area. It produced a thick cloud of black smoke, which changed to white following burnout as an auxiliary system pumped a flow of carbon dioxide into the motor to quench its internal fire. It delivered up to 2.9 million pounds of thrust, with an average of 2.45 million over the test, compared to predictions of 3.1 and 2.4 million. The propellant burning rate was slightly slower than anticipated, and the burning time of 124 seconds was about four seconds longer than planned. A change in the propellant recipe promised to produce the proper burn rate, by adding more oxidizer.[44]

The thrust vector control worked as planned, and the hardware was in good condition following the test. However, the ignition was more rapid than predicted, with the peak rate of thrust rise being more than twice the allowed limit of 150,000 pounds during ten milliseconds. The solution appeared to lie in modifying the main propellant charge of the igniter, to produce a less intense flame. The igniter for DM-2, the next prototype motor slated for test, was modified accordingly.

The DM-1 firing also disclosed problems in the thermal protection of the nozzle. Its thick pads of carbon fiber were intended to ablate, to erode during a burn; they were used

Fig. 39. *Test of Thiokol's Solid Rocket Motor in Utah. (NASA, courtesy Dennis R. Jenkins)*

only once and replaced after every flight. They needed adequate thickness and could not erode too rapidly. However, the test showed excessive erosion in areas that were protected with silica fiber, such as the outer nozzle. These were given carbon fiber for DM-2, which was more robust. The nozzle boot, which protected the flexible bearing, had plies of rubber separated by sheets of Teflon. The Teflon also gave way to sheets of carbon fiber, to make the boot more flexible.[45]

Meanwhile, work proceeded on DM-2. Preparations began in June with the assembly of the steel casting segments. Two of them were filled in August, and the last of them, the aft section, was cast in December. The firing took place in mid-January 1978, and the measured parameters again showed success: peak thrust of 2.85 million pounds, average of 2.5 million, duration of 129 seconds. This was 5 seconds longer than planned and resulted from the winter weather, which chilled the charge of propellant.[46]

Although the basic design was clearly suitable, DM-2 led to further engineering changes. Within the steel casings, the weight of rubber insulation and polymer liner fell from 23,900 pounds in DM-1 to 19,000 for DM-3, as examination of the casings showed that the reduced thickness still gave adequate protection. The modified igniter reduced the peak rate of thrust rise from 359,000 pounds in ten milliseconds to 273,000, but this still was nearly twice the allowed limit. Again the flame from the igniter appeared to be too intense, and again the design of its main propellant charge was modified. Further changes, within the nozzle, added margins of safety. The carbon-fiber layer at the front, which had been thickened slightly following the test of DM-1, now went from 3.38 inches to 4.16 inches in depth. The boot also became thicker, while the use of carbon cloth phenolic continued to spread.

As with the SSME, the transition to testing of full-scale motors greatly lessened the need for component testing, while exercising the components under physically realistic conditions. For instance, the problem of excessively rapid ignition clearly required both an igniter and a motor, and not an igniter alone. However, parachute development still had to proceed separately, and the drop tests from the B-52 took place during 1977 and 1978, in parallel with the work at Thiokol.[47]

At NASA-Marshall, other work in parallel featured structural and vibration tests. These used the casings of DM-1 and DM-2, which had been designed for reuse. Thiokol personnel broke them down into casting segments and filled them anew, this time with inert propellant that lacked oxidizer, then sent them off by rail. The first such segment went to Marshall in September 1977. Here, for the first time, loaded casting segments were stacked vertically, in the fashion of working SRBs. Tests with these units began a year later and continued into early 1979, with the orbiter Enterprise in the starring role.[48]

During 1978, after the DM-2 firing, Thiokol prepared for DM-3 and DM-4. A shuttle's twin SRBs required closely matched burn rates, and company managers believed they could achieve this using dual casting, which would fill two segments simultaneously from the same propellant mixers. The Air Force notes somewhat delicately that "due to the accidental loss of a mix building during operations on another contract, the dual casting method was shelved, and the more conventional method of filling one segment at a time was used."

This was not the only mishap. Some dual casting did take place, and while filling two center segments, problems with a vacuum bell allowed air to enter, producing a cast segment that contained bubbles, voids, and porosity. There was no hope of using it, and this segment was scrapped. A replacement case suffered serious damage during subsequent handling operations, and while it is not easy to do harm to half-inch-thick steel, this case also went for scrap. Next, a major nozzle component was destroyed when a lifting device failed. Another failure took place in the cables of a gantry crane, dropping a concrete weight into a storage pit. These accidents, coupled with the earlier ones, led NASA-Marshall to send a group of specialists to review Thiokol's in-house handling procedures. Fortunately, this rash of misfortunes already was at an end.[49]

Developmental problems remained. As assembly of DM-3 neared completion, study of the casing of DM-2 disclosed an area of high erosion near the end of a casting segment. To prevent the ends of these segments' propellant charges from burning, it was necessary to apply a layer of inhibitor, a noncombustible substance molded to each charge. It appeared prudent to protect the casing of DM-3 by adding more inhibitor and to lengthen its rubber insulation as well. This meant dismantling DM-3 for rework, which delayed its firing from July to October 1978. It too was successful.

The thrust vector control of DM-3 mounted flight-type actuators, where previous tests had used hardware from the first stage of the Saturn V. DM-3 also used the nozzle with robust thermal protection, which worked well. However, the igniter again produced an overly rapid thrust rise. Indeed, the peak rate of increase now was even higher than in DM-2, at 290,000 pounds in ten milliseconds compared with the 273,000 of that earlier test. The changes to the igniter had the welcome consequence of virtually eliminating its combustion instability. However, there now appeared no clear route to an acceptable peak rate of thrust increase, and it was clear that this rate had to be rapid, or the twin SRBs might violate the restriction on thrust imbalance. A new igniter modification returned its design more nearly to that of DM-1, and management decided to live with the associated fast thrust rise, at least until better understanding was in hand.[50]

The firing of DM-4, in February 1979, closed out the developmental series on a high note. Ahead lay three qualification firings, during the subsequent year, which tested the SRM in its flight configuration.[51] Assembly of flight hardware was already under way, with fabrication of components for the first nozzle having begun during 1978. Again there were difficulties, as tape-wrapped parts experienced high scrap rates because they showed delamination following curing. The problems were traced to tape-wrap mandrels and to improperly designed tooling. Engineers prepared new tools and changed the wrapping technique, and the yield of good components rose markedly.

Other activity initiated the routine transport by rail of finished casting segments. Thiokol used eight heavy-duty flatcars, enough for two complete SRBs at one segment to each car. The firm sent a complete casing to NASA-Marshall for structural testing there and then followed with two empty and two filled sets for use with Enterprise. It also conducted railroad coupling tests, using one segment on its flatcar. These determined the speeds at which coupling and uncoupling could be accomplished, without causing damage.

The last four-segment sets went off to Marshall in September and October 1978. Each segment in turn rode the fifty-eight-wheel transporter over an eighteen-mile road that had been reinforced for these heavy loads. The route led to a railhead at Corinne, Utah. With the era of the shuttle at hand, historians of the West could recall that this town had flourished amid an earlier national effort.[52]

In 1869, during construction of the transcontinental railroad, it had been a Hell on Wheels town. This was a place of gamblers, prostitutes, and sellers of whiskey, who stayed there for a while before moving on to the next such location, farther down the line. Long after Corinne became nothing more than a station stop, guidebooks continued to regale travelers with tales of sin and debauchery. Yet as this lurid past faded from memory, the contrast remained instructive. It showed that as the nation acquired the means to deal with challenges that were increasingly demanding, it also became more genteel. Well before the advent even of airplanes, let alone of the shuttle, Hell on Wheels towns had become nothing more than memories from a colorful past.[53]

A century after completion of that transcontinental route, the railroad still was a mainstay. Thiokol used it anew during 1979, shipping the first flight motors to Cape Canaveral on 30 August and 8 November. Here indeed was a milestone, for with the development and qualification nearing completion, the track out of Corinne showed green lights for flight operations. Within the VAB at the Kennedy Space Center, work crews proceeded to stack the arriving segments and to build two complete SRBs, set side by side with a space between them. That space was reserved for another major element of the space shuttle: the External Tank.[54]

The External Tank

With its length of 155 feet and diameter of 27.5 feet, the External Tank was by far the largest cryogenic stage ever built. Had it been seven feet taller, its outer shell could have completely enclosed the entire Saturn I launch vehicle. Its hydrogen tank was the tallest and most voluminous vessel ever constructed as part of a liquid-fueled launch vehicle. Yet the ET was more than a container of propellants; it was the structural backbone of the entire space shuttle. It mounted both the orbiter and the SRBs, with a strong beam running across its internal diameter to take the thrust of those boosters.

Like the SRB, the ET drew extensively on recent experience. The first two stages of the Saturn V, the S-IC and S-II, had also amounted to large tanks—and had actually been wider, with diameters of 33 feet. The S-IC was nearly as long as the ET, at 138 feet, and thus was even more voluminous, though its fuel, kerosene, was less demanding than liquid hydrogen. These Saturn stages had played major structural roles as well. The S-IC carried the weight of the entire stack. The S-II, hydrogen-fueled and with nearly two-thirds the ET's capacity, supported the fully fueled S-IVB stage along with the Apollo moonship.[55]

But as the SRB was more than a scaled-up booster from the Titan III, the ET was not merely an elongated S-II. The S-II and S-IVB mounted their engines at the bottom and placed their heavy tankfuls of liquid oxygen at the bottom as well. This reduced structural weight by eliminating the need for the hydrogen tankage to support those masses of oxygen. It also reduced the stages' moment of inertia, making them more responsive when steering by gimbaling their engines.

Within the ET, the weight of oxygen was six times that of the hydrogen, and the oxygen tank went to the front. The reason was that the SSMEs did not thrust along the centerline of the ET, Saturn-style, but were mounted within the orbiter, well to one side. With a rear-mounted oxygen tank, the shuttle following SRB separation would have resembled a large boulder with rocket engines firing parallel to its surface. Rather than accelerating the vehicle, the off-center SSMEs would have tended to spin it around, with this tank being the center of mass.

It was essential that the SSMEs' thrust vector pass through the shuttle's center of gravity. With a rear-mounted oxygen tank, the SSMEs would have had to gimbal at very large angles. The forward-mounted oxygen tank moved this center well to the front, greatly easing the requirements for gimbaling. It also necessitated a beefed-up ET that could support that up-front weight, while making the shuttle less responsive by increasing its moment of inertia. Indeed, the postseparation shuttle resembled a barbell, with a mass of oxygen at one end and the SSMEs at the other. This was awkward, but it was acceptable; the alternative of a rear-mounted oxygen tank was not.[56]

The ET resembled the S-II in its thermal protection, for both placed insulation on the exterior of the tankage. The S-II used it to reduce boil-off to acceptable levels, with this stage remaining cold to the touch. It was not a problem when ice formed on this exterior; the ice acted as additional insulation, then fell away amid the vibration of liftoff. But for the ET, falling ice was impermissible. It could damage the delicate thermal-protection tiles of the adjacent orbiter. The insulation on the ET, therefore, had to be thicker to prevent ice from forming in the first place.

When the ET went into production, its insulation was a spray-on foam, similar to a grade that had seen use on the S-II. James Odom of NASA-Marshall, the ET project manager, notes that this tank "has about a third of an acre of surface on it. And every square inch of that surface is covered with insulation. We put it on in what we call a 'barber pole' process: we rotate the tank past a spray gun. It looks just like a paint sprayer. You rotate the tank as you translate the spray gun. This process puts a spiral pattern on it."[57]

In addition, the ET needed more than insulation; it called for ablative thermal protection. This was not because the shuttle was to race upward at hypersonic speed while still in the atmosphere. It stemmed from shock-impingement heating, at merely supersonic velocity. Supersonic flight produces shock waves; when these strike an adjacent surface, they produce an intense rate of localized heat transfer. In Odom's words, "Now, the vehicle is aerodynamically extremely dirty. At the front edge of the tank, that has its own very clean aerodynamics. As you come further back, you pick up the two SRBs, and you're going to have shock waves coming off the two SRBs and then just a little further you've got the nose of the orbiter. Then you drop down a little further, you've got all the structural attachments that's right in the slip stream that's between these big bodies. Where you have these shock impingements, you will actually sweep large areas where the heating rates may be up to forty BTUs per square foot per second." This was less than one-tenth the heat load on a nose cone re-entering the atmosphere at the modest speed of 10,000 ft/sec. But the ET was to be built of thin aluminum sheets, and it needed this protection.[58]

The ET also differed from the S-II in cost control. The S-II was part of Apollo, which had drowned its problems in money. But the ET needed low cost, and for a time it seemed it would achieve this. The cost per tank in the production program, estimated at $1.8 million initially, was just $500,000 higher in 1978. Indeed, the ET was to call for a true program of aerospace production, for the shuttle was to fly up to sixty times per year, and every mission needed its own tank. This contrasted with the orbiter, SRB, and SSME, which were reusable and hence were built only in limited numbers.[59]

"There was nothing really challenging technologically in the tank," recalls James Kingsbury, a senior manager at NASA-Marshall. "The challenge was to drive down the cost." His fellow managers had built both the S-IC and S-II; they were prepared to draw on their experience in seeking low cost for the ET, in both design and production. Like those stages, the ET was to be built of welded aluminum. The selected alloy promised low manufacturing costs. Bulkhead designs cut the parts count and emphasized producibility, reducing the amount of welding. Simplified structural components promised costs as low as one-fifth of counterparts from the S-IC. Other changes cut the cost of machining, such as adopting spray-on insulation. Adroit choices of welding methods and standardization of procedures promised higher welding speeds and reduced set-up time. The overall design stressed simplicity, placing the SRB thrust beam between the tanks and deleting a retro-rocket that had initially been planned for its nose.[60]

Good tooling also helped. This called for more than lathes and drill presses; it meant construction of enormous jigs, frameworks used in aligning parts for assembly. James Odom notes that "in aircraft programs you'll build a prototype on relatively 'soft' tooling, or developmental tooling, and then once you have completed the concept verification, then you will design and build the production tools for the high rate." He continued,

> *What I wanted to do was to build the test tanks on the same tooling that I was going to be building the flight tanks, because I wanted to be sure that what I tested, I could build. I took the risk and put a $200 or $300 million investment into tooling up front that normally gets invested later in a program. Fortunately, it paid off for me because I didn't change the tooling after that first time. I just added more of the same kind. We put a lot of emphasis on keeping our tool design concurrent with our flight hardware design. And, to me, the benefit from it is it forces you to design in more manufacturability in your initial tool design.*[61]

In 1976, with assembly of the first tanks imminent, the Michoud facility counted nearly seven hundred large tools and thirty-four manufacturing fixtures. Some of these had dimensions of 35 by 180 feet, noteworthy examples being trim-and-weld installations for the hydrogen tank. One of them trimmed the edges of skin panels and welded them both inside and outside, forming cylindrical sections some 20 feet long. An assembly fixture then received these sections, joining all of the hydrogen tank's assemblies and domes in one set of welding operations.

"We have, as best we can tell, the largest weld tool in the world," Odom boasted in 1984. "We put the whole hydrogen tank together; it's just like a great big lathe. We start off with a dome, and then we drop an approximately 20-foot cylinder in and just keep dropping these cylinders in and welding them together until you get to the end and then you put another dome on the other end. That tool, then, is about fifty yards long."[62]

Designers also learned from experience with the Saturn V and other large launch vehicles. These had shown "geysering," which occurs in long vertical tubes filled with liquid oxygen. If heat leaks into such a pipe, it can vaporize some of this liquid, which forms a bubble. As the bubble grows, it displaces the weight of some of the liquid, reducing the pressure below and allowing more oxygen to vaporize. The bubble fills much of the tube, expanding until it reaches the top, where it erupts into the tank and forms a geyser.

The liquid oxygen within the tank, denser than water, now can fall past the bubble to refill the emptied duct. This can cause a damaging water-hammer effect, which would be particularly severe in the ET. Its main oxygen line was eighty-four feet long and seventeen inches across; geysering could produce water-hammer pressures above 800 psi. In addition, gas fills the space in the tank above the oxygen. The spray from the geyser would cool this gas and cause it to condense, reducing the tank's internal pressure and leading to tank collapse.

Geysering thus had the potential to destroy both the oxygen tank and its main feed line, with the danger being exacerbated because this tank was at the top. However, the cure was simple: a four-inch pipe adjacent to the main line. Helium, rising within that pipe, promoted a recirculation that drew cold liquid oxygen from the bottom of the tank into the main duct. This cold liquid did not form bubbles; that happened only when liquid oxygen in the main line was close to its boiling point. The antigeysering system was tested separately, and it worked.[63]

The first complete ET went to the SSME test center in Mississippi in September 1977 for use with the three-engine cluster. In addition, that tank underwent testing in its own right, as it took on propellants for the first time. In initial exercises more than 1.5 million pounds of liquid oxygen and liquid hydrogen were loaded and off-loaded safely. There were no observable leaks from any system. Propellant temperatures remained near predictions, and boiloff rates were as expected. This showed that the insulation was performing appropriately. The antigeysering arrangements were given their own workouts, and Lawrence Norquist, a manager at Michoud, described their performance as "especially gratifying."

The hydrogen and oxygen tanks faced each other across a cylindrical enclosure known as the intertank. Leaks from those vessels had the potential of forming dangerously combustible mixtures; hence the intertank was fitted with a purge system. In tests, hazardous gas concentrations dropped to zero within ten minutes after initiating the intertank purge, again as predicted. This validated that system as well.[64]

The second and third ETs were slated for structural and vibration tests. The Michoud Assembly Facility lacked the specialized stands and equipment for this work, but these existed at NASA-Marshall, as legacies of the Saturn V program. The big tanks traveled to Marshall aboard barges, which made their way up the Mississippi and Tennessee Rivers. Motorists drive that distance in a few hours, via interstate highway from New Orleans to Huntsville, Alabama, but the trip by barge took eleven days.[65]

Four sets of structural tests took place at Marshall, from April 1977 to November 1979. They dealt with the intertank, the hydrogen tank, and the oxygen tank, with the latter undergoing two series of tests. Each such series took months, with the hydrogen-tank exercises covering over a year during 1978 and 1979. The intertank came first; it was a structural cylinder that supported the oxygen vessel, holding up to 1.33 million pounds of this liquid. The intertank also faced stresses due to thermal gradients because it had a very cold propellant container at each end. The engineers at Marshall reproduced those thermal gradients by using nonflammable liquid nitrogen, thus avoiding the use of dangerous rocket propellants, and got good results. The loads came from hydraulic jacks that applied torsion, bending moment, and compression.

The second round of tests placed the oxygen tank atop that intertank. The shuttle was to accelerate at up to three g's. To simulate the g-forces associated with that oxygen vessel, engineers filled it with driller's mud, used in the oil industry. It had nearly twice the density of liquid oxygen. Overall, the testing demonstrated that the ET and its component tanks could withstand 140 percent of their flight loads.

But in working with the hydrogen tank, nothing was light enough to take the place of liquid hydrogen. In the words of James Odom,

> *We tested a hydrogen tank with roughly a quarter of a million pounds of liquid hydrogen in it. Now, if you have a rupture there, that could get very exciting. So to do that, the only facility in the country that could put the loads on and accommodate this large hydrogen tank was our test stand here at MSFC that did the development and acceptance testing of the S-IC. We literally hung this tank in the test stand up on the large load ring that had taken these seven and a half million pound loads on the Saturn V, and literally took the tank and pushed on it just like it sees its loads in flight with a quarter of a million pounds of hydrogen on board. Now, that will keep you up nights when you're doing that.[66]*

No such heroics marked the vibration tests, which went forward concurrently at Marshall using the third prototype ET. It sufficed to fill the oxygen tank with various levels of water

Fig. 40. *Transport by barge of an External Tank. The tugboat's elevated pilot house prevents the ET from obstructing the view. (NASA)*

and to simulate the light liquid hydrogen with nothing more than air. These tests mated the ET to Enterprise and subsequently added dummy SRBs, creating a vehicle that for the first time looked like a complete space shuttle.

The structural testing applied loads, weights, and pressures. The vibration tests used acoustically powered shakers of 150 and 1,000 pounds force, which used rods to transmit their oscillations to hard points in the mated vehicle. A crane lifted Enterprise into the test facility by hoisting it into the air using cables, where it hung suspended like a gaffed fish. After being mated to the ET, that same crane lifted the vehicle anew, allowing its vibrations to duplicate those of the shuttle in free flight. But with the SRBs, the combination was too heavy for that crane. Engineers allowed the vehicle to rest on flexible air bags, which gave similar freedom of movement.

The shakers oscillated at frequencies from 1.5 to 50 hertz, exciting structural frequencies as low as 2 hertz. *Aviation Week* described the tests:

Fig. 41. *Hoisting Enterprise into position at NASA-Marshall for structural tests. (NASA, courtesy Dennis R. Jenkins)*

During shuttle flight, if sensors pick up what appear to be vehicle trajectory changes but in fact are deflections of the shuttle structure, the flight control system would steer the shuttle out of its proper trajectory to satisfy the errant sensors. Because such deflections could be oscillatory in nature, continual adjusting by the flight control system to follow structurally induced sensor inputs could result in an unstable or overstressed vehicle that would break up in flight.

In most if not all tests, the test officials will have data in front of them describing the computer-predicted response for each mode. During tests, they will try and get the spacecraft "ringing" into a particular mode, such as giving the solid rocket boosters a rolling moment relative to the external tank.

Engineers will tune the shaking of the shuttle to reach a predicted point. If the mode is not reached at that predicted point, they will continue to tune the ringing of the vehicle until they reach that particular mode. The difference in what it takes to reach a particular mode versus what was predicted will be important data.

"Ringing" represented resonance, a match between the excitation frequency and the vehicle response that showed itself as a marked increase in the amplitude of this response. Through such matches, investigators learned the vehicle's natural frequencies and modes of oscillation.

During similar tests of the less complex Saturn V in 1966 and 1967, engineers had found that a planned placement of flight control sensors on the S-IVB could have led to the destruction of the vehicle in flight because of flexing modes found at that location. Similarly, the work with the shuttle found enhanced oscillations around the planned locations of rate gyros on both the SRBs and the orbiter. Those instruments measured rates of rotational movement. The data from these tests allowed designers to move those gyros to safer locations.[67]

These initial ETs were not flight-rated, and problems with their thermal protection delayed production of the first flight models. Richard Foll, Martin Marietta's director of engineering at Michoud, describes the ablative coating as "a mixture of two GE resin silicones, has some carbon black in it, has some cork in it, has phenolic balloons." At room temperature it could stretch like Silly Putty, but at around -200 degrees Fahrenheit it turned into a brittle glass. Foll adds that "if you would take a vehicle, chill it down with this material applied to the outside, and then stretch it, that you've got a very brittle material that you are stretching."[68]

This was no matter to treat casually, for Odom notes that if it came off, "in even relatively small pieces, like six to ten inches in diameter," shock-impingement heating during ascent "can get enough heat onto the aluminum tank that's underneath, to melt it, and then you can lose the whole vehicle. And there's no nondestructive evaluation techniques that you can go in and test after you have put it on to really know that it's on, and it's adhering."[69] The solution lay in stretching the ablative material before applying it to the surface. It could retain this stretch while warm and even after turning to glass when cold. Then, when the skin of the

ET flexed and stretched under its loads, it relieved the built-in stress within the ablator and did not apply new stress that could cause it to fracture and perhaps to flake off.[70]

Spray-on application of the ablator, blowing it directly onto the tank, lay in the future. Instead the ablator was to be bonded to the surface, which was a labor-intensive process. Initial estimates indicated that only modest areas of the ET would need it, but ongoing studies of shock-impingement heating, which included work in wind tunnels, increased the areas to be covered by as much as 50 percent. The move to a spray-on process sought to save time in production, as Martin Marietta requested funds for a suitable installation. It was expected to be in service by 1981, in time to support a stepped-up production program. But in the meantime, there was nothing to do but rely on the bonding process, which added 46 days to the time needed to fabricate a tank at Michoud.

These production problems affected the delivery schedule. The first three prototypes had left Michoud between September 1977 and March 1978 for use in SSME testing and in the structural and vibration work at NASA-Marshall. The delivery date for the first flight tank, initially set for December, slipped to February 1979 and then slipped further. Finally, on 29 June, NASA formally accepted this tank at Michoud. This cleared it for shipment by barge to Cape Canaveral. It arrived a week later, with the first two flight SRBs following during subsequent months.[71]

Storable Propellants and On-Orbit Maneuvering

Within the field of liquid rocketry, an important branch features the use of storable propellants. In contrast to cryogens such as liquid oxygen, storables remain liquid at room temperature. It is possible to load them within a rocket and to hold them in the tanks for long periods. Many successful combinations are also "hypergolic," igniting on contact when they mix in a combustion chamber. A particular choice, monomethyl hydrazine as fuel with nitrogen tetroxide for the oxidizer, gave the basis for the shuttle's OMS and RCS. This was appropriate, for in contrast to the SSME, which burned its cryogens immediately after liftoff, these systems had to remain ready for use throughout a mission. The two systems complemented each other. Indeed, at the rear of the orbiter they were mounted within the same installations, known as OMS pods, and drew their propellants from the same tanks.[72] Twin OMS engines, each rated at 6,000 pounds of thrust, injected the shuttle into its orbit following SSME shutdown and ET separation. Those same engines enabled this vehicle to maneuver while in orbit, perhaps to rendezvous with a target. At the end of the mission, the OMS fired again and reduced the orbiter's speed, permitting reentry.

The RCS provided attitude control, while working in tandem with OMS. Prior to the deorbiting OMS burn, RCS thrusters turned the orbiter to fly tail-first. Following that burn, they

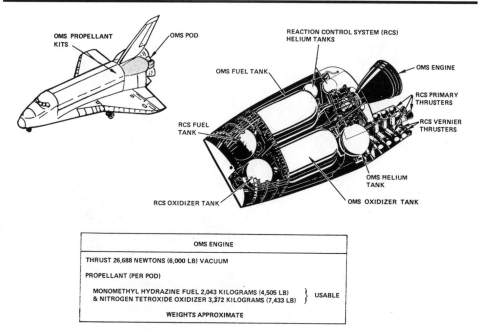

Fig. 42. *Orbital Maneuvering System. (NASA)*

reoriented the orbiter into a nose-high attitude for atmosphere entry. They allowed the orbiter to execute small motions in any direction, which were essential for docking following rendezvous; if your car had a similar system, you could pull alongside a parking space and then enter it by scooting sideways. The RCS also provided safety, for while either OMS engine could perform the de-orbit maneuver, there was enough on-board propellant to allow RCS thrusters to fill this role if both OMS engines were to fail.[73]

Though McDonnell Douglas built the OMS pods, Aerojet Liquid Rocket Company crafted the engines, having won the development contract in February 1974.[74] This award reflected the company's strong leadership in the field of storable liquid propulsion. Indeed, the firm traced its beginnings to the invention of this form of rocketry, having grown out of research at Caltech immediately prior to World War II. Aerojet developed and produced large engines of this type for the Titan II and Titan III. More particularly, it built the Service Propulsion System (SPS), the main engine of the Apollo spacecraft.

This SPS was critical indeed, for if it failed to ignite following a moon landing, it would have left its astronauts stranded in lunar orbit, with no way to return home. The spacecraft carried no second engine; in the words of astronaut Frank Borman, "It simply has to work at that point." Aerojet met this challenge by seeking the greatest possible simplicity in the SPS

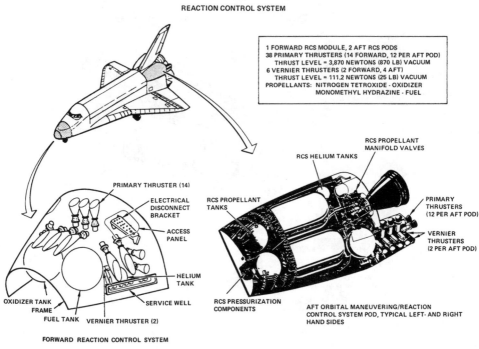

REACTION CONTROL SYSTEM

1 FORWARD RCS MODULE, 2 AFT RCS PODS
38 PRIMARY THRUSTERS (14 FORWARD, 12 PER AFT POD)
THRUST LEVEL = 3,870 NEWTONS (870 LB) VACUUM
6 VERNIER THRUSTERS (2 FORWARD, 4 AFT)
THRUST LEVEL = 111.2 NEWTONS (25 LB) VACUUM
PROPELLANTS: NITROGEN TETROXIDE - OXIDIZER
MONOMETHYL HYDRAZINE - FUEL

Fig. 43. *Reaction Control System. (NASA)*

design and did the same with the OMS engine.[75] This engine contrasted sharply with the SSME. The latter was complex indeed, with turbopumps, preburners, and their associated valves—all of which offered rich opportunities for serious problems during development. The OMS engine, like the SPS, dispensed with all that. Their function, after all, was to feed propellants into the main combustion chamber, and the OMS engine did this by relying on pressurized tanks. Such tanks can be heavy, but those of the OMS system were modest in size, while the OMS engines operated at a low chamber pressure, 125 psi. This simplicity meant that there was no elaborate start sequence, no Rube Goldberg ignition arrangements. You opened a couple of valves; helium pressure in the tanks did the rest, and away you went.

The OMS engines were smaller than the SPS, delivering less than one-third of that earlier engine's 20,500 pounds of thrust. But they had to last much longer. The SPS had been rated for only twelve and a half minutes of life and counted as rugged when one test model held out for thirty minutes. The OMS engine was to serve for a hundred shuttle missions and to achieve a total life of fifteen hours.[76]

The SSME faced similar demands, and the OMS engine adopted its technology in its regenerative cooling. The SPS had used no such cooling, protecting its thrust chamber with an

200

ablative liner. The OMS thrust chamber was stainless steel with slotted grooves milled into its periphery; electrodeposited nickel formed a surrounding close-out. Fuel, flowing within those channels, carried away the heat. The thrust chamber was the size of a large vase, ten inches across and two feet long.[77]

While the OMS engine was simple in design and used familiar technology, it faced the issue of combustion instability. Earlier engines had suppressed this by adding baffles to the face of the injector. They reduced engine performance; they also were hard to cool and tended to crack. Aerojet avoided baffles and sought instead to promote stable combustion with "acoustic cavities." The field of acoustics deals with sound waves and pressure disturbances; the SSME had similar cavities for the same reason. These were small depressions within the thrust-chamber wall that damped pressure oscillations. The successful use of such cavities demanded much work with injectors, and Aerojet pursued the rapid assessment of a number of injector designs.

Former practice called for machining channels for fuel and oxidizer flow into injector plates, which were welded together. "To make a conventional injector took about three months," David Winterhalter, chief of space shuttle propulsion at NASA Headquarters, told *Aviation Week*. The new process used platelets 0.0008 inches thick. Each platelet was photo-etched, being covered with a protective layer of photoresist and exposed to light through a patterned mask. The light weakened portions of this photoresist; etching then removed the underlying metal, for a platelet design that matched the mask. Eight such platelets, joined by diffusion bonding, formed a complete injector plate. "It's one simple process," said Winterhalter. "You can make hundreds of identical plates and change the injector pattern easily if required."[78]

Extensive testing was a specialty of the house at Aerojet; the SPS had fired some 3,200 times before qualifying for flight. Studies of combustion stability used bombs, small explosive charges placed within the combustion chamber in deliberate attempts to disrupt the flow. Initial work indeed found instability, which called for consideration of changes in both the injector plate and the acoustic cavities. The evaluation of new designs for the OMS engine called for 350 tests, many of which had two bombs, to give more data.

In operational use, the OMS engine was to avoid being sensitive to changes in chamber pressure or oxidizer-to-fuel mixture ratio. It was to accommodate a range for these parameters. The test conditions went well outside the anticipated operating limits and led to selection of suitable designs. This brought a prototype engine that was further scrutinized, in 225 tests that used 343 bombs. This work with injectors delayed their development by six months, but Aerojet chose a final design in October 1975. Engine development then went forward.[79]

This effort included a great deal of component testing, for the engine's modest pressures and thrust levels brought similar modesty in the component test facilities. In testing full-scale engines, more than five-sixths of the firings were at other than nominal operating conditions, thus demonstrating the robustness of the design.[80] The tests showed robustness in other ways as well. The OMS engines were to be installed only a few feet from the SSMEs, and Aerojet had to prove that the SSMEs' loud roars would not cause damage. Company engineers ran an OMS engine repeatedly while a Titan III first-stage engine boomed away on the same test rig. Yet even this was not loud enough, so they repeated this test at Johnson Space Center, using the largest horn available at that center for studies of acoustics.[81]

In mid-1978 Dan David, the program manager at Aerojet, wrote that "development testing of the OMS Engine is complete. The development program can best be summarized by stating that the program did not face any 'show stoppers.'"[82] Because the OMS engines were to fire only in space or in near-vacuum, within the upper atmosphere, the tests did not take place at sea level; all firings used altitude chambers. Aerojet had facilities that simulated the rarefied pressure of seventy thousand feet. A subsequent test series was conducted at NASA-Johnson's White Sands Test Facility in New Mexico during the last four months of 1978. It exercised a prototype OMS pod, complete with tanks, valves, and regulators.[83]

By mid-1979, with qualification testing under way, the program recorded more than 1,600 firings. These counted thrust chamber assemblies as well as complete engines. Another Aerojet manager described them as "anticlimactic, not only because the tests have gone smoothly, but also because nearly all the tests had been performed previously at the development level." He noted that "extensive testing was done at the component level, and all major and most minor problems were solved long before development testing at the engine level. Most of the testing was done at the extreme limits of the required operating range rather than at the nominal design point."

Qualification testing continued during 1979, certifying the engines for use in piloted flight. Meanwhile, McDonnell Douglas built OMS pods, with engines included, that were meant for actual use. These went to the Kennedy Space Center, where in May 1980 they were installed on the orbiter Columbia, which was being readied for the first flight of a shuttle into space.[84]

An OMS pod also held RCS thrusters: twelve that each had 870 pounds of thrust, along with two "vernier," or fine-adjustment, units rated at 25 pounds. The nose mounted fourteen more primary thrusters, for a total of thirty-eight, along with two additional verniers. They were built by the firm of Marquardt, a company that specialized in small rocket engines and had developed similar thrusters for the Apollo moonship. The vernier was an uprated version of a thruster built for the Manned Orbiting Laboratory. The primary had considerably more thrust than its Apollo counterparts, reflecting the two-hundred-thousand-pound weight of a shuttle in orbit.[85]

There was columbium on Columbia; the OMS and RCS engines made good use of this metal, also called niobium. Its melting point, 4,500 degrees Fahrenheit, is one of the highest known, considerably higher than that of iron. Uncooled columbium formed the outer nozzle of the OMS engines, withstanding temperatures up to 2,400 degrees. The RCS thrusters used this metal for the thrust chamber itself, cooling it with a film of fuel. Those chambers suppressed combustion instabilities with acoustic cavities, again as with OMS.[86]

The development of the RCS thrusters proceeded straightforwardly. Test units reached White Sands late in 1977, with a major program of firings running from May to November 1978. Within this developmental program, the primary thrusters fired more than fourteen thousand times, mostly in short pips of less than a second; the verniers burned nearly one hundred thousand times. These tests exercised the units over a broad range of mixture ratios and propellant delivery pressures and temperatures. Qualification testing began in January 1979, with flight units being delivered to Kennedy Space Center in April.[87] Then in early 1980, tests of complete RCS systems, including flight tankage, revealed a problem.

The tanks relied on pressurized helium to feed propellants to the thrusters, and in zero-g, this gas could enter an engine and cause it to misfire. An on-board computer then might shut down that thruster and remove it from service, just when it would be needed. To prevent this, the tanks held long channels that were protected with stainless-steel mesh, finely woven screening that filtered out gas and allowed only fuel and oxidizer to flow through the passages to the RCS engines. The tests of 1980 showed that the simultaneous firing of several thrusters could produce a sudden drop in tank pressure that would cause the screens to break down, allowing helium to reach those engines.

There was no easy solution, and NASA coped with this by writing a software command to prevent the computer from firing more than three thrusters simultaneously in any one system. This allowed adequate on-board thrust for all standard maneuvers, but it was insufficient for an abort with return to launch site. For such an emergency, the orbiter was to fire as many as seven thrusters. Managers coped with this as well, by loading the aft RCS tanks until they were completely full, leaving no room for any helium within those tanks. However, this too was merely an expedient, for the system had not been designed for this. The first several missions flew with such "overfilled" tanks, while researchers continued to seek a permanent solution. They developed sophisticated mathematical models of propellant behavior and benchmarked these models with tests that used tanks fitted with special instruments.

This work showed that a four-thruster capability was feasible using normally filled aft tanks. It also showed that even without overfill, firing seven thrusters during abort would cause them to ingest only minimal amounts of gas. In 1983, two years into flight operations, NASA deleted the requirement for overfill and accepted the RCS system as certified for use in all circumstances.[88]

Auxiliary Power Units

The orbiter resembled a large aircraft not only because it had wings and a vertical fin but also because it needed on-board power to move its control surfaces and to operate its landing gear and brakes. Like other such aircraft, the shuttle relied on conventional hydraulic systems, each with its own pump and power source. Standard design practice calls for several such systems; for example, the Boeing 747 has four of them, all independent, with any of them capable of providing flight control if the others should be damaged. The shuttle had three.[89]

Although its hydraulic lines and actuators followed familiar designs, the power source was new. It was an APU rated at 135 horsepower that used hydrazine, N_2H_4, as a monopropellant that drove a turbine. The hydrazine decomposed as it flowed over a catalyst, releasing heat and forming hydrogen, nitrogen, and ammonia. Though these gases were combustible, they emerged from the catalytic bed at 1,700 degrees Fahrenheit and 1,170 psi, which was enough to drive the turbine without combustion. The turbine wheel, 5.25 inches in diameter, rotated at 72,000 rpm. It powered a hydraulic pump by means of a gearbox, and the pump put pressure on the fluid. This pressure then moved the actuators.[90]

Despite its modest size and power rating, the APU turbine showed several of the same development problems that plagued the much larger turbines on the SSME. Yet the APU had to meet a particularly tight schedule, for it had to be ready for use within Enterprise during the Approach and Landing Tests in 1977, nearly four years before a shuttle would fly to orbit. This meant that the builder of the APU, Sundstrand, faced pressures of its own. However, the APU did not attempt to push technology to the limit, as with the SSME. Its technical problems generally did not result from extreme stresses or temperatures; rather, they reflected the need to learn how to use meticulous care in design and fabrication. Some issues were simply accepted without resolution, as when the turbine housing developed cracks due to thermal stresses. These were in noncritical areas and did not affect APU operation, while two test APUs ran successfully with cracked housings for more than seventy hours. Managers endorsed the housing as qualified for twenty hours of service life, even with cracks. The housing would then be replaced as part of scheduled maintenance.

Turbine blades also showed cracks, at both the blade tips and roots. Beveling the tips solved the first problem; it diminished fatigue that had caused the cracks and physically removed the metal that was subject to fracture. The blade-root difficulty proved to result from stresses in corners, which were relieved by making them rounder. Other problems arose in welding a shroud, a narrow ring that surrounded the tips. The shroud developed its own cracks, while some welds lacked strength. Engineers changed the shroud material from Hastelloy X to Inconel 625, with the latter showing increased strength and better weldability. They also introduced a precisely controlled electron-beam welding process. These changes brought a trustworthy turbine.

Other issues arose in developing a valve to control the flow of hydrazine into the gas generator, with its catalytic bed. The flow was not continuous; rather, the valve injected fuel in spurts, which led to large pressure fluctuations. It tended to leak and showed limited life due to wear and breakage of the valve seat, made of tungsten carbide. The cure was not simple; it emerged through considerable work in redesign, stress analysis, and manufacturing.

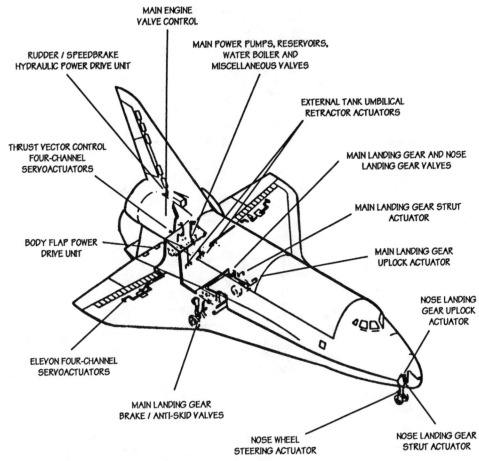

Fig. 44. *Orbiter systems actuated by hydraulics. (NASA, courtesy Dennis R. Jenkins)*

Machining of the seat used an electrodischarge process that left a recast layer. This layer was irregular and porous and led to cracks. Honing the seat with diamond slurry removed it, strengthening the seat against failure. Careful attention to internal corners and interfaces removed sources of stress. The tungsten carbide was sensitive to many solvents and other fluids; design changes proved necessary to give it protection. A review of APU development notes that "the machining and manufacturing process turned out to be almost an art, and all seats and poppet assemblies were manufactured in a small, one-man shop."[91]

The fuel pump, which supplied hydrazine, had gears that quickly showed wear. Redesigned gears used many small teeth rather than a few large ones and minimized the sliding contact that promoted wear. The new gears also were made from a very hard tool steel that

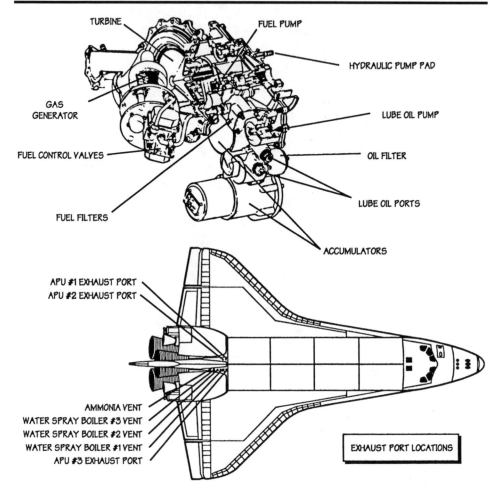

Fig. 45. *Auxiliary Power Unit with exhaust ports. (NASA, courtesy Dennis R. Jenkins)*

could resist abrasion. The initial pump design had an unacceptably short service life, but the modified version offered an effectively unlimited life.

Of these problems, the turbine-wheel cracks had the potential to produce outright shutdown of an operating unit. The others acted more to reduce the APU's rated life. Hence, even though the APU was still in development during 1977, it was possible to fit Enterprise with prototype units and to use them during the ALT flights. Six such units, including spares, operated for a total of thirty-five hours.

The most significant problem arose during the first captive-active flight, with astronauts on board while this orbiter remained mated to its 747. One of the three installed APUs developed a fuel leak that was large enough for a pilot in a chase plane to observe. The leak came from a fuel pump seal that had previously shown problems. The seal had used a bellows to assure a tight fit; a redesign replaced the bellows with a flexible seal and stopped the leaks. Through such changes, a reliable APU design emerged.[92]

This same APU was selected for the SRB, to power the hydraulic system that gimbaled its movable nozzle. Here it was called the Hydraulic Power Unit (HPU). The power requirement was the same as for the orbiter, 135 horsepower, while the service demands were considerably less stringent. On the orbiter, each APU was to run for 82 minutes during a mission, operating through liftoff and ascent, de-orbit, reentry, and landing. The HPU had to operate for only 2.4 minutes, during ascent. Orbiter versions were to serve for forty missions and fifty operating hours; the SRB required only twenty flights and two hours.

These reduced requirements permitted a simplified design. While the basic HPU remained the standard one from Sundstrand, it deleted several appurtenances of the gearbox. On the orbiter, these promoted extended run times; on the SRB, they were not needed. The gas-generator valve, a complex component that had given much trouble in development, gave way to a less sophisticated counterpart that cost only one-fourth as much to purchase. This search for simplicity carried over to prelaunch check-out, for there was much interest in testing the HPU without a full hot-fire exercise. This proved to be achievable using gaseous nitrogen, which rotated the turbine at speeds up to 76 percent of the operational rate. This cold-gas system also permitted check-out of the complete system for thrust vector control, which demanded up to 26 horsepower.

However, the SRB imposed greater demands in actual use. Its HPU faced considerably harsher vibrations, being mounted immediately adjacent to the solid rocket motor. It had to withstand the shock of impact into the ocean and then had to survive a bath in salt water for up to a week, while the SRB was recovered and towed to shore. Hence the HPU was hardened, made more resistant to damage. For instance, its mounts acted as vibration isolators and actually reduced the vibration loads to levels below those of the orbiter.[93]

Development testing of the SRB thrust vector control system took place at NASA-Marshall. The first set of tests, in July 1976, used an orbiter APU and components that either were off the shelf or were built in-house. Flight-configured hardware entered development testing during 1977; a formal qualification program followed, from November 1977 to June 1979. However, while this work featured hot firings of HPUs, it treated the thrust vectoring as a system in its own right, separate from the SRB. Tests with SRB motors took place in parallel, with additional flight hardware being shipped to Thiokol in Utah for use during the firings of DM-3 and DM-4 in October 1978 and February 1979. Three subsequent tests with qualification motors, during the following year, gave final assurance that the HPU was ready for flight.[94]

When Columbia flew into space in April 1981 it mounted an impressively large array of propulsion systems. It had three SSMEs, each with turbopumps that pushed the limits of the feasible. The combined power of all three sets of pumps could drive the aircraft carrier *Nimitz* at flank speed.[95] There were two SRBs, each with a pair of HPUs that included their own turbines. Three APUs, with turbines, rode aboard the orbiter. Two OMS engines, along with forty-four RCS thrusters, primary and vernier, completed the list. The shuttle had liquid rocket engines of both the cryogenic and storable-propellant types, as well as two of the largest solid rocket motors ever built. At liftoff, all this equipment had to be ready to work properly or the launch would be scrubbed.

CHAPTER SIX

Thermal Protection

"Re-entry is perhaps one of the most difficult problems one can imagine," Theodore von Kármán, the dean of American aerodynamicists, wrote in 1956. "It is certainly a problem that constitutes a challenge to the best brains working in these domains of modern aerophysics."[1] At that time, with the Air Force's big missile programs under way, the challenge lay in protecting a small ICBM nose cone. Fifteen years later, the space shuttle offered considerably greater difficulties. It was vastly larger; its thermal protection had to be reusable, and this thermal shield demanded both light weight and low cost. Fortunately, by 1970 there were several promising approaches, both for initial or interim use and for the long term.

Early Developments

The problems of atmosphere entry have been strongly experimental in nature, calling for extensive research and developmental testing in ground facilities, along with flight tests. Yet one of the most important early contributions, in 1953, was entirely an exercise in applied mathematics. This was the work of two investigators at NACA's Ames Aeronautical Laboratory, H. Julian Allen and Alfred Eggers, who addressed the issue of the optimum shape for reentry. Conventional thinking held that hypersonic reentry demanded the ultimate in needlenose sharpness. For the Atlas ICBM, initial concepts resembled a church steeple with a flagpole. Allen and Eggers showed that this shape in fact was the worst, for it would lead to the highest rates of heat transfer. Instead of being sharp, the nose cone was to be blunt, like a sphere rather than a long pointed spike.[2]

Early support for this finding came in 1954, amid feasibility studies of the X-15. This hypersonic rocket plane could not simply point its nose in the direction of flight during reentry, like a dive bomber. With its streamlined shape, it would enter the dense lower atmosphere at too high a speed, encountering excessive heat loads as well as aerodynamic forces that would cause it to break up. Analysis showed that the plane should enter with its nose high, presenting its flat undersurface to the air. It then could lose speed in the upper atmosphere, reducing the aerodynamic loads while easing the heating. In the words of John Becker, the study leader, "It became obvious to us that what we were seeing here was a new manifestation of H. J. Allen's 'blunt body' principle. As we increased angle of attack, our configuration in effect became more 'blunt.'" Decades later, the shuttle reentered using the same nose-high attitude, and for the same reason.[3]

The first engineering solution to the reentry problem was the heat sink. In the X-15 this took the form of a skin and structure of Inconel X-750, a temperature-resistant nickel steel. Heat came in at the surface and was absorbed within the thickness of the skin, without raising its temperature beyond acceptable limits. The Thor missile used heat sink as well; with a range of fifteen hundred miles, its nose cone was of heavy copper.[4] But copper heat sink was inadequate for an ICBM, with a range of six thousand miles and considerably higher velocity at reentry.

When von Kármán made his comment, three years after Allen and Eggers wrote their paper, no good solution was in view. Yet amid the fast pace of technical advance, a valuable new approach quickly emerged: ablation. George Sutton, a physicist and mechanical engineer at General Electric, was its inventor and early proponent. When he presented initial experimental results at a symposium in June 1957, George Solomon, an Air Force advisor, rose to his feet and stated that this was the solution to the problem of thermal protection.[5]

Ablation of a nose cone resembles the charring of wood as it gives off gases and forms charcoal. The heat of reentry causes the ablative material to decompose, forming a porous carbon-rich char along with gases produced by pyrolysis. The char has low heat conductivity and protects the underlying material, though it oxidizes slowly as reentry proceeds. The gases flow outward, forming an additional protective layer and carrying away heat.

The materials typically are composites, with temperature-resistant fibers—carbon, glass, graphite—in a matrix of plastic such as a phenolic or epoxy resin. Carbon cloth phenolic, used in the SRB, is an example; it is formed to the desired shape by tape wrapping. Other ablative heat shields are fabricated by molding the fibrous resin at high pressure followed by curing. The cured resin produces the carbon char, which has little mechanical strength. The fibers hold it to the underlying unheated substrate.[6]

Ablatives offered a much better solution than copper heat sink, being much lighter while withstanding reentry at considerably higher speeds. They swept the field in the world of long-range missiles then went on to provide heat shields for the Mercury, Gemini and Apollo spacecraft. Apollo's role was particularly demanding, for when returning from the moon, its spacecraft entered the atmosphere with twice the kinetic energy of a return from low orbit. This high kinetic energy resulted from the energy of escape velocity being twice that of low orbit.

210

After 1970, as NASA looked ahead to the shuttle, ablatives did not appear promising as thermal protection. They could serve only for single flights and would then require costly replacement, whereas the shuttle was to be reusable. There also was concern over the weight of ablatives when covering large areas. They indeed had an advantage, for they worked. Yet although ablatives had offered bright hope in 1957, by 1970 they appeared as no more than a backup, an interim approach to permit the shuttle to fly while NASA went forward with better concepts.[7]

High on the list was the hot structure, which was considerably more complex than the X-15's heat sink. It called for a primary or load-bearing structure that had some heat resistance but could not directly face the temperatures of reentry. The vehicle's skin would consist of thin, shingle-like plates, free to expand or contract due to the temperature changes. These had to withstand aerodynamic loads; they were not to blow off like roof shingles in a hurricane. However, they were expected to lose strength at elevated temperatures.[8]

These plates were to be fabricated from metals and alloys chosen for heat resistance. Lacking heat-sink capacity, they would reradiate absorbed heat, with their high temperatures resulting from equilibrium between heat absorption and loss by radiation. Insulation, placed beneath these shingles, would protect the underlying primary structure by retarding the inward flow of heat. By the time this heat began to penetrate in earnest, the vehicle was to be out of the danger zone, slowing after reentry and entering a realm of cooler temperatures.

The first major use of hot structures came on the Air Force's Dyna-Soar, a small piloted craft that resembled the eventual space shuttle and that was to ride to orbit atop a Titan III. Its primary structure was René 41, a nickel-chromium alloy that also contained cobalt and molybdenum. Originally developed for use in jet engines, it retained useful strength at 1,800 degrees Fahrenheit. Skin panels could accommodate temperatures nearly a thousand degrees higher; they were made of columbium alloy with insulation underneath. Wing leading edges, which were to reach 3,000 degrees, used molybdenum alloy. The nose cap, rated at 4,300 degrees, was zirconia. This high temperature reflected the relative sharpness of the nose, for the Dyna-Soar was too small a craft to have room for genuine bluntness.

Like the leading edges that were immediately adjacent, the insulation had to withstand 3,000 degrees Fahrenheit as well. The solution was Q-felt, a quartz fiber. This material also protected the hydraulic system, which used hydraulic fluid as a coolant. The pilot compartment and on-board electronics demanded additional cooling and used a wall moistened with water. A water-glycol coolant loop carried internal heat to a heat sink that used liquid hydrogen.

Defense Secretary Robert McNamara canceled Dyna-Soar in December 1963, with its first flight still in the future. The project nevertheless reached an advanced state of development and played a key role in bringing hot structures from the realm of initial experiments into working technology. In 1958, prior to this program, the state of the art was 600 degrees

Fahrenheit for skin panels, leading edges, and primary structure, using stainless steel and titanium. Dyna-Soar raised the temperature limits virtually beyond recognition.

It spurred the construction of ground-test facilities that could test and characterize its refractory metals at their elevated temperatures. These alloys oxidized when hot; they were given protective coatings, which demanded appropriate methods for application. Other coatings blackened the exterior panels to improve their ability to radiate heat. Designers had to allow for the tendency of columbium to creep or stretch slowly when hot and for the brittleness of molybdenum. Metallurgists developed heat-treatment techniques that promised to alleviate this brittleness, while engineers fabricated forgings of molybdenum, along with rivets, nuts and bolts of this metal.[9]

This technology also had a thorough workout in a small unpiloted vehicle called ASSET. It did not reach Dyna-Soar's orbital speeds, but it imposed similar heat loads by executing long hypersonic glides in the upper atmosphere at speeds as high as 19,500 ft/sec. Its shape resembled Dyna-Soar's, as did its in-flight temperatures, its hot-structure design, and its selected materials. Six ASSETs flew from Cape Canaveral between 1963 and 1965, five of them successfully. One was recovered from the ocean, giving specialists a close look at a hot structure that had survived its trial by fire.[10]

Hot structures elicited strong interest during subsequent years amid space shuttle design studies. Ablatives offered one-time-only use, but hot structures held the promise of being reusable, for they were designed not to oxidize or erode during entry. Their use of skin and leading-edge panels also promoted maintainability, for damaged panels might readily be located and replaced. But as space shuttle concepts took form, during the late 1960s, hot structures faced strong competition from a third approach to thermal protection: the use of reusable surface insulation (RSI). This gave rise to the "tiles" of the final shuttle design.[11]

There are several types of such insulation. The shuttle type came along only after much research. Like the panels of a hot structure, these tiles were to reradiate their incoming heat, to maintain a surface temperature that was high but nevertheless in equilibrium. Like those panels, tiles also showed enough intrinsic heat resistance to withstand temperatures in the thousands of degrees. But while hot structures separated the panels from the underlying insulation, in a tile they were part of the same block of material, which was bonded to the vehicle skin.

Tiles retained the reusability and maintainability of a hot structure, for they were not consumed during reentry and could be replaced when damaged. They also brought new advantages. They needed no protection against oxidation, for their materials were already fully oxidized. They offered light weight, while their insulating properties were superb. In Dyna-Soar, Q-felt insulation had only modest effectiveness; that is why the primary structure was René 41. For the shuttle, RSI insulated well enough to permit a primary structure of aluminum. This metal has no heat resistance worth mentioning; it loses strength at temperatures of a few hundred degrees. But it has long been the most familiar of aerospace materials, the lowest in cost, and the easiest to work with.

RSI grew out of ongoing work with ceramics for thermal protection. Ceramics had excellent heat resistance, light weight, and good insulating properties. But they were brittle, and they cracked rather than stretched in response to the flexing under load of an underlying metal primary structure. Ceramics also were sensitive to thermal shock, as when heated glass breaks when plunged into cold water. This thermal shock resulted from rapid temperature changes during reentry.[12]

Dyna-Soar gave an early example of the use of ceramics, in the nose cap. Its primary structure was graphite reinforced with silicon. Tiles of sintered zirconia covered this structure and were held in place with thick pins of zirconia. Designers expected that this ceramic would crack and reinforced the zirconia with platinum-rhodium wire. A cracked tile or pin then could continue to hold together after a fashion and would not break apart completely.[13]

The background to the shuttle's tiles lay in work dating to the early 1960s at Lockheed Missiles and Space Company. Key people included R. M. Beasley, Ronald Banas, and Douglas Izu. The point of departure involved the use of matted fibers of ceramic rather than monolithic blocks. A Lockheed patent disclosure of December 1960 gave the first presentation of a reusable insulation made of ceramic fibers for use as a reentry vehicle heat shield. Initial research dealt with casting fibrous layers from a slurry and bonding the fibers together.

Related work involved filament-wound structures that used long continuous strands. Silica fibers showed promise and led to an early success: a conical radome of thirty-two-inch diameter built for Apollo in 1962. Designed for reentry, it had a filament-wound external shell and a lightweight layer of internal insulation cast from short fibers of silica. The two sections were densified with a colloid of silica particles and sintered into a composite. This gave a non-ablative structure of silica composite, reinforced with fiber. It never flew, as design requirements changed during the development of Apollo. Even so, it introduced silica fiber to the realm of reentry design.

Another early research effort, Lockheat, fabricated test versions of fibrous mats that had controlled porosity and microstructure. These were impregnated with organic fillers such as Plexiglas (methyl methacrylate). These composites were not ablative, for the filler did not char. Instead it evaporated or volatilized, producing an outward flow of cool gas that protected the heat shield at high heat-transfer rates. The Lockheat studies investigated a number of fibers, including silica, alumina, and boria. Researchers constructed multilayer composite structures of filament-wound and short-fiber materials that resembled the Apollo radome. Impregnated densities were 40 to 60 lb/ft^3, the higher number being close to the density of water. Thicknesses of no more than an inch gave acceptably low back-face temperatures during reentry simulations.

This work with silica-fiber ceramics was well under way at the time of John Glenn's first flight to orbit, in February 1962. By 1965, a specific formulation of bonded silica fibers was ready for further development. Known as LI-1500, it was 89 percent porous and had a density of 15 lb/ft^3, one-fourth that of water. Its external surface was impregnated with filler to a predetermined depth, again to provide additional protection during the most severe reentry heating. By the time this filler was depleted, the heat shield was to have entered a zone of more moderate heating, where the fibrous insulation alone could provide protection.

Initial versions of LI-1500, with impregnant, were intended for use with small space vehicles similar to Dyna-Soar that had high heating rates. Space shuttle concepts were already provoking attention—the January 1964 issue of the trade journal *Astronautics & Aeronautics* presents the thinking of the day—and in 1965 a Lockheed specialist, Max Hunter, introduced an influential configuration called Star Clipper. His design called for LI-1500 as the thermal protection.

Like other shuttle concepts, Star Clipper was to fly repeatedly, but the need for an impregnant in LI-1500 compromised its reusability. But in contrast to the earlier reentry vehicle concepts, Star Clipper was large, offering exposed surfaces that were sufficiently blunt to benefit from the Allen-Eggers principle. They had lower temperatures and heating rates, which made it possible to dispense with the impregnant. An unfilled version of LI-1500, which was inherently reusable, now could serve.

Here was the first concept of an external insulation, bonded to the vehicle skin, that could reradiate heat in the fashion of a hot structure. However, matted silica fiber by itself is white and has low thermal emissivity, making it a poor radiator of heat. This meant excessive surface temperatures that called for thick layers of the silica insulation, adding weight. To reduce the temperature and the thickness, the silica needed a coating that would turn it black, for high emissivity. It then would radiate well and would stay cooler.

The selected coating was a borosilicate glass, initially with an admixture of Cr_2O_3 and later with silicon carbide, SiC, which further raised the emissivity. Glass and silica are both silicon dioxide, SiO_2; this assured a match of the coefficients of thermal expansion of the coating and substrate, to prevent the coating from cracking under the temperature changes of reentry. The glass coating could soften at very high temperatures, to heal minor nicks or scratches. It also offered true reusability, surviving repeated cycles to 2,500 degrees Fahrenheit.

A flight test came in 1968, as NASA-Langley investigators mounted a panel of LI-1500 RSI to a reentry test vehicle, along with several candidate ablators. This vehicle carried instruments, and it was recovered. Its trajectory reproduced the peak heating rates and temperatures of a reentering space plane such as Star Clipper. The LI-1500 test panel reached 2,300 degrees and did not crack, melt, or shrink. This proof-of-concept test pointed to the plausibility of high-emittance reradiative tiles of coated silica for thermal protection.[14]

214

Yet for the nose and leading edges, silica RSI was not enough; a reentering shuttle would need a thermal protection with even more temperature resistance. Carbon was an obvious choice. It had withstood extreme temperatures as early as 1942, when vanes of graphite dipped into the rocket exhaust of the V-2 and deflected this flow of hot gas to provide thrust vector control. Graphite had formed the primary structure of the Dyna-Soar nose cap. Graphite also served as the leading edges of ASSET, withstanding 3,000 degrees Fahrenheit.

In contrast to dense refractory metals such as molybdenum, graphite was lighter than aluminum. It oxidized when hot, as you may demonstrate in your back yard with a charcoal grill, but like those metals it could be protected with antioxidation coatings. Nevertheless, despite its track record, graphite failed to enter the mainstream of thermal protection. It was brittle, damaged easily, and did not lend itself for use with thin-walled structures.[15] The development of a better carbon began in 1958, with Vought Missiles and Space Company in the forefront. It went forward with support from the Dyna-Soar and Apollo programs and brought the advent of an all-carbon composite consisting of graphite fibers in a carbon matrix. Existing composites had names such as carbon-phenolic and graphite-epoxy; this one was carbon-carbon.

Carbon cloth gave a point of departure, produced by oxygen-free pyrolysis of a woven organic fiber such as rayon. Sheets of this fabric, impregnated with phenolic resin, were stacked in a mold to form a lay-up and then cured in an autoclave. At that point, one had a shape made of laminated carbon cloth phenolic, an ablative material closely resembling the nozzle lining of the shuttle's SRB. Further pyrolysis converted the resin to its basic carbon, yielding an all-carbon piece that was highly porous due to the loss of volatiles. It needed densification, which was achieved through multiple cycles of reimpregnation under pressure with an alcohol, followed by further pyrolysis. These cycles continued until the part had the specified density and strength.

Carbon-carbon retained the desirable properties of graphite in bulk: light weight, temperature resistance, and resistance to oxidation when coated. It had some strength along with a low coefficient of thermal expansion, which assured excellent resistance to thermal shock and to the stresses produced by temperature change. It had much better damage tolerance than graphite, and its carbon-phenolic lay-ups could readily be formed in advance to a desired shape.[16]

Researchers at Vought conducted exploratory studies during the early 1960s, investigating resins, fibers, weaves, and coatings. In 1964 they fabricated a Dyna-Soar nose cap of carbon-carbon; the program had been canceled, but this exercise permitted comparison of the new nose cap with its predecessor of graphite and zirconia tiles. In 1966 this firm built a heat shield for the Apollo afterbody, behind the main ablative shield. A year and a half later, the company constructed a wind-tunnel model of a Mars probe that was designed to enter the atmosphere of that planet.[17]

None of these exercises approached the full-scale development that Dyna-Soar had brought to hot structures. As with the concurrent work on RSI, they definitely were in the realm of the preliminary. Even so, by 1970 it was clear that one could arrange the thermal-protection technologies in an order that matched diminishing experience with increasingly attractive potential.

Ablatives continued to reign, representing a technology that not only had flown with re-peated success but also had protected astronauts even during the demanding return from the moon. However, for the shuttle they offered only interim use. Hot structures were in the fore-front among the reusable approaches, reflecting the experience of Dyna-Soar and ASSET. RSI and carbon-carbon were still in their infancy, but promised a new generation of materials hav-ing particular effectiveness. This was the state of the art as NASA turned to the shuttle.[18]

Protecting the Shuttle

During 1969 an initial round of space shuttle studies went forward at several major contrac-tors. They were exercises in preliminary design, evaluating alternatives and conducting trade studies for the purpose of choosing a preferred configuration. The chosen approach was a two-stage, fully reusable design, which reflected the preference of Maxime Faget, director of research and engineering at the Johnson Space Center. At that early date, the contractors' work included careful consideration of RSI but showed a decided preference for hot struc-tures in designing their orbiters.[19]

McDonnell Douglas proposed an orbiter with titanium primary structure, thermal pro-tection being provided by panels of columbium, nickel-chromium, and René 41. This study also considered the use of tiles made of a "hardened compacted fiber," which was unrelated to Lockheed's RSI. However, they did not recommend this approach, for those tiles were heavier than plates or shingles of alloy, and less durable. North American Rockwell offered a titanium structure that used RSI on the bottom, where temperatures were high, along with metallic shingles.

Lockheed proposed an aluminum structure; its orbiter thus resembled a conventional air-plane. This company was home to both Star-Clipper and LI-1500, and its shuttle designers accordingly followed the lead of Max Hunter by considering the use of the latter for thermal protection. They concluded, though, that this carried high risk. An alternate TPS appeared preferable, using metallic shingles that resembled those of McDonnell Douglas. General Dy-namics came in with a primary structure of aluminum along with some titanium. Thermal protection was a hot structure: columbium alloy on the bottom, titanium on the cooler sides and top.

216

Significantly, these concepts were not designs that these companies were prepared to send to the shop floor and to build immediately. They were paper vehicles that would take years to develop and prepare for flight. Yet despite this emphasis on the future, and notwithstanding the optimism that often pervades such preliminary-design exercises, not one company—not even Lockheed—was willing to recommend RSI as the baseline. It lacked maturity, with hot structures standing as the approach that held greater promise.[20]

Nevertheless, everyone was aware that RSI indeed might have its day as a consequence of continuing research and development. Lockheed continued to stand in the forefront of this work, amid ongoing activity at its Palo Alto Research Center. Investigators succeeded in further cutting the weight of RSI by raising its porosity from the 89 percent of LI-1500 to 93 percent. The material that resulted, LI-900, weighed only nine pounds per cubic foot, one-seventh the density of water.[21]

There also was much fundamental work on materials. Silica exists in crystalline forms: quartz, cristobalite, tridymite. These not only have high coefficients of thermal expansion but also show sudden expansion or contraction with temperature due to solid-state phase changes. Cristobalite is particularly noteworthy; above 400 degrees Fahrenheit it expands by more than 1 percent as it transforms from one phase to another. Silica fibers for RSI were to be glass, an amorphous rather than a crystalline state having a very low coefficient of thermal expansion and an absence of phase changes. The glassy form thus offered superb resistance to thermal stress and thermal shock, which would recur repeatedly during each return from orbit.[22]

The raw silica fiber came from Johns-Manville, which produced it from high-purity sand. At elevated temperatures it tended to undergo "devitrification," transforming from a glassy to a crystalline state. Then, when cooling, it passed through phase-change temperatures and the fiber suddenly shrank, which brought large internal tensile stresses. Some fibers broke, producing internal cracking within the RSI and degradation of its properties. This would grow worse during subsequent cycles of reentry heating.

To prevent devitrification, Lockheed worked to remove impurities from the raw fiber. Company specialists raised the purity of the silica to 99.9 percent while reducing contaminating alkalis to as low as 6 ppm. Lockheed did these things not only at the laboratory level but also in a pilot plant. This plant took the silica from raw material to finished tile, applying 140 process controls along the way.

Established in 1970, the pilot plant was expanded in 1971 to attain a true manufacturing capability. Within this facility, Lockheed produced tiles of LI-1500 and LI-900 for use in extensive programs of test and evaluation. The company turned them out reliably and in quantity. In turn, the general availability of these tiles encouraged their selection as the prime candidate for shuttle TPS, in lieu of a hot-structure approach.[23]

General Electric also became actively involved, studying types of RSI made from zirconia and from mullite, $3 \, Al_2O_3 \cdot 2 \, SiO_2$, as well as from silica. The raw fibers were commercial grade, with the zirconia coming from Union Carbide and the mullite from Babcock and Wilcox. Devitrification was a problem, but whereas Lockheed had addressed it by purifying its fiber, GE took the raw silica from Johns-Manville and tried to use it with little change. The basic fiber, known as Microquartz, already had served as insulation on the X-15 and other craft. It contained nineteen different elements as impurities. Some were present at a few parts per million, but others—aluminum, calcium, copper, lead, magnesium, potassium, sodium—ran from 100 to 1,000 ppm. In total, up to 0.3 percent was impurity. General Electric treated this fiber with a silicone resin that served as a binder, pyrolyzing the resin at high temperatures. This transformed the fiber into a composite, sheathing each strand with a layer of amorphous silica that had a purity of 99.98 percent and more. This high purity resulted from that of the resin. The amorphous silica bound the fibers together while inhibiting their devitrification. General Electric's RSI had a density of 11.5 lb/ft³, midway between that of LI-1500 and LI-900.[24]

These developments drew interest at NASA-Ames and NASA-Langley, which had long been in the forefront of work on reentry and thermal protection. With space shuttle studies under way, it quickly became clear that the shuttle's requirements differed markedly from those of Apollo. The peak heat transfer rate was lower, but the exposure was more prolonged, reaching into the hundreds of seconds. A reentering shuttle was to face an airflow with local speeds up to Mach 10, in contrast to the transonic flows over the nearly flat heat shield of Apollo. A shuttle's thermal protection had to qualify for a hundred reuses, in contrast to the once-only practice of Apollo. Because a shuttle was to be large in size, much of the heat transfer would result from turbulent flow, which carried heat more strongly than the laminar flows of the small Apollo spacecraft.

Ablative materials had been easy to characterize, but the reusability of a shuttle's hot structure or RSI demanded more. Such a thermal-protection system might degrade slowly through dozens of simulated flights—and then give way entirely. Such systems therefore demanded precise measurements of changes in thickness, composition, strength, crystallinity and crystal size, and optical properties. In testing ablative specimens, one could learn a great deal by observing their temperature and rate of erosion. The shuttle called for more: scanning electron microscopy, X-ray fluorescence, X-ray diffraction, spectral reflectance, methods for evaluating changes in strength.[25]

Test facilities raised issues of their own. Early research had simulated the hot and high-speed flows of reentry by placing samples of material in the exhaust of a rocket engine. However, this technique could not reproduce the rarefied flows of the upper atmosphere; the exhausts were too dense. These rockets had given way to several types of hot-gas wind tunnels. The arc jet, a standard, used an intense electric arc to heat its airflow, reaching temperatures

218

above 10,000 degrees Fahrenheit.[26] In 1970 NASA had arc jets rated at 20 megawatts, with electric power for a town of seven thousand. Even this was not enough, though, for shuttle thermal-protection studies demanded larger test specimens. For the ablatives of Apollo, it had sufficed to use samples the size of postage stamps. The shuttle test program required full-scale tiles and insulated hot-structure panels, which were several inches across.

At Ames the need for new test facilities suited the wishes of the center director, Hans Mark, who wanted to construct such installations. Howard Larson, head of the Thermal Protection Branch, proposed to build a new supersonic wind tunnel of the "vitiated air" type, achieving high temperatures by burning methane in its airflow. Its test gas then would consist of the products of combustion. His division chief, Dean Chapman, added his own proposal as he recommended a 60-megawatt arc jet with three times the power of existing ones. Hans Mark warmly endorsed both concepts and won preliminary approval from NASA Headquarters in 1971. But the director of Langley, Edgar Cortright, objected to having Ames receive the vitiated-air facility. He argued that such a wind tunnel would better suit the research objectives of his own center, which already had the nation's largest installation of this type, reaching Mach 7 within an eight-foot test section. Cortright won his point and took Larson's installation for his own. However, Ames received authorization to proceed with the big arc jet. This fitted that center's capabilities, for Ames already possessed NASA's largest direct-current power source. The new facility entered service in April 1975, in time to contribute to the mainstream development of shuttle thermal protection.[27]

The plans for new test installations, in 1970, complemented new NASA-sponsored work that addressed a broad range of issues in thermal protection. The funding for this work in fiscal year 1970 came to nearly $4 million, which was no mean sum; the centerpiece of NASA's shuttle work that year was a pair of detailed studies of two-stage, fully reusable configurations, each funded at $8 million, at North American Rockwell and McDonnell Douglas. Nearly two-thirds of the new thermal-protection funding went for carbon-carbon and RSI, but NASA also allocated close to $1 million to address some well-defined problems involving hot structures and advanced ablatives.[28]

Work on ablatives, funded at $425,000, sought low cost and light weight. The standard method of fabrication, used in Apollo, placed the ablator within an open-faced fiberglass honeycomb to prevent the material from cracking and to retain the char during reentry. There now was interest in replaceable heat-shield panels, held in place with nuts and bolts; these fasteners were to lie at the bottom of deep holes in the ablator, filled with plugs of that material. A new ablator, phenolic nylon, was becoming available in densities as low as 12 lb/ft^3, putting it on a par with RSI. It promised low cost along with high effectiveness. It could absorb 8,000 BTU/lb, enough to boil eight times its weight in water.

Even in the dollars of 1970, $425,000 did not go far. It nevertheless sufficed to fund a small coterie of investigators, who dealt with specific issues: fabrication of low-cost panels, relaxing some nonessential standards in quality control, wind-tunnel and arc jet tests in existing facilities.[29] Work on hot structures, funded at $525,000, showed its own focus by emphasizing superalloys. These materials, based on iron, nickel, and cobalt, drew on the metallurgy of jet engines and were close to being state-of-the-art for shuttle thermal protection. The most attractive included nickel-based Inconel 718 and René 41 along with cobalt-based Haynes-25 and -188. The latter gave improved oxidation resistance when hot. However, superalloys generally could resist oxidation and corrosion, even without protective coatings.

For use in the shuttle, there nevertheless was much to learn concerning their engineering details. Their upper temperature limits were not defined, while the cobalt-based alloys needed study of their high-temperature susceptibility to creep and to oxidation under stress. Panels of superalloy needed seals, to prevent hot gas from penetrating. Insulation, such as Microquartz with the very low density of 3.5 lb/ft^3, had to be packaged and attached to the panels or the primary structure. In turn, this packaging had to withstand high temperatures and prevent absorption of moisture. The panels had to hold amid the loud noise of the shuttle's rocket engines. They also needed leeway for thermal expansion; yet they could not undergo aerodynamic flutter, which would weaken them through fatigue.

NASA was not going to address all these issues during 1970, not with only half a million dollars, but this sum at least funded the fabrication and test of additional hot-structure articles. Even so, the superalloys were limited to 1,800 degrees Fahrenheit and below, which meant they could cover only about two-thirds of the orbiter. The hot undersurface, exposed to the onrushing airflow, needed more.[30] Hence there was interest in refractory metals that could allow hot structures to reach beyond the limits of the superalloys. Tantalum promised a limit of 2,800 degrees Fahrenheit, but it needed protection against oxidation and no suitable coatings were known. Columbium, that mainstay of Dyna-Soar and ASSET, could reach 2,400 degrees repeatedly in routine use. Appropriate coatings were available, making this metal suitable for the undersurface. Still it faced technical issues that resembled those of the superalloys, including demonstration of a hundred reuses while surviving the acoustic as well as the thermal environment. In 1970, facilities for such testing did not exist, but again NASA expected to learn more by using existing arc jets and wind tunnels.[31]

Carbon-carbon needed its own coatings. By itself it oxidized rapidly above 1,500 degrees, but when protected it could withstand 3,000 degrees and more. This material had its start in a graphite-cloth lay-up impregnated with resin, and one could mix powdered silicon into the resin used with the outer plies. When heated, the powder reacted chemically to form silicon carbide, an effective oxidation inhibitor. Other carbide-forming elements were also of interest, as were refractory oxides that could serve as overlays.[32]

During 1970 and 1971 NASA's contractors carried through detailed design studies that included selection of materials for both primary structure and thermal protection. North American and McDonnell Douglas dealt with the two-stage fully reusables, while Grumman, funded at $4 million in a search for lower-cost designs, proposed a two-stage configuration that placed the orbiter's hydrogen fuel in expendable tanks.[33] In 1969, amid the initial round of studies, the four participating firms had split their choices evenly between titanium and aluminum for the primary structure, while showing a decided preference for hot structures. The new round of studies maintained this preference.

All three contractors specified primary structures of titanium for their orbiters. They also proposed to use titanium hot structures for the upper fuselage and the top side of the wings. This placed them in a position to draw on the technology of the SR-71 aircraft, which had been built largely of titanium. They also used carbon-carbon for the nose cap. For the rest of the orbiter, the choices were considerably more eclectic.

Grumman and North American selected carbon-carbon for the wing leading edges, but McDonnell Douglas preferred coated columbium. The companies split three ways in dealing with other areas of the surface. North American selected a mullite RSI for the undersurface and the sides of the forward fuselage. For those same regions, McDonnell Douglas chose the nickel superalloy Hastelloy-X. Grumman used René 41 for the fuselage sides, Haynes 188 for the undersurface.[34]

The reports documenting these design decisions came out in mid-1971. By then it already was clear that NASA needed new configurations that could cut sharply into the cost of development, and within weeks the contractors did a major turnabout. All three went over to a primary structure of aluminum. They also abandoned hot structures, on which they had lavished so much attention. They turned instead to the use of phased technology, with interim systems that could get the shuttle up and flying, to be followed by subsequent systems that would give the definitive shuttle. Ablatives were to provide the interim thermal protection, with RSI coming in later.[35]

What brought this dramatic change? The advent of RSI production was critical. This drew attention from Max Faget, who had crafted initial concepts of the two-stage fully reusable and went on to offer a succession of preliminary orbiter designs that helped to guide the work of the contractors. His most important concept, designated MSC-040, came out in September 1971 and served the contractors as a point of reference. It used aluminum and RSI.[36]

"My history has always been to take the most conservative approach," Faget declares. Everyone knew how to work with aluminum, but titanium was literally a black art. Much of the pertinent shop-floor experience had been gained within the SR-71 program and was classified. Few machine shops had pertinent background, for only Lockheed had constructed an airplane—the SR-71—that used titanium hot structure. For the superalloys and columbium, the situation was worse, because these metals had been used mostly in turbine blades. Lockheed had encountered serious difficulties as its machinists and metallurgists learned to use titanium.[37] The shuttle program faced tight cost constraints, and no one wanted to risk an overrun while metalworkers wrestled with the problems of other new materials.

Titanium offered a potential advantage through its temperature resistance; hence its thermal protection might be lighter. This apparent weight saving was largely lost, however, because of a need for extra insulation to protect the crew cabin, payload bay, and on-board systems. Aluminum could compensate for its lack of heat resistance because it has higher thermal conductivity than titanium. The loss of a tile would not necessarily lead to a burn-through, for the aluminum could spread the heat across a wide area.

Designers expected to install RSI tiles by bonding them to the skin, and for this, aluminum had a strong advantage. Both metals form thin layers of oxide when exposed to air, but that of aluminum is more strongly bound. Adhesive, applied to aluminum, therefore held tightly. The bond with titanium was considerably weaker and appeared likely to fail in operational use at around 500 degrees Fahrenheit. This was not much higher than the limit for aluminum, 350 degrees, showing again that titanium's heat resistance did not lend itself to operational use.[38]

The complexity of hot structures also militated against them. Their mechanical installation called for a plethora of clips, brackets, stand-offs, frames, beams, and fasteners. Structural analysis posed a formidable task. Each of the many panel geometries needed its own analysis, to show with confidence that the panels would not fail through creep, buckling, flutter, or stress under load. Yet this confidence might be fragile, for hot structures had limited ability to resist overtemperatures. They also faced the continuing issue of sealing panel edges against ingestion of hot gas during reentry.

NASA-Langley had worked to build a columbium heat shield for the shuttle and had gained a particularly clear view of its difficulties. It was heavier than RSI but offered no advantage in temperature resistance. In addition, coatings posed serious problems. Silicides (silicon compounds) showed promise of reusability and long life, but they were fragile and easily damaged. A localized loss of coating could result in rapid oxygen embrittlement at high temperatures. Unprotected columbium oxidizes readily, and above the melting point of its oxide, 2,730 degrees Fahrenheit, it can burst into flame.[39] "The least little scratch in the coating, the shingle would be destroyed during reentry," says Faget. Charles Donlan, the shuttle program manager at NASA Headquarters, made a similar comment in 1983: "We learned that the metallic heat shield, of which the wings were to be made, was by no means ready for use on a returnable spacecraft. The slightest scratch and you are in trouble."[40]

In this fashion, having taken a long look at hot structures, NASA did an about-face as it turned to the RSI that Lockheed's Hunter had recommended as early as 1965. Then in January 1972 President Nixon and his budget director George Shultz endorsed a space shuttle program that was to build it properly from the start, avoiding the use of interim systems. Out went the plans for phased development; the baseline design now was to use RSI and carbon-carbon from the outset. Within days after Nixon's decision, NASA's Dale Myers, the associate administrator for manned space flight, spoke to a lunar science conference in Houston and stated that the agency had made the basic decision to use RSI.[41]

This decision nevertheless left important issues unresolved. Though RSI now was in the forefront, it would not win by default; if its development was to encounter unusual difficulty, NASA still could tide the program over by using ablatives. A year later, Eugene Love, a specialist in space shuttle technologies at NASA-Langley, discussed their status at the annual meeting of the American Institute of Aeronautics and Astronautics: "Ablators are baselined as a confident fallback solution (temporary) for both leading edges and large surface area, should development of the baseline approaches lag. Overall technology progress in ablators in the past four years has resulted in major reductions in the weight and cost. In particular, the cost reductions and innovative refurbishment methods have combined to place ablators in a comforting backup position for the shuttle."[42]

The winning shuttle proposal, from North American, continued to specify carbon-carbon for the nose cap and leading edges along with mullite RSI for the undersurface and forward fuselage. These design features were held over from the fully reusable orbiter of the previous year. Although most primary structure was aluminum, that of the nose was titanium, with insulation of zirconia lining the nose cap. The wing and fuselage upper surfaces, which had been titanium hot structure, now went over to an elastomeric RSI consisting of a foamed methylphenyl silicone, bonded to the vehicle in panel sizes as large as thirty-six inches. This RSI gave protection up to 650 degrees Fahrenheit.[43]

Still, was mullite RSI truly the one to choose? It came from General Electric and had lower emissivity than the silica RSI of Lockheed but could withstand higher temperatures. Yet the true basis for selection rested on the ability to withstand a hundred reentries, as simulated in ground test. NASA conducted these tests during the last five months of 1972, using facilities at its own Ames, Johnson, and Kennedy centers, with support from Battelle Memorial Institute.

The main series of tests ran from August to November and gave a clear advantage to Lockheed. That firm's LI-900 and LI-1500 went through one hundred cycles to 2,300 degrees Fahrenheit and held to specified requirements for thermal conductivity and back-face temperatures. The mullite showed excessive back-face temperatures and high thermal conductivity, particularly at high temperatures. As testing increased in severity, the mullite also developed coating cracks and gave indications of substrate fracture.

The tests then introduced acoustic loads; each cycle of simulation now subjected the RSI to the loud roars of rocket-powered flight along with the heating of reentry. LI-1500 continued to show promise. By mid-November it demonstrated the equivalent of twenty cycles to 160 dB, the acoustic level of large launch vehicles, and 2,300 degrees Fahrenheit. A month later, NASA conducted what Lockheed describes as a "sudden death shootout": the contending materials went into a single large twenty-four-tile array at NASA-Johnson for renewed thermal-acoustic tests. After twenty cycles, only Lockheed's LI-900 and LI-1500 tiles remained intact.

Lockheed won the thermal-protection subcontract in 1973, with NASA selecting LI-900 as the baseline RSI. The firm responded with preparations for a full-scale production facility in Sunnyvale, California. In addition, LI-1500 demonstrated a hundred cycles to 2,500 degrees Fahrenheit and survived a thermal overshoot to 3,000 degrees as well as an acoustic overshoot to 174.5 dB. With this, NASA had a suite of thermal-protection materials that appeared ready for mainstream development and eventual operational use.[44]

Test and Development

As with other elements of the shuttle, the fabrication of the thermal protection took place amid considerable care. In fashioning the nose cap of carbon-carbon, the Vought Corporation used a Union Carbide carbon cloth impregnated with Hexcel phenolic resin. The lay-up ran from nineteen plies in skin surface areas to twice that number at the rear, where the cap would be mounted to attachment rings. It was cured in an autoclave at 350 degrees Fahrenheit, inspected with X-rays or ultrasonics, then held for seven days in an electric furnace at 550 degrees. Pyrolyzation came next. The part went into a retort that was packed with coke, which absorbs oxygen, and kept at 1,500 degrees for seventy hours. This converted the phenolic resin into porous carbon. Workers then impregnated the component with furfuryl alcohol, cycled it anew through the autoclave and electric furnace, and pyrolyzed it once again. Twice more it was treated with alcohol, heated, and pyrolyzed, each time gaining strength and density as the carbon became less porous. These treatments raised its strength from 3,000 to 18,000 psi.

The finished carbon-carbon had a coefficient of thermal expansion an order of magnitude lower than that of most metals, which reduced thermal stresses. It also had the happy property of actually gaining strength with temperature, being up to 50 percent stronger at 3,000 degrees Fahrenheit than at room temperature. For oxidation protection, it was packed in a mixture of silicon carbide, alumina, and metallic silicon, then baked in an argon atmosphere at 3,000 degrees for eight hours. This converted the outermost millimeter of surface to silicon carbide, SiC. This layer contained tiny cracks, which were filled through impregnation under pressure with tetraethyl orthosilicate. A final stage of heating deposited silica within the pores and cracks, as a further shield against oxidation. Treatment with a sealant finished the work.[45]

Manufacture of the silica tiles was more straightforward, at least in its basic steps. The base material consisted of short lengths of silica fibers of 1.5-micron diameter. A measured quantity of fibers, mixed with water, formed a slurry. The water was drained away; workers added a binder of colloidal silica, then pressed the material into rectangular blocks that were ten to twenty inches in dimension and more than six inches thick. These blocks were the crudest form of LI-900, the basic choice of RSI for the entire shuttle. They sat for three hours to allow the binder to jell, then were dried thoroughly in a microwave oven. The blocks moved through sintering kilns, which baked them at 2,375 degrees Fahrenheit for two hours, fusing binder and fibers together. Band saws trimmed distortions from the blocks, which were cut into cubes and then carved into individual tiles using milling machines driven by computer. The programs contained data from Rockwell International on the desired tile dimensions.

Next the tiles were given a spray-on coating. After being oven-dried they returned to the kilns for glazing at temperatures of 2,200 degrees for ninety minutes. To verify that the tiles had received the proper amount of coating, technicians weighed samples before and after the coating and glazing. The glazed tiles then were made waterproof by vacuum deposition of a silicon compound from Dow Corning, while being held in a furnace at 350 degrees Fahrenheit. These tiles were given finishing touches before being loaded into arrays for final milling.[46]

Although the basic LI-900 material showed its merits during 1972, it was another matter to produce it in quantity, to manufacture tiles that were suitable for operational use, and to provide effective coatings. To avoid having to purify raw fibers from Johns Manville, Lockheed asked that company to find a natural source of silica sand with the necessary purity. The amount needed was small, about twenty truckloads, and was not of great interest to quarry operators. Nevertheless, Johns Manville found a suitable source, in Minnesota.

Problems then arose in maintaining adequate purity during manufacture at that firm's plant in Waterville, Ohio. Initial trials showed that residue from other products was contaminating the shuttle-grade silica. Lockheed sent employees to Waterville to improve the fibers' purity. Johns Manville cleaned its equipment thoroughly, relined some of its furnaces, and used isolation rooms to prevent contamination from other grades of fiber that could float through the air.

Other problems arose when shaping the finished tiles. Initial plans called for a large number of identical flat tiles, varying only in thickness and trimmed to fit at the time of installation. However, flat tiles on the curved surface of the shuttle produced a faceted surface that promoted the onset of turbulence in the airflow, resulting in higher rates of heating. The tiles then would have had to be thicker, and this threatened to add weight. The alternative was an external RSI contour closely matching that of the orbiter's outer surface. Lockheed expected to produce thirty-four thousand tiles for each orbiter, grouping most of them in arrays of two dozen or so and machining their back faces, away from the glazed coating, to curves matching the contours of the shuttle's aluminum skin. Each of the thirty-four thousand tiles was to be individually numbered, and none had precisely the same dimensions. Instead, each was defined by its own set of dimensions. This was costly, but it saved weight.

Numerically controlled milling machines did the work. Lockheed purchased three small-bed versions for carving individual tiles, and two large-bed types to shape the inner surfaces on tile arrays. These machines were special in their own right, for they used no oil, which could impair the coating process. "Each tile has an individual program," Kevin Forsberg, Lockheed's thermal-protection project manager, told *Aviation Week*. "It is a horrendous job. The manufacturing people were appalled." A particular issue involved replacing damaged tiles: "When two hundred tiles are identical, you can make 230 and if some are damaged, you have spares. If each is unique, you can't afford to make dupes."[47]

Difficulties also arose in the development of coatings. The first good one, LI-0042, was a borosilicate glass that used silicon carbide to enhance its high-temperature thermal emissivity. It dated to the late 1960s; a variant, LI-0050, initially was the choice for operational use. This coating easily withstood the rated temperature of 2,300 degrees Fahrenheit, but in tests it persistently developed hairline cracks after twenty to sixty thermal cycles. This was unacceptable; it had to stand up to a hundred such cycles. The cracks were too small to see with the unaided eye and did not grow large or cause tile failure. Still, they would have allowed rainstorms to penetrate the tiles during the weeks that an orbiter was on the ground between missions, with this water adding to the launch weight.[48]

Help came from NASA-Ames, where researchers were close to Lockheed both in their shared interests and in their facilities being only a few miles apart. Howard Goldstein at Ames, a colleague of the branch chief Howard Larson, set up a task group and brought in a consultant from Stanford University, which also was just up the freeway. They spent less than $100,000 in direct costs and came up with a new and superior coating called reaction-cured glass. Like LI-0050, it was a borosilicate, consisting of more than 90 percent silica along with boria (boron oxide) and an emittance agent. The agent in LI-0050 had been silicon carbide; the new one was silicon tetraboride, SiB_4. During glazing it reacted with silica in a way that increased the level of boria, which played a critical role in controlling the coating's thermal expansion. This coating could be glazed at lower temperature than LI-0050, reducing the residual stress that led to the cracking.[49] SiB_4 oxidized during reentry, but in doing so it produced boria and silica, the ingredients of the glass coating itself.

When we look at a shuttle orbiter we see black tiles covering the undersurface and forward fuselage and white tiles elsewhere. They all were designed as standard LI-900 with its borosilicate coating, but the black ones had SiB_4 and the white ones did not. This additive actually had a disadvantage in cooler areas of the shuttle, for it absorbed solar heat. White tiles reflected the sun, to reduce on-orbit system temperatures. The black tiles were usually 6 by 6 inches with thickness from 0.5 to 3.5 inches, as required. White tiles generally were 8 by 8 inches and 0.2 to 1.0 inches thick.[50]

As insulation, they were astonishing. You could heat one in a furnace until it was white-hot, remove it, allow its surface to cool for a couple of minutes—and pick it up by its edges using your fingers, its interior still at white heat. Nevertheless, they lacked structural strength and were brittle. They could not be bonded directly to the orbiter's aluminum skin, for they would fracture and break because of their inability to follow the flexing of this skin under its loads.[51] Designers therefore placed an intermediate layer between tiles and skin, called a strain isolator pad (SIP). It was a felt made of Nomex nylon from Du Pont, which would neither melt nor burn. It had useful elasticity and could stretch in response to shuttle skin flexing without transmitting excessive strain to the tiles. Workers bonded tiles to SIP and SIP to skin by using adhesive, a room-temperature vulcanizing methylphenyl silicone. This retained resiliency to temperatures as low as -170 degrees Fahrenheit, which kept it from becoming brittle when in shadow during an orbital flight.

This adhesive was a close chemical relative of the foamed RSI that Rockwell International had initially planned to use on the coolest portions of the orbiter, below 650 degrees. However, the Nomex felt had some temperature resistance and proved superior for this purpose. In this fashion, the SIP extended to cover areas that used no tiles, including the wing upper surface, upper fuselage sides, payload bay doors, and OMS pods. It was installed in blankets, three by six feet, and glued in place using the standard SIP silicone adhesive.[52]

Testing of tiles and other thermal-protection components continued through the 1970s. NASA-Ames was particularly active, where this work was part of a surge of effort that dealt broadly with the entire shuttle. NASA planned to spend thirty-five thousand wind-tunnel hours through fiscal 1979, of which 40 percent were to be accomplished at Ames. During 1973 and 1974, nearly half of Ames's shuttle wind-tunnel work was conducted within a single facility, a 3.5-foot hypersonic tunnel that used interchangeable nozzles to operate at Mach 5, 7, and 10. It ran two shifts per day, conducting about ten tests during each sixteen-hour day, with this pace continuing for several years. *Aviation Week* described the overall activity as "an unprecedented concentration of resources on a single vehicle." A branch chief stated that the rapid pace of work meant a new way of life, for the atmosphere had changed from a somewhat relaxed exploration of research problems to that of a production facility where everyone had to meet deadlines and keep the schedule from slipping.

Ames also had its Unitary Plan wind tunnels, with three test sections driven by a common power plant that ranged from Mach 0.7 to Mach 3.5. A low-speed tunnel had a forty- by eighty-foot test section and accommodated a model orbiter that was forty-four feet long, with enough detail to assess the influence of the tiles on the vehicle's aerodynamics during approach and landing. Ames also had arc jets, including the new 60-megawatt facility, with nozzles that could create a free jet at Mach 6 and flow in ducts at Mach 3.5.[53]

A particular challenge lay in creating turbulent flows, which demanded close study because they increased the heat-transfer rates many times over.[54] During reentry, hypersonic flow over a wing is laminar near the leading edge, transitioning to turbulence at some distance to the rear. No hypersonic wind tunnel could accommodate anything resembling a full-scale wing, and it took considerable power as well as a strong airflow to produce turbulence in the available facilities. Even the 60-megawatt arc jet could not do this with its free jet. Nevertheless,

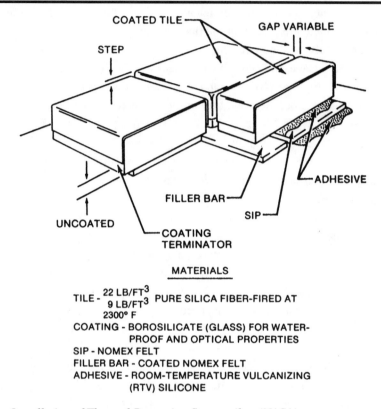

Fig. 46. Installation of Thermal Protection System tiles. (NASA)

Ames succeeded in creating such flows. Researchers used a 20-megawatt arc jet that fed its flow into a duct that was nine inches across and two inches deep. The narrow depth gave a compressed flow that readily produced turbulence, while the test chamber was large enough to accommodate panels with size of eight by twenty inches. This facility supported the study of coatings that led to the use of reaction-cured glass. Tiles of LI-900, six inches square and treated with this coating, survived a hundred simulated reentries at 2,300 degrees, in turbulent flow.[55]

The Ames 20-megawatt free jet arc facility made its own contribution, in a separate program that improved the basic silica tile. In response to excessive temperatures, these tiles failed by shrinking and becoming denser, which brought unacceptable changes in their dimensions. Investigators succeeded in reducing the shrinkage by raising the tile density and by adding silicon carbide to the silica, rendering it opaque and reducing internal heat transfer. This led to the introduction of a new grade of silica RSI with density of 22 lb/ft^3, having greater strength as well as improved thermal performance.[56]

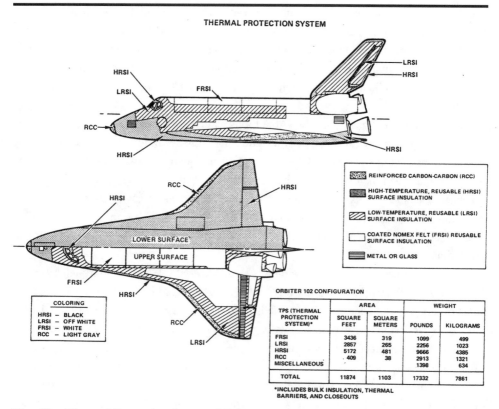

THERMAL PROTECTION SYSTEM

REINFORCED CARBON-CARBON (RCC)

HIGH-TEMPERATURE, REUSABLE (HRSI) SURFACE INSULATION

LOW-TEMPERATURE, REUSABLE (LRSI) SURFACE INSULATION

COATED NOMEX FELT (FRSI) REUSABLE SURFACE INSULATION

METAL OR GLASS

COLORING	
HRSI	BLACK
LRSI	OFF WHITE
FRSI	WHITE
RCC	LIGHT GRAY

ORBITER 102 CONFIGURATION

TPS (THERMAL PROTECTION SYSTEM)*	AREA		WEIGHT	
	SQUARE FEET	SQUARE METERS	POUNDS	KILOGRAMS
FRSI	3436	319	1099	499
LRSI	2857	265	2256	1023
HRSI	5172	481	9666	4385
RCC	409	38	2913	1321
MISCELLANEOUS			1398	634
TOTAL	11874	1103	17332	7861

*INCLUDES BULK INSULATION, THERMAL BARRIERS, AND CLOSEOUTS

Fig. 47. *Thermal Protection System. (NASA)*

The Ames researchers carried through with this work during 1974 and 1975, with Lockheed taking this material and putting it into production as LI-2200. Its method of manufacture largely followed that of standard LI-900, but whereas that material relied on sintered colloidal silica to bind the fibers together, LI-2200 dispensed with this and depended entirely on fiber-to-fiber sintering. LI-2200 was adopted in 1977 for operational use on the shuttle, where it found application in specialized areas. These included regions of high concentrated heat near penetrations such as landing-gear doors, as well as near interfaces with the carbon-carbon nose cap where surface temperatures could reach 2,600 degrees Fahrenheit.[57]

Other NASA centers had their own test facilities. Langley came close to matching Ames, with a 20-megawatt arc jet, a Mach 10 continuous-flow hypersonic tunnel, and a transonic tunnel with eight-foot test section. Vitiated-air installations, a specialty at Langley, included the new Thermal Protection Structural Test Facility and the existing High Temperature Tunnel, with Mach 7 capability and its own eight-foot section.[58] Other arc jets, rated at 5 and 10 megawatts, were at the Johnson Space Center. The 10-megawatt installation, known as the Atmospheric Reentry Materials and Structures Evaluation Facility (ARMSEF), dated to 1968 and saw extensive use in developing every type of thermal protection used on the orbiter. It conducted materials studies for shuttle thermal protection as early as 1969, with Apollo still very much under way, while its pace of activity stepped up after 1977 and continued even after the first flight to orbit.[59]

Rockwell International relied on NASA for its wind tunnels, but it had enough test equipment to make it a major center in its own right. It had four separate acoustic facilities along with three installations that heated tiles intensely using quartz lamps. NASA's Langley and Johnson centers had their own acoustic and heating capabilities, while Vought had in-house graphite heaters to test its carbon-carbon. Rockwell also had an environmental laboratory that subjected tiles to rain, salt spray, and high humidity as at Cape Canaveral. In addition, Rockwell, Vought, and NASA-Dryden all were ready to carry out structural tests of thermal-protection elements.[60]

Testing proceeded in four overlapping phases. Material selection ran through 1973 and 1974 into 1975; the work that led to LI-2200 was an example. Material characterization went concurrently and extended midway through 1976. Design development tests covered 1974 through 1977; design verification activity began in 1977 and ran forward through subsequent years. Materials characterization called for some ten thousand test specimens. This work used statistical methods to determine basic material properties. These were not the well-defined properties that engineers find listed in handbooks; they showed ranges of values that often formed a Gaussian distribution, with its bell-shaped curve. This activity addressed such issues as the lifetime of a given material, the effects of changes in processing, or the residual strength after a given number of flights. A related topic was simple but far-reaching: to be able to calculate the minimum tile thickness, at a given location, that would hold the skin temperature below the maximum allowable.

At Ames material characterization was handled largely by the 20-megawatt aerodynamic test facility. It had a wide variety of nozzle arrangements, making it one of the most versatile facilities within that center. Investigators supplemented their wind-tunnel measurements with X-ray diffraction to measure crystallization, trace-chemical analysis, and determination of optical properties such as emissivity.[61]

Carbon-carbon required its own tests. Though coated to resist oxidation, it oxidized slowly at high temperature and lost mass. It was necessary, therefore, to define the oxidation rates and the effect on mechanical properties. Air-oxidation tests, performed over a wide range of pressures and temperatures, used arc jets, radiant heaters, and furnaces. Other properties depended on the number of carbon cloth plies and the temperature, and it took some twenty-two hundred tests to obtain statistically valid characteristics that were suitable for use in design.[62]

Design development tests used only 350 articles, but spanned four years because each of them required close attention. An important goal involved validating the specific engineering solutions to a number of individual thermal-protection problems. The nose cap and wing leading edges, made of carbon-carbon and subjected to the highest temperatures, held a large degree of concern. Their attachments were exercised in structural tests that simulated flight loads up to design limits, with design temperature gradients. Other tests, conducted within an arc-heated facility, determined the thermal responses and hot-gas leakage characteristics of interfaces between the carbon-carbon and RSI. Installations of RSI brought their own issues, particularly where they were penetrated. The undersurface had landing-gear doors, which needed careful tests to guard against leakage of hot gas. RCS thrusters had their own penetrations, as did flash evaporators and urine dumps. The payload-bay doors had their own seals. The crew access door required its own assurances, for it was located on the lower forward fuselage in an area of high temperature.[63]

Design development testing also addressed basic issues of the tiles themselves. There were narrow gaps between them, and while Rockwell had ways to fill them, these gap-fillers required their own trials by fire. A related question was frequently asked: what happens if a tile falls off? A test program addressed this—and found that in some areas of intense heating, the aluminum skin indeed would burn through. The only way to prevent this was to be sure that the tiles were firmly bonded in place, and this meant every one of them in the critical areas.[64]

Design verification tests used fewer than fifty articles, but these often represented substantial portions of the orbiter. An important test article, evaluated at NASA-Johnson, reproduced a wing leading edge and measured five by eight feet. It had two leading-edge panels of carbon-carbon set side by side, a section of wing structure that included its principal spars, and aluminum skin covered with RSI. It had insulated attachments, internal insulation, and interface seals between the carbon-carbon and the RSI. It could not have been fabricated earlier in the program, for its detailed design drew on lessons from previous tests. It withstood simulated air loads, launch acoustics, and mission temperature-pressure environments, not once but many times.

Large surface panels survived their own verification tests. A flat panel, four by five feet, had aluminum skin supported by structural stringers and covered with RSI. Such test items could verify the planned life of a hundred missions and did this for demanding areas that included the undersurface. Like the NASA-Johnson leading edge, these panels were too large for arc-heated wind tunnels; they took their workouts in radiant-heating facilities.[65]

Continued testing brought a continuing flow of engineering modifications. The layout of the white tiles and nylon-felt RSI changed after 1975, while ablatives made a minor comeback as they found use in the elevons. The orbiter had inboard and outboard elevons rather than single units that would span the wing trailing edges. Late in development, thermal analysis and wind-tunnel testing indicated that the areas between the elevons, midway along the span, would reach 3,200 degrees Fahrenheit—far too high for any RSI, even LI-2200. NASA responded by turning to its old standby, the ablative of the Apollo heat shield. Small panels of this ablator, placed at the facing edges of the elevon segments, solved the problem, even though the panels had to be replaced after every flight.[66]

The testing ranged beyond the principal concerns of aerodynamics, heating, and acoustics. There also was concern that meteoroids might not only put craters in the carbon-carbon but also cause it to crack. At NASA-Langley, researcher Donald Humes investigated this by shooting small glass and nylon spheres at target samples using a light-gas gun driven by compressed helium. Helium is better than gunpowder; it can expand at much higher velocities, and Humes's projectiles flew at speeds near 5.5 kilometers per second. Meteoroid impact speeds would be up to ten times higher, but Humes assumed that particles with equal energy create equal damage.

The kinetic energy of his spheres varied from 0.2 to 74 joules. At 3 joules, the crater penetrated the oxidation-resistant surface layer of the material and reached the carbon interior. An 11-joule collision produced an impact crater on the front surface and a spallation on the back surface, although there was no hole through the material. The specimens were completely penetrated at 34 joules and higher. The material proved resistant to cracking. Humes added that carbon-carbon "does not have the penetration resistance of the metals on a thickness basis, but on a weight basis, that is, mass per unit area required to stop projectiles, it is superior to steel."[67]

Yet amid all the advanced technology of arc jets, light-gas guns, and hypersonic wind tunnels, one of the most important tests was also one of the simplest. It involved nothing more than taking tiles that were bonded with adhesive to the SIP and to underlying aluminum skin and physically pulling them off.

The Shuttle Comes Unglued

It was no new thing for people to show concern that the tiles might not stick. In 1974 at NASA-Ames a researcher noted that aerodynamic noise was potentially an important problem and told *Aviation Week,* "We'd hate to shake them all off when we're leaving." At NASA-Johnson during 1975 the ARMSEF arc jet saw extensive use in lost-tile investigations. Some laboratory tests at Rockwell showed weak bonds, but this was attributed to faulty specimens used in the tests, and there was little follow-up.[68]

There was reason to believe that the forces acting to pull off a tile would be as low as 2 psi, some 70 pounds for a tile six inches square. This was low indeed; the adhesive, SIP, and RSI material all were considerably stronger. As a consequence, the thermal-protection testing gave priority to thermal rather than to mechanical work, essentially taking it for granted that the tiles would stay on. However, the attachment of the tiles to the shuttle lacked adequate structural analysis and did not take account of peculiarities in the components. The SIP had some fibers oriented perpendicular to the cemented tile undersurface. The tile was made of ceramic fibers, with these fibers concentrating the loads. This meant that the actual stresses they faced were substantially greater than the average.[69]

Columbia, orbiter OV-102, was the first to receive working tiles rather than the Styrofoam substitutes of Enterprise. Columbia also was slated to be the first into space. It underwent final assembly at the Rockwell facility in Palmdale during 1978. Check-out of on-board systems began in September and installation of tiles proceeded concurrently. This work was scheduled for completion in January 1979, with Columbia to be rolled out from its plant on 23 February.[70]

Mounting the tiles was not at all like laying bricks. Measured gaps were to separate them; near the front of the orbiter, they had to be positioned to within 0.017 inches of vertical tolerance to form a smooth surface that would not trip the airflow into turbulence. This would not have been difficult if the tiles had rested directly on the aluminum skin, but they were separated from that skin by the spongy SIP. The tiles also were fragile. An accidental tap with a wrench, a hard hat, even a key chain could crack the glassy coating. When that happened, the damaged tile had to be removed and the process of installation had to start again with a new one.

The SIP for a particular tile was an individual pad cut to size and somewhat smaller than the tile proper, leaving room around the edges. Filler bars took up that room; these were strips of SIP material that lay at the base of the narrow gaps between adjacent tiles. Installation began by painting the orbiter's aluminum skin with a corrosion inhibitor of green epoxy. A blueprint-like guide, generated by a drafting machine at Palmdale and printed on transparent Mylar plastic, indicated where the filler bars were to be laid out. An instrument called a comparator verified the tile alignment. A vacuum seal locked the comparator to the aluminum surface, and a technician checked the tile fit by using light indicators. The gaps were critical; on the forward lower fuselage, where heating was intense, they could not be closer than 0.017 inches, or tiles might strike or rub against each other due to flexing of the vehicle during re-entry. They could not be farther apart than 0.030 inches or too much heat could reach the skin. The accuracy of a tile fit had to be as great as 0.002 inches. If a fit lacked precision, the tile might be sanded to bring it within the tolerances.

After confirming an accurate fit the installer attached the filler bars to the aluminum using adhesive. The tile went to a separate room, where its SIP pad was bonded into place with more of that glue. Tile and pad were weighed to ensure that they had received the correct amount of adhesive; then they were left to cure at room temperature for several hours. Meanwhile, both NASA and Rockwell inspectors checked and recorded each operation. A

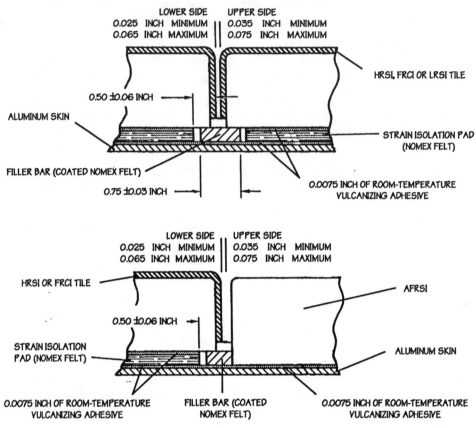

Fig. 48. *Thermal Protection System interfaces. (NASA, courtesy Dennis R. Jenkins)*

worker applied adhesive to the tile cavity on the side of the orbiter, to a depth of 0.005 to 0.007 inches. Mechanical or vacuum devices held the tile in place. It cured for four days, then was given a pull test and a measurement of gap widths. With this, it was ready.[71]

The tiles came in arrays, each array numbering about three dozen tiles. It took 1,092 arrays to cover this orbiter, and NASA reached a high mark when technicians installed 41 of them in a single week. Rockwell had 150 people on the job, some of whom worked at the company's Downey plant, while others were at McDonnell Douglas in St. Louis. These people worked on special projects, applying tiles to the forward reaction control module in Downey and to the OMS pods in St. Louis. Proceeding in parallel with the main activity in Palmdale, they shaved two to three months off the production schedule.

Some of these workers had experience in related fields such as glass fiber bonding, and program officials tried to hire people with such backgrounds. Many others lacked such ex-

234

perience, and Rockwell set up a training program, with two weeks in the classroom and two to three weeks of on-the-job training. The program included certification in various skills, lessons on thermal-protection materials, and methods of handling the delicate tiles so as to avoid damaging them. People learned to apply adhesive in ways that would make the tiles stick.[72]

The schedule came unglued before the tiles did. If everything had proceeded in a logical fashion, the Palmdale workers would have carried through the complete job of tile installation before sending Columbia off to Kennedy Space Center for its first flight. This did not happen, for reasons that stemmed from continuing delays in the planned date of that flight. During much of 1978 NASA and Rockwell looked ahead to a first launch of Columbia in the latter part of 1979. To prepare for this, Rockwell recruited and trained a work force of eleven hundred that was to serve as the ground crew. But late in 1978, in quick succession, the SSME experienced failures of the heat exchanger and main oxygen valve while under test, with the latter failure destroying an engine and delaying further firings of the three-engine cluster. Thoughts of a 1979 launch date soon went out the window, and Rockwell found itself with a Kennedy Space Center work force that had little to do.[73]

Rather than lay them off and then hope to rehire them in due course, Rockwell officials elected to turn some of these people into tile installers. Columbia was to go to the Cape, ferried atop its 747, with some seven thousand of its thirty-four thousand tiles missing. It also lacked its main engines and much of their insulation, Auxiliary Power Units, on-board computers, and fuel cells. Columbia needed about six months to complete its assembly, and members of the launch crew were in a position to help. They also would have a live shuttle orbiter to work with, allowing them to exercise their ground equipment and operate the on-board systems. This work would keep everyone busy during 1979.

This orbiter spent an additional week in Palmdale while workers filled the gaps, where tiles were missing, with temporary versions made of Styrofoam. This gave a smooth exterior surface, reducing aerodynamic loads during the ferry flight. On 8 March, in a replay of the prelude to the 1977 flights of Enterprise, Columbia rode the ninety-wheel trailer from Palmdale to Edwards Air Force Base. It was hoisted onto the back of the 747, which made a seventeen-minute test flight the following day. It all looked familiar: tail cone on the orbiter, Fitzhugh Fulton once again at the 747's controls.[74]

The temporary tiles were attached with double-backed tape; those that faced severe aerodynamic pressure had adhesive as well. In addition, there was tape on the surface of the orbiter, to smooth the airflow and to guard against seepage of water. Engineers had tested this surface tape on a T-38 aircraft; it had shown no sign of coming loose. But as the 747 left the runway, ground controllers noted debris on the concrete. Two T-38 chase aircraft pilots reported that tape was flapping from the orbiter's fuselage. Following the landing, it became clear that some sections of tape had worked loose during the flight and were buffeted by the airflow against adjacent tiles, loosening some and damaging others. Donald Slayton, the

NASA manager for orbital flight tests, said that about three dozen temporary tiles were lost. Another four or five permanent tiles fell off due to whipping from the tape and because of excess aerodynamic pressure resulting from loss of temporary ones that lay immediately ahead and had acted as a shield. Significantly, no permanent tile, bonded in place with adhesive, had failed due to weakness of the adhesive itself when facing normal loads.

NASA got rid of all the tape, including that on the surface, and reinforced the temporary tiles by gluing them all into place. The event nevertheless brought a public-relations flap, for Rockwell's press officials failed to make clear the difference between the plastic and the RSI tiles, which made it easy for people to believe that the real ones had fallen off. But when Columbia reached the Cape two weeks later, visual inspections showed that a number of tiles indeed had come loose. This meant that trouble with the tiles was only beginning.[75]

The thermal-protection staff initially numbered 260, but during their first three weeks, they succeeded in applying only about 200 new tiles. Quickly, the number to be installed began to grow. In April the count stood at 8,000; by early May it was 10,800. Rockwell responded by hiring several hundred additional people from the local area to help with the work. Even so, at midyear the rate of tile installation was only 300 per week, with the program calling for up to 500 per week to hold at least a modest chance of launch before April 1980. "The magnitude of the job is far greater than any of us anticipated," said Donald Phillips, the shuttle mission manager at Kennedy Space Center.[76]

There were other problems. The temporary tiles had been glued in place using their own adhesive, which overlapped and covered portions of the permanent tiles. The excess adhesive had to be removed, which called for much scraping and brushing. Workers had drilled holes in the temporary tiles to allow the adhesive to penetrate to the base, but the drills had scratched the orbiter's aluminum skin, and the scratches had to be repaired. In addition, the temporary tiles proved to have been attached so firmly that they were difficult to remove. Rockwell handed out plastic tools to cut them away, but their removal proved time-consuming.[77]

Worse news came around midyear, as detailed studies showed that in many areas, the combined loads due to aerodynamic pressure, vibration, and acoustics would produce excessively large forces on the tiles. Work to date had treated a 2-psi level as part of normal testing, but the new data showed that only a small proportion of the tiles already installed faced stresses that low. The breakdown was as follows:

Number of Tiles	Force Level (psi)
3,400	2 or less
6,300	2 to 6.5
3,000	6.5 to 8.5
5,500	8.5 to 13
247	13 and above

The first group appeared safe, but the rest were open to question. NASA officials already knew that the tiles were weak, for pull tests had shown that many of them broke away from their adhesive at only about 60 percent of their published tensile strength. Together, these findings meant that the basic usefulness of the tiles as thermal protection was suddenly in doubt.[78]

This loss of strength applied to both the LI-900 and LI-2200 tiles. Work at Rockwell gave representative numbers for the components of the thermal-protection system and for the complete system (table 6.1). For both grades of RSI, the strength of the combination, as mounted to the skin of the shuttle, was only about half that of the RSI itself, which was the weakest of the group.[79] Superficially, it appeared that there was no real problem, that LI-900 could continue to serve in most areas while LI-2200 might substitute in the limited regions that called for higher strength. However, installed tiles needed strength beyond their calculated forces to provide a margin of safety. The strength of the LI-900 combination, given as 11.7 psi, actually was not a fixed value. It had a range that showed considerable scatter, with 11.7 psi as the mean. Each tile had its own value, and many tiles were less. Nor could NASA turn to LI-2200 and use it freely; it would add weight.[80]

What caused the weakness? The fault lay in the nylon-felt SIP, which had been modified by "needling" to increase its through-the-thickness tensile strength and elasticity. This was accomplished by punching a barbed needle through the felt fabric, some one thousand times per square inch, which oriented fiber bundles transversely to the SIP pad. Tensile loads applied across the SIP, acting to pull off a tile, were transmitted into the SIP at discrete regions along those transverse fibers. This created localized stress concentrations where the stresses were as much as 1.9 times the mean value. These local areas failed readily under load, causing the glued bond to break.

Rockwell and NASA responded by demanding their own margin of safety. No installed tile was to be trusted, not even the ones in the 2-psi zones. Instead, technicians were to use vacuum suction cups to test each tile as an individual unit, applying 1.25 times the calculated flight force. Sensitive microphones, mounted to a tile, listened for the snapping sounds that signaled incipient failure. To remain in place, a tile had to survive this test without cracking. Any that cracked would be removed.[81]

There also was a clear need for means to increase the strength of the tiles' adhesive bonds. Initial thinking called for placing a thin layer between tile and SIP, with the glue on both sides. Aluminum foil had its proponents; it adhered well to the adhesive. But it would expand and contract readily with temperature changes, and even though it was thin, it might impose unwanted loads on a tile. Sheets of carbon-polyamide composite avoided the thermal problem and gave an initial solution.[82] The definitive solution came during October. It was known that LI-2200, which was denser, gave a stronger bond. The new method consisted of modifying a thin layer at the bottom of each tile to give it the qualities of this material. The process, called densification, used Du Pont's Ludox with a silica slip. Ludox was colloidal silica

Table 6.1. Strength of Standard Thermal Protection System

Material	Strength (psi)
LI-900 RSI	24
0.16-inch SIP	41
Adhesive	480
Combination	11.7
LI-2200 RSI	60
0.09-inch SIP	68
Adhesive	480
Combination	30.2

Source: Cooper and Holloway, "Shuttle Tile Story," *Astronautics & Aeronautics,* Jan. 1981, 29.

stirred into water and stabilized with ammonia; the slip had fine silica particles dispersed in water. The Ludox acted like cement; the slip provided reinforcement, in the manner of sand in concrete.

The tiles had been waterproofed during manufacture, and workers started by applying isopropyl alcohol as a wetting agent. They then painted the mix of slip and Ludox onto the backs of tiles and set them to dry for twenty-four hours in air and for two hours in an oven at 150 degrees Fahrenheit. Each tile was then weighed to learn whether it needed more Ludox. The dried tile, made gray with pigment for identification, was waterproofed anew using Dow Corning's standard silicon compound, Z-6070. It then was ready for reinstallation on Columbia, again by being glued into place atop an intermediate pad of SIP.[83]

Densification restored the lost strength, as shown in new data (table 6.2).[84] The work of reinstallation proceeded in parallel with the tests of tile strength, and *Science* gave a picture of the activity:

> *In a large hangar at the Kennedy Space Center in Cape Canaveral, Florida, still encased in a cocoon of metal pipes and walkways, sits the space shuttle Columbia, the most advanced hybrid airplane-space capsule ever developed. Amid a high-pitched yowl from huge fans, several hundred workers imported from California hustle to and from the Columbia, each carrying small boxes containing a single tile to be fixed to its hull. Others gaze at intricate, checkered maps of the shuttle's skin, tacked onto long rows of portable corkboard. Still others tinker with wires and small parts within the shuttle's belly, beneath a lighted sign that blazes, "Vehicle Powered." Men and women carrying clipboards rush about, watching small groups of people either affixing tiles or attempting to pull them off with small arrays of vacuum pumps.*[85]

Fig. 49. Thermal-protection tiles on the underside of the forward fuselage. (NASA, courtesy Dennis R. Jenkins)

The hangar was the Orbiter Processing Facility (OPF), which adjoined the immense Vehicle Assembly Building. The second bay of the OPF became a center for tile densification. The tile work force numbered 940 in early 1980, many being in their late teens and early twenties. They worked three shifts, with about twenty large trailers parked nearby to provide support—including food service and rest rooms.

The work went badly during 1979, for as people continued to install new tiles, they found more and more that needed to be removed and replaced. In June NASA and Rockwell had estimated that they had to fit 10,500 into position. At year's end they had applied 12,000 tiles, newly installed or remounted. A month later, only 1,200 of the June tally had not been mated—and the count of tiles to emplace was up to 13,100, as ongoing pull tests continued to disclose weak ones.

Orderly installation procedures broke down. Rockwell had received the tiles from Lockheed in arrays and had attached them in well-defined sequences. Even so, that work had gone

Fig. 50. *Thermal-protection tiles on the underside of a wing. (NASA, courtesy Dennis R. Jenkins)*

slowly, with 550 tiles in a week being a good job. Now, however, Columbia showed a patch-work of good ones, bad ones, and open areas with no tiles. Each individual tile had been shaped to a predetermined pattern at Lockheed, using that firm's numerically controlled milling machines. But the haphazardness of the layout made it likely that any precut tile would fail to fit into its assigned cavity, leaving too wide a gap with the adjacent ones.

Many tiles therefore were installed one by one, in a time-consuming process that fitted two of them into place and then carefully measured space for a third, designing it to fill the

Table 6.2. Strength of Densified Thermal Protection System

Material	Strength (psi)
0.16-inch SIP, undensified LI-900	11.7
0.16-inch SIP, densified LI-900	22.6
0.09-inch SIP, undensified LI-2200	30.2
0.09-inch SIP, densified LI-2200	46.3

Source: Holloway, "Shuttle Tile Story," *Astronautics & Aeronautics,* Jan. 1981, 30.

space between them. The measurements went to Sunnyvale, California, where Lockheed carved that tile to specifications and shipped it to Kennedy Space Center. Hence each person took as long as three weeks to install four tiles. Densification also took time; a tile removed from Columbia for rework needed fourteen days until it was ready for reinstallation.[86]

New procedures sometimes brought new trouble. There was much interest at Rockwell in using ultrasonic instruments to make absolute determinations of the strength of tiles, a quantity that correlated closely with the speed of sound within the RSI material. A staff scientist recommended a commercially available ultrasonic unit that he had used successfully in studies of wood and concrete. Initial use with tiles looked promising. By February 1980 the test equipment was ready at Kennedy Space Center, complete with trained operators. Most tiles required minimum strength of 13 psi, or minimum ultrasonic velocity of 1,710 ft/sec. But whereas direct tensile tests of tiles had shown that only about 1 percent of them had strength below this acceptance level, rejection rates based on velocity measurements approached 10 percent. The overriding concern was not to accept any weak tiles, and many were removed even though they indeed were sound. It took months to learn how to use the ultrasonic data properly and to reduce the rate of false rejections.[87]

How could these problems have been avoided? They all stemmed from the fact that the tile work was well advanced before NASA learned that the tile-SIP-adhesive bonds had less strength than they needed. The analysis that disclosed the strength requirements, varying with location on the surface of the shuttle, was neither costly nor excessively demanding. It might readily have been in hand during 1976 or 1977. Had this happened, Lockheed could have begun shipping densified tiles at an early date. Their development and installation would have occurred within the normal flow of the shuttle program, with this change amounting perhaps to little more than an engineering detail.

The reason this did not happen was far-reaching, for it stemmed from the basic nature of the program. The shuttle effort followed "concurrent development," with design, manufacture, and testing proceeding in parallel rather than in sequence. This approach carried risk, but the Air Force had used it with success during the 1960s. It allowed new technologies to enter service at the earliest possible date. But within the shuttle program, funds were tight. Managers had to allocate their budgets adroitly, setting priorities and deferring what they could put off. To do this properly was a high art, calling for much experience and judgment, for program executives had to be able to conclude that the low-priority action items would contain no unpleasant surprises. The calculation of tile strength requirements was low on the action list because it appeared unnecessary; there was good reason to believe that the tiles would face nothing worse than 2 psi. Had this been true, and had the SSMEs been ready, Columbia might have flown by mid-1980. It did not fly until April 1981, however, and in this sense, tile problems brought a delay of close to a year.

The delay in carrying through the tile-strength computation was not mandatory. Had there been good reason to upgrade its priority, it could readily have been done earlier, in a more timely fashion. The budget stringency that brought this deferral (along with many others) thus was false economy par excellence, for the program did not halt during that year of launch delay; it kept on writing checks for the contractors and employees. The missing tile-strength analysis thus ramified in its consequences, contributing substantially to a cost overrun in the shuttle program.[88]

Yet while NASA might be locking the barn door after the horse was stolen, during 1979 that agency gave the same intense level of attention to the tiles' mechanical problems that it had previously reserved for their thermal development. Wind tunnels entered the effort: a facility with eight-foot test section at NASA-Langley, another of eleven feet at NASA-Ames, and a big one of sixteen feet at the Air Force's Arnold Engineering Development Center. NASA-Dryden also dealt with specialized topics by flying RSI panels on two fighter planes, an F-104 and F-15. These simulated the aerodynamic loads of launch, executing flight profiles that subjected their tiles to 1.4 times their design pressures.[89]

Densification solved the overall problem of the tiles. It was simple in concept, but difficult to implement because it had to be applied on a large scale and with close attention to detail. The wind-tunnel and flight-test work gave their own simple solutions, which were much easier to put into practice because they were used only in limited areas. For instance, one small group of tiles bordered the windshield, overhanging its edge with extensions that reached downward and nearly touched the windows. Tests showed that the airflow could exert pressures on the overhang that tended to peel these tiles off. They needed extra strength, and this came from their grain structure. Internal layers of silica fibers generally ran parallel to the skin, in an arrangement that minimized heat transfer. New windshield tiles, with grain running perpendicular to the skin, had more strength while continuing to give acceptable thermal performance. As a final touch, technicians glued the overhang to the window, sealing it against the airflow.

The front of the OMS pods proved to flex excessively, which could cause them to break and even fall off. The solution was to "dice" their tiles, carving them into small pieces that easily followed the deflections. Another problem arose on the body flap, which extended to the rear of the fuselage below the main engines. Tiles of LI-2200 at the flap's corners were highly stressed, with SIPs that tended to fail. They obtained support from specialized screws with broad threads that could hold within the delicate silica of these tiles.[90]

The work appeared to be in hand by May 1980 as NASA anticipated a launch date of March 1981. Then Rockwell officials proposed densifying all tiles, regardless of their required strength. The tiles already were rated as capable of withstanding 125 percent of their calculated loads, and this new effort would raise this margin to 175 percent. But it also required some 9,000 additional tiles to undergo densification. Alan Lovelace, NASA's deputy administrator, said that this could delay the first flight by two and a half months, while adding as much as $500 million to the total program.

NASA and Rockwell officials met at the end of May and elected to hold their decision until July to see how matters would develop. By then the situation was clearer, for 1,700 tiles still were below 125 percent, based on the most recent loads analysis. Another 1,700 lay in areas of substantial densification, and it appeared prudent to densify these as well. A study of combined loads, resulting from vibration, acoustics, and aerodynamic shock waves, identified 1,276 other tiles that needed this treatment, for a total of nearly 4,700. This was only half the earlier number, but still it meant that Columbia needed several additional months of tile work.[91]

The effort continued to follow the pattern of three steps forward and two steps back. For a while, more tiles continued to be removed than were put on in a given week. People took notice during a week in early September, when 738 went on and 307 were removed, either for densification or replacement by a different tile, for that week showed a net gain of 431. Even so, the end was in sight, with 26,281 in place out of 30,922 required, leaving 4,741 to go. Two months later this number was well below a thousand, with technicians mounting more than five hundred per week.[92]

A key event lay immediately ahead: the roll-out of Columbia from the OPF into the Vehicle Assembly Building, where it was to be mated with its ET and twin SRBs and prepared for launch. During the spring of the previous year, before the main tile problems had come to light, the schedule had called for this to occur on 24 November 1979, with first flight penciled in for May or June 1980. Late in July 1980, at a space shuttle program review, NASA set 23 November as the date for roll-out, a delay of one year almost to the day. First flight was set for late March 1981, but officials declared that a slip into April was highly possible.[93]

This time the schedule went as planned. Late on 24 November, on the anniversary of that earlier planned date, a tow vehicle pulled Columbia into the VAB, as a large crowd watched and cheered. This orbiter had spent twenty months within the OPF, and the tiles still needed minor touch-up and detail work. But the SRBs and ET were already in place. Within two days, Columbia was hard-mounted to the latter, forming a live shuttle in flight configuration. Kenneth Kleinknecht, the manager of Columbia at NASA-Johnson, put it succinctly: "The vehicle is ready to launch."[94]

CHAPTER  SEVEN

The Orbiting Airplane

With its rocket propulsion and its thermal protection, the space shuttle broke new ground. It also had on-board temperature control and, to protect its crew, environmental control and life support. It used the Apollo launch facilities at Kennedy Space Center, relying strongly on support from its launch control center. Yet it was more than an orbiting spacecraft; it also was an airplane. It had wings, a tail, and control surfaces. It also had a suite of on-board avionics that was advanced for its day and could execute an automatic landing approach. The aluminum structure and landing gear drew on familiar themes in aircraft design. It nevertheless remained extraordinary, for nothing like it had ever been built.

Aerodynamics and Structures

The shuttle was one of the last major aircraft to rely almost entirely on wind tunnels for studies of its aerodynamics. There was much interest in an alternative: the use of supercomputers to derive aerodynamic data through solution of the governing equations of airflow, known as the Navier-Stokes equations. Solution of the complete equations was out of the question, for they carried the complete physics of turbulence, with turbulent eddies that spanned a range of sizes covering several orders of magnitude. But during the 1970s, investigators made headway by dropping the terms within these equations that contained viscosity, thereby suppressing turbulence.[1]

People pursued numerical simulation because it offered hope of overcoming the limitations of wind tunnels. Such facilities usually tested small models, which failed to capture important details of the aerodynamics of full-scale aircraft. Other errors arose from tunnel walls and model supports. Hypersonic flight brought its own restrictions. No installation had the power to accommodate a large model, realistic in size, at the velocity and temperature of reentry.[2]

By piecing together results from specialized facilities, it was possible to gain some insights into flows at near-orbital speeds. The shuttle reentered at Mach 27. NASA-Langley had a pair of wind tunnels that used helium, which expands to very high flow velocities. These attained Mach 20, Mach 26, even Mach 50. However, their test models were only a few inches in size, and their flows were very cold and could not duplicate the high temperatures of atmosphere entry. Shock tunnels, which heated and compressed air using shock waves, gave true temperature up to Mach 17 while accommodating somewhat larger models. However, flow durations were measured in milliseconds.[3]

During the 1970s the largest commercially available mainframe computers included the Control Data 7600 and the IBM 370-195.[4] These sufficed to treat complete aircraft—but only at the lowest level of approximation, which used linearized equations and treated the airflow over an airplane as a small disturbance within a uniform free stream. The full Navier-Stokes equations contained sixty partial derivatives; the linearized approximation retained only three of these terms. It nevertheless gave good accuracy in computing lift, successfully treating such complex configurations as a shuttle orbiter mated to its 747. The next level of approximation restored the most important nonlinear terms and treated transonic and hypersonic flows, which were particularly difficult to simulate in wind tunnels. The inadequacies of wind-tunnel work had brought such errors as misprediction of the location of shock waves along the wings of the C-141, an Air Force transport. In flight test this plane tended to nose downward, and its design had to be modified at considerable expense.

Computers such as the 7600 could not treat complete aircraft in transonic flow, for the equations were more complex and the computation requirements more severe. They nevertheless made important contributions in the study of wings. HiMAT, a highly maneuverable NASA experimental aircraft, flew at Dryden and showed excess drag at Mach 0.9. Redesign of its wing used a transonic-flow computational code and approached the design point. The same program, used to reshape the wing of the Grumman Gulfstream, gave considerable increases in range and fuel economy while reducing the takeoff distance and landing speed.[5]

During the 1970s, NASA's most powerful computer was the Illiac IV, at Ames Research Center. It used parallel processing and had sixty-four processing units, achieving speeds up to twenty-five million operations per second. Built by Burroughs Corporation with support from the Pentagon's Advanced Research Projects Agency, this machine was one of a kind. It entered service at Ames in 1973 and soon showed that it could run flow-simulation codes an order of magnitude more rapidly than a 7600. Indeed, in its performance it foreshadowed the Cray-1, a true supercomputer that became commercially available only after 1976.

The Illiac IV was a research tool, not an instrument of mainstream shuttle development. It extended the reach of flow codes, treating three-dimensional inviscid problems while sup-

246

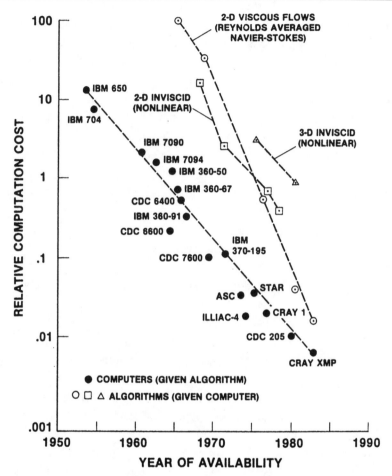

Fig. 51. *Development of capability in computational fluid dynamics. Improvements in algorithms matched those in computers. (NASA)*

porting simulations of viscous flows that used approximate equations to model the turbulence.[6] In the realm of space shuttle studies, Ames's Walter Reinhardt used it to run a three-dimensional inviscid code that included equations of atmospheric chemistry. Near peak entry heating, the shuttle would be surrounded by dissociated air that was chemically reacting and not in chemical equilibrium. Reinhardt's code treated the full-scale orbiter during entry and gave a fine example of the computational simulation of flows that were impossible to reproduce in ground facilities.[7]

Such exercises gave tantalizing hints of what would be done with computers of the next generation. Still, the shuttle program was at least a decade too early to use computational simulations both routinely and effectively. NASA therefore used its wind tunnels, and at times it turned to those of the Air Force as well. Hypersonic wind tunnels rely on hot air at high pressure, and in August 1975 the 3.5-foot facility at NASA-Ames suffered an explosion that blew off the roof of its building. It had been conducting space shuttle tests, and NASA moved these tests to the Air Force's Arnold Engineering Development Center while the Ames installation was under repair.[8]

The wind-tunnel program gave close attention to low-speed flight, which included approach and landing as well as separation from the 747 during the 1977 flight tests of Enterprise. As early as 1975, Rockwell built a $1 million model of the orbiter at 0.36 scale, lemon yellow in color and marked with the blue NASA logo. It went into the forty- by eighty-foot test section of Ames's largest tunnel, which was easily visible from the adjacent freeway. It gave parameters for the astronauts' flight simulators, which previously had used data from models at 3 percent scale. The big one had grooves in its surface that simulated the gaps between thermal-protection tiles, permitting assessment of the consequences of the resulting roughness of the skin. It calibrated and tested systems for making aerodynamic measurements during flight test and verified the design of the elevons and other flight-control surfaces as well as of their actuators.[9]

Other wind-tunnel work strongly influenced design changes that occurred early in development. The most important was the introduction of the lightweight delta wing late in 1972, which reduced the size of the SRBs and chopped a million pounds from the overall weight. Additional results changed the front of the ET from a cone to an ogive and moved the SRBs rearward, placing their nozzles farther from the orbiter. The modifications reduced drag, minimized aerodynamic interference on the orbiter, and increased stability by moving the aerodynamic center aft.

The activity disclosed and addressed problems that initially had not been known to exist. Because both the SSMEs and SRBs had nozzles that gimbaled, it was clear that they had enough power to provide control during ascent. Aerodynamic control would not be necessary, and managers believed that the orbiter could set its elevons in a single position through the entire flight to orbit. However, work in wind tunnels subsequently showed that aerodynamic forces during ascent would impose excessive loads on the wings. This required elevons to move while in powered flight to relieve these loads. Uncertainties in the wind-tunnel data then broadened this requirement to incorporate an active system that prevented overloading the elevon actuators. This system also helped the shuttle to fly a variety of ascent trajectories, which imposed different elevon loads from one flight to the next.[10]

A problem of similar importance arose in January 1980. Work at NASA-Langley disclosed that the orbiter had far less directional stability at Mach 6, during descent, than had been anticipated. This stemmed from its nose-high attitude and proved to occur from Mach 8 to 2.

Fig. 52. *0.36-scale model of the shuttle orbiter in a wind tunnel at NASA-Ames. (NASA, courtesy Dennis R. Jenkins)*

It appeared that at high angles of attack the vehicle produced vortices that interacted with the vertical tail to diminish stability, causing the shuttle to sideslip with potential loss of control.

An on-board flight control system was to steer the orbiter during its descent. To do this successfully, it needed accurate detail on the vehicle's responses to movement of its elevons and rudder. NASA responded by setting up a task force with people from Rockwell, Langley, and NASA-Johnson, a group that included aerodynamicists and specialists in flight control and flight simulation. Continuing wind-tunnel tests produced data that quickly went into flight simulators, which tested methods for regaining adequate margins of control. A change in the flight path during descent allowed Columbia to avoid the worst instability, maintaining control as it slowed through Mach 6.[11]

Much wind-tunnel work involved issues of separation: Enterprise from its carrier aircraft, SRBs from the ET following burnout. At NASA-Ames a fourteen-foot transonic tunnel

249

investigated problems of Enterprise and its 747. Using the same equipment, engineers addressed the separation of an orbiter from its ET. Ordinarily this was supposed to occur in near-vacuum, but it posed aerodynamic problems during an abort.

The SRBs brought their own topics. They had to separate cleanly; under no circumstances could a heavy steel casing strike a wing. Small solid rocket motors, mounted fore and aft on each booster, were to push them away safely. It then was important to understand the behavior of their exhaust plumes, for these small motors were to blast into onrushing airflow that could blow these plumes against the orbiter's sensitive tiles or the ET's delicate aluminum skin. Wind-tunnel studies helped to define appropriate angles of fire. These studies also showed that a short, sharp burst from the motors was best. Their design called for them to burn for 0.68 seconds with thrust of 21,000 pounds, as necessitated by the casings' heavy weight.[12]

Prior to the first orbital flight in 1981, the program racked up 46,000 wind-tunnel hours, consisting of 24,900 hours for the orbiter, 17,200 for the mated launch configuration, and 3,900 for the carrier aircraft program. During the nearly nine years from contract award to first flight, this was equivalent to operating a facility sixteen hours a day, six days a week. The program used 101 models: 45 for aerodynamics, 34 for heat transfer, and 22 for issues of structural dynamics. Investigations spanned the full range of flight regimes, with 38 percent of the orbiter test hours at subsonic speeds, 44 percent in the transonic realm, and 18 percent for hypersonic flight.

During the first year of the space shuttle program, into early 1973, wind tunnel activity emphasized configuration development and refinement of the design. A second phase of tests ran through 1975. Rockwell devoted the majority of its efforts to the Approach and Landing Tests and to the 747 as a carrier aircraft. Other work evaluated stability and control characteristics across the entire Mach range and studied control surface effectiveness and hinge moments. Related efforts addressed issues of vehicle aerodynamics that dealt with the needs of the flight control system, which was to execute an automated atmosphere entry and descent. A third phase began in early 1976 and continued through first flight, five years later. It worked to verify the predicted aerodynamic characteristics of the final shuttle design, from Mach 0.2 to Mach 15. It conducted tests along the flight trajectory that featured small increments in Mach number, angle of attack, angle of sideslip, and control surface position. Other work sought to minimize model-to-model and tunnel-to-tunnel discrepancies. This phase conducted 28 test series and used 14 wind tunnels.[13]

Structural tests complemented the activity in aerodynamics. The mathematics of structural analysis was well-developed, with computer programs called NASTRAN that dealt with strength under load while addressing issues of vibration, bending, and flexing. The equations of NASTRAN were linear and algebraic, which meant that in principle they were easy to solve. The main problem was that there were too many of them, for the most detailed mathematical model of the orbiter's structure had some fifty thousand degrees of freedom. Analysts introduced abridged versions that cut this number to one thousand and then relied on experimental tests for data that could be compared with the predictions of the computers.[14]

There were numerous modes of vibration, with frequencies that changed as the shuttle burned its propellants. Knowledge of these frequencies was essential, particularly in dealing with problems of "pogo." This involved a longitudinal oscillation like that of a pogo stick, with propellant flowing in periodic surges within its main feed line. Such surges arose when their frequency matched that of one of the structural modes, producing resonance. The consequent variations in propellant-flow rate then caused the engine thrust to oscillate at that same rate. This turned the engines into sledgehammers, striking the vehicle structure at its resonant frequency, and made the pogo stronger. It weakened only when consumption of propellant brought a further change in the structural frequency that broke the resonance, allowing the surges to die out.

Pogo was common; it had been present on earlier launch vehicles. It had brought vibrations with acceleration of nine g's in a Titan II, which was unacceptably severe. Engineering changes cut this below 0.25 g, which enabled this rocket to launch the manned Gemini spacecraft. Pogo reappeared in Apollo, during the flight of a test Saturn V in 1968. For the shuttle, the cure was relatively simple, calling for installation of a gas-filled accumulator within the main oxygen line. This damped the pogo oscillations. However, design of this accumulator called for close understanding of the pertinent frequencies.[15]

The shuttle structural test program began in 1970, at NASA-Langley. Initial work used a one-eighth-scale model. Within its ET water served in lieu of liquid oxygen, as these fluids had similar density. The liquid-hydrogen tank was filled with lightweight plastic pellets that had the same low density as this fuel. The model SRBs used standard solid propellant, with its oxidizer replaced by a salt to make the mixture inert. By varying the loadings of the "propellant," it was possible to simulate conditions at different moments during ascent. Analysis using NASTRAN disclosed as many as two hundred modes of vibration having frequencies below 20 hertz. Studies using the one-eighth-scale model showed that in general, predicted and measured frequencies agreed to within 10 percent. However, the relatively small size of this model meant that its construction was somewhat simplified, which raised questions as to the adequacy of the simulations.[16]

Work with Enterprise could have permitted study of this issue using a full-scale orbiter, but it would not be available until relatively late in the development program. The desire for better data therefore led to the ¼-Scale Model Program, initiated early in 1975 as a joint activity of NASA-Johnson and Rockwell International. Its model was scaled directly from working blueprints; this permitted accurate duplication of details of the primary load-bearing framework. The selected scale of one-fourth met size restrictions of available test facilities and was large enough to replicate structural joints. Fidelity in replicating these joints was essential in obtaining trustworthy data, but the one-eighth-scale model had not done this. The quarter-scale model captured these structural details far more effectively, using screws and rivets with diameters as small as 0.039 inches.

In the one-quarter model the "propellants" of the SRBs and the ET oxygen tank were the same as in the one-eighth-scale model. However, for the hydrogen tank the quarter-scale ET now used nothing more than air. With liquid hydrogen having only one-fourteenth the density of water, air proved adequate in place of the earlier plastic pellets. Test conditions simulated a shuttle at liftoff, at maximum aerodynamic pressure during ascent with the SRBs partly empty, at SRB burnout with these solid motors being completely empty, and during subsequent boost with the SRBs having fallen away. Other tests involved an empty SRB and ET, each considered as a separate element, as well as the orbiter by itself, both with and without a simulated forty-five-thousand-pound payload.

The work dealt with 281 vibrational modes. Half of these showed agreement of measured frequencies to within 5 percent of the calculated values; another one-fourth of these modes showed correlation within 10 percent. However, the final one-fourth of the modes were off by more than 10 percent, and a few were in error by 30 percent and more. This highlighted deficiencies in the mathematical models, which required further study. Still, the one-quarter-scale ET gave particularly good correlations, raising analysts' confidence in the accuracy of their computer model. This allowed a reduction in the scope of subsequent vibration testing of the full-sized ET.[17]

The most important tests used actual flight hardware: the orbiter Enterprise; STA-099, a full-sized structural test article that later became the Challenger; prototypes of the ET and its tanks; and SRBs that contained inert propellant. At the Palmdale plant, prior to its roll-out in September 1976, Enterprise was held horizontally and subjected to vibration tests. The results were not far removed from the mathematical predictions (table 7.1).

In 1978, following the Approach and Landing Tests, Enterprise went to NASA-Marshall, where the structural work included studies of the ET. For vibrational tests, engineers assembled a complete shuttle in flight configuration by mating Enterprise to an ET and a pair of dummy SRBs. Again, there was good agreement between theory and experiment (table 7.2). The tests showed good agreement for mode shapes and frequencies below 10 hertz. The correlations between test and analysis were of lesser quality between 10 and 20 hertz and deteriorated further at higher frequencies. However, the most important modes were below 10 hertz, which allowed investigators to conclude that their mathematical models were suitable for flight certification.[18]

One problem that these models addressed came at liftoff. The ignition of the three SSMEs imposes a sudden load of more than a million pounds of thrust. This force bends the SRBs, placing considerable stress at their forward attachment to the ET. If the SRBs were to

Table 7.1.   Vibration Data: Orbiter

Mode Description	Predicted Frequency (Hz)	Test Frequency (Hz)
Fuselage bending	5.09	5.97
Wing bending	7.86	7.31
Vertical fin bending	4.15	3.80
Vertical fin torsion	17.20	14.27

Source: Chaffee, *Conference,* 329.

ignite at that moment, their thrust would add to the stress. To reduce the force on the attachment, analysts took advantage of the fact that the SRBs would not only bend but would sway back and forth somewhat slowly, like an upright fishing rod. The strain on the attachment would increase and decrease with the sway, and it was possible to have the SRBs ignite at an instant of minimum load. This called for delaying their ignition by 2.7 seconds, which cut the total load by 25 percent. The SSMEs fired during this interval, which consumed propellant, cutting the payload by six hundred pounds. Still, this was acceptable.[19]

While Enterprise underwent vibration tests, STA-099 showed the orbiter's structural strength by standing up to applied forces. Like a newborn baby that lacks hair, this nascent form of Challenger had no thermal-protection tiles. Built of aluminum, it looked like a large fighter plane. For the structural tests, tiles were not only unnecessary; they were counterproductive. They had no structural strength of their own that had to be taken into account, and they would have received severe damage from the hydraulic jacks that applied the loads and forces.

STA-099 and Columbia had both been designed to accommodate a set of loads defined by a database designated 5.1. In 1978 there was a new database, 5.4, and STA-099 had to withstand its loads without acquiring strains or deformations that would render it unfit for flight. Yet in an important respect, this vehicle was untestable; it was not possible to validate the strength of its structural design merely by applying loads with those jacks. The shuttle structure had evolved under such strong emphasis on saving weight that it was necessary to take full account of thermal stresses, resulting from temperature differences across structural elements during reentry. No facility existed that could impose thermal stresses on so large an object as STA-099, for that would have required heating the entire vehicle.

STA-099 and Columbia had both been designed to withstand ultimate loads 140 percent greater than those of the 5.1 database. This safety factor of 140 percent was standard at NASA-Johnson; it had applied as well to the spacecraft of Mercury, Gemini, and Apollo. However, the structural tests on STA-099 now had to validate this safety factor for the new 5.4 data-

Table 7.2. Vibration Data: Complete Shuttle

Mode Description	Predicted Frequency (Hz)	Test Frequency (Hz)
SRB roll (symmetric)	1.90	2.05
SRB roll (antisymmetric)	1.88	2.08
Orbiter bending	2.97	3.24
Wing bending	6.70	6.43

Source: Chaffee, *Conference,* 333.

base. Unfortunately, a test to 140 percent of the 5.4 loads threatened to produce permanent deformations in the structure. This was unacceptable, for STA-099 was slated for refurbishment into Challenger. Moreover, because thermal stresses could not be reproduced over the entire vehicle, a test to 140 percent would sacrifice the prospect of building Challenger while still leaving questions as to whether an orbiter could meet the safety factor of 140 percent.

NASA managers shaped the tests accordingly. For the entire vehicle, they used the jacks to apply stresses only up to 120 percent of the 5.4 loads, rather than 140 percent. When the observed strains proved to match closely the values predicted by stress analysis, the 140 percent safety factor was deemed to be validated. In addition, the forward fuselage underwent the most severe aerodynamic heating; yet it was relatively small. It was subjected to a combination of thermal and mechanical loads, which simulated the complete reentry stress environment in at least this limited region. Additional investigations tested the ultimate strength and fatigue resistance of specified critical components. Finally, STA-099 was given a detailed and well-documented post-test inspection.

The work got under way in August 1978. A 430-ton steel framework surrounded the vehicle as jacks applied up to 256 loads, distributed over 836 application points. The activity involved thirty-nine test conditions, which simulated thirty-two critical design criteria. Instruments measured stiffness as well as strength. The thermal tests used electrically heated blankets to warm the forward fuselage, reproducing the temperatures of a few hundred degrees that would result from heating through the RSI. In a separate test, the fuselage was given loads that simulated the impact of the nose landing gear on a runway. Computed stresses proved to be conservative, for whereas there was good agreement between the measured and predicted longitudinal skin stresses, circumferential skin stresses were 30 percent less than calculated. Hardware modifications were minor. The forward reaction-control-system tank-support structure proved to lack lateral stiffness and failed in test but was redesigned and accepted. Following these tests, STA-099 was marked for refurbishment as the flight vehicle Challenger, to join Columbia as part of the growing fleet.[20]

Fig. 53. *STA-099 being readied for structural tests at Lockheed. (Rockwell International, courtesy Dennis R. Jenkins)*

Fuel Cells for Power

The orbiter's internal systems ran on electric power, which came from fuel cells. A fuel cell runs electrolysis in reverse, for while electrolysis passes an electric current through water to break its molecules into hydrogen and oxygen, fuel cells allow hydrogen and oxygen to combine so as to produce water plus electric current. The principle was discovered by Sir William Grove in 1839, but the first practical fuel cell was demonstrated only in 1959, its inventors being Francis Bacon and J. C. Frost of Cambridge University. Even then, however, fuel cells offered the promise of producing considerably more power than an equal weight of batteries. NASA sponsored early research and development and used them on both the Gemini and Apollo spacecraft.

A single fuel-cell unit has a maximum theoretical voltage of 1.23 volts, though internal losses within the shuttle's cells reduced this to 0.90 volts. High currents require large surface

Fig. 54. *STA-099 within its test rig. (Lockheed Martin, courtesy Dennis R. Jenkins)*

area of the electrodes, which therefore are porous; useful voltages are attained by stacking a number of units in series. The main components are an anode for the hydrogen, a cathode for the oxygen, and an electrolyte. The anode holds a catalyst, which dissociates hydrogen molecules into protons and electrons. The electrons produce the electric current, while the chemical reaction takes different forms depending upon the choice of electrolyte. It may be either acidic or alkaline; in either case, the net result is that ions flow through it, from one electrode to the other, and react to produce water. The electrons also flow between the electrodes, but not by a direct route. Their path takes them through the electric circuitry, to provide on-board power.[21]

Although the Apollo fuel-cell designs were frozen relatively early in that program, NASA's Manned Spacecraft Center pursued a succession of development programs that continued to advance the technology. Initial goals included longer life along with an easing of Apollo-era operational constraints, and early support went to Allis-Chalmers and General Electric. GE used an acidic electrolyte, in the form of a solid polymer. Allis built an alkaline fuel cell that held its electrolyte within a matrix of highly porous asbestos, the cancer-causing properties of which were not known at the time. The Allis concept was chosen for use in Apollo Applications, which developed into Skylab, and for the Air Force's Manned Orbiting Laboratory. The MOL was canceled in 1969, while Skylab fell prey to budget cuts and saved money by using existing Apollo fuel cells. These events led Allis to drop its work in this field. But GE remained active and was joined by the Power Systems Division of United Technologies Corporation, which was working on an alkaline design similar to the one that Allis had abandoned.

In 1970 the space shuttle was on the horizon. People appreciated that it would need far more power than Apollo, thus calling for further advances in fuel-cell technology. The work broadened to include storage vessels, with Beech Aircraft building test articles. Engineers at the Manned Spacecraft Center had Lockheed study an integrated system that could supply hydrogen and oxygen not only to fuel cells but also to a cryogenic OMS while providing additional oxygen for the flight crew to breathe. MSC also carried through a second round of demonstration projects, awarding contracts to GE and to UTC. General Electric further extended the life by humidifying the hydrogen gas, while UTC introduced an improved alkaline electrolyte matrix. The new fuel cells were to advance beyond those of Apollo by offering reusability. Both companies built engineering models as small stacks with four to six fuel-cell units, each of which went through a five-thousand-hour test program.

Work within NASA complemented that of the contractors. Designers had expected to use a platinum-palladium catalyst, but research at NASA-Lewis led to a gold catalyst, which became standard. It gave increased efficiency and brought a significant drop in the system weight, for two stacks of fuel-cell units, rather than three as originally required, could meet the needs of a shuttle mission. Another improvement came from MSC. It starved the cells of oxygen after shutdown by supplying an inert gas to the cathode, with hydrogen on the anode, while continuing to draw power from the system. This procedure prevented a small but unrecoverable voltage loss that accompanied existing shutdown methods. It reduced the rate of performance degradation and made possible accurate performance predictions.

General Electric and UTC completed their fuel-cell technology programs in 1973, and William Simon, a specialist in this field at NASA-Johnson, assesses what was learned:

Through these programs, major technological barriers were overcome and the fuel cell and cryogenic storage system was placed in the enviable position of being one of the few systems truly ready, from a technology standpoint, for the development program. The development effort was reduced to solving engineering problems, though not at all insignificant, for which no major scientific breakthroughs were required. This state of technological preparedness earned for the power generation and reactant storage system the distinction of being the only subsystem which reached the end of the originally defined Shuttle development program on schedule and within the allocated budget.[22]

United Technologies was selected to develop the flight fuel cells, which were to be alkaline, using a solution of potassium hydroxide as the electrolyte. The storage of reactants followed Apollo practice by holding the hydrogen and oxygen in tanks as supercritical gases. They were cooled to cryogenic temperatures and held at high pressure, but the temperatures were not low enough for liquefaction, thus avoiding problems of separating gas from liquid in zero gravity.[23]

The fuel-cell design called for two power-producing stacks, each with thirty-two units. An early prototype, designated DM-2A, had a single flight-configured stack along with a nonflight accessory section. It went under test at NASA-Johnson's Thermochemical Test Facility with the goal of providing insight into long-term performance and degradation. It ran successfully for five thousand hours at an average load of 4.5 kilowatts and with an average loss of less than 1 volt. Test engineers mapped its operating characteristics over a full range of flight conditions. They also evaluated power-up capabilities, responses to transient electrical loads, purge requirements, and instruments for diagnosing system health, along with procedures for start-up and shutdown.

Three engineering problems arose in the course of system development. The first brought difficulties in removal of water; it first showed up in 1975, when a hydrogen pump seized up during vibration testing. A failure investigation revealed that the impeller had rubbed against its housing; it also found contamination inside the pump. A redesigned pump successfully completed the vibration test as well as other tests, but the following year brought further problems. The pump repeatedly surged and stalled, either discharging sudden bursts of hydrogen or delivering little if any. This proved surprisingly hard to resolve. Engineers tried pump purging and new techniques for feeding in hydrogen, but it took three successive redesigns before these malfunctions went away. In time, however, a substantially modified pump successfully completed a two-thousand-hour test and qualified for flight.

The second important development problem involved plates that formed channels for the reactant gases. These were of magnesium, chosen for its light weight, and needed protection against corrosion due to water in the stack. The protection came from a thin layer of nickel, and in accordance with standard industrial practice, the magnesium was first dipped in a chemical solution to promote the adhesion of an intervening layer of copper, which bonded to the nickel. However, the nickel layers showed plating defects in the form of tiny blisters on the surface that exposed the underlying magnesium. Investigation showed that the standard plating process, though well established for other uses, was inadequate for this specific application. The solution called for better understanding of the sources of defects, which led to improvements in process control, paced by better inspection techniques. The problem did not go away, but these changes raised the proportion of good plates to acceptable levels.[24]

As with the APUs, prototypes of the fuel cells were available as early as 1976 and were installed aboard Enterprise and used during the Approach and Landing Tests. However, these flights' durations were brief, and the system needed further development to qualify for repeated use during long orbital missions. An important step toward that goal came during 1978, when a full flight-configured system operated at NASA-Johnson. It had two thirty-two-cell power-producing stacks and was essentially identical to the operational version. It arrived at Johnson after having run for some two thousand hours and rolled up another thousand hours during the tests, which included short-term high-power levels, cool-down after landing, and long periods of open circuit.

During 1979 and 1980 the third significant engineering problem came to the forefront. It arose within a qualification unit, which was to deliver power levels of between 2 and 12 kilowatts, according to a specified cycle, for 2,000 hours. Some 600 hours into the exercise, the system's voltage fell below the minimum requirement of 27.5 volts during a transition between those limiting power levels. Engineers continued the test to 2,061 hours, with the voltage during this power-up transient falling to 25 volts. Then they dismantled the fuel cell for analysis and study.

The anode, or hydrogen electrode, had shown worse voltage decay than the cathode at high power levels. The post-test tear-down showed white deposits of calcium silicate on the anode, which were judged to be a primary contributor to the loss of performance. The system still was using asbestos as the matrix that held the electrolyte, and studies of asbestos from other units showed that this material contained as much as several percent of calcium. Tests showed that it tended to make its way to the anode, where it prevented that electrode from working properly.

It took a year of intensive effort to address this, as people from NASA-Johnson, Rockwell, and United Technologies all contributed. Two new processes gave the solution. The first one treated the electrode surfaces so as to achieve greater uniformity, thus minimizing cell-to-cell performance variations. This procedure was already in hand, for it was in use within the NASA-Lewis fuel-cell effort, and had also been applied within a UTC fuel cell developed for the U.S. Navy. The second process leached the asbestos to remove calcium impurities, by reacting the calcium in the matrix with an acid and rinsing away the soluble reaction product.[25]

The final fuel-cell plant was more than forty pounds lighter than its Apollo counterpart, while delivering eight times the average power. The most advanced batteries of the day would have required about ten times as much weight; off-the-shelf batteries that were readily available would have weighed up to twenty-five times more. Table 7.3 compares the fuel cells and tanks of Apollo and the shuttle. The tank held more oxygen than could react with the available hydrogen. The extra oxygen was for the crew to breathe and hence was formally a part of the life-support system.[26]

Table 7.3. Fuel-Cell Characteristics, Apollo and Shuttle

	Apollo	Shuttle
Fuel-cell unit		
Output (kW)		
Minimum–maximum	0.6–1.4	2–12
Average	0.9	7
Voltage	27–31	27.5–32.5
In-flight restart?	No	Yes
Restarts allowed	None	50, no maintenance
		125 with maintenance
Powerplant life (hours)	400	2,000, no maintenance
		5,000 with maintenance
Weight (lbs.)	245	202
Specific weight (lb/kW)	270	29
Current density (A/ft^2)	90	230
Cost (millions of 1982 dollars)		
Development	151	22.6
Per production unit	3.0	2.2
Fuel-cell tankage		
Tank capacity (lbs.)		
Hydrogen	29	92
Oxygen	330	781
Tank weight (lbs.)		
Hydrogen	91	227
Oxygen	80	215
Cost (millions of 1982 dollars)		
Development	37.3	14.4
Production set (two tanks)	2.0	1.6
Tank reusability	None	100 missions

Source: Data from Chaffee, *Conference,* 716–17.

Environmental Control and Life Support

For astronauts, the availability of oxygen is certainly a matter of life and death. The analyst James Oberg, writing of the Soviet space station Salyut 4, notes that problems within such on-board systems can tax the heroism of spacefarers, even when these difficulties do not immediately threaten their lives.

Salyut 4 was a riposte to America's Skylab, which hosted three astronauts for eighty-four days. It reached orbit just after Christmas 1974, subsequently housing two cosmonauts who set a record for their country by remaining aboard for twenty-nine days. Two new crew members, Pyotr Klimuk and Vitaly Sevastyanov, arrived in May 1975. They reached an orbiting station whose on-board systems were beginning to deteriorate. Life aboard Salyut demanded control of the humidity, for moisture in a man's breath totals several pounds per day, which had to be removed from the air. This required nothing more than cold plates to condense the moisture, but on Salyut 4 they did not work properly.

After a month in space, the cosmonauts had plenty to complain about. "The windows are still fogged over," one of them remarked. "The green mold is halfway up the wall now. Can't we come home?" The answer was *nyet:* like soldiers in the Great Patriotic War, they were to stay at their post and do their duty for the motherland. The green mold continued to spread, while Klimuk and Sevastyanov continued to ask if they could come down. Finally, late in July they received the order they wanted and returned safely. They set a new Soviet record of sixty-three days in space, but it was clear that the systems of Salyut needed more work.[27]

Aboard the shuttle, the life-support system provided flight crews with an on-board atmosphere virtually identical to the air we breathe. It had sea-level pressure, 14.7 psi, and contained 20 percent oxygen and 80 percent nitrogen. Canisters of lithium hydroxide absorbed carbon dioxide from the astronauts' respiration. Oxygen, stored in tanks, thus passed into the cabin, through the crew members' lungs, and into the canisters. Water for the astronauts came from the fuel cells; a dehumidifier within the cabin removed excess moisture and pumped it into a holding tank. The orbiter lacked on-board shower facilities; to compensate in part for this, pumps circulated the air through filters, which removed bacteria and odors.[28]

Astronauts stored their food in stowage bins and prepared it by using a galley. The shuttle lacked an on-board refrigerator, but canned goods were in the cupboard. Many choices were dehydrated, promoting storability while cutting weight and volume. To reconstitute your dinner, you pushed a blunt-nosed needle through a diaphragm in the food package to inject water. Other foods—turkey, beefsteak, ham, chicken à la king, freeze-dried strawberries, cheese spread, peanut butter, chocolate brownies, cocoa powder—were irradiated with gamma rays from radioactive cobalt-60, which killed germs.

The galley, built by General Electric, resembled a soft-drink dispenser, being molded to fit the curving cabin wall. It offered hot and cold water, an oven with a forced-air heater, and storage space for trays, wet wipes, drink cups, and similar items. Scissors were standard table utensils, used to cut open the food packages, but crew members could eat the contents using conventional tableware.[29] The galley also included a zero-g wash station. This was a spherical container twelve inches across, made of transparent Lexan plastic. You pushed your hands and forearms through openings in the sides; a nozzle with selectable water temperature allowed for wetting and scrubbing. Near the bottom a vacuum port provided an outlet.

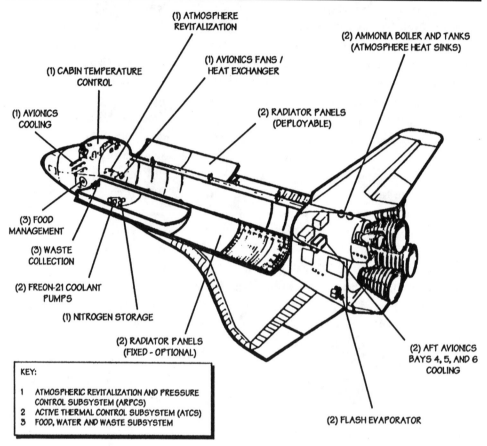

Fig. 55. *Environmental Control and Life Support System. (NASA, courtesy Dennis R. Jenkins)*

The menu provided up to twenty-nine entrées, ten vegetables, ten soups, nine fruits, five salads, and fourteen beverages—all nonalcoholic. *Design News* provides an example of *la cuisine de l'espace:*

> *Les Hors d'Oeuvres: La Salade au Saumon*
> *La Soupe: La Soupe aux Dindes*
> *Le Poisson: La Langouste au naturel*
> *L'entrée: Le Biftek Almondine*
> *Les Fruits: Les Apricots à la Maison*

All these items, including the lobster (langouste), were available from the on-board stores.[30]

The shuttle also had a Waste Collector System, or toilet. The zero-g commode of *2001: A Space Odyssey* came with a long list of instructions, but the shuttle's unit was much simpler. It resembled a standard toilet; you sat on it with a seat belt and a fixed foot support. A unisex urinal served both men and women. It used a funnel attached to a hose, with airflow carrying its liquid to the holding tank.

You defecated into a porous bag, with a current of air carrying the solid waste toward its bottom. The filled bag went into a storage area that was depressurized, allowing the waste to dry. The airflows came from the cabin air and picked up odors and germs galore, but again a filter of activated charcoal removed these contaminants. The filtered air then returned to the cabin, none the worse for its experience. Following a flight, the toilet came out as a unit and went back to its manufacturer, where technicians removed its contents and cleaned it for further use.[31]

To provide personal comfort, the shuttle's on-board systems called for a number of motors, fans, separators, instruments, and controls, many of which introduced problems of design and development. Apollo and Skylab had used analogous systems, but as with everything else aboard the shuttle, its environmental control and life support demanded very high reliability and long operational life. No one wanted to abort a mission because green mold was halfway up the walls.

The Atmospheric Revitalization Subsystem provided the crew with air to breathe. Its motors and pumps were critical. Their bearings had to operate for ten thousand hours, while withstanding vibrations that imposed strong oscillatory forces. These vibrations resulted from sudden flows of water or urine into centrifugal separators, which produced off-balance conditions. Resistance to corrosion was also of high importance.

Motors relied on ball bearings that were sealed and lubricated with grease. These gave operating lives far in excess of the design requirement. A cabin fan accumulated more than fifty-six thousand hours in operation; an avionics fan, used in cooling on-board electronics, topped twenty-eight thousand. However, pumps presented their own requirements. Engineers selected designs with immersed motors, which use the pumped liquid for its coolant. These saved money and weight, while providing good efficiency. Ball bearings now were undesirable, for they developed excessive friction when immersed. The solution lay in sleeve bearings, cylinders of carbon that fitted snugly around the shaft. These worked well whether wet or dry, allowing a water pump to run for forty-two thousand hours in ground test. This represented nearly five years of continuous operation.

Water separators were key elements of both the urinal and the dehumidifier. Their liquids arrived as drops or globules entrained within a flow of air, which came from an internal fan. The airflow, laden with liquid, entered a centrifugal separator, a chamber that rotated at 5,900 rpm and turned on the same shaft as the fan. This high speed compressed the air at the chamber periphery, generating pressure that pushed the accumulating liquid into a narrow tube, through which it flowed to the holding tank.[32]

The urinal and dehumidifier brought specialized issues. For the urinal, astronauts faced the problem of the "last drop," which could fill a tablespoon. It consisted of urine that dribbled from the urethra with a falloff in bladder pressure at the end of urination, to cling to the urethral opening in zero-g. Airflow within the urinal provided suction, and a port at the top of the collection funnel drew the last drop into the system. The user came away feeling clean and sanitary.[33] The dehumidifier faced a somewhat similar issue. It used a condenser, a chilled plate on which moisture could condense. The apparatus then needed means to collect the condensate and transfer it into the water separator. Wicks were undesirable; they were susceptible to clogging by collecting dust, and they needed prewetting when dry. Instead, the shuttle relied on a novel component called a slurper. This was a channel with small holes to let in the water, and with a hydrophilic or water-attracting coating on the surface surrounding the holes. The fan within the separator provided suction through the holes, and the slurper acted as a wick that would not clog or demand prewetting. It handled up to four pounds per hour of condensate.[34]

Water separation and collection was one function of the environmental control; other functions dealt directly with the cabin atmosphere. A pressure gauge was an instrument of considerable importance, and an aneroid barometer offered a point of departure. This barometer uses a sealed chamber with a flexible metal bellows that follows pressure changes by moving inward or outward. An aircraft altimeter measures pressure and presents it as an altitude, by arranging for the bellows to move needles on the instrument face. The shuttle introduced the issue of how to use this bellows to generate an electrical signal that could trigger caution and warning devices. The solution involved mounting a fine wire to the bellows and holding it in tension within a magnetic field. An oscillating current within the wire, tuned to its natural resonant frequency, caused it to vibrate. A drop in the cabin pressure brought a change in the wire's tension and hence in this frequency. Electronic circuitry detected this and retuned the oscillation to restore the resonance. It was easy to measure the new frequency, which gave both the pressure and its rate of falloff. The result was a versatile electronic instrument that had a rated life of twenty thousand hours and that avoided the gears and rotating parts of a conventional altimeter.

Oxygen and nitrogen flowed from tanks at pressures up to 3,300 psi, and valve-position indicators gave close readings that permitted precise control. The indicator placed two high-strength samarium-cobalt magnets on a valve stem. These produced a measurable voltage difference across a current-carrying metal strip, because of the electromagnetic Hall effect. A transducer or sensor used this voltage change to determine valve position to a precision of 0.003 inches. This was three times the accuracy of mechanical indicators, which allowed lower flow settings. The new indicator used solid-state electronics and did not require penetration of the valve pressure wall. In addition, the absence of moving parts eliminated failures due to wear.

Flow sensors also were electronic, while measuring flow rates for oxygen and nitrogen from 0 to 5 pounds per hour. The sensor allowed most of the gas to pass through a relatively large pipe. This channel held several thicknesses of woven stainless-steel wire mesh that created a pressure drop. A fine tube pierced this filter; a small portion of the main flow went through this tube and followed the pressure change. The sensor measured flow rate by arranging for the flow through the tube to cool two coils of wire, thus changing their electrical resistance. These coils wrapped around the tube and were electrically heated; each of them served as part of a balanced electrical circuit. When the gas stream cooled the tube, it changed the balance, which permitted measurement of the flow. The system triggered warning signals if the flow rate dropped below 4.9 pounds per hour, allowing astronauts to make good use of their valves.

In addition to sensing and controlling the total pressure in the cabin, a separate instrument determined the partial pressure or component of pressure due to oxygen. This guarded against the possibility that crew members might asphyxiate in an atmosphere of steady and normal pressure that was almost entirely nitrogen. This oxygen sensor carried over from earlier programs, having accumulated thousands of hours of flight time during the Skylab and Apollo-Soyuz missions. It provided the control signal to maintain proper oxygen levels, and again it did this electronically. The sensor used an electrochemical cell somewhat resembling a fuel cell. It had a pair of electrodes immersed within an alkaline electrolyte, with a gas-permeable membrane. Oxygen in the cabin air penetrated the membrane and reached the sensing electrode, with a catalyst that ionized the oxygen molecules. These flowed through the electrolyte to the second electrode, which was of copper and formed an oxide. The oxidation released a flow of electrons, which provided the output signal. If the oxygen partial pressure fell off, the electron current diminished as well, allowing the system to add more oxygen under automatic control.

The copper electrode was consumable and was subject to depletion, but the rated operating life topped six thousand hours, even when the cabin air was mildly oxygen-rich. An initial calibration could last through the months of operation. Background gases did not affect the sensor, for only oxygen produced the flow of electrons. The sensor also responded quickly and accurately, tracking a sudden change in oxygen partial pressure in thirty seconds or less, to a precision of better than 90 percent.[35]

Oxygen meant fire hazard, and the orbiter had extinguishers with Halon gas in each of three avionics bays. It also had smoke detectors, which detected the fine particulates released by a flame. An initial design collected such particles on an oscillating quartz crystal, which vibrated at a precise resonant frequency. Though the absorbed smoke had minimal mass, it could change the mass of the crystal and bring a shift in the frequency, allowing it to sound an alarm. However, dust in the air collected as well, and the crystal lost its usefulness within a relatively short time. The selected alternative used a small quantity of a radioactive element, americium-241, with a half-life of 458 years. It emitted ionizing radiation that turned air molecules into charged particles, which formed an ion current within a chamber. Smoke particles absorbed some of the ions, reducing this current, with the falloff in current being proportional to the concentration of smoke. When it fell below a programmed threshold, the unit triggered a fire alarm, allowing the Halon extinguishers to flash into action.

In addition to providing oxygen and safeguarding against its dangers, the life-support system obtained potable water from the fuel cells. These cells operated at 150 degrees Fahrenheit and 60 psi, and the water that formed at the anodes was full of dissolved hydrogen. This gas came out of solution at cabin pressure, 15 psi, and it had to be removed. The system did this by using another Apollo technique. Water from the cells, flowing at twenty-three pounds per hour, passed through tubes of palladium-silver with interiors that had been treated with a catalyst. The tubes' exterior was exposed to the vacuum of space. The alloy was permeable to hydrogen, which diffused through it while the water remained behind. The final design removed the free gas and left only small quantities of dissolved hydrogen, which was acceptable. A separator unit weighed little more than three pounds and was no larger than a loaf of bread; yet it played an essential role in providing water that was fit to drink.[36]

The environmental control incorporated an Active Thermal Control Subsystem. It featured an array of coolant lines, with individual loops serving particular items of equipment. Heat exchangers transferred the heat to two main Freon ducts, which carried it to a succession of heat sinks. These rejected the heat, with specific sinks disposing of it prior to liftoff, in space, and during reentry and landing.

The cabin atmosphere used water coolant to control the cabin's internal temperature in the fashion of air conditioning. Water also cooled the condenser plates of the dehumidifiers, while some electronic equipment was mounted on cold plates that were chilled with Freon. The fuel cells combined hydrogen with oxygen in a reaction that was tantamount to combustion; it produced heat, which demanded specialized loops. Payloads within the cargo bay had two separate ducts that respectively used Freon and water. Hydraulic systems also used Freon, to control the temperature of the hydraulic fluid. Dual Freon pipes, operating simultaneously to provide redundancy, transported heat from its sources to its sinks. The sources—cabin, payload, fuel cells, hydraulics—delivered their heat to the Freon using heat exchangers. Four different heat-sink devices then were necessary to provide heat rejection during all mission phases.

On the launch pad and during the countdown, the orbiter relied on ground-support equipment. A heat exchanger, permanently installed within the vehicle, plugged into a refrigeration system on a ground cart. This cart pumped chilled Freon into the shuttle using coolant lines with quick disconnects. The on-board heat exchanger was dead weight, for it had to fly to orbit and yet was used only on the ground. To cut this weight to a minimum, the servicing equipment provided a high rate of coolant flow and a particularly cold temperature.

266

During the powered flight to orbit, the vehicle initially used no heat rejection, relying on its slow rate of temperature rise to prevent damage due to retained heat. Water evaporation provided the next heat sink. The Freon in the central loops had to stay below 40 degrees Fahrenheit, and the evaporator was to operate at 35 degrees. It did this by relying on "flash evaporation," at pressures of 0.1 psi or less. At such low pressures, water boils just above the freezing point. The shuttle reached altitudes with correspondingly low air pressure some two and a half minutes into a flight and then was free to employ this technique.

The fuel cells produced more water than the crew could use, and the excess was available for additional flash-evaporation cooling during the mission. However, this was merely a supplement. The main on-orbit heat sink was a set of radiators mounted to the inside of the payload bay doors. Those doors opened soon after reaching orbit and remained open until late in a mission. When they closed, prior to reentry, the shuttle turned again to flash evaporation.

Following reentry, during the last ten minutes of flight and the first few minutes after landing, the orbiter used a fourth heat sink: a boiler that evaporated ammonia, NH_3, and vented it overboard. Water was unsuitable; below 140,000 feet it boiled at too high a temperature. Ammonia was toxic, but it boiled at −28 degrees Fahrenheit at sea level, giving an excellent heat sink. It was an effective evaporant; when it vaporized, it absorbed nearly 600 BTU per pound. This limited the required weight of the ammonia supply, which had to ride with the shuttle to orbit and back. The ammonia boiler served as the heat sink until a ground crew could bring up a refrigeration cart, to off-load the task of cooling from the shuttle itself.[37]

For on-orbit cooling, the payload bay doors came from Rockwell's Tulsa, Oklahoma, division and were two of the few structural elements that were not made of aluminum. They had initially been slated to use this metal, but the need to reduce weight brought a 1974 decision to fabricate them of graphite-epoxy composite. At the time, they were the largest structures made of this material. It brought a weight savings of nine hundred pounds, or 23 percent. The doors were built to withstand acoustic loads of 163 dB during launch, on-orbit temperatures from −170 to +220 degrees Fahrenheit, and reentry temperatures of 350 degrees, as limited through use of thermal protection. They were designed for a hundred flights and a ten-year service life.[38]

The radiators received Freon from the main coolant loops, which followed a circuit that took them through heat exchangers having successively higher operational temperatures. The hydraulics exchanger was the hottest, at 160 to 180 degrees Fahrenheit. The radiators disposed of heat at rates up to 100,000 BTUs per hour, cooling the Freon to 40 degrees. Vought Systems Division was the contractor, assembling the radiator panels by attaching tubes to the face sheets of aluminum honeycomb panels. Honeycomb formed the core, like the meat in a sandwich, with these sheets on either side resembling bread. Freon in the tubes warmed the sheets, which rejected the heat as thermal radiation.

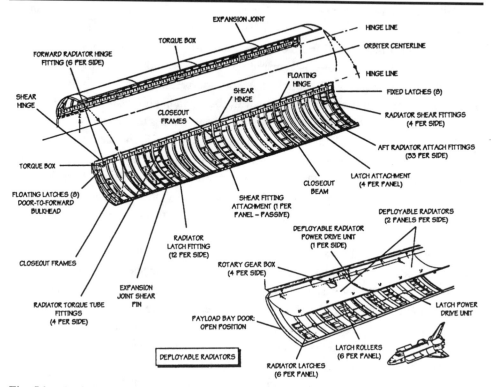

Fig. 56. *Payload bay doors. (NASA, courtesy Dennis R. Jenkins)*

Each of the two main loops served its own door, with each door mounting up to four radiator panels. The two forward units had hinges and could swing outward to radiate heat from both sides; their coolant tubes therefore were set more closely together. Each door also had a permanently installed third panel, fixed to its inner surface and radiating from one side only. A fourth panel could be added as an option, if a mission was to produce more heat.[39]

Flash evaporation, which supplemented the on-orbit radiators, was an innovation in the use of water as a heat sink. Water outperformed ammonia, absorbing more than a thousand BTUs with each pound that boiled; indeed, it was the best evaporant available. In addition, the flash evaporator gave a useful method for expelling excess water when the on-board storage tanks filled to overflowing, as they took in this excess from the fuel cells. The evaporator converted this water to steam and allowed it to flow through a nozzle at high velocity, to stream away from the vehicle. This prevented the formation of ice crystals that could contaminate the nearby environment and that formed copiously if the water was simply pumped into the vacuum of space.

Previous spacecraft had also used water evaporators for heat rejection. Mercury, Gemini, and the Apollo command module used wick-fed boilers. The Saturn V and the Apollo lunar module and space suit relied on porous plates that accumulated ice and allowed it to sublimate, passing directly from solid to vapor in vacuum. However, these devices lacked long life and had limited response and heat load range. The flash evaporator thus evolved as a new and advantageous approach. It sprayed water on the walls of a chamber that was heated by the Freon coolant, the chamber being maintained at low but not zero pressure. This allowed the water to evaporate at the desired temperature of 35 degrees Fahrenheit. Though simple in concept, it required close control. The water entered in pulses, and it was essential that each such discharge evaporate completely. The pressure in the chamber dropped as each spurt of vapor tailed off in its flow through the nozzle. If this pressure fell sufficiently, water left in the chamber could freeze. It then would form an insulating layer that resisted sublimation. The next water spray would not evaporate but would freeze as well, thickening the ice. Soon the evaporator would be completely clogged.

Investigators at Rice University, only a few miles from NASA-Johnson, carried out basic research on the physics of the flash-evaporation process, and engineers at Johnson tested a flash evaporator within a large evacuated facility. Using this Space Environmental Simulation Laboratory, NASA managers arranged for the evaporator to operate under flight conditions. They intentionally flooded the system's chamber and allowed it to freeze. Then a backup control allowed the Freon temperature to rise from 40 to 62 degrees, within the evaporator. This initiated sublimation of the ice, which increased the internal pressure. The ice thawed rapidly, restoring the evaporator for further use. This showed how a flight crew could deal with a freeze-up while in orbit, if it were to occur.[40]

Auxiliaries: Landing Gear, Remote Manipulator, Space Suits

During a mission, the shuttle made routine use of its fuel cells, life support, and environmental control. It also carried equipment that saw use only at specialized moments. The wings and engines were prime examples; another was the thermal protection. Others included landing gear, astronauts' pressure suits, and the Remote Manipulator System, a long mechanical arm that deployed and retrieved payloads.

The X-15 used a wheeled nose gear, but the shuttle was the first space vehicle to mount aircraft-type wheels and support struts for the main gear. B. F. Goodrich, the subcontractor that built the tires and brakes, worked vigorously to reduce their weight. Landing speed was an important design issue, for the shuttle was to touch down at 245 to 260 miles per hour, some 100 miles per hour faster than a commercial airliner. An airliner's tires typically endure some one hundred landings before being changed. For the SR-71, which touched down at up to 250 miles per hour, tires held up to fewer than ten landings. The shuttle's tires were initially rated for six such reuses.[41] These tires were smaller and less numerous than those of other aircraft. This subjected them to overloading. An airliner such as the Lockheed L-1011, which imposed a similar load, had tires of 52-inch diameter. Those of the shuttle's main gear were 44.5 inches. The SR-71, weighing 172,000 pounds at takeoff, had main gear with three-wheel trucks. The orbiter was heavier, but its trucks had only two wheels.[42]

The shuttle's tires held 315 psi of pressure in the main gear and 300 in the nosewheels. Conventional aircraft tires had a normal pressure loss of two to three percent per day, but for the shuttle, this was totally unacceptable. From the time the orbiter was mated to the ET until the moment of touchdown, a duration lasting up to several months, there was no possibility of servicing or reflating the tires. They therefore used a heavy inner liner that maximized air retention, cutting the leakage to less than 0.1 percent per day. Prior to flight, technicians measured the leak rate of each tire and gave them overinflation based on the anticipated time for launch-pad check-out and for the mission.

Aircraft during development verify their tire designs with taxi tests, including aborted takeoffs, and with landings. The shuttle followed this procedure, at least to a degree, by using prototype tires during the Approach and Landing Tests. The main gear had dimensions of 44.5 by 16 inches with twenty-eight plies, a measure of strength. Following these tests, the size remained unchanged but the design was upgraded to thirty-four plies. This would handle the heavy load of the most demanding situation: a 240,000-pound orbiter, still with its payload, landing after an abort with return to the launch site.

These tires were not steel-belted radials; they needed more freedom to flex. They obtained it from a bias-belted design that used nylon cord. To test them, B. F. Goodrich turned to Wright-Patterson Air Force Base, which had a suitable dynamometer. It caused tires to roll on a rotating cylinder, ten feet in diameter. The cylinder's surface was smoother than the concrete of a runway, but such tests were harder on tires than actual landings. There was increased stress from additional bending that occurred as the tire tread conformed to the curvature of the dynamometer. The tire's inflation pressure also had to be increased, to reproduce the peak sidewall deflection.

The Air Force facility was more diverse in its capabilities and more flexible than any other tire-testing center in the nation, including that of Goodrich itself. It was automated and had computer control; an operator could type in the specifications for a new test before the tire had time to cool down. A hydraulic system duplicated high-speed landings by stroking the tire against the rotating cylinder, raising the load from zero to 150,000 pounds in 0.1 second. The shuttle's tires also conducted simulated touchdowns in crosswind, which required the vehicle and its landing gear to yaw.[43]

The landing-gear struts functioned as shock absorbers, following standard practice. These struts cushioned the vertical sink rate, with a maximum allowable rate of 9.6 ft/sec.

The orbiter touched down with its nose high; when it rotated forward and brought down the nose wheels, the front strut absorbed that sudden load in turn. The struts contained hydraulic fluid along with nitrogen gas, which was separated from the fluid by a diaphragm. This prevented the gas from dispersing into the fluid in zero-g and assured proper shock strut performance.

The wheels also had brakes, which also came from B. F. Goodrich. Each wheel had its own disk brake, with beryllium rotors to absorb the frictional heat and carbon composite linings to provide the friction. This use of beryllium had recently become standard, for brake heat sinks of this metal had seen service on the Pentagon's C-5A and F-14 aircraft. Carbon composite was new but useful, offering great wear resistance, light weight, high strength, and good temperature resistance.

The brakes had limited life, paralleling that of the tires. Dynamometer tests showed that the brakes could meet the specification of five normal stops, with the brake assembly absorbing 36.5 million foot-pounds of energy, along with one emergency stop at 55.5 million, following an abort with return to launch site. The brakes obtained certification as early as August 1977, nearly four years before STS-1. Even so, that mission, along with many that followed, avoided the fifteen-thousand-foot strip at Kennedy Space Center, landing instead at Edwards Air Force Base, where the runways were longer still. It took flight experience, followed by a major redesign and upgrade of the brakes, before they had enough stopping power to permit the shuttle to land routinely at the Cape.[44]

A second auxiliary system, the Remote Manipulator, grew out of an international venture with Canada. As early as 1970 NASA and the State Department invited that nation, along with Europe, Japan, and Australia, to participate in the shuttle program. In Europe, this initiative led to Spacelab; in Canada it brought the Manipulator. By deliberate choice, neither project held the urgency of the main shuttle program. The shuttle could still conduct an active flight schedule if Spacelab encountered delays, while the robot arm remained a paper program until 1975. Moreover, the shuttle could fly initially without it. Still, such manipulators were part of the plan from its outset; artwork of 1972 shows an orbiter in space with two such arms. Informal contacts with Canada dated to 1973, when NASA's Office of Manned Space Flight sent a delegation from Rockwell to spend a week in that country and to learn whether it could take on the task of development. The Rockwell group determined that Canadian industry indeed was ready.

A Canadian governmental interagency review concluded that the National Research Council of Canada would be the responsible organization, with that country's National Aeronautical Establishment (NAE) managing the development. In June 1974 Frank Thurston, director of the NAE, visited NASA Headquarters and proposed formally that Canada would develop the Manipulator System with its own funds, provided that NASA would purchase and pay for the production units. An initial estimate put the cost of development at $30 million.[45]

271

Arnold Frutkin, NASA's assistant administrator for International Affairs, wrote that "the Canadian concept is to parallel the Spacelab agreement with ESRO almost exactly." That agreement set pro-rata shares for the contributions from each of the European member nations, based on their gross national products. With Canada's GNP being some 9 percent as large as the combined total of the ESRO states, it was prepared to spend as much as $35 million, which provided financial leeway.

Frutkin recommended that NASA accept Thurston's offer. "We could reduce our future Shuttle funding requirements by upwards of $30 million," he noted. He also saw political benefits that reinforced those of the Spacelab collaboration: "If we can combine European and Canadian inputs to the STS of very nearly half a billion dollars in dollar equivalent, we will have done a great deal to consolidate the stabilization of the program, both with respect to the Administration and the Congress."[46]

Leroy Day, NASA's deputy director of the shuttle program, took the role of Thurston's counterpart. He led a group that visited Canada in the fall of 1974, traveling to industrial firms and receiving briefings. This NASA inquiry built on the earlier evaluation of capabilities performed by Rockwell. Day then recommended to Fletcher that NASA should negotiate a formal memorandum of understanding (MOU) with the Canadians, a document that would define the two nations' commitments and responsibilities.[47]

The shuttle's manipulator concept took shape within a broader field of remotely controlled manipulators, which were useful in nuclear power plants and underwater applications. The firm of General Electric had a strong interest in this technology, and a company official, Otto Klima, argued that Canada was placing itself in a position to build similar devices for nonspace purposes. He declared that this would grow into a billion-dollar business and that NASA should not help Canada emerge as a competitor. Lobbyists from GE presented this viewpoint on Capitol Hill and won support from senators that included Hugh Scott, the Republicans' leader.

NASA's George Low noted "strong Congressional pressure . . . to keep us from giving this business to the Canadians." Fletcher looked into the matter; in Low's words, "they had finally reached the conclusion that Klima was wrong, that the investment that NASA was making was not so large as to put the US out of business and the Canadians into business, and we should go ahead with the Canadians. Gus Weiss for the WH agreed with this position." Fletcher met with Senator Lawton Chiles, an influential Democrat who agreed with GE, with Low noting that "if we lost Chiles over this issue, we would lose much more than the $30 million." However, Fletcher's position prevailed.[48]

Leroy Day led the team of NASA negotiators that drafted the MOU, with this group having members from NASA-Johnson and from the Legal and International Affairs Offices of the Washington headquarters. They hammered out most of the MOU's provisions in Ottawa during April 1975, adding further details during subsequent weeks. Representatives of both nations signed the document in July. The development contract came to $75.8 million, paid in full by Canada, with an additional $66 million in U.S. dollars for purchase by NASA of production units. The Canadian contribution reflected a more accurate assessment of project requirements and was more than twice the initial offer. Ottawa nevertheless accepted this expense, anticipating that it would boost that nation's position in aerospace and advanced technology. NASA also gave a sweetener, assuring Canada of preferred access to the shuttle for its payload sponsors.

In Toronto, Spar Aerospace Products took a sole-source contract from the National Research Council, based on previous experience with similar manipulators. Another Toronto company, Dilworth, Secord, Meagher and Associates, devised the "end effector," the robotic grasping device that would attach to the end of the arm, to hold and release payloads. In Montreal, CAE Electronics built controls and displays within the shuttle orbiter, while RCA handled the project's electronic interfaces with the rest of the shuttle. Not all the elements of the system displayed the maple leaf; the Canadians did not hesitate to turn to the Yankees. General Dynamics in San Diego, which had experience with graphite-epoxy, used this composite to fabricate the arm booms, though Spar provided the specifications.[49]

The final design amounted to a human arm that was fifty feet long. A shoulder joint mounted it to the left upper wall of the payload bay, moving the entire arm in pitch and yaw but not in roll. An elbow joint, at the end of the sixteen-foot upper arm, flexed the lower arm about a single axis, like your own elbow. A set of wrist joints moved the end effector in pitch, yaw, and roll, giving it considerably more freedom of motion than a person's hand. Closed-circuit television cameras, at the elbow and wrist, permitted close monitoring of the arm's activity. Within the orbiter, an aft mission station stood at the rear of the flight deck. A display and control panel provided information to the mission specialist, who served as the operator. This crew member controlled the arm by using two hand controllers, or joysticks. One extended the arm and moved its end in any desired direction. The second operated the end effector, moving the wrist in all three axes.[50]

The arm was built for use in zero-g, which raised questions as to how to test it on the ground. Spar Aerospace used an air bearing floor that allowed the arm to move freely and without friction. "These tests will give us some idea of how the arm works in each of the six degrees of freedom," G. M. Lindberg, the project manager at the National Research Council, told *Aviation Week* in 1978. "While the tests will confirm that the system can perform, they won't give us a totally definitive evaluation. We still will rely on mathematical models for a large part of system evaluation."[51]

Development proceeded smoothly. A NASA internal memo noted that "all issues have been resolved at the Day-Thurston level; no issues have gone to the Administrator." In particular, early in 1979 these officials agreed on terms for production, which led to a separate contract between NASA and Canada. The Remote Manipulator did not fly on STS-1, but it

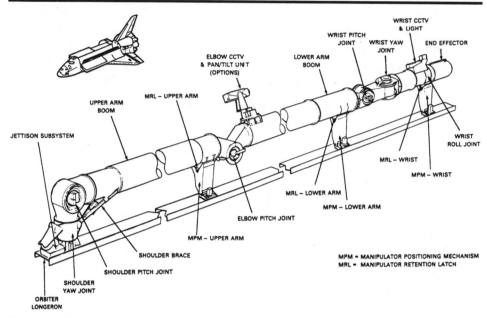

Fig. 57. Remote Manipulator System. *(Spar Aerospace, Ltd.)*

flew in November 1981 on STS-2, the second flight of Columbia. It went on to become a standard element of on-orbit shuttle operations.[52]

This arm provided one means whereby astronauts could literally extend their reach. A mobile and self-contained space suit was another, for it allowed crew members to address particular tasks using their own eyes and hands. The two systems complemented each other. An astronaut at the end of the arm was the most versatile of end effectors, while this Manipulator offered a safe method to hold him securely and to place him in appropriate locations while wearing his space suit. NASA had a long involvement with such suits. Astronauts wore them aboard the X-15, Mercury, Gemini, and Apollo. During Gemini missions they repeatedly left their spacecraft to perform "extravehicular activity" (spacewalks) and to learn how to work in zero-g. Pressure suits also were essential elements of Apollo and Skylab. When Neil Armstrong took his one small step, he did it while wearing a space suit.[53]

One would not compare such a garment to a skin diver's scuba gear and wet suit. It was a spacecraft in its own right, with a life-support system. The shuttle's cabin provided a mix of oxygen and nitrogen at sea-level pressure. But the space suits had to flex and bend in the face of forces produced by their internal pressure. This dictated low pressure, 4.3 psi. To prolong astronauts' stay times in vacuum, they provided oxygen only. Two tanks within the backpack stored this gas, at 1,000 psi; there was enough for seven hours. As exhaled breath

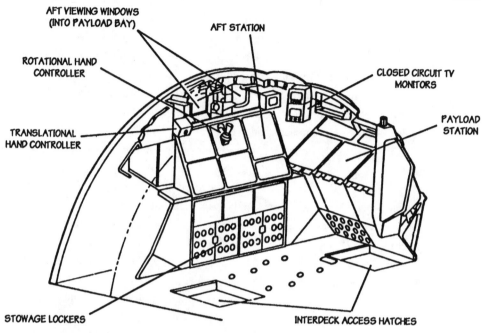

Fig. 58. *Orbiter aft mission station. (NASA, courtesy Dennis R. Jenkins)*

contains a considerable amount of unmetabolized oxygen that can be inhaled anew, the suit used recirculating ventilation. Lithium hydroxide removed carbon dioxide from the exhalate; activated carbon absorbed odors. As on the shuttle, the space suit oxygen supply replenished only the portion of this gas that was irretrievably converted to CO_2 and absorbed.

The shuttle's suit also incorporated active cooling while controlling the humidity of the recirculating oxygen, using methods that closely resembled those of the orbiter's life support system. Within the suit, an astronaut wore a flexible garment resembling a set of long johns, fitted with tubes for cooling water. A water sublimator and heat exchanger, similar to those of the Apollo suit, cooled the water as well as the oxygen. To remove moisture from the oxygen, the stream of exhalate was cooled in the sublimator, producing condensate that accumulated on surfaces treated with a hydrophilic coating. A slurper drew the water into a rotating unit that spun at 19,000 rpm, duplicating the centrifugal separator of the shuttle's environmental control. As on the orbiter, a tube touched the surface of the separated water and drew it off, allowing it to flow either into a storage tank or directly to the sublimator. By using metabolic water, the space suit reduced its need for stored water, cutting the size and weight of the backpack.

In addition to removing moisture from the oxygen, the suit also addressed a longstanding issue by extracting bubbles of air from the cooling water. These bubbles arose because the liquid cooling garment was only partly filled with water and because dissolved gases came out of solution during depressurization. The suit removed these bubbles by allowing the water to flow through a fine-mesh screen, which held back this air, while permitting the water to pass through freely. The air, mixed with a small flow of water, went into the rotating water separator. When it came out, it mixed with the oxygen.

The suit also carried self-contained caution and warning. This improved on that of Apollo, which had relied on ground controllers who monitored conditions using telemetry. The shuttle's suit had a microprocessor along with five kilobits of memory, placing all information directly with the user. A twelve-character display used light-emitting diodes. It presented a list of check-out procedures, tracked the supply of oxygen and water, and noted out-of-limit conditions and corrective actions. A silver oxide-zinc battery powered this system, while also providing power for the fluid circulation equipment and for the suit's radio.

Each Apollo suit had been custom-tailored to fit an individual astronaut. The shuttle's was built to serve a wide range of sizes, for both genders. The range reached from the fifth percentile for women to the ninety-fifth percentile for men, including all people except the tiniest and largest. The suit thus fitted women barely five feet in height, slender and petite, as well as men standing well over six feet tall with broad shoulders and robust chests. To do this, designers built it from components: helmet, a hard upper torso that carried the backpack, upper and lower arms, gloves, waist, a leg section that resembled a pair of pants, and boots. Each component came in a range of sizes, which could be assembled to fit a particular person. NASA thus exchanged custom tailoring for off-the-rack ensembles in offering its lineup of fashions.

The shuttle suit was much easier to put on and to check out than its Apollo counterpart. That one took half an hour to don and as much as an hour to check, often with help from a second crew member. For the shuttle suit, these activities took fifteen minutes or less and involved only the user. This quick responsiveness allowed engineers to dispense with an Apollo requirement whereby that program's suit had to be compatible with both the spacecraft life-support system and with its own backpack. This had compromised the design by requiring a detachable backpack, with multiple hoses, electric cables, and connections for oxygen, cooling water, and emergency pressurization. But the shuttle's suit was a completely autonomous unit, independent of the orbiter. This permitted a convenient layout that mounted the backpack permanently to the upper torso, integrating the suit with its life support, controls, and displays. If the shuttle were to lose cabin pressure, astronauts expected to save their lives by donning their suits in timely fashion, with this integration of the systems contributing to the suits' ease of use.

In the movie *2001: A Space Odyssey,* the HAL 9000 computer went berserk. It caused a robot to turn on the astronaut Frank Poole, played by the actor Gary Lockwood, cutting his air hose and hurling him into space to die. That fictional space suit followed Apollo's, which used exposed hoses that indeed could have ruptured, leading to an astronaut's death. The shuttle's suit eliminated this hazard by using internal channels and ducts. It also provided more upper-body mobility than the Apollo suit, as it used gimbaled constant-volume joints in the shoulders, wrists, and waist. These avoided taxing the user's strength, for they did not compress the internal atmosphere.[54]

Yet in a very real sense, though an improvement over the space suits of Apollo, the shuttle suit stood in that program's shadow. Like the spacecraft of Mercury and Gemini, Apollo craft used a pure oxygen atmosphere when in space, at 5 psi of pressure. This simplified the design of the life-support system, which avoided the complexities of using a two-gas system, with nitrogen as well as oxygen, to simulate air at normal pressure. In turn, the reduced oxygen pressure of Apollo cut the weight of its spacecraft because their structures were built to hold less internal pressure. But Apollo's single-gas oxygen system led to a disastrous fire in a spacecraft in 1967, which killed the astronauts Gus Grissom, Edward White, and Roger Chaffee.

That fire halted the program in its tracks, forcing a major redesign. Engineers meticulously substituted fireproof materials and introduced a two-gas system for Apollo ground tests.[55] After that, there was no possibility that the shuttle would use a pure oxygen system, and in this respect its cabin atmosphere was a reflection of its time. The shuttle introduced its own extensive fireproofing while using a two-gas cabin atmosphere, closely resembling the air we breathe, to eliminate the fire hazard of pure oxygen. This belt-and-suspenders approach ensured that whatever harm might strike the shuttle, at least it would not suffer another Apollo fire.[56] But this cabin atmosphere brought significant difficulty in using the shuttle's space suit.

In using pure oxygen at reduced pressure, that suit followed the practice of Gemini and Apollo. In those programs, with the suit atmosphere matching that of the cabin in both pressure and composition, an astronaut could step easily into space, without delay. But the shuttle's cabin atmosphere brought a considerable disadvantage in using the suit, for this air held substantial amounts of nitrogen at sea-level pressure. The suit, again, provided pure oxygen at a pressure that was much lower. To prevent nitrogen from bubbling into the blood and causing the bends, an astronaut had to breathe pure oxygen at sea-level pressure for more than three hours before passing through the airlock into vacuum. This was highly inconvenient, but there was no alternative. It was one more legacy of the Apollo fire.

Hamilton Standard held the $27.5 million contract for the suit, developing and building the backpack and integrating it with the rest of the system. This firm also supplied a "walk-around bottle," a portable system that provided oxygen for prebreathing during preparations for egress into vacuum. A subcontractor, ILC Industries, built the soft parts of the suit.[57]

Under a separate contract, Martin Marietta provided a 330-pound rocket pack called the Manned Maneuvering Unit (MMU). Its background included a maneuvering unit developed for the Air Force's Manned Orbiting Laboratory, which never flew. The MMU was a variant of a similar unit that Martin had devised during the early 1970s and that astronauts had tested during the Skylab program. Its user wore the new one like a second backpack. Nitrogen, stored at 3,000 psi, gave a total velocity change of some seventy feet per second; a stabilization system enabled the user to fly straight. The MMU had a specific purpose: to assist a free-flying astronaut in the recovery of a satellite. This meant grabbing it, stopping its rotation, and moving it into the shuttle's cargo bay.[58]

The MMU improved on its Skylab counterpart by introducing fingertip control of motion. This contrasted with the earlier use of a hand-grip control, which was tiring when wearing gloves in vacuum. An MMU could achieve controlled rotation in pitch, roll and yaw, complementing its ability to translate in any direction; on-board gyroscopes supported an automatic attitude-hold capability, freezing a user's orientation. Electrical outlets enabled an astronaut to operate power tools, a portable light, cameras, and instrument monitoring devices; a pair of silver-zinc batteries provided on-board power for the MMU. It could range up to 450 feet from its orbiter.

Though approved for development in 1975, the MMU remained in the design definition stage until funding became available in 1979. Astronauts trained to use it by practicing in a simulator, which incorporated a moving-base carriage that could translate and rotate in any direction, as well as a large-screen television display. The MMU first flew to orbit aboard the tenth shuttle flight, in February 1984, with the astronaut Bruce McCandless posing for a memorable photo as he wore a unit on his back while floating freely in space.[59]

The MMU was one of a number of systems that enabled astronauts both to work effectively and to live in some semblance of comfort. One would not say that the shuttle included all the comforts of home; after all, it lacked showers. But it offered fresh, clean air with ample water and electric power, and an acceptable toilet. Space suits, MMUs, and the robot arm gave astronauts an admirable set of tools. There also was a pervasive concern with safety, for the vehicle as well as for its crew.

Safety and the Shuttle

Within the shuttle program, there was ongoing concern for the lives of its astronauts. The 1967 Apollo fire continued to drive home its lessons. Four years later, astronauts aboard Apollo 13 had barely avoided their own deaths after an on-board explosion disabled their spacecraft. The Soviets had their own cautionary tales. They had lost the cosmonaut Vladimir Komarov, also in 1967, through nothing more than a fouled parachute that failed to deploy. The crew of Soyuz 11 subsequently died as well, asphyxiating in space when a faulty valve brought a loss of the cabin atmosphere.[60]

For the space shuttle, Level I requirements from NASA Headquarters stated an explicit concern for safety: "The orbiter vehicle shall provide a safe mission termination capability through all mission phases." This requirement covered launch preparations, prior to liftoff, as well as ascent to orbit. It carried weight during selection of the prime contractor in 1972. Lockheed, which competed for this award, lost points because it left a sixty-five-second gap during ascent, with no provision for abort.[61]

On the launch pad, a slide-wire system offered means for quick escape. Five wires led from the launch tower to ground bunkers some twelve hundred feet away. Each wire supported one basket; with each basket carrying two crew members, as many as ten people could slide to safety. An outward opening side hatch, within the orbiter, also assisted rapid emergency egress.

Initial missions counted as test flights. As in any high-performance aircraft, crew members were to have ejection seats. Subsequent flights were to delete these seats, for as a 1976 statement noted, "Ejection seats would not be practical for operational missions that will carry a large number of passengers who have had no flight experience." Still, these seats provided additional safety for the lives of the early flight crews.[62]

The system also offered means to abort a flight at any time after liftoff. The SRBs, burning during the first two minutes, were to serve as abort motors by carrying the orbiter and ET to a high altitude. At any time during the next two and a half minutes, following burnout of these solid boosters, the orbiter could draw on the propellants in the ET to return to the launch site. If an SSME had shut down in flight, the orbiter could use its surviving engines as big retro-rockets, to slow the vehicle to a halt at high altitude and then to accelerate backward, again to land at the launch site.

Following those first four and a half minutes, the shuttle still could use its surviving main engines to abort into a very low orbit, from which it would reenter and land following a single loop around the earth. If an emergency such as leaking cabin pressure brought a need for immediate return, or if two SSMEs went out, the craft could cross the Atlantic and land on a runway in Spain or western Africa.[63]

With provisions for abort being mandatory, a second Level I requirement dealt with redundancy in design: "The redundancy requirements for all flight vehicle subsystems (except primary structure, thermal protection system, and pressure vessels) shall be established on an individual subsystems basis, but shall not be less than fail-safe. "Fail-safe" is defined as the ability to sustain a failure and retain the capability to successfully terminate the mission."[64] In practice, this requirement led to many key shuttle subsystems appearing on the vehicle in numbers of two or three.

Components such as the SRB, intended only for brief use, sometimes called for no more than dual redundancy, with a prime subsystem and a backup. Thus, the thrust vector control on each solid booster had two independent hydraulic systems, each with its own HPU. Dual redundancy also extended to the payload bay doors, which had independent electrical and mechanical actuators. The electrical system was the prime, but if it failed, a crew member could don a space suit and close the doors mechanically.

Many critical systems went further and were "fail-operational."[65] This meant continuing a mission on failure of a prime and terminating it safely on failure of the backup. In some instances, dual redundancy achieved this. The SSME's Honeywell electronic controller had dual-redundant circuitry, with two channels for every critical function. But if both channels went out, their engine would shut down safely. The orbiter had two OMS engines; if it lost one, the other could gimbal to thrust through the vehicle's center of mass and conduct the de-orbit maneuver. But if an orbiter lost both OMS, it still could de-orbit by using its small attitude-control thrusters in lieu of them.[66]

Still, there were a number of situations where nothing less than true triple redundancy could guarantee fail-operationality. Hence the orbiter had three SSME engines. The orbiter used its hydraulics on numerous occasions during a flight; it therefore had three independent hydraulic systems and three APUs, in contrast with two of each in the SRB. Any one hydraulic system in the orbiter, with its APU providing hydraulic pressure, could do a complete job of actuating the orbiter's controls. In similar fashion, on-board electric power came from three fuel cells, each of which was connected to one of three independent electrical buses.[67]

Level I requirements amounted to standing orders that were broad in scope, giving general guidelines of similar breadth. Other broad guidelines came from a standing committee, the Aerospace Safety Advisory Panel (ASAP), which had been chartered by Congress to make recommendations to the NASA administrator. It had been active during earlier piloted programs, including Apollo and Skylab. It maintained its involvement during the shuttle program.

This panel had important similarities to the committee headed by MIT's Eugene Covert, which reviewed the SSME project in the late 1970s. Both groups took their membership from distinguished technical leaders; during the mid-1970s the members of ASAP included a president of Lockheed, the director of the Defense Nuclear Agency, and the head of NASA's Kennedy Space Center. Consultants included William Mrazek, a former director of engineering at NASA-Marshall.

Both groups also emphasized the quick look and the annual report. The Covert panel spent no more than a few days at particular sites such as Rocketdyne; its two reports came out about a year apart. ASAP made similar visits to contractors and to other centers, supplementing inspections by the full committee with more detailed fact finding by specialized teams and individuals. But whereas the Covert committee was an ad hoc group that made its reports and then disbanded, ASAP was a permanent body. Its members held six-year terms; they visited plants and facilities at least once a year and issued their reports annually.[68] They dealt with broad issues, leaving project managers to address the details. For instance, in March 1977 its recommendations concerning the SRB included the following:

The nozzle bearing boot, although it has passed some tests, is not out of the woods as yet. There are concerns with regard to assuring that the maximum material temperatures are not exceeded during the firing time and that no splits or openings occur allowing hot gas flow inside the bearing.

The Auxiliary Power Unit has experienced some "under performance" tests which require a reexamination and review to define the manner in which the performance and reliability of these important units can be upgraded.

The panel avoided micromanagement. This same report of March 1977 discussed the SSME, which at that time was beset by difficulties. Even so, the panel made "no specific recommendations at this time" because "the engine development problems are well recognized by the proper levels of management and solutions are being sought and evaluated."[69] These were the working levels, which dealt with day-to-day issues of design and test. Here the activity featured ongoing concern with failure modes analysis, which led to identification of critical items. Such analysis sought to identify all the ways an item of equipment could fail and to document the potential effects on the system of each failure mode. Failures ranked as Criticality 1 if they threatened the lives of crew members or the existence of the vehicle, Criticality 2 if they threatened the mission, and Criticality 3 for anything less.

Failure modes analysis required up-to-date design data both for equipment and for their functional relationships to the rest of the system. Engineers sought to list all conceivable malfunctions and to describe both intermediary and ultimate effects of failure on other parts of the whole. They ranked each failure mode by noting its severity and its consequences to the vehicle, its astronauts, and the mission.[70]

This work often brought the identification of hazards, but it was not always either feasible or desirable to render them safe. When the probability of danger appeared low, sometimes a hazard could be classed as an "accepted risk."[71] An important example, early in the program, involved thrust termination on the SRB. The issue was whether to provide a method to shut down the burning of an SRM in the event of an emergency.

It was easy to stop a liquid-fueled engine; it was simply a matter of turning valves to interrupt propellant flow. Solid boosters required more heroic measures. The plan called for blowing off ports in the front of the motor, allowing hot gas to escape from both ends and lowering the internal pressure, thus extinguishing its burn. Studies showed that thrust termination would impose a heavy jolt upon the shuttle in flight. To prevent this sudden mechanical shock from producing structural damage, the orbiter needed more strength—and this extra robustness threatened to cost 19,600 pounds in additional weight. This was unacceptable. Fortunately, it proved feasible to delete the thrust termination, thus avoiding the weight penalty, and to treat this deletion as an accepted risk. Experience had shown that large solid motors were inherently reliable. For the shuttle, thrust termination appeared useful only in highly unusual circumstances: loss of thrust from two or more SSMEs, loss of thrust vector control in two or three of these engines, or loss of thrust vector control on two or more axes of the SRBs. Accordingly, in April 1973 NASA officials deleted the requirement for thrust termination.[72]

As an ongoing activity, as shuttle development proceeded, the Safety Division at NASA-Johnson provided a continuing review of specific issues in design and operation. Working with the prime contractor Rockwell International and with NASA project offices, this division noted safety concerns and presented them to the Major Safety Concerns Screening Board within the Safety, Reliability, and Quality Assurance office. Individual members of this board cited the safety issues that concerned them, compiling them into a report that was updated every three months. A typical report presented some two dozen "open safety concerns" that were being addressed, and a similar number of "closed safety concerns" that had been resolved. The report also summarized the accepted risks.

For instance, the SSME incorporated a heat exchanger, which used engine heat to produce a flow of gaseous oxygen that pressurized the oxygen tank within the ET. At Rocketdyne, a failure modes analysis identified this heat exchanger as posing a Criticality 1 hazard, for a leak from its heat-exchange coil could allow hydrogen to enter that oxygen tank. This would form a dangerously explosive mixture that could cause the ET to blow up, with loss of both crew and vehicle. Rockwell International performed a hazards analysis and concluded that the risk was acceptably low. NASA-Marshall nevertheless recommended a design change. At NASA-Johnson, this change was disapproved by the Level II Program Requirements Control Board (PRCB), which dealt with the overall shuttle design. This board concluded that in light of the Rockwell analysis, the baseline heat exchanger design was adequate. Periodic leak checks appeared sufficient to forestall the danger.[73]

Acceptance of such risks took place above the contractor level, within the program. The PRCB had the appropriate authority. So did the Space Shuttle Program Manager, at Level II. Some risks were accepted at the project level, by the Orbiter Level III Configuration Control Board or by the Orbiter Project Office at JSC. On the other hand, there were issues that required a decision at Headquarters.[74] On-orbit rescue was an example. This was a Level I requirement, which stated, "To fulfill the space rescue role the Space Shuttle System shall have the capability to launch within 24 hours after the vehicle is mated and ready for transfer to the pad." However, this implied a need to have two orbiter vehicles actively available, which would not be possible during the initial flights. To prevent a program delay Headquarters officials decided in May 1974 to waive this requirement, at least during those early flights, and to treat the lack of rescue capability as an accepted risk. This was possible because the likelihood of an on-orbit failure requiring rescue appeared very low.[75]

Elements of shuttle safety thus included Level I requirements for abort and for system redundancy, recommendations from the ASAP, failure modes and criticality analysis, and formal procedures for mitigating hazards and for selecting accepted risks. However, another approach to safety was conspicuous by its absence: probabilistic risk assessment. In the words of a NASA paper titled "Product Assurance for Shuttle," "The requirement for numerical goals, apportionments and predictions has been eliminated. Past experience indicates that complex math models tend to lose credibility and serve no practical purpose."[76]

This aversion to quantitative risk studies drew on experience during Apollo, where at times they actually proved detrimental to success. NASA contracted with a branch of General Electric to conduct a full numerical probabilistic risk assessment of the likelihood of landing astronauts safely on the moon and returning them to Earth. The GE study indicated the chance of success was less than five percent. NASA administrator Thomas Paine kept this under wraps, fearing that such pessimistic predictions could harm the program. The actual results of Apollo showed six completely successful lunar landings in seven attempts, with the seventh—Apollo 13—nevertheless rescuing its astronauts and saving their lives. With this, the agency turned away from quantitative statistical goals and assessments.

"Statistics don't count for anything," NASA's head of reliability for Apollo told *IEEE Spectrum* in 1989. He was Will Willoughby, who went on to become director of reliability management for the U.S. Navy. "They have no place in engineering," he continued. He asserted that risk is minimized not by statistical test programs, but by "attention taken in design, where it belongs."[77]

This attitude was not new. During Project Mercury, it was necessary to man-rate the Atlas ICBM that served as the booster, and the senior statistician for Atlas put it bluntly: "Reliability is a property of the equipment which must be designed into the equipment by the engineers. Reliability cannot be tested into a device and it cannot be inspected into a device; it can only be achieved if it is first designed into a device." The Mercury program used redundancy to achieve reliability.[78] The shuttle did the same, through its emphasis on fail-safe and fail-operational design.

One sees what this meant in considering the test and qualification requirements for the SSME. These began with Design Verification Specifications, which gave a basis for deriving the developmental test program. Whenever an engine went on a stand for a firing, whether in Mississippi or in California, it had ten to twenty test objectives, several of which were to support a particular specification. Qualification for flight carried two requirements: a total of sixty-five thousand seconds of single-engine operation and successful completion of two cycles of certification tests on each of two engines. The first criterion came from NASA's John Yardley and followed military precedent by being equivalent to forty shuttle missions. The second stipulation was outlined in the Contract End Item Specification and resulted from cooperative work involving Rocketdyne and NASA-Marshall. The two-cycle requirement exposed each certification engine to ten thousand seconds of operation, which was tantamount to some twenty flights. This meant that the test engines were obligated to demonstrate a substantial fraction of their rated life: fifty-five missions, or twenty-seven thousand seconds.[79]

While accumulating the sixty-five thousand seconds, the engine design showed large improvements in mean time between failures and in mean time between component replacements. However, Rocketdyne's Bob Biggs notes that the program of development testing imposed no numerical goals or specifications for these parameters. During the certification testing, any failure in the course of a cycle meant that the failed engine had to start anew, from the beginning of that cycle. Even so, such failures did not count as a black mark against the reliability of the SSME design.[80]

Yet if quantitative risk analysis could lead to unwarranted pessimism, lack of such analysis appears to have encouraged an attitude that conclusions from such studies could readily be ignored or fudged. This was particularly true for the SRB. Between 1979 and 1982, the J. H. Wiggins Company of Redondo Beach, California, conducted a study of shuttle safety, published in three volumes. The work focused particularly on the risk of losing a shuttle vehicle during launch. The Wiggins assessment reviewed the history of other solid rocket motors and showed that they had undergone an average of one catastrophic failure for every thirty-four to fifty-nine launches. A later study, prepared by Teledyne Energy Systems of Timonium, Maryland, reported similar findings and proposed "a failure rate of around one in a hundred."

The sponsoring NASA organization, the Space Shuttle Range Safety Ad Hoc Committee, rejected the Wiggins conclusions. It took the view that the Wiggins study included data from solid motors developed ten to twenty years earlier, but that the SRB, as a more modern design, would also be more reliable. The committee therefore assigned a failure probability of one in a thousand for each SRB. Moreover, because the SRB design benefited from improvements in design and manufacturing that had contributed to its man-rating, the committee lowered the probability of failure to one in ten thousand. That would have allowed NASA to launch a shuttle vehicle every week for a hundred years before experiencing a Challenger-like catastrophe.[81]

In fact, Challenger blew up on the shuttle's twenty-fifth flight, for a disastrous failure of one out of the first fifty SRBs.[82] The Wiggins study had hit the probability dead-on; it could hardly have been more accurate if its authors had used a crystal ball. Accordingly, it is appropriate to revisit the cause of that accident and to view it in the light of the concepts of redundancy and accepted risk.

The SRM was built of propellant-filled segments that fitted together at joints. Putty sealed the joints, but it had minimal ability to resist a motor's strong internal pressure, so each joint was sealed using two O-rings. This represented dual redundancy, in line with the Level I requirement. At the outset, in 1973, NASA's James Fletcher had praised this design: "The Thiokol motor case joints utilized dual O-rings and test ports between seals, enabling a simple leak check without pressurizing the entire motor. This innovative design feature increased reliability and decreased operations at the launch site, indicating good attention to low cost

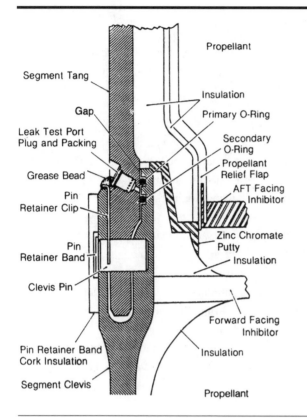

Propellant

Segment Tang

Insulation

Gap

Primary O-Ring

Leak Test Port
Plug and Packing

Secondary
O-Ring

Propellant
Relief Flap

Grease Bead

AFT Facing
Inhibitor

Pin
Retainer Clip

Pin
Retainer Band

Zinc Chromate
Putty

Insulation

Clevis Pin

Forward Facing
Inhibitor

Pin Retainer Band
Cork Insulation

Insulation

Segment Clevis

Propellant

Fig. 59. *Detailed cross-section of a Solid Rocket Motor joint. The interior of the SRM is to the right. (NASA)*

DDT&E and production."[83] Following the loss of Challenger, a commission headed by Secretary of State William Rogers examined the accident closely and issued a report. It noted that many people had shared Fletcher's sanguine attitude:

> *The Space Shuttle's Solid Rocket Booster problem began with the faulty design of its joint and increased as both NASA and contractor management first failed to recognize it as a problem, then failed to fix it and finally treated it as an acceptable flight risk.*
>
> *Morton Thiokol, the contractor, did not accept the implication of tests early in the program that the design had a serious and unanticipated flaw. NASA did not accept the judgment of its engineers that the design was unacceptable, and as the joint problems grew in number and severity NASA minimized them in management briefings and reports. Thiokol's stated position was that "the condition is not desirable but is acceptable."*

> *Neither Thiokol nor NASA expected the rubber O-rings sealing the joints to be touched by hot gases of motor ignition, much less to be partially burned. However, as tests and then flights confirmed damage to the sealing rings, the reaction by both NASA and Thiokol was to increase the amount of damage considered "acceptable." At no time did management either recommend a redesign of the joint or call for the Shuttle's grounding until the problem was solved.*[84]

Responding both to the accident and to the Rogers Commission report, NASA implemented a program of sweeping change. Quantitative risk estimation made a comeback, as the agency set up a risk management program. Its manager, Benjamin Buchbinder, told *IEEE Spectrum* that probabilistic hazard analysis had real value "in understanding the system and its vulnerabilities." Such studies could contribute to assessing the interactions of subsystems, the effects of human activity and environmental conditions, and in detecting common-cause failures. Environmental issues were important; cold weather prior to the launch of Challenger had stiffened the O-rings and led to the vehicle's loss.

The SRBs were extensively redesigned, with particular attention being paid to the joints. Engineering changes included addition of a third O-ring, thus achieving triple redundancy. The modifications worked; after the return to flight in 1988, the shuttle flew without incident. At a high cost, which included the lives of all seven members of the Challenger flight crew, NASA succeeded in removing the SRM joint design from its list of accepted risks.[85]

CHAPTER  EIGHT

The Electronic Shuttle

The Boeing 747 and the C-5A were among the first aircraft to use digital computers, special-purpose systems that supported these planes' inertial guidance.[1] The space shuttle went much farther, becoming very nearly a computer peripheral. Pilots did not fly it in traditional stick-and-rudder fashion. Instead they moved hand controllers that sent commands to central computers, which moved the elevons and fired the reaction-control thrusters. These computers also generated data that appeared on cathode ray tube displays on the flight deck, simplifying the cockpit layout. The main engines operated under direct computer control. In addition, ground-based systems checked out the shuttle prior to launch, conducted the countdown, and ordered liftoff.[2]

Defining the Requirements

When compared with previous programs such as Skylab and Apollo, the shuttle's on-orbit operations were to be considerably more complex and extensive, and were to support a much wider variety of payloads. In addition, the orbiter was to make precisely controlled and un-powered runway landings. These requirements supplemented a longstanding NASA rule that a mission must be aborted unless at least two means of returning safely to Earth were available. These stipulations shaped the design approach.

In previous programs, safe return meant a parachute landing in the ocean, close to a recovery ship. This simple procedure led to relatively simple backup systems; they lacked the performance of the primary operational systems, but complied with the mission rules. For the shuttle, however, the maneuvers of atmosphere entry through final approach and landing imposed performance requirements that were demanding indeed. Therefore, backup systems with degraded performance were unacceptable.

Aborted missions were part of the program's contingencies, but it also was unacceptable to execute an abort in response to a single failure. The shuttle was to fly up to sixty times per year, and it needed reliability to attract its clientele. NASA therefore imposed a new and comprehensive requirement, whereby failures in avionics systems were to provide fail-operational capability. The on-board electronics were to remain fully capable of performing the mission after any single failure, and of returning safely to a runway after any two failures. This stipulation, along with the need for fully capable backup systems, dictated a need for multiple and independent hardware installations to ensure reliability through redundancy.

This raised the issue of failure detection in components. The industry had experience with built-in test equipment, but such circuits brought their own problems. They often had failed in use, leading to false conclusions as to the operability of a unit. A preferred method of fault detection, selected for the shuttle program, called for "voting" by comparing the operational data produced by several devices or subsystems operating in parallel. This approach set the requirement for redundancy. Two parallel units were too few; either of them might be wrong, with no way to tell. Three units gave a minimum; two good ones could "outvote" and identify a failed one. Four parallel units then accommodated a second failure in the same area, meeting the requirement for fail-operational capability. This combination of fault isolation and fault tolerance led to quadruple redundancy, which became prevalent in the shuttle's avionics.

Autonomous operation was a third requirement, mandated by the Air Force and accepted by NASA as a design goal. This meant that the flight crew was to manage the on-board systems to a larger degree than in earlier spacecraft. Such vehicles had relied on telemetry from on-board subsystems, which fed into Houston's Mission Control. For the shuttle, much of this data was to remain within the orbiter, with the flight computers processing it and displaying it to the crew in usable forms. In addition, many aspects of system management had to be placed under computer control, to avoid presenting crew members with excessive workloads.

Flight control imposed its own requirements. The orbiter was an unstable airplane; it could not be flown manually even for short periods, either during ascent or reentry, without full-time stability augmentation. This ruled out the possibility of direct flight control using a conventional stick and rudder with hydraulic boost. In addition, the shuttle was to cover an extraordinarily broad range of flight regimes, from hypersonic reentry in the upper atmosphere to subsonic approach and landing. It also was to use both aerodynamic control surfaces and reaction-control thrusters. The different control modes and flight regimes required different control laws, which were mathematical equations that related pilot inputs to control responses.

The elevons and other surfaces had a great deal of control authority, which they needed to meet the vehicle's requirements. This brought a situation wherein a hard-over command to a control actuator, issued in error and remaining in effect for as little as 0.01 to 0.4 seconds, could cause structural failure and loss of the orbiter. Therefore, the flight control system had to prevent actuator hard-overs no matter what elements might fail. In addition, the short reaction time ruled out any direct role for the flight crew in reacting to system failures and dictated a fully automatic response.

These considerations led to a decision to use digital flight control. Such systems had been used with great success during Apollo, though no airplane had yet flown with such control until 1972, at the start of the shuttle program. But digital aircraft control was in active development at that time, and offered particular advantage for the shuttle. Such a system readily accommodated stability augmentation and automated response to faults. It also allowed a computer to hold all control laws in memory. During each distinct flight phase, the computer would select and implement the appropriate set of laws, blending reaction controls and aerodynamic control surfaces for the best effect.[3]

The shuttle also broke new ground in its on-board data systems. Conventional avionics relied on wire bundles, with each separate wire dedicated to a single signal or function, but the orbiter was large in size, with numerous sensors, control effectors, and other devices distributed all over the vehicle. The alternative was the data bus, which resembled a telephone line that serves a number of users. Each sensor or instrument had its own identifying "telephone number," which allowed the computer to respond appropriately. TRW estimated in 1971 that data buses would save twenty-five hundred pounds on the orbiter while requiring 500 watts less power. The technology was in its infancy at that time. Nevertheless, it received a significant boost in 1975, when the Air Force issued MIL-STD-1553, a set of standards for operational data buses. Shuttle data buses were not 1553-compliant, but strongly resembled systems that met the 1553 requirements in full.

In sum, the requirements for shuttle electronics included the following:

No degraded backup systems
A fail-operational approach
Use of operational data to detect and isolate failures
Quadruple redundancy to isolate a second failure
Automatic failure detection and recovery for time-critical functions
Full-time digital flight control
Data buses to save weight; computer-driven cockpit displays to enhance vehicle autonomy and reduce flight crew workload.[4]

The On-Board Computers

It was clear at the outset that computers would serve such purposes as flight control, guidance and navigation, and data processing. Designers initially considered options that placed the space flight and aerodynamic ascent and reentry functions in different machines, or that

289

separated guidance and control from other data-processing operations. However, this brought difficulties in attempting to interconnect and use multiple complexes of machines, which might be of different types, installed in different numbers. This drove the configuration to the selected choice: a single suite of general-purpose computers that would serve all required uses.

During the early 1970s standardized "black boxes," often with multiple circuit boards, played the role of today's Intel Pentium and other microprocessors. At that time, only two avionics computers under development appeared potentially suitable for the shuttle. The Singer-Kearfott SFC-2000 was a candidate for the B-1 bomber; the IBM AP-101 was a derivative of that firm's 4 Pi series, which drew on technology from the System/360 commercial program of the mid-1960s. IBM 4 Pi computers flew aboard Skylab and were selected for the F-105 and B-52 aircraft. Both the Singer-Kearfott and IBM machines required extensive modification, and NASA chose IBM.[5] This drew on a long involvement with that company, which had built flight computers for all Saturn-class launch vehicles as well as for Gemini and Skylab.[6] The Skylab experience was particularly pertinent. Two 4 Pi units, a primary and a backup, gave NASA its first flight experience with a true dual-redundant computer system aboard a spacecraft. The twin computers were part of Skylab's Attitude and Pointing Control System. It controlled the attitude of this orbiting laboratory about all three axes and kept it locked on the sun, which provided all on-board power through solar panels.

The two 4 Pis were identical, and either was fully capable of performing all computations on its own. The system carried redundancy management software, which could detect deviations from specific performance criteria and automatically command the primary computer to hand over control to the backup. NASA had used custom-built computers aboard the Gemini and Apollo spacecraft, but the 4 Pi units of Skylab were tested, off-the-shelf technology. The successful use on Skylab of the dual computers, along with the redundancy management, opened the door to the far more extensive use of the similar AP-101 units aboard the shuttle.[7]

Standard considerations of redundancy meant that the orbiter would have at least four main computers, designated AP-101B, and NASA soon added a fifth. Initially it was intended to provide additional computational capacity, thus addressing uncertainty as to future needs. In particular, it would off-load functions from the four other computers that were not critical to flight, including management of the payloads and of vehicle systems during the mission in orbit. However, the role of this fifth computer soon expanded.

During preparation for the Approach and Landing Tests, a review panel concluded that the use of a single set of flight-control software represented an unnecessary risk. The flight computers all were to run the same code, which would contain errors; these computers therefore would all make the same mistakes. Standard error-detection techniques would not find

Table 8.1. Comparison of Three Computer Systems

	Saturn I-B, V	Skylab	Space Shuttle
Year available	1964	1971	1974
Computer functions	Prelaunch and orbital check-out; vehicle sequencing; guidance, trajectory insertion	Attitude control; redundancy management; generating signals for control and display	Prelaunch checkout; system and payload management; guidance, navigation, flight control during ascent, orbit, and entry
Human-computer interfaces	Through ground control computer only, before launch	Display and control panel	CRT data display; keyboard data entry; hand controls; flight display; manipulator control
Arithmetic	Fixed point	Fixed point	Fixed and floating point
Word length (bits)	Data: 26 Instruction: 13	Data: 16 Instruction: 8 and 16	Data: 16 and 32, fixed point; 32 and 64, floating point Instruction: 16 and 32
Available instructions	18	54	154
Computing speed (ops/sec)	11,300	60,000	480,000, fixed point 325,000, floating point
Special features	None	Interrupt provision	Microprogramming, high-level language, 19-level interrupt, 24 general registers
Memory Type	Ferrite core	Ferrite core	Ferrite core, memory chips
Capacity (words)	16,384	16,384	40,960 (1975) 106,496 (c. 1980)

Table continues

291

Table 8.1. *Continued*

	Saturn I-B, V	Skylab	Space Shuttle
Access time (microseconds)	10	1.2	0.375
Size (ft³)	2.4	2.2	0.87
Weight (lbs.)	75	97.5	57.9
Power (W)	150	165	350

Source: Data from Cooper and Chow, "Development," 6, 7, 10–11.

---

them, for these techniques assumed that at least two of the AP-101B units would give correct and valid results. Enterprise was to rely entirely on its digital flight control in steering to its landings; there was no manual backup. If undetected software problems caused it to tumble out of the sky, the flight crew members might save their lives using ejection seats with parachutes, but would have to watch helplessly while their craft crashed onto the desert floor.

The solution called for a backup flight control system with a completely independent set of software that would run on the fifth computer. This backup code came from Rockwell International, whereas the primary flight-control code was written by IBM. The backup was less complex and less capable than the primary programs, but it could land Enterprise safely in the event of a severe failure within those programs. The initial set of backup routines included twelve thousand words of executable instructions, half of which dealt with flight control as such while the other half addressed ground check-out and built-in test of the computer. However, the backup flight system grew to become an important entity in its own right, with some one hundred thousand words.[8]

Any one of the five AP-101B units could serve as the fifth or backup machine, with an astronaut making the designation through a keyboard entry. The selected unit responded by requesting the backup software from memory, and then would operate on standby. During normal operations, with the shuttle under control of the primary system, the backup monitored and received data from all primary computers and their sensors. This enabled the backup system to keep itself current, and to maintain the ability to take control of the shuttle at any time.[9]

Data buses provided close interconnections between the five computers and their input-output devices, and the shuttle's sensors and on-board systems. Each bus had a clock rate of one megabit per second, with the data being transmitted using pulse-code modulation. This provided a rapid stream of the binary digits 0 and 1, which were encoded as pulses. A high-tech form of internal communications, it had a clear resemblance to the Morse code of nineteenth-century telegraphy, the pulses replacing that era's dots and dashes.

Twenty-eight data buses served the complete shuttle (table 8.2). Data transfers were formatted in a standardized code, which provided multiple checks of the correctness of words and messages. All data traffic over the buses was initiated by command words sent by the computers. To transmit data, perhaps to mass memory or to a flight-deck display, the computer sent a command that indicated the imminence of the transmission, followed by the required number of data words. To receive data, the machine sent a command requesting the type and amount. A terminal unit, serving the appropriate display or subsystem, then responded with the desired number of words. The expected word count had to be correct, or the entire message would be rejected.

The system used three types of computer memory: chips, ferrite core, and magnetic tape for mass storage. This reflected a conservative use of existing technology, in the magnetic cores, while taking advantage of the burgeoning promise of microelectronics, with its emphasis on single-chip devices of unparalleled power and capacity. The consequence was a computational architecture that offered considerably more flexibility than that of Skylab, which had used similar central processors.[10]

Ferrite cores were rings of iron-rich ceramic, about a millimeter in diameter. Each core lay at the intersection of two wires, which threaded through its hole. Numerous wires formed a grid that resembled window screening, with a core at each such intersection. Pulses of current, sent through two selected wires, singled out their particular core and magnetized it, with its magnetic north pole either pointing up or down. This corresponded to the two binary digits. Computers could read data from the cores as well as write onto them, and for more than twenty years ferrite-core memory dominated the industry. Jay Forrester of MIT, an early leader in computers, had introduced this technology around 1952. This form of memory caught on quickly because it was far more compact than the alternative, which used individual vacuum tubes as memory elements. It also was vastly more reliable and less demanding of power, and generated far less heat. Even the rise of the transistor, as a discrete circuit element, did not challenge it. Core memory stood its ground until the advent of the single-chip semiconductor memory, during the 1970s.

A 1962 survey of computers by the author John Pfeiffer described a "typical computer array" as "made up of 4,096 cores (a 64 by 64 square) on a flat wire grid which, in the words of one designer, 'resembles a repair patch on a copper screen.' Such arrays are stacked one above the other like the floors of a building." An experimental computer of the day had a core memory with thirty-eight 4,096-bit arrays arranged in such stacks, which fitted "into a little

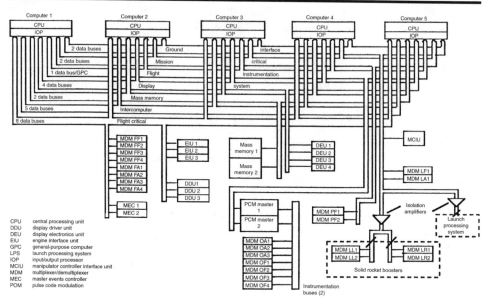

Fig. 60.   *Shuttle data bus networks. (NASA, courtesy Dennis R. Jenkins)*

blue box no larger than a bedtime radio set. Other machines have memory units containing millions of bits, or millions of cores."[11]

The shuttle initially had 40 kilowords of core memory in each of its five main computers, a kiloword (kword) being 1,024 32-bit words. This proved inadequate; the capacity increased to 104 kwords in early flight versions. Each AP-101B was packaged in two boxes, one holding the central processor with 80 kwords of memory, the other containing the input-output processor and the remaining 24 kwords of core. The four primary computers all held the same data in memory. Hence the effective capacity of the system, the four plus the fifth, was 208 kwords, or 832 kilobytes, with 1 byte being 8 bits.[12] By today's standards, this is minuscule. The Apple Power Macintosh, a recent personal computer, readily accommodated 128 megabytes of main memory, some 150 times more. Indeed, it would have taken the full core capacity of all five computers, working independently, merely to play some of the popular video games of the 1990s.[13]

However, the term "core memory" had a double meaning, for it referred to the memory's role as well as to its technology. This was the main memory, offering fast response and random access even while being limited in capacity. For mass storage, computers had long relied on tape drives, which could store immense amounts of data even while being sharply limited in speed by the slow rolling of the tape reels. Popular images of computers often fea-

294

Table 8.2. Allocation of Shuttle Orbiter Data Buses

Number of Buses	Function
8	Flight-critical services: guidance and navigation, flight control, sequencing, main engine
5	Intercomputer data transfer
2	Mass-memory interface
4	Flight deck displays
5	Pulse-code modulation master unit interface (one dedicated to each computer unit)
2	Systems management; payload operational sequence
2	Ground interface; remote manipulator

Source: NASA SP-504: Hanaway and Morehead, *Space Shuttle Avionics System,* 1989, 30.

tured such revolving reels. They stored large databases and lengthy programs, supplying information and code to the core memory as needed, to be held for immediate use.

The shuttle used two identical tape units, each with storage capacity of 134 megabits, or 4.2 million 32-bit words. Thus these tape drives, which served in lieu of floppy-disk units, each had one-eighth the capacity of the cited Apple Macintosh. Nevertheless, this mass memory served a number of purposes. It stored prelaunch test routines, diagnostic test programs for fault isolation, formats for flight-deck displays and for downlink data transmissions, program and data segments to be loaded into core during specific mission phases, system software, and duplicate copies of on-line programs for initial loading, reloading, or reconfiguration of the computers, as in choosing a backup.[14]

The core and mass memories provided the principal forms of data storage, and the shuttle also used early types of memory chip. In 1973 they were available with 4,096 bits of read-only memory and 1,024 of random-access memory. Two years later, these numbers quadrupled.[15] Such chips permitted microprogramming: the use of small stand-alone routines, residing in read-only memory. Designers used microprogramming in both the central processor and the input-output processor. This gave considerable flexibility, implementing a more comprehensive repertoire of instructions and accommodating changes in both the instruction set and the system architecture.

The input-output processor (I/O) was an important unit in its own right. It was the same

295

size as the central processor unit (CPU). The latter performed all arithmetical operations. The I/O controlled the data buses as they served the shuttle at large. It also performed data buffering, or temporary accumulation. It had its own instruction set; its data processing programs were independent of those in the CPU. By placing the I/O and CPU in separate boxes, engineers facilitated their design and development while simplifying subsequent maintenance.[16]

The advent of the shuttle as a program, which received Nixon's approval in 1972, coincided in time with the advent of the earliest microchips. The shuttle's use of computers took form amid the revolution in microelectronics that transformed the field during the 1970s, and it is appropriate to ask why NASA did not take more advantage of this. The answer lies in appreciating that the microchips of the 1970s were severely limited in their capabilities. Though widely touted as wonders of that era, they lacked the power to handle the demands of the space shuttle. The first of them, the Intel 4004, was a single-chip microprocessor dating to 1971. It used four-bit words, in a minimal arrangement that allocated one word to each decimal digit, such as 5 or 8. A few months later, Intel brought out an eight-bit device, the 8008, that offered somewhat more power and greater flexibility. Other firms followed with their own eight-bit chips, beginning in 1973. Together, these chips established eight bits as the standard for that time. They also gave the technical base for the first personal computers.

In 1976 Enterprise was in final assembly at Palmdale. In that year the firm of Fairchild Semiconductor brought out the F8 microcomputer. It was a technical tour de force, for its computer power matched that of ENIAC, the world's first general-purpose electronic digital computer. Built in 1945, ENIAC had filled a large room with its circuitry; the F8 put all its circuits on a single board. It was three hundred thousand times smaller than ENIAC, used fifty-six thousand times less power, and had a mean time to failure measured in years rather than hours. Even so, by the mid-1970s the world had moved well beyond ENIAC. A 1975 article in *Scientific American* stated, "A good example of an application for which microcomputers are ideally suited is a system designed to control automobile traffic lights. Thousands of such controllers are currently being installed." It was a long step upward from traffic-light control to the space shuttle, and even the F8 had far less power than the shuttle needed.[17]

The shuttle's computer system was a long-lead item. It needed an early start, which meant an early commitment to hardware. It had to be ready in time for the Approach and Landing Tests of 1977. Lead-time considerations then compelled NASA to rely on the state of the art as of the early 1970s. This ruled out microprocessors, which then were in their infancy. Yet while good eight-bit microchips lay in the future, the IBM AP-101 was available and suitable. The IBM designers made no attempt to place a central processor on a single microchip, or to fit a complete computer, with memory and input-output devices, onto a single circuit board. They willingly spread the central processor across several boards, with the complete system filling two boxes. In exchange, they gained power and flexibility.

In 1976 sixteen-bit chips were available; a single word represented integers as high as 65,535. The shuttle used floating-point word lengths of up to sixty-four bits, handling the equivalent of integers as large as 1.84×10.[18] The F8 could add two eight-decimal-digit numbers in 150 microseconds, giving an upper limit of 6,700 addition operations per second in fixed point. The shuttle's computer was rated at 480,000 operations per second in fixed point and 325,000 in floating point—with its operand numbers being enormously larger.[19]

In its relationship to computers as well as in its cabin atmosphere, the shuttle was a child of its technological era. The world at large had a strong interest in supercomputers, with good ones arriving late in the decade. Even so, the shuttle program was ten to fifteen years too early for these systems to make serious contributions in computational aerodynamics, leaving NASA to rely on its wind tunnels. Similarly, despite that era's vast excitement over microchips and personal computers, the shuttle effort came along too soon for these forms of electronics to contribute. In important respects, the AP-101B was a minicomputer of the late 1960s.

Minicomputers had offered some of that decade's important innovations. They came from companies such as Digital Equipment and Data General. They put serious computing power in the hands of individual scientists and small companies, at affordable prices. The Digital Equipment PDP-8 of 1965 was one of the important early models; it was the size of a two-drawer file cabinet and cost eighteen thousand dollars. It used integrated circuits that packaged fewer than a dozen transistors onto a single chip, advancing beyond the discrete components of the early 1960s but falling well short of the large-scale integration of Intel.[20]

Minicomputers flourished into the 1980s. A widely read book of 1982, Tracy Kidder's *Soul of a New Machine,* gave an intimate view of the development of an advanced model at Data General. This technology, not that of microchips, provided the technical base for the AP-101B.

Software

The system grew to include more than 270 components, depending on the mission, and used some five hundred thousand words of software code. The work of programming and debugging was greatly eased by use of a high-level language called HAL/S, which resembled the widely used Fortran.[21]

Computer languages exist at several levels, and one may illustrate this with an example. Consider a short subroutine that executes the following: "Given two positive numbers X and Y, subtract the smaller from the larger; add 5 to the result; then return to the main program." On an Intel 8080 microprocessor, one could encode this in three ways:[22]

Machine Language	Assembly	High-Level Language
00 100 001		
00 000 010	ABS LXIH 2 7	ABS: PROCEDURE;
00 000 111		
01 111 110	MOVA, M	If X > Y THEN Z = X − Y + 5;
		OR ELSE Z = Y − X + 5;
00 100 011	INXH	RETURN;
10 010 110	SUBM	END ABS;
11 110 010		
00 001 011	JP LOC	
00 000 011		
00 101 111	CMA	
00 111 100	INRA	
11 000 110	LOC ADI 5	
00 000 101		
00 100 011	INXH	
01 110 111	MOVM, A	
11 001 001	RET	

Machine language, written in binary, requires the programmer not merely to state the arithmetic operations but to define locations in memory and to specify the flow of data from one such location to another. Its statements control the work of the computer's logic gates. It is the language of a machine's operating system, which defines its fundamental instructions and operations, but it is clearly too long and unwieldy for ordinary use.

Assembly language replaces the binary statements of machine code with short statements of text. These often have a mnemonic character, making them suitable for the programmer's memory as well as the machine's. Thus, in the example, MOVM, A and MOVA, M move data between locations in memory, while SUBM subtracts. There is one-to-one correspondence between assembly statements and their counterparts in machine language; a specialized routine called an assembler carries out the translation.[23]

Much of Apollo's software was written in assembly, and while it worked, it still required programmers to pay meticulous attention to the machine instructions and to the addresses of items of data in memory. The shuttle was to have up to thirty times more data and lines of programming, and there was obvious advantage in allowing programmers to dispense with the details of data flow and storage. To do this, NASA turned to a high-level language that resembled ordinary algebra. As the example illustrates, there is no simple connection between statements in such a language and corresponding statements in assembly or machine code. The language in the example is PL/M; the same is true of Fortran, Ada, C++, and other popular choices. The computer carries through the translation using a compiler, which automatically deals with the bookkeeping of information storage and transfer. Compiled code again is written in machine language, but now the machine itself, not the programmer, does the writing.[24]

NASA's language of HAL/S was a shuttle-oriented variant of an existing language, HAL. It came from Intermetrics, a firm in Cambridge, Massachusetts, founded by five programmers from MIT who had written software for Apollo. The name honored Hal Laning, a professor at that university who had worked with Jay Forrester and whose contributions to computer language laid important groundwork for HAL.

High-order languages carry a disadvantage, for their compiled code lacks the efficiency that can be attained by experienced programmers working in assembly. When speed and memory are at a premium, as was true for the shuttle, the need to make optimum use of these resources can militate in favor of assembly, despite its inconveniences. NASA conducted a test, with different teams of programmers writing code in assembly and in HAL/S, and found that use of the latter imposed a loss in efficiency of only about 10 percent. This was acceptable, for the high-level language promised increased programmer productivity, better program maintainability, and enhanced visibility into the software. The operating system remained in machine code, but NASA directed the use of HAL/S for the rest of the software.[25] In contrast to existing languages such as Fortran, HAL/S was tailored specifically for use with aerospace avionics. It allowed programmers to define priority levels and scheduled computational tasks accordingly. Long-running processes of low priority, such as navigation, used data that did not change greatly from one second to the next. These were interrupted repeatedly to allow higher-priority processes, such as flight control, to execute.

The shuttle's computers received and responded to data in real time, and HAL/S enhanced the ability of programmers to write routines that did this. It promoted time-sharing, whereby the computers could run several programs concurrently, using interrupts to switch between them. This allowed the shuttle's numerous automated systems to respond to rapidly changing events without interfering with each other, as they shared the computers' resources.[26]

Redundancy management was an important task for the software. The four main computers operated independently, yet all were expected to give the same output at the same time. This took effort to achieve, for when two or more computers operate at several priority levels and are subject to interrupts, the precise order of processing and hence the numerical outputs of these systems are typically not the same.

Even when these computers represent the same model from the same manufacturer, their speeds of execution and the timing of an interrupt-invoking process are not identical. Consider two computers that both run programs A and B, switching between them by means of interrupts. Computer 1 runs code A, stores a value of the variable X, and receives an interrupt that switches it to run code B, which reads and uses that value. Computer 2 also runs code A, but at a slightly slower pace, and receives the interrupt, switching it to run code B, before reaching the programmed instruction to store the value of X. Code B, running in computer 2, again reads and uses X—but its X represents an earlier value that differs from the one in computer 1. This process, called "slivering," allows identical programs in duplicate computers to see different data. In this example, computer 2 will need a second interrupt, returning it to code A, before it can store an updated value of X. It then will need a third interrupt before it can return to code B and pass on that value. By then, computer 1 probably is working with data in its own code B that are different indeed.

The shuttle used several rules to maintain identical data in all central processors. Input transactions used a "command and listen" technique, with one computer requesting the data and all receiving the reply. This ensured that all would receive identical input data, but it left open the possibility of an input error. The computers guarded against this by exchanging status information following each input. If any computer detected an error, none would use the particular data in which the error occurred. The system also protected data transfers from the input-output processors into the central processors. It did this by having the central processor halt momentarily until a transfer was complete.

These rules dealt with input data; the system also prevented slivering and protected data when passed between different programs and priority levels. The key was synchronization of the computers, which occurred nearly five hundred times per second. Each computer sent a synchronization code and read the synch lines from the others to see whether they were sending the same code. If any were not, the others would wait until every computer was sending the proper code. This wait lasted 0.004 seconds, which was large enough to allow for normal variations in processing speeds. After that, any computer that was not sending the proper code had failed to synch. It was dropped from the redundant suite of machines.

To protect interprogram data, the basic method was "synch and disable." It disabled the interrupts, preventing code A in the example from storing X in one computer until it was ready to store the same X in all computers. Only then would the interrupts be reenabled. A separate method protected data by locking it away, preventing access by any other program. The data would be unlocked only after the new value of X was stored in all computers.[27]

In addition to ensuring integrity of data, redundancy management also had to assure integrity of the computers themselves. A failed computer was to be identified to the crew using self-test techniques that detected failures at least 96 percent of the time. The failure of one machine could not cause another one either to identify itself as failed, or to generate incorrect output. The crew could reconfigure the redundant computer suite and shut down all transmissions from a failed unit, then restore such a unit to service where practical.

All computers were to give the same outputs. The selected basis of comparison was to sum the outputs to be transmitted over data buses during one computational cycle, and to compare the sums. Such a sum represented a word in memory, and the words in the separate computers had to match, bit for bit. If any particular machine failed this test twice, it was identified as a failed unit. This comparison-of-output technique detected most failures. Even so, identification was not certain if the faulty unit was simply notified that it had failed. Therefore, properly running computers in the set needed a means of forcing the faulty computer to indicate itself as having failed. To do this, the system used special hard-wired logic as part of its redundancy management, with the integrity of this logic being assured through separate test programs in each computer.[28]

Redundancy management also dealt with unusual cases where conventional voting, or comparison of outputs, could fail. If the four computers gave a vote of 2 to 2, the system would display warning messages to the crew and recommend use of the backup flight system. If two computers failed, comparison of outputs from the other two allowed detection of a disagreement, but could not determine which of them was giving the invalid result. Each computer therefore carried self-test programs as well as built-in test equipment.[29]

Similar considerations applied to managing redundancy in the sensors and other onboard electronics. Here the data was measured rather than computed. The system was to compare redundant data, discard any value that diverged beyond an acceptable threshold, and select the middle value of three good inputs, or the average if only two were available. However, this again called for meticulous care.

Fault detection had to minimize the likelihood of false alarms, while maximizing the probability of detecting an erroneous signal. These requirements were totally contradictory and conflicting. As a compromise, the system accepted performance within three statistical standard deviations of normal, a criterion that rejected only 0.26 percent of measured values. This brought the problem of determining the values of the standard deviations, most of which had to be derived analytically because of insufficient test data.

Other measures sought to ensure that compared items of data had been taken at the same time and had not been corrupted when digitized and transmitted over data buses. The system had to identify a faulty unit over the complete range experienced in the data, account for any expected unique or peculiar behavior, and use built-in test equipment when only two redundant units were still operating. It had to accommodate start-up and shutdown, degrade as harmlessly as possible, and provide the flight crew with status information, warning messages, and opportunities to intervene as appropriate.

The computer complex interfaced with 38 subsystems on the orbiter and four on the SRBs. Each subsystem and replaceable unit had its own requirements for redundancy management, which had to be treated individually. At times these requirements cut across several subsystems, with the Reaction Control System being a case in point. The redundancy management logic had to detect and isolate thrusters that were failed off, failed on, and leaking. Depending on the type of failure detected, it had to issue commands to appropriate valves to prevent loss of propellant or other dangerous conditions. The RCS included propellant tanks, pressurization systems, manifolds, valves, and forty-four thrusters for flight control. The thrusters were divided into four groups, any two of which could maintain vehicle control about all axes. Each of the thruster groups and associated valving was managed by one of the four redundant avionics assemblies. The system also had three electric power buses, a dual instrumentation system with sensors, and displays and controls for the crew. The redundancy management system had to function amid all this complexity.[30]

After ensuring that the computers and on-board systems were operating properly, the system then had to fly the shuttle as an airplane. To do this, it relied on Primary Avionics System Software (PASS), with requirements that reached well beyond those of Apollo. The Apollo computers were mostly on the ground, at Mission Control in Houston. The shuttle, on the other hand, was to have far more on-board autonomy. In principle, it could fly on its own like an airplane, dispensing with Mission Control. (In practice, though, it used the Houston facilities in Apollo's manner.)[31]

The software for STS-1 in 1981 had eight separately executable programs, all sharing a common operating system. These programs were stored on mass-memory tape and loaded into the on-board computers at crew request. The cited lengths (see table 8.3) included the operating system.[32] This software supported three primary preflight and in-flight applications. The first was Guidance, Navigation, and Flight Control. It determined vehicle position, velocity, and attitude; managed subsystem redundancy; and provided the crew with the displays and data necessary to control the avionics. This software also issued all engine and aerodynamic surface commands from liftoff through touchdown and roll-out. Software supporting this application was resident in five of the eight cited programs. In particular, these codes included the control laws that flew the vehicle during descent. This suite of laws used reaction-control thrusters exclusively during early entry, then gradually blended in the aerodynamic control surfaces as they became effective. Roll control using elevons came first; pitch came next, then yaw, controlled with the rudder. Around Mach 3.5, the orbiter went over completely to aerodynamic control, with the thrusters being totally deactivated.

The second major software application, Systems Management, monitored the performance of orbiter subsystems. It provided fault detection and announced anomalies to both crew and ground. The third application, Vehicle Checkout, permitted the testing and certification of subsystems during preparation for flight, prelaunch countdown, and while in orbit just prior to reentry.[33]

Four facilities supported software development. The Software Development Laboratory was the principal center for debugging, integrating, and verifying the on-board codes. IBM,

Table 8.3.   Software for STS-1

Program	Functions	Length (32-bit kwords)
Preflight initialization	Check-out of subsystems and of data processing system hardware and interfaces	72.4
Preflight guidance, navigation and control check-out	Calibration and alignment of inertial measurement units; check-out of control system and display	81.4
Ascent/aborts	Terminal launch countdown; guidance, navigation, flight control	105.2
On-orbit	Guidance, navigation, flight control	83.1
On-orbit check-out	Guidance, navigation, flight control; check-out of entry subsystems	80.3
On-orbit systems management	Subsystem monitoring; fault annunciation; payload bay door operation; antenna management	84.1
Entry	Guidance, navigation, flight control	101.1
Mass-memory utility	Mass-memory alterations	70.1

Source: AIAA Paper 81-2135, fig. 1.

the software contractor, built it; it was located at NASA-Johnson, where IBM ran it under contract to NASA. This center assembled all software elements into the mass-memory tape that loaded the on-board tape units. The second facility, Shuttle Avionics Integration Laboratory (SAIL), also at NASA-Johnson, served as a simulation center. It brought together avionics hardware, flight software, and ground systems, which were integrated and tested in accordance with flight procedures. NASA and Rockwell International used this center for avionics verification and for certification of the hardware and software. The Flight Systems Laboratory was located at the Rockwell plant in Downey, California. Its capabilities and its role in the shuttle program resembled those of SAIL. However, whereas SAIL emphasized prelaunch, ascent, aborts, and orbit insertion, the Downey center dealt with on-orbit activities and with de-orbit, entry, and landing.

The fourth facility, the Shuttle Mission Simulator (SMS), was NASA's main training installation for flight crews. Located also at NASA-Johnson, it gave realistic environments that trained crews in all mission phases from prelaunch through landing and roll-out. For STS-1

303

training, it offered two flight simulators, both of which employed actual shuttle data-processing hardware. The "moving base" simulator, installed in 1977, provided cockpit motion by moving in any direction while rolling about all three axes. There was also a "fixed-base" simulator. Both provided extensive visual displays along with a direct link to the Mission Control Center, also at NASA-Johnson. A. J. Macina, an IBM senior systems analyst, wrote that "although not directly involved in software testing, the SMS does provide an invaluable benefit in this area. Its extensive program of crew training under off-nominal conditions tends to subject the software to a wide variety of stress conditions. This provides added confidence in the performance and capabilities of the system."[34]

Author Dennis Jenkins relates the development of PASS, the complete suite of flight programs:

> After the requirements were considered firm enough to actually start coding, IBM engineers had to go through one of the most rigorous processes known to software engineering. Following an inspection of the requirements documents, a design specification was generated using a Process Design Language (PDL) that described the specification in algorithmic form. This PDL was critically inspected by members of the software development team, as well as outside parties. Next, the design was translated into HAL/S source code by a programmer, who then performed first-level testing of the software to ensure that the logic produced the desired results. This was followed by another formal inspection of the code in which a group of analysts and engineers "walked through" the code line-by-line to compare it to the original design specification and a checklist of standards. Errors detected at this stage were logged to a data base which was later used as a quality assurance tool in determining the most prevalent causes of errors, and help to determine how to eliminate them in the future. IBM figures that 85 percent of software errors were eliminated by this process, a remarkably high percentage based on industry averages. After this process, a software module was subjected to a rigorous eight-level testing process conducted by an independent IBM organization (i.e., not directly connected to the software development organization). After passing the eighth and final test, the software was considered to be "man-rated." To support the development and testing of PASS, IBM relied on a collection of Shuttle-specific software tools involving more than 2 million lines of code.[35]

Programs used in such critical systems as air traffic control cost some fifty dollars per line of code, as written, documented and tested. Such programs averaged ten to twelve errors per thousand lines of software, and to NASA this was unacceptable. It paid one thousand dollars per line and cut the defect rate by two orders of magnitude, to 0.11 errors per thousand lines.

IBM carried out twenty-four interim releases of PASS software from October 1977 to STS-1 in April 1981. The ninth release in December 1978 gave full software capability to support the first flight. Nevertheless, additional releases proved necessary to accommodate continuing changes in requirements and to correct discrepancies.[36]

While IBM was working on PASS, Rockwell proceeded with the backup flight system. Initially it was simple indeed; prior to the Approach and Landing Tests, an early version was intended merely to stabilize the vehicle long enough for the crew to bail out. This backup software was intended only for the ALT series; it would be deleted once confidence in the primary flight software was in hand. However, this primary software had five hundred thousand words of code; it would take up to ten thousand years to test every possible set of logical branches. The need for backup flight software, operating during both ascent and descent during flights to orbit, thus reappeared. It had to accommodate abort options, which included once around the earth followed by reentry. The backup code now needed navigation and guidance, and quickly it loomed as a major software effort in its own right. From an initial twelve thousand words, it grew to nearly one hundred thousand.

Rockwell was not IBM; it lacked the in-house staff and facilities needed for large-scale software development. The company therefore took the role of prime contractor for the backup system, providing the operating system, the displays, and the overall integration. Intermetrics wrote code for system management, sequencing, uplink, and ground check-out. The Charles Stark Draper Laboratory, another offshoot of MIT, programmed the guidance, navigation, and flight control.

With this division of responsibility established, matters went downhill quickly. Draper needed detailed descriptions of the operating system; then, when these were not available, its programmers had to fall back on developing their best guesses. Intermetrics wrote sequencing codes but found that the sequences depended on guidance and navigations events, for which Draper had been unable to do the coding. Rockwell could not give proper attention to the operating system because its staff was spending most of the time resolving problems of integrating the three sections of software.

New simulation techniques helped, allowing Draper to run code successfully using its version of the operating system. The next step was to integrate the software from Draper, Intermetrics, and Rockwell into a single tape for verification at Downey, and to compare the output with equivalent data from the tests at Draper. However, continuous comparison was not possible at first because no automated plotting programs were available. Edward Chevers, a specialist at NASA-Johnson, notes that during early use of simulations in testing, the backup system "was a virtual basket case. When output data were compared, there were times when it was impossible to determine whether they were from the same program."

With time, people developed better understanding of the differences in output results, while more of the testing process became automated. This permitted comparisons between the backup and primary systems. One test case showed close agreement—but both proved to be wrong! This led the engineers to test additional cases. The backup flight system also underwent extensive testing at the Flight Simulation Laboratory in Downey and at SAIL in Houston. These tests generally went well; although a few of them did not, programmers traced the causes and established reasons for shortfalls in code performance. Prior to STS-1, enough data was in hand to give high confidence that the backup system would be ready for use, if the necessity arose.[37]

Avionics

The shuttle's digital flight control was an advanced form of fly-by-wire, a term that dated to 1956, when *Aviation Week* used it in writing about research at Wright-Patterson Air Force Base. In an initial experiment, during the mid-1960s, investigators fitted a B-47 bomber with a rate gyro to sense motion, a set of actuators, and an analog computer. This system controlled only the pitch axis, and operated in parallel with the standard mechanical system. Pilots took off under mechanical control and then switched to fly-by-wire.

Its success sparked additional Air Force interest, which grew further during the Vietnam War. The Air Force found itself losing substantial numbers of F-4 and F-105 fighters due to enemy fire that hit vital flight-control mechanisms. Other planes, hit similarly, made it back to base because the damage had spared their own mechanical control systems. Fly-by-wire now drew interest because it offered the prospect of multiple wire paths to the control surfaces. Battle action might cut several such wires in a redundant set, while leaving one still intact—and that would suffice for a safe return from combat.

Rechristened the Survivable Flight Control System, fly-by-wire took to the sky anew in April 1972 within a modified F-4. Analog flight computers from Sperry commanded the control surfaces; pilots from the Air Force, Navy, Marines, and NASA flew this plane at Edwards Air Force Base. This was one of the first aircraft to fly without mechanical flight controls. Its fly-by-wire system had been planned as a safety measure, but its control laws also made the plane more responsive and more maneuverable. The flight-test program ran for thirteen months.[38]

NASA also liked fly-by-wire. This agency had used this form of digital control to command the reaction-control rockets of Gemini and Apollo. With the shuttle in view during the early 1970s, it wanted to extend this technology to aircraft. Further, it went the Air Force one better by using digital rather than analog systems. NASA worked with an F-8C Navy fighter at its Flight Research Center, initially installing a surplus Apollo guidance computer with an inertial measurement unit that had been built for use in the piloted Lunar Module. The F-8C made its first digitally controlled flight in May 1972, less than a month after the initial flight of the Air Force's modified F-4.

Beginning in 1976 this F-8C served as a test bed for shuttle hardware. The single Apollo computer gave way to three IBM AP-101 units, with software that implemented the voting procedure if their results disagreed. The software also tested redundancy management for the shuttle, issuing orders to synchronize the three computers and marking one of them as a failed unit if it could not do this within four milliseconds. The F-8C gave greater realism than a flight simulator, and assisted in the development of control laws. These took form as algebraic equations, programmed in software, that converted pilots' movements of the cockpit controls into deflections of the shuttle's elevons or rudder.[39]

Conventional aircraft controls placed a stick or yoke with a wheel in the cockpit for roll and pitch, with rudder pedals for yaw. Hydraulic boost systems included elaborate sensing devices to provide feedback to the pilots, requiring that they exert greater forces on the controls to produce larger deflections of the aerodynamic surfaces and to achieve rapid response or abrupt maneuvers. The shuttle kept the hydraulics, which it needed to apply large forces to the control surfaces, but made no attempt to provide feedback. It also kept the rudder pedals. These operated the rudder by using fly-by-wire during atmospheric flight, then steered the nosewheel and applied the brakes on the main wheels when on the runway.

Instead of a stick or yoke, the flight crew used translational and rotational hand controllers. The latter was a joystick; you grasped it in your hand, turning it in roll, pitch, and yaw, and controlling vehicle attitude accordingly. During ascent, the SSMEs and SRBs gimbaled under automatic control, but astronauts could override this and steer using the hand controllers. These same controllers gimbaled the OMS engines, commanded the reaction-control thrusters, and executed rendezvous maneuvers. During reentry, they controlled these thrusters as well as the elevons, in accordance with the programmed control laws.[40]

Flight-control specialists could change these laws, thereby modifying the vehicle response, by writing new code. In this fashion, they dealt with a "pilot-induced oscillation" (PIO) that Enterprise displayed during the final flight of the Approach and Landing Tests. A PIO somewhat resembles the zigzag course of a motorist who turns the wheel rapidly from side to side, making the car swerve to left and right like a skier doing a slalom. For Enterprise, it arose because there was a small but noticeable time delay between the moment a pilot moved the hand controller and when the craft could begin to respond. Writer and computer specialist James Tomayko notes that if the delay was too long, the pilot could lose patience and move the joystick even more. By then the first command was in process, producing a larger effect and causing an overshoot. Seeing this, the pilot could react by using the hand controller to give an opposite command—which could result in an overshoot in the opposite direction. The resulting oscillation, such as an up-and-down motion in pitch, could grow quickly and cause the vehicle to crash.

For Enterprise the cure emerged from a flight-research program that used the NASA-Dryden F-8C. The solution consisted of new software instructions that enabled the computer to recognize the sharp back-and-forth movements of the hand controller that would bring on the PIO. The new code prevented the flight computer from blindly following the pilot's ac-

307

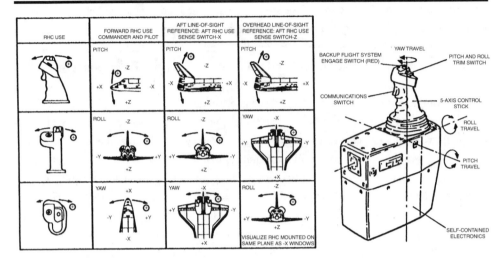

Fig. 61. *The rotational hand controller, or joystick. (NASA, courtesy Dennis R. Jenkins)*

tions. Instead, this computer was to command a smooth response, as if that motorist's steering mechanism prevented a slalom by responding sluggishly to his wild movements of the wheel. This solved the PIO problem for the shuttle, with the new software being installed aboard Columbia for its first flight to orbit. This work also contributed to aviation as a whole, for it introduced new methods for preventing PIO that applied to other fly-by-wire craft.[41]

Within the shuttle, standard quadruple redundancy applied to the control surface actuators. It was essential to prevent a hardover command to a control surface. The actuator did this by receiving four electrical inputs, one from each on-board computer. Each input drove its own hydraulic device; the sum of these four devices' hydraulic pressures then drove the main hydraulic actuator. If one of these inputs opposed the other commands, as by erroneously calling for a hardover, this set up a force fight among the four devices. The opposing input would be overpowered, three to one, and the sum of the pressures again sufficed to drive the main actuator correctly.[42]

It took more than hydraulics to steer the shuttle properly; like any launch vehicle, it needed on-board guidance. During the 1950s, when the Pentagon had pursued its big missile programs, inertial guidance had loomed as one of the more difficult technical issues. But in the 1970s, suitable sensors were available off the shelf. The shuttle used two Ball Brothers star trackers, which measured azimuth and elevation of bright stars. Doors protected them from the atmosphere during ascent and entry; automatically operating shutters gave protection against viewing the sun while on orbit. The computer software stored a catalog of a hundred stars, allowing observation in virtually any orbital attitude or location. These star trackers aligned a set of three inertial measurement units, which supplied data on vehicle attitude and acceleration. These units, from Singer-Kearfott, combined that firm's KT-70 Gyroflex platform, used on the A-7E attack jet, the P-3C antisubmarine warfare aircraft, and the SRAM attack missile, with the SKN-2600 electronics package that had been selected for the F-16 fighter.

The orbiter also mounted four Bendix accelerometers and four Northrop rate gyro assemblies, with the latter serving for attitude control. It proved difficult to find acceptable locations for these gyros, and the vibrational test program at NASA-Marshall contributed substantially to the eventual solution. Ideally, these instruments would have been mounted at the center of gravity, to measure true rigid-body rotation. The aft bulkhead of the payload bay was close to this center of gravity, and designers initially mounted a rate gyro assembly to each of this bulkhead's four corners. Ground vibration testing subsequently found local resonances that made these positions unacceptable. The mounting locations were changed twice before an acceptable position emerged, at the center base of the bulkhead.[43]

Data derived from these sensors was presented on the flight-deck displays, which were generally conventional in character. These instrument panels included more than twenty-one hundred switches and controls, with many additional functions being operated through keypads. This spaceship's cockpit instruments included attitude indicators, horizontal situation indicators that resembled artificial horizons, Mach indicators that gave accurate readings at high angles of attack, altitude and vertical velocity indicators, flight control position indicators, reaction control system activity lights, and a g-meter. Cathode ray tube displays, three on the forward instrument panel and one in the mission specialist station, were the most innovative features. Each video screen measured five by seven inches and had its own keyboard; switches allowed crew members to select guidance and navigation, systems management, or payload as major functions. These displays could access any data in the computers.

These videos represented a step beyond conventional flight-deck layouts, where all the instrument displays were electromechanical and each instrument served a dedicated purpose. Such flight decks were severely crowded because they had to present all items of data that a flight crew might ever need. During the 1980s, commercial airliners installed "glass cockpits" that relied primarily on video displays; even standard instruments such as artificial horizons became images on a cathode ray tube. Because each such display could serve many functions, the character of the flight deck changed dramatically as it merely presented the data that the crew needed at a given moment, with the rest being held in reserve. The shuttle flight deck was not a glass cockpit, at least not during the 1970s and 1980s, but its use of general-purpose video displays was a significant step in that direction.[44]

The flight deck, which served as the on-board control center, also functioned as the communications center. Basic requirements called for voice channels up and down, an uplink command channel, a downlink telemetry channel, two-way Doppler to determine velocity for navigation, a ground ranging capability, and provisions for encryption and decryption of voice and data to meet Air Force needs for security. In addition, there were a number of specialized requirements.

The communications system had to accommodate voice, command, and data traffic with existing ground-based tracking stations: NASA's Ground Spaceflight Tracking and Data Network, the Air Force's Space Ground Link System. It also had to be able to use NASA's Tracking and Data Relay Satellite (TDRS) system, in geosynchronous orbit. Downlink data included real-time operational telemetry, television, and wide-band data from payloads and experiments. Because the orbiter was to fly as an airplane, it had to communicate with air traffic control. The system needed all-attitude operation, voice and data links to a free-flying astronaut, text and graphics uplinks, tracking of satellites for rendezvous, and extensive remote control from the ground to reduce flight crew workload. In addition, all antennas had to be either flush-mounted to the skin, under the thermal protection, or else deployable on orbit and retractable for ascent and entry.[45]

NASA's TDRS was innovation in its own right. It was to replace many NASA ground stations, serving as a communications satellite system for use with low-orbiting spacecraft. These were to point their antennas upward, communicating with the TDRS spacecraft, which would relay their data to Earth. Late in 1976, NASA awarded the contract for this program to a consortium of Western Union Space Communications, TRW, and Harris Electronics. The $796 million award was the largest NASA contract to that date for an automated space project. Initial plans called for two spacecraft to ride Atlas-Centaur rockets to geosynchronous orbit during 1979, with the system becoming operational in 1980. This did not happen; the first TDRS did not reach orbit until 1983. The next one was lost aboard Challenger in 1986, but the complete three-satellite constellation finally entered service in 1989. Until then, NASA had to continue to operate its ground stations.[46]

Nevertheless, the prospect of TDRS allowed shuttle designers to install a dual-purpose antenna. Communications with TDRS called for data rates as high as fifty megabits per second; transmission frequencies had to be correspondingly high, which allowed the use of a small antenna that still gave a well-collimated beam. This parabolic antenna, three feet in diameter, deployed from within the payload bay. It also served as an on-board radar, tracking satellites at distances of up to twelve nautical miles and determining range, range rate, angle, and angle rate. Astronauts had to interrupt their TDRS communications to use this antenna in its radar mode, but this saved the weight of a second antenna and its associated electronics.[47]

When communicating with a ground station, the shuttle used its radio link for navigation. The station was at a known location, and the radio had a sharp frequency. Receivers, both at the station and on the orbiter, measured the frequency shift produced by the Doppler effect, and converted this shift into an accurate time history of range rate. This data permitted determination of the shuttle's position and velocity, or, equivalently, its orbital elements. Knowledge of these elements then allowed prediction of the vehicle's flight path in space.

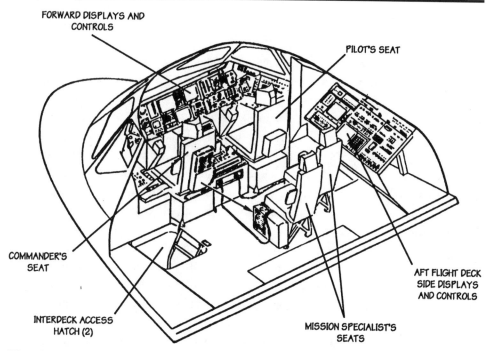

FORWARD DISPLAYS AND
CONTROLS

PILOT'S SEAT

COMMANDER'S
SEAT

AFT FLIGHT DECK
SIDE DISPLAYS
AND CONTROLS

INTERDECK ACCESS
HATCH (2)

MISSION SPECIALIST'S
SEATS

Fig. 62.   *Orbiter flight deck. (NASA, courtesy Dennis R. Jenkins)*

Air-traffic control brought its own issues. Collision avoidance was not one of them; NASA simply barred all traffic from the vicinity of a launch or landing. But the shuttle faced significant issues in finding its runway and executing a final approach.

Following reentry, the orbiter surrounded itself with a sheath of ionized gases that resulted from atmospheric heating. This brought a radio blackout. On emerging from the blackout, this vehicle needed a navigation aid that could steer it toward the runway. The solution lay in the Pentagon's Tactical Air Navigation (TACAN) system, with a transponder on the ground. In response to transmissions from the shuttle, it gave both bearing and range, with the latter found by bouncing a signal pulse off this station and measuring the time of flight. TACAN was certified only for altitudes up to 50,000 feet, and the shuttle needed 140,000. Nevertheless, standard TACAN proved suitable, and three off-the-shelf transceivers went aboard the orbiter.[48]

Yet while TACAN could allow the shuttle to navigate and to find its way to the airstrip, it was unsuitable for final approach. The vehicle needed equipment that used radio to mark the specific direction in the air that led to the runway. The Air Force used Ground Controlled Approach, with radar at an airfield and a ground controller issuing instructions. It was like being blindfolded and having someone call out directions so that you could walk a straight line. This was unsuitable for NASA, which wanted the shuttle to execute automated approaches, following a radio beam to the runway under computer control.

This was possible in principle with the Instrument Landing System, which had been the civilian world's standard bad-weather landing system since 1948. It used two beams transmitted by separate antennas. The localizer beam showed the way to the airport; the glideslope beam pointed into the sky at an angle of three degrees above the horizontal, marking an electronic ramp for landing approaches. Common practice called for airliner pilots to receive the ILS signals using on-board radios and to fly their planes accordingly, by hand. Automated landing approaches also were feasible. Nevertheless, ILS had its own disadvantages. Its beams tended to bend or wiggle when some of their energy bounced off large buildings. At Chicago's Midway Airport, Runway 22L lacked ILS because its signal would reflect from the Sears Tower, producing high levels of interference. At Kennedy Space Center, the Vehicle Assembly Building was one of the world's largest structures. The shuttle's landing strip stood virtually in its shadow, where it faced similar interference. In addition, ILS offered only a single straight-in flight path; its transmitters amounted to a lighthouse that kept its beam fixed, requiring pilots to follow its direct line. This was unsuitable for the shuttle. A returning orbiter needed a range of glideslopes as well as freedom to maneuver before turning onto final approach, immediately prior to landing.

The solution called for a microwave landing system. Microwaves had much higher frequency than the transmissions of ILS; they could be focused more sharply, and were much less prone to interference from reflections. Their landing approaches were free of bends and wiggles; pilots said that a microwave landing was like flying down a wire. In addition, the short wavelengths of microwaves permitted compact antennas that could scan through the nearby airspace to cover a considerable volume. This gave a broad range of glideslopes and approach azimuths.[49]

The microwave beams were flat, thin, and triangular. The localizer beam represented a thin wedge resting on its side; the glideslope beam extended from side by side and angled above the horizon. Both beams scanned an orbiter five times per second, the localizer sweeping from side to side and the glideslope oscillating up and down. Together they swept through a volume of airspace that had the shape of a tall pyramid resting on its side, with the apex at the runway. The system covered a range of forty degrees in azimuth and thirty in elevation.[50]

The planned approach began following emergence from the radio blackout, with the orbiter navigating by TACAN, inertial guidance, and barometric altitude. At fourteen thousand feet it entered a wide circular turn, picking up the microwave signals and aligning with the runway. The on-board computers engaged an autoland system; under automatic control, the shuttle executed a straight-in heading at a glide-path angle of twenty-one to twenty-four degrees. The landing proceeded rapidly, taking seventy to eighty seconds from initiation of autoland to touchdown while covering eight to ten miles.

The vehicle velocity was typically 290 knots, with the aim point four thousand feet in front of the runway threshold. At nineteen hundred feet in altitude the crew retook control and pulled up, transitioning to a glideslope of three degrees. Relying on the energy of their flight speed, they held that three-degree slope, deploying the landing gear at five hundred feet. Below two hundred feet, an on-board radar altimeter provided altitude data. The shuttle crossed the threshold at sixty feet in wheel height, flared to a path angle of one-half degree, then touched down twenty-five hundred feet inside the runway. The speed at touchdown was to be 170 knots.[51]

Like the on-board computers, the microwave system drew extensively on existing designs. The contractor, AIL Division of Cutler-Hammer, built the ground equipment as a modification of a landing system developed for the air force of Sweden. Airborne electronics were adapted from a landing system built for the U.S. Army, picking up azimuth and elevation signals and interrogating a ground receiver that served as distance measuring equipment. Installations were initially slated for Kennedy Space Center and for Edwards AFB, where a version of this microwave system participated in the Approach and Landing Tests.[52]

The combination of on-board computers, software, and advanced avionics paced the state of the art. Similar equipment entered use with the F/A-18 fighter, which went into operational service in 1980. But a 1989 NASA review noted that "even today" the shuttle's electronics "remains the most sophisticated, most advanced, most integrated system in operational use." It represented the first successful attempt to implement a comprehensive fail-operational redundancy concept in an avionics system. It also introduced complex techniques for redundancy management.

It used digital data buses to perform flight-critical functions. Its software was written in a high-level language. It made extensive use of flight software programs stored in a tape memory, thus expanding the effective size of its computer memory. It relied on digital fly-by-wire in operational atmospheric flight, not merely in test. It placed multifunction cathode ray tube displays on its flight deck, and integrated flight control with the rest of its avionics. In addition, while taking advantage of current developments in on-board computer applications, the space shuttle program also advanced the application of computers to ground checkout and preparation for launch.[53]

The Launch Processing System

Countdowns predated the use of computers during preparation for launch. Some veterans still remember the 1950s, which featured programs such as the Thor missile. Weeks of preparations culminated in a "hot count," a countdown preceding a static firing on the launch pad, followed a few days later by the actual launch. The count took four hundred minutes and featured a thousand separate events, including checks of equipment on the ground as well as aboard the Thor. When problems showed up, the count would hold until they were fixed, and that could take a while.

The ground crew put power into the missile and proceeded with check-out, with the guidance system causing many of the delays. During one long hold, over a difficulty that was not in his area, the propulsion specialist Bill Ezell of Rocketdyne went out and played nine holes of golf, only to return and find that nothing had changed. Some people took the long delays as opportunities for a catnap; they dozed in a car or lay down on a drawing board in a room that adjoined the launch center. But if the problem really was serious, the test conductor would scrub the launch. "Some counts went for thirty hours before we canceled," Ezell remembers. Then, with the fault finally repaired, the count would resume again—right from the top, at T−400, with plenty of opportunity for more to go wrong.

Jim Dorrenbacher, a Thor man at the contractor, Douglas Aircraft, recalls that preparation for a launch "involved a lot of what-ifs. For instance, if there was a problem, you had to see if it was in the vehicle—or the test equipment. You always wondered if you were doing it right. And there were never enough people with the background and experience." As the count neared zero, "you watched the pressure gauges, the power supply, the electronics gauges—those were tense moments. Everybody holding their breath and hoping that nothing wiggled." The test conductor faced a row of controllers, each with his own console. At the appropriate moment, still watching the gauges, each in turn gave a "go" sign by making a thumbs-up motion. The test conductor, said to have the "golden thumb," brought it down hard on the firing button, and the engine ignited with a roar.[54]

Launch vehicles with inertial guidance carried an on-board computer, which converted measured accelerations into velocity and position. The first Saturn I vehicles inherited their guidance packages from the earlier Jupiter missile and used manual check-out in the style of Thor, though now an automatic sequencer gave the firing command rather than a test conductor. An improved Saturn I, built for flight to orbit, used a new guidance system with a digital computer from IBM. The presence of this computer in the launch vehicle led directly to the use of another one for check-out. This trend started when NASA-Marshall, builder of the Saturn I, bought an RCA 110 computer to communicate with that on-board computer. At first the 110 merely handled communications and switching; engineers continued to activate procedures and conduct tests manually. But the 110 could do more, for it was one of the first priority-interrupt computers on the market. It permitted the division of a program into several sections, each with a different priority level, and allowed an engineer to switch immediately from one test to another during operations.

In 1962 IBM proposed that Chrysler Corporation, the manufacturing arm of NASA-Marshall, could use the 110 to conduct some check-out tests automatically, at the factory. Chrysler did this, using a new computer language, HYLA. In addition, several computers from Packard Bell entered service. They had their own language, SOL; they shared a common memory, and conducted further automated check-outs of Saturn I elements.[55]

Manual check-out meant that operators scanned gauges and dials, flipped switches, obtained responses, and judged whether observed readings were within acceptable limits. Automated check-out meant that at least to some degree, a computer could step through a sequence of measurements and compare them against preprogrammed standards. It saved time by eliminating much of the need for a test engineer to verbally verify a checklist over a telephone link, when working with another engineer at the launch pad. Automation also assured repeatability of a test or procedure, duplicating the exact conditions of a failure and reverifying a system after components had been replaced.[56] Still, manual check-out was what people knew and trusted; automated check-out entered Saturn I operations only in increments. Ten of these vehicles flew between 1961 and 1965. Flight Five, in January 1964, was the first orbital mission; it introduced the on-board IBM computer and the on-line 110 in the blockhouse. Subsequent flights added new automated guidance system tests, while the launch crew installed the first cathode ray tubes for data display. Check-out of the first stage went from fully manual in Flight One to 60 percent automated in Flight Nine. But computers did not replace the manual procedures; the launch team was ready to revert at any time to hands-on operation.

Even so, it was clear from the start that the Saturn V would be far too complex for manual work. In developing that rocket's automated procedures, the Saturn I-B program served as a test bed. The RCA 110, fitted with more memory and renamed the 110A, became a workhorse. Ludie Richard, a senior manager at NASA-Marshall, placed a 110A in the blockhouse of Launch Complex 34, home of the Saturn I-B, with this computer controlling a second 110A at the pad. This prefigured the planned Saturn V arrangements, which called for similar use of dual 110As.[57] The Saturn I-B first flew during 1966, and a 110A provided the data links. All test transmissions went through this computer; if it failed, the entire check-out would stop. But Hans Gruene, head of Launch Vehicle Operations, refused to place sole reliance on the computer for check-out. While the 110As indeed could conduct check-out procedures, Gruene insisted that his launch crews were to retain the option of reverting to manual test, as with the Saturn I. This reflected more than conservatism; it resulted from significant difficulties in writing computer programs that could carry out the test procedures. IBM initially did the programming, writing code in machine language. However, those programmers worked largely by translating English-language written instructions, and these did not spell out every detail; many contingencies depended on engineers' intimate knowledge. The programmers thus lacked full awareness of launch procedures, while Gruene's staffers were in no position to write code.

NASA's Ludie Richard came to the rescue. He appreciated that if test engineers were to embrace automated check-out, they would indeed have to write the pertinent computer pro-

315

grams. His staff responded by inventing ATOLL: Acceptance Test or Launch Language. The initial Saturn I-B launch, early in 1966, used only half a dozen ATOLL routines, but later flights in this series showed increasing numbers as launch engineers took advantage of this language's user-friendly features.

The Saturn V built on these developments. Its computer systems embellished those of the Saturn I-B, placing a 110A in each firing room at the launch control center. This moon rocket rode a mobile platform to the launch pad, and each platform mounted its own 110A. This permitted check-out in the Vehicle Assembly Building as well as at the pad.[58] The 110A served for countdowns of the complete Saturn V. It also saw service in check-out of its S-IC first stage and its Instrument Unit. However, the second and third stages—the S-II from North American, the S-IVB built by Douglas—used a different arrangement. Both burned liquid hydrogen, mounted the J-2 engine, and ignited at altitude. Both stages used a different check-out computer, the Control Data CDC-924A, which gave added flexibility and used newer computer technology to provide additional test capability.[59]

Separate events brought automated check-out to piloted spacecraft. As early as 1961, Project Mercury engineers found that they had their hands full in dealing with fewer than a hundred telemetered measurements. Studies of Apollo projected an increase to 2000, which meant that automated check-out was inevitable. Two engineers, Thomas Walton and Gary Woods, demonstrated that digital equipment could display data that previously had appeared on dials and strip charts. This won them permission to look for a computer that could handle the data in real time. They chose the Control Data CDC-168; Walton then persuaded the Gemini Project Office to use two of them to check out that spacecraft.

This raised the question of whether a single set of check-out equipment might serve both the Saturn V and its Apollo moonship. Managers from NASA-Marshall and from Cape Canaveral met during 1963 and decided not to do this; these vehicles' check-out arrangements would remain separate. This brought the advent of ACE, Acceptance Checkout Equipment, with its CDC-168s. The first version went on line late in 1964, with subsequent versions supporting Apollo and Skylab.[60] The final Saturn V check-out system tested 2,700 discrete signals and determined its overall health by receiving 150,000 signals per minute. ATOLL also flourished. There were 21 programs in this language on the first Saturn V launch in November 1967, 43 on Apollo 11, and 105 on Apollo 14. The flexibility of ATOLL made it particularly useful for Saturn systems that changed from one mission to the next.[61]

During check-out an engineer selected a desired program by typing an instruction using a keyboard. The programs were stored on tape; using a video monitor, the computer displayed the status of that routine and stated when it was loaded from the tape and ready to run. If the computer found a hardware problem, it gave the engineer a choice: terminate that program and select another; go back for another review of the original options; return to the step that produced the fault; disregard the problem and proceed to the next test step. For instance, if the routine was to check a propellant system, one step in the sequence had the computer issue a command to an instrument mounted on a valve that controlled an engine's ratio of fuel to oxygen. The instrument signaled the position of the valve, open or shut, and the system checked this response and store it in memory. As the test proceeded, the results appeared on the engineer's display and were printed out and recorded on tape.

Automated check-out was something of a mixed blessing. As Wernher von Braun put it, "The more you know about your vehicle, the greater the likelihood that you do discover something that is not operative and as a result you cannot launch. So it is like the man seeing his doctor every two weeks. We have assurance that we won't launch a sick bird. But at the same time, the probability of hitting a red line while we are on the count is vastly increased."[62]

Not all on-board systems lent themselves to automated check-out. Propellant leaks still required a visual search; mechanical failures were best diagnosed through personal inspection.[63] However, designers succeeded in building suitable check-out equipment for the subsystems that used electrical actuation and measurement, allowing the majority of check-out tests to proceed under computer control. Check-out of the Saturn V, including its Instrument Unit, was 90 percent automated; for the last twenty hours of a countdown, this number was 85 percent.[64]

Automated check-out also saved time while allowing engineers to increase the scope of their tests. The technical staff of Douglas used manual techniques on early S-IV stages, which flew as second stages of the Saturn I. The work required twelve hundred hours per stage. Its subsequent counterpart, the S-IVB of the Saturn I-B and Saturn V, used a single J-2 engine where the S-IV had flown with six RL-10s. Managers nevertheless did not simplify their preflight procedures; to the contrary, they increased the amount of testing by 40 percent. However, they also replaced the manual check-out with fully automated procedures—and cut the check-out time to five hundred hours. Computer-assisted check-out also helped the Saturn family attain a near-perfect flight record, for all Saturn I, Saturn I-B, and Saturn V launch vehicles lifted off successfully and flew without a disabling failure.[65]

The space shuttle's automated check-out arrangements emerged from this background. The shuttle brought new requirements, with speed of launch preparation being at the top. It had taken as much as half a year to make a Saturn V ready for flight. For the shuttle, this was to drop to two weeks. At the Kennedy Space Center, Tom Walton took the lead in planning a new Launch Processing System (LPS) that would support such a schedule.

Apollo had used five major systems for check-out, control, and monitoring, each with its own unique hardware, software, operational procedures, and maintenance. Walton anticipated that his LPS would take shape as a greatly enlarged ACE, a unified system that could deal with the entire vehicle. Nor was he willing to farm out its development to a contractor; he headed his center's Design Engineering Directorate, and he was prepared to do the work in-house. He won support from the center director, Kurt Debus, in 1972.

This broke with the practice of Apollo. During that program, the NASA centers responsible respectively for the Saturn V and the Apollo moonship also designed the control and ground support systems used at the Cape. NASA-Marshall thus developed check-out and launch equipment for its moon rocket, while the Manned Spacecraft Center did the same for Apollo's piloted spacecraft. This division of responsibility militated in favor of crafting ACE as a separate system. Indeed, when NASA's Joseph Shea went to Houston in 1963 to take over the management of Apollo, he transferred Walton's ACE group from Cape Canaveral and took it as part of his own center.

For LPS, however, Walton and Debus wanted the Kennedy Space Center to act as its own prime contractor. The center had the necessary staff and facilities, or could readily acquire them. The new approach meant that people who would operate the LPS would also be involved in its design, and would be thoroughly familiar with it. This contrasted with the Apollo approach, which had required Kennedy personnel to maintain and operate systems such as ACE that had been designed in Houston or at NASA-Marshall.

As an initial exercise, Walton's staff built a small model of a liquid hydrogen loading facility, with real valves and tanks and with colored water simulating this propellant. Using a Digital Equipment PDP 11/45 minicomputer, they wrote software that used a video display to present a schematic view of the pipes, tanks, and valves, with pressures and other parameters appearing next to appropriate components. This system transferred fuel to a model of the shuttle, under software control, with an engineer monitoring flow rates and pressures at the console.

In 1972 Robert Thompson, head of the shuttle project office at the Manned Spacecraft Center, selected the LPS concept of NASA-Kennedy over an alternative from his own center. In the words of James Tomayko, "Thompson judged Kennedy's to be the best proposal, but he also thought it more efficient for NASA to develop the Launch Processing System where it would eventually be used, and by the people who would use it."[66]

A colleague of Walton, Frank Byrne, laid out the basic architecture. He broke with the practice of Apollo, which had relied principally on a limited number of mainframes. These lacked the memory to hold all the check-out programs, which had to be loaded and run in a series. Instead, Byrne introduced the general use of minicomputers, placing some forty such machines within each firing room, including one at each console. Individual areas of effort—engines, propellants, avionics—could have their own consoles, each with its own disk drive and with appropriate software to run the test sequences. This brought the important advantage of parallel testing, saving time during check-out by allowing several sets of procedures to go forward simultaneously.[67]

The consoles and minicomputers were key elements in the Checkout, Control and Monitor Subsystem (CCMS), which was the heart of the LPS. In essence, the CCMS comprised the equipment within a firing room. It could be switched to serve an orbiter during check-out and postflight activity, within the Orbiter Processing Facility. It could also connect to the shuttle during preparation for flight, within the VAB or at the launch pad. The CCMS operated in real time, with all its minicomputers drawing data from a common memory. However, it handled only the most current data, for this memory held just 128 kilobytes. Hence the system also needed mass storage, to hold information from prior tests.

Apollo test data had been recorded and stored on magnetic tapes, which were filed away within an archive. Finding and displaying particular items of information was a time-consuming process, akin to searching for individual documents in a technical library. LPS introduced the Central Data Subsystem (CDS), built around two pairs of Honeywell 66/80 computers along with numerous hard-disk drives. These stored more than test results; they also held failure reports and descriptions of equipment configurations.

At the time of its installation, the CDS filled two-thirds of an acre within the Launch Control Center. Its two-hundred-megabyte drives gave it one of the world's largest "disk farms." You would think you were in the household appliance department of Sears if you had been there, for each disk drive was about the size of a kitchen dishwasher and looked like a washing machine, though they were painted blue rather than white. This was not a real-time system; it was as slow as the Internet, and like the Internet, it required users to wait for up to several minutes to retrieve specific items of data. In time the CDS overflowed, requiring old data to go back onto tapes in the archive. But during preparations for STS-1 and for the next few flights, the CDS greatly speeded access to engineering information.

A third element of the LPS, the Record and Playback Subsystem (RPS), addressed the unwillingness of launch controllers to fully trust their computers. Plenty of hard-bitten veterans had fond memories of the days of the golden thumb, and they wanted to see raw data in familiar formats such as strip charts, rather than the processed data of the CCMS. The RPS accommodated such wishes. It also allowed engineers to retrieve the unprocessed information, to verify that the CCMS had correctly interpreted it.[68]

The LPS brought other improvements. Apollo check-out systems had used display monitors but presented their data as rows of numbers, in formats that made it difficult for users to gain an overview of an entire system. No individual could deal with more than a fragment; it took hundreds of people in a firing room to count down and launch a Saturn I-B or Saturn V. Photos of the era show them sitting side by side in front of long rows of consoles, within an electronic dystopia that might have suited Orwell's *1984*.

LPS enhanced the reach of programs, cutting the number of controllers by enabling each of them to take broader responsibility. A CCMS firing room had only fifteen consoles, each with three large color displays. These introduced diagrams and color displays, like those of the model propellant-loading facility.[69] A new language, GOAL (Ground Operations Aerospace Language), made it easier still to write high-level computer code, as an example will illustrate. Here are steps taken from printed liquid-hydrogen procedures and their equivalent in GOAL:

Open transfer line valve GLHK1221E.

OPEN <GLHK1221E>;

a. In the event the open indication is not verified by GLHX1223E or GLHX1233E in 25 seconds, select secondary mode. In the event that the open indication is still not verified by GLHX1223E or GLHX1233E in 25 seconds, notify the console that the transfer line valve did not open.

VERIFY <GLHX1223E> IS OPEN OR <GLHX1233E> IS OPEN WITHIN 25 SECS THEN GO TO STEP 100;

OPEN <GLHK1231E>;

VERIFY <GLHX1223E> IS OPEN OR <GLHX1233E> IS OPEN WITHIN 25 SECS THEN GO TO STEP 100;

b. Energize GLHK1231E for secondary mode.

Monitor GLHK1229E and GLHT1227A transfer line liquid and temperature probes are set within one minute.

RECORD TEXT (XFER LINE VLV DID NOT OPEN) TO <PAGE A> BLINK RED;

TERMINATE;

After a stable wet indication on GLHX1229E and/or −423 °F on GLHT1227A, wait five minutes before continuing.

STEP 100 VERIFY <GLHX1229E> IS WET OR <GLHT1227A> IS LESS THAN OR EQUAL TO −423 DEGF WITHIN 1 MIN THEN GO TO STEP 200;

RECORD TEXT (XFER LINE LIQ SENSOR AND XFR LINE TEMP FAILED TO GIVE PROPER RESPONSES) TO <PAGE A> BLINK RED;

TERMINATE;

STEP 200 DELAY 5 MINUTES;

(*Source:* Henry C. Paul, CASI 77A-35323, p. 7-2)

IBM developed GOAL, intending it to be largely self-documenting and as close to the English language as possible. The first formal training course for test engineers, which included the study of GOAL, ran for only three weeks of half-day classroom instruction. Even so, many of these engineers remained reluctant to write their own code, and NASA set up a separate organization at Kennedy to take their requirements and translate them into GOAL.[70]

Procedures written in this language controlled everything from the simple opening of a valve to the sequence of activities that loaded propellants into the ET. A representative CCMS operation filled the ET with 140,000 gallons of liquid oxygen, transferring it from a storage tank a third of a mile away. This called for only a single console, C3, in the firing room. At the time of STS-1 its operator was Jimmy Rudolph. He began by running several programs to verify that the system was ready to begin. These programs established that all "exception monitor limits" were set to standby, all system measurements were being reported and all mechanical valves were being cycled to determine their readiness to operate. The CCMS did these things without human intervention, finishing within about ten minutes.

With verification complete, Rudolph awaited a "go" signal from the test conductor. When it came, he could respond by pushing a single button marked "Fill." This initiated the loading of the liquid oxygen, which continued automatically until completion. It started with a ten-minute chilldown of the pump and pipes in the storage area, to prevent this liquid from flashing into gas. The shuttle also underwent chilldown of its main engines and of the oxygen tank within the ET. These operations used cold gaseous oxygen that had boiled off from the liquid in the storage tank. Next came a slow filling of the ET tank until it was 2 percent full, further chilling the pipes and valves while preventing a damaging water-hammer effect, or surge of excess pressure, from developing within the pipes. Rapid pumping followed, filling this tank to 93 percent, with the CCMS then topping off the tank, again slowly, to 100 percent. Sensors atop the tank measured the amount; GOAL codes monitored the sensors, with programmed algorithms removing ambiguity due to sloshing of the liquid during the fill.

With the shuttle sitting on its pad, some oxygen boiled off within the ET, to be vented overboard. The CCMS monitored this boil-off and replenished it, ensuring that the tank stayed full. This continued until nine minutes prior to ignition, with the ET tank then being sealed, allowing the boil-off to pressurize it for launch. Following liftoff, other CCMS codes automatically safed the ground support equipment.[71]

From the outset, in 1972, it was clear that the LPS would have to handle at least as many data parameters as Apollo. The final Apollo missions had used some ninety-three hundred data parameters; initial estimates put the number for the shuttle near eighty-five hundred, though this grew to forty thousand by STS-1. As with Apollo and earlier programs, CCMS was to check out the ground equipment as well as the shuttle proper. The requirements for this task were comparable to those of checking out the entire orbiter.

Development of the LPS went forward smoothly. In March 1973 NASA-Kennedy's Design Engineering Directorate set up a concept-development group. Its first major set of specifications, the "LPS Concept Description Document," came out in October. A detailed set of functional requirements was in hand during April 1974. A month later, NASA awarded an $11.5 million contract to IBM to provide systems engineering and software development. This preceded the procurement of hardware and software, and permitted an integrated system design. It contrasted with previous programs, in which managers had first selected the hardware and then had to fit the software to a system that had been committed in advance.

Other contracts followed. The firm of Modular Computer Systems was chosen in June 1975 to provide minicomputers for the CCMS. The initial order called for sixty Modcomp II/45 machines for $4.2 million. Two months later, Martin Marietta received a $22.8 million award to provide the consoles and their supporting electronics. The fourth major contract went to Honeywell in December for its mainframes.

That firm's 66/80 computers reached the Launch Control Center early in 1976. In November Martin delivered the first consoles. These enabled IBM to write operational software; by February 1977 enough such software was in hand to allow system users to begin programming in GOAL. Firing Room 2 became operational in September 1977, with Firing Room 1 following in February 1978. Use of GOAL flourished. The staff wrote 1,381 programs in this language, totaling fourteen million words.[72]

Managers still liked hands-on involvement for some activities. This was particularly true at the Hypergolic Maintenance Facility, an operational center at the Cape that dealt with the OMS and reaction control systems. As late as 1998 the manager, Retha Hart, noted that to pressurize some tanks, "the engineer has to sit there and watch the pressure as it increases, then manually cut it off. Then let it go a little longer and again manually cut it off. He has to keep watching it as it builds up." Nevertheless, as the date approached for the shuttle's first flight, such vestiges of manual check-out were very much the exception. At the Cape, as well as in flight, the world of the shuttle was largely one of computers.[73]

CHAPTER NINE

The Program Struggles

No one ever built the space shuttle. With some fifty thousand people working within this program at its peak, none of them could ever take more than a tiny part of it as their personal concern. Instead, during the years of development, members of project staffs addressed a succession of small, well-defined tasks. Like the Space Shuttle Main Engine, made up of numerous parts that fitted together, the program as a whole amounted to an assembly of its own large number of parts. Program management then consisted of defining them, linking them in a logical order, tracking their costs, and measuring performance.

Schedule, Performance, and Cost

The process began with NASA's Requests for Proposal, soliciting bids for major system elements such as the orbiter or the SSME. Such documents presented requirements for contractor management systems. A selected contractor had to show that its management arrangements indeed complied with these requirements, and there was a strong tendency to use existing management systems that were already in place. At NASA Headquarters, Howard Roseman was the chief of program evaluation. He wrote that government officials expected to show "understanding of the way the contractor best conducts his business. His procedures, his reporting and control methods, and his preferred measurement techniques will be studied by NASA. It is very likely that these existing methods will already satisfy government requirements."[1]

A winning contractor received a Statement of Work, issued by NASA. It contained a Work Breakdown Structure (WBS) as a principal element. According to Dale Myers,

> It clearly defines what is expected of them by dividing their total tasks into logical
> and manageable elements, each of which may represent an element of hardware, a
> service or a function to be performed. For each element of the WBS discrete sched-

ules, cost targets and organizational responsibilities are assigned and the contractor is held accountable for their accomplishment. The Work Breakdown Structure is a "family tree" of activities, structured to make it possible to identify all program elements and their relationship to the program organizational structure. It represents the primary means for integrating the many thousands of tasks which must be accomplished.[2]

The WBS also provided a framework for parceling out work within a contractor's organization. Starting at the program level, corporate planners successively divided the effort into low-level increments that represented the way in which the work was to be performed. These increments defined manageable units of work, described for purposes of planning and control of cost, schedule, and technical performance.

The contractor aligned or integrated its organizational structure with the WBS so as to assign responsibility for the tasks to be performed. With functional responsibilities having been established, managers could further subdivide the effort into "work packages," with these packages being presented to the performing departments. Each work package described the task to be performed, identified the responsible organization, authorized expenditure of funds and resources, and identified budget and schedule constraints. Frederick Peters, chief of schedule analysis at NASA-Johnson, described these packages as "the basic building blocks used for detailed planning, assignment and control of the Project's performance."[3]

As a separate set of documents, contractors also received a sixteen-volume compilation from the program office at NASA-Johnson: JSC 07700, *Level II Program Definition & Requirements.* It gave a basis for project management, providing a compendium of specific requirements, directives, procedures, agreements concerning interfaces, and information regarding capabilities of systems. Volume 3 dealt with program planning and analysis, including schedules.

Program management, as discussed previously, amounted to a dialogue. NASA raised issues or defined requirements, and contractors answered with responses that could guide the process of decision making. The WBS, together with guidelines set forth in the JSC compilation, were intended to elicit a response from each prime contractor in the form of a set of program logic networks. In Myers's words, "while the WBS essentially describes *what* has to be done, the logic networks describe *how* and *when* the work must be accomplished."

Logic networks were schedules, prepared by a contractor, defining the detailed sequence and pace of the tasks laid out in the WBS. Such a network took form as a diagram resembling a flow chart used in computer programming, with this "flow chart" describing a program to be executed, not by a computer, but by a corporate organization. Logic networks presented the sequence and interrelationship of all project activities, along with the significant resulting events or milestones.

324

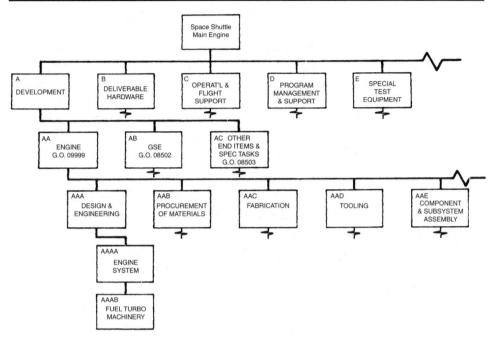

Fig. 63. *Work Breakdown Structure for the Space Shuttle Main Engine. (Dale Myers)*

The three NASA program levels each defined their own milestones. Level I, at Headquarters, gave those that this agency had committed to achieve in statements to Congress. Level II milestones came from the JSC program office; milestones at both levels were imposed upon the projects. An example was a date for delivery of flight-rated SSME engines. NASA project managers, at Level III, created their own milestones to monitor progress against those from the higher levels.[4]

Budgeting was important, with a contractor's work packages providing a basis. Each work element had its own budget, including overhead. Internal corporate accounting methods gave records of accrued costs. To track running costs in the course of ongoing work, contractors implemented the Performance Measurement System. It compared actual costs to those projected within the WBS; it also tracked schedules.

Every increment of effort, from a project down to the level of a work package, had a schedule of milestones representing what was to be accomplished. Each increment also had its budget, showing funds to be allocated from one month to the next, and one could match the planned expenditures to the milestones. This gave rise to three types of cost: (1) Actual Cost, the cost actually incurred or accrued to date; (2) Budgeted Cost for Work Performed (BCWP), the planned cost in the budget for achieving the milestones actually reached to date;

325

and (3) Budgeted Cost for Work Scheduled (BCWS), the planned cost in the budget for achieving the milestones planned to date.

To track the costs, managers kept a running tally over time of the following quantity: Cost Variance = (BCWP − Actual)/BCWP. Expressed as a percentage, this quantity gave values that one could plot over time to display a cost trend. Cost overruns stood out sharply in such a curve, with the trend serving to predict whether ultimate costs would meet those of a budget.

Costing also provided a basis for measuring schedule delays, which meant that the basis for determining such delays was in dollars rather than weeks or months: Schedule Variance = (BCWP − BCWS)/BCWS. Hence this quantity tracked the budgeted costs for milestones planned and achieved to any date, rather than the milestones themselves. Again, a curve of values, plotted over time, showed whether its corresponding project element was lagging.[5]

This system readily accommodated technically risky projects, including the SSME. These had particularly large numbers of milestones, to allow for closer monitoring of their schedules. They also had ample amounts of slack in the schedules, which took the form of time allotted between scheduled and latest-allowable dates for the milestones.

A related arrangement, the Technical Performance Measurement System, dealt explicitly with achievement of technical goals. It compared achieved values of performance, measured over time, with values specified by contract. Again, the emphasis was on quantitative determinations that permitted the portrayal of trends. The system called for identification and measurement of critical technical parameters, which permitted evaluation of variances and construction of plots showing changes in these variances over time.

For example, the orbiter was to achieve specified values of the following nine parameters, for which the contractor, Rockwell International, was responsible:

Payload to orbit, pounds
Lateral range (crossrange), nautical miles
External tank separation and disposal
Touchdown speed, knots
Intact abort coverage, percentage of flight to orbit
Time for emergency egress of astronauts
Ferry range, nautical miles
Turnaround time, weeks
Cost per flight

The latter was particularly important, for NASA had promised the OMB in 1972 that it would achieve a value of $10.45 million per flight in 1971 dollars. An entire requirements document within the Level II compilation, volume 16 of JSC 07700, addressed the issue of controlling this single parameter of cost per flight.[6]

With contractors using the Performance Measurement System to track their budgets, NASA exercised similar care in crafting its own budgets. The process began at the contractor level, with these companies presenting their requirements to their Level III project managers. All Level III managers had the responsibility for predicting their own cost requirements, including financial reserves to provide a cushion, and for submitting budgets. These went through a detailed review, within the Level II program management and then at Headquarters, with the federal budget for the space shuttle program being updated twice a year. After a budget plan had been approved in Washington, funds were then authorized directly to the Level III project managers, with the Level II program manager at JSC being responsible for assuring that all projects had the means to cover their budgeted requirements.[7]

Significant technical tools supported these management processes. Logic networks existed not only at the corporate level but at the project and program level, with the latter, at Level II, giving a master logic network. It gave a basis for schedule development, again at the program-wide Level II, and for construction of subordinate logic networks at the projects' Level III. The Level II logic network, maintained by the Program Office at JSC, used PERT II. PERT, Program Evaluation and Review Technique, dated to the Navy's Polaris program of the 1950s. NASA's version, written in the Fortran computer language, had been developed in 1965 for use on the IBM 7094 computer and in 1979 was adapted for use on the powerful IBM 370 series. PERT saw extensive use during the Apollo program and returned to service during development of the shuttle.

The shuttle program as a whole was far too complex for any one person to comprehend in its entirety. Still, although no single person could construct its complete logic network, it was possible to treat contractor-level networks as fragments of the whole, or "fragnets." A fragnet constituted any meaningful portion of a project network. The use of standardized nomenclature throughout the shuttle program meant that PERT could identify terms that were common to two or more fragnets. Then, with all such common terms being noted across the entire array of fragnets, PERT automatically linked them to form the master network.

These common terms represented interfaces, not only between program components such as the ET and SRB but also between organizations. The master logic network, at Level II, gave expected and allowed dates for planned events: an important test, for instance, or a technical review. In this manner, PERT supported the development of project schedules. PERT also accommodated the display of new fragnets, as subsets of the master network that crossed the boundaries of the earlier fragnets that had come from the contractors.[8]

PERT offered one set of management tools. Another set took form as NASA's Management Information Centers (MICs). These amounted to "war rooms," centralized locations where current information was available that covered a broad range of topics in depth. Such centers existed at Headquarters, in the JSC Level II program office, in the Level III project offices at the Johnson, Marshall, and Kennedy centers, and at the prime contractors: Rocketdyne, Rockwell International, Thiokol, and Martin Marietta, home of the ET.

The shuttle's MICs grew out of Apollo-era Program Control Centers at similar locations, which supported the Saturn V. Marshall's center had been rimmed with sliding status charts, including an enormous PERT chart that covered eight hundred entities and summarized ninety thousand events taking place around the country. Double screens accommodated briefings both with slides and with viewgraphs. Closed-circuit television linked this center to KSC and Houston. The center supported teleconferencing, with microphones in the ceiling that picked up people's voices where they sat. During such teleconferences, attendees could see the viewgraphs at other locations, from copies had been transmitted by fax. The MICs of the shuttle program were similar. They had devices for data retrieval and display, along with data and voice communication links. They supported their own teleconferences, using microphones, loudspeakers, and telephone lines to transmit briefing information. They did not offer the advantages of the Internet, which had not yet been invented. Still, they eliminated much need for participants to travel to meetings while widening the number of attendees.[9]

Shuttle program management thus resembled a well-oiled machine, deploying sophisticated resources for planning, budgeting, performance measurement, and presentation of information. However, the grease for this machine was money. When it was lacking, project managers could state precisely what they wanted to do, without necessarily being able to do it. NASA's Howard Roseman notes that "where requirements exceed available funds, the Program Manager will assess the most critical project needs and divide available resources according to program priorities." In choosing those priorities, however, the importance of certain issues could be overlooked. That is how people missed the problem of the thermal-protection tiles, which might have been caught much earlier and far less expensively.

In 1972 NASA had promised the OMB to carry through the development of the shuttle for $5.15 billion in 1971 dollars. This was another critical budget goal with its own separate document—volume 15 of the compilation JSC 07700—to guide managers in its achievement.[10] Still, this estimate proved to be low, and there were reasons. Humboldt Mandell, a NASA management analyst, notes "enormous pressure at the beginning of a program to estimate the actual costs to be lower than historical trends might indicate. Lower estimates simply increase the probability that the program will overrun its costs. Program managers feel that they will be able to do better than their predecessors, and they often are willing to assume high risks in initial program estimates to help sell the program in the political arena." Mandell quoted a 1962 statement by David Novick, a leader in modern cost estimation: "The incentives to estimate low are much greater than the penalties, if indeed there are penalties." Program planning, Mandell added, was also significant: "Programs with longer definition phases have proven to have the least cost and schedule overruns." The shuttle's definition phase was vanishingly short, with the promised cost of $5.15 billion coming virtually at the same time as NASA's decision to use solid boosters. Hence this estimate lacked a base of study in depth, to make it realistic.[11]

Management issues also were pertinent. In 1971, at the start of the shuttle program, NASA had conducted an extensive series of interviews that sought advice on how to avoid problems. Interviewees included Lockheed's Clarence "Kelly" Johnson, director of this company's Skunk Works that built the SR-71, which to this day remains the most advanced airplane ever flown. George Schairer, a vice president at Boeing, recounted the lessons learned from experience with commercial airliners. Robert Widmer, a vice president at General Dynamics, summarized similar experience with the Air Force's B-58.

An important recommendation called for management through small, hand-picked government program offices and contractor teams. Lockheed's Skunk Works was a classic example. In 1965, during the SR-71 program, it had only fourteen departments and no more than five resident customer representatives. Its Air Force project office was similarly small, with only 35 people. The Navy's Polaris program had been managed from a highly compact Special Projects Office—and it took the program from initial approval to operational missiles at sea in less than four years.[12]

The shuttle's counterpart, if any, was the Level II program office at JSC. Its manager, Robert Thompson, concerned himself largely with the overall configuration and with interfaces. In addition, the Level III project office for the orbiter, headed by Aaron Cohen, was also at JSC. Thompson gave considerable leeway to NASA-Marshall within its own domain, which included the SSME, SRB, and ET. This accommodated another conclusion from the 1971 interviews: to allow contractors maximum autonomy. However, the author Dennis Jenkins notes that the Level II office had a tendency to impose new requirements on the lower levels—without providing additional funds. The tight budgets that resulted saved money in the short run but invited delays, and these delays boosted expenditures in the longer term.

Other problems arose from concurrent development. In no way did the shuttle use off-the-shelf main engines or thermal protection; both were new, and both pressed the state of the art. They had to be developed concurrently with the orbiter, not only as projects in their own right but as elements of the complete shuttle system. The problems and delays of the thermal-protection tiles arose accordingly.[13]

Such issues affected other programs as well as the shuttle itself. Events showed that cost overruns could pile up until a project became too expensive and had to be canceled. This happened with the proposed Inertial Upper Stage. Unmet milestones could accumulate until only a massive injection of new funds could save a project. This happened with a key shuttle payload, the Space Telescope. Indeed, it came close to happening with the shuttle itself.

Shuttle Upper Stages

The shuttle was born amid high hopes. It took initial form amid NASA's far-reaching plans of 1969, which aimed at piloted missions to Mars by the mid-1980s. Space stations, thickly sown between the earth and the moon, were to play essential roles in preparing for these missions. The shuttle was to support the stations as a logistics vehicle, flying to low Earth orbit.

The shuttle could greatly extend its usefulness by launching spacecraft and satellites as well. However, much of this traffic required higher orbits, such as the geosynchronous orbit used in communications. The shuttle alone could not fly so high, but it might carry an upper stage that would reach such orbits. Still, amid the optimism of the early 1970s, NASA's planners aimed at nothing so simple as boosting satellites to geosynchronous orbit on one-way trips, as was already the practice. These people had their eyes on a space tug. In a typical mission, it would carry a spacecraft to geosynchronous altitude, inject it into orbit, then leave it and rendezvous with a second satellite. It would link up with that satellite, fire its engine anew, and return with this spacecraft to low orbit, where the shuttle would retrieve the mated pair. Alternately, the tug might carry a robotic manipulator and service a geosynchronous spacecraft by installing a new module, before returning to the shuttle as its mother ship. It indeed would return; the tug was intended to be reusable.[14]

NASA-Marshall conducted initial studies of such concepts. The work did not aim at detailed design, but rather sought to understand how increasingly capable tug concepts might accomplish a successively broader range of missions. A 1969 concept called for a tug that amounted to a shortened S-IVB stage from the Saturn V. It was to ride into space aboard that moon rocket, use the space station for its base, use a shuttle for propellant resupply, and carry a crew module. It would range as far as the moon, for which it was to carry landing legs.[15] But during 1970, as budget cuts hit home, NASA abandoned its plans both for a space station and for continued use of the Saturn V. The tug now became a true shuttle upper stage, to be carried within its payload bay. New concepts proposed a reusable Centaur, to burn hydrogen for high energy, as well as a reusable Agena, burning storable propellants that promised safety and simplicity in design.

In 1972, with NASA's budget offering little leeway, Marshall officials considered that they might be trying to bite off more than they could chew. It appeared prudent to hedge their prospects by commissioning studies of existing upper stages as low-cost stopgaps, allowing the shuttle to launch geosynchronous missions while awaiting the true space tug of the future.[16] The tug was a NASA concept, but this agency was willing to invite others to share its work. NASA already was preparing to give Spacelab to Europe and soon would give the shuttle manipulator arm to Canada; might someone else take on the project of a shuttle upper stage? The obvious candidate was the Air Force. The point man in the ensuing discussions was Malcolm Currie, the director of defense research and engineering (DDR&E).

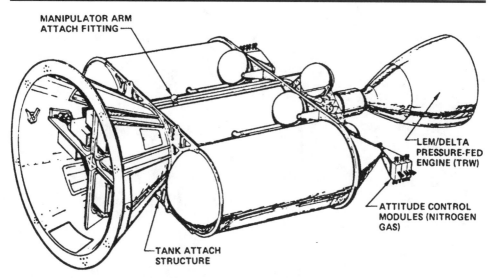

MANIPULATOR ARM
ATTACH FITTING

LEM/DELTA
PRESSURE-FED
ENGINE (TRW)

ATTITUDE CONTROL
MODULES (NITROGEN
GAS)

TANK ATTACH
STRUCTURE

Fig. 64. A space tug concept derived from a Delta upper-stage engine and associated tankage. (TRW)

George Low met with Currie in September 1973. Their main topic of discussion was aeronautics facilities, but Low also raised the matter of the tug. He said that NASA would find it highly desirable for the Air Force to fund and manage its development, not only because of NASA's budget problem but to draw NASA and the Pentagon closer within the overall shuttle program. Low wrote in a memo, "Currie is not at all adverse to this idea and felt that if the tug were to be a minimum tug to meet Air Force requirements, then an Air Force development with DOD funding would not be out of the question."[17]

Low and Currie met again a month later and reached a tentative agreement. The Defense Department indeed would develop an interim upper stage for the shuttle, and would do this by modifying an existing expendable upper stage. This version would not retrieve spacecraft in orbit but was to be designed from the start for reusability. The program was to face a cost ceiling of $100 million, but if it could not build a reusable stage for this amount, then the stage would be expendable. The Air Force secretary, John McLucas, endorsed this agreement. Currie wrote that the Air Force would use the new stage "with the Space Shuttle during the period when high priority military payloads will be transitioning from launch on current expendable vehicles to launch on the Shuttle." It also would "permit NASA to develop, at a more leisurely pace, an optimum upper stage with full capability."[18]

Still, Low, Currie, and McLucas resembled two-star generals. They could recommend, but others at higher levels would decide. The shuttle proper had required the endorsement of

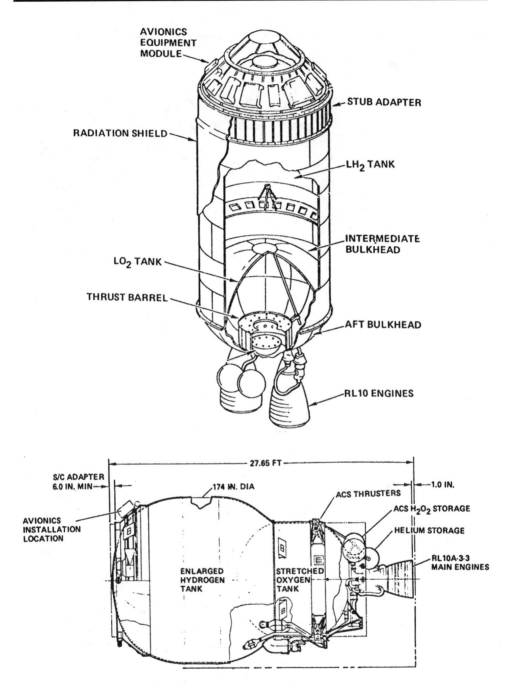

Fig. 65. *Centaur upper stage (top) with space tug concept designed as a derivative. (General Dynamics)*

Nixon himself, and the new Interim Upper Stage (renamed Inertial Upper Stage in 1977) needed its own support. Currie took an important step in February 1974, as he and Walter LaBerge, the assistant Air Force secretary for research and development, met with the deputy secretary of defense, William Clements. Currie phoned Low the next day, and Low wrote that the meeting had been "very positive. Clements quickly realized that for a relatively small investment in dollars, the DOD can get a large piece of the action." He decided that the Pentagon indeed would develop the IUS.[19]

Not all of Clements's colleagues agreed. In May Currie warned Low that they faced opposition from Leonard Sullivan, the assistant secretary for program analysis and evaluation. In Low's words, "Apparently none of us in NASA, in DDR&E, and in the Air Force had reckoned with Sullivan or even remembered that he existed. He had not been involved by any of us in any Shuttle discussions and apparently resents this. But more importantly, he is involved in DOD Systems Acquisitions Reviews, and in the last one on the Shuttle, he presented some very negative results of studies that he had performed. His conclusion is that the Shuttle has no economic or other benefit to the DOD and, therefore, the DOD should not support it financially." Sullivan had the ear of the secretary of defense, James Schlesinger, and asserted that Schlesinger "would just as soon not fund any part of the Shuttle," not even the IUS. He might "just barely go along" on the IUS, but only because he "does not want to be the member of the Administration who scuttles the Shuttle."[20]

Further meetings ensued at the Pentagon, and Sullivan won an important point as his colleagues agreed that the Defense Department would not pay for the procurement of the orbiters that it would use. NASA would pay to build them and then would allow the Air Force to use them freely as elements of a national system that was to be made available to all. Sullivan gave ground as well, for Currie and LaBerge convinced him to go along with an informal DOD position that gave full support to the IUS.[21]

On 11 July senior Pentagon officials agreed that they would indeed commit to developing the IUS. Four days later, Fletcher and Low met with Schlesinger and Clements to discuss the shuttle. Schlesinger confirmed that this agreement constituted the decision. He then stated that he considered the Pentagon to be a partner with NASA on the shuttle. Low shared the news with Currie and asked him to send Clements a copy of a study on uses of the shuttle. In August Clements sent Fletcher a letter of his own: "Once the Shuttle's capabilities and low operating cost are demonstrated we expect to launch essentially all of our military space payloads on this new vehicle and phase out of inventory our current expendable launch vehicles."[22]

These expendables might be slated for phaseout, but both NASA and the Air Force expected them to leave a legacy in the form of an existing stage that would serve as the IUS. New use for such upper stages was an old story. Agena, which first flew in 1959 atop the Thor, later was adapted for Atlas and Titan III-B. Centaur, built initially as an upper stage for

Atlas and first flow in 1963, was being readied for the Titan III-E.[23] It was natural to envision that the IUS would evolve in similar fashion.

Early thoughts focused on four existing liquid-propellant stages. Table 9.1 illustrates their capabilities when carrying payloads to geosynchronous orbit atop Titan III launch vehicles.[24] NASA preferred Centaur, which burned hydrogen. Its high energy and large payload capacity made it ideal for planetary missions, which were important in the agency's future plans.[25] The Air Force had Transtage, the standard third stage of the Titan III, and Agena. Both burned storable liquid propellants. They already had seen much use with existing payloads, and would make it easy to switch from Titan III to shuttle—or to keep Titan III, if events were to dictate such a decision.

Boeing had a fifth alternative in its solid-fueled Burner II.[26] By itself it was too small for consideration, but Boeing officials argued that an enlarged version merited consideration alongside the liquid-fueled competitors. They won their point and took a study contract. Similar contracts went to Lockheed for Agena, to Martin Marietta with Transtage, to McDonnell Douglas for Delta, and to General Dynamics, home of Centaur. Each of these five studies cost from $635,000 to $640,000. They ran from October 1974 to midyear in 1975.[27]

"NASA is involved in study and assessment activity and will be consulted before a stage is selected," Brig. Gen. Henry Stelling, an Air Force space director, told *Aviation Week*. He also noted pointedly that "the baseline approach and the approach that we used to justify the Air Force budget for the IUS is one that will satisfy only DOD requirements." George Low had his own view: "We would either like to get the high performance of the Centaur stage or the simplicity of the solid stages. The Transtage, which may well be the Air Force's number one candidate, has the disadvantages of both and the advantages of neither. Its integration into the Space Shuttle payload bay would be almost as difficult as that of the hydrogen Centaur stage, would require a great deal of provisioning for venting and dumping, and its performance characteristics are no better than those of the solid stage."[28]

The Air Force's Walter LaBerge was to make the decision, and on 4 September 1975 he announced his choice. The IUS would be expendable and was to use solid propellants. Such a vehicle promised to be simpler, more reliable, and safer than a liquid-fueled IUS, and could also offer lower costs by being easier to handle and simpler to maintain. NASA held off until 3 September before recommending solids, for its officials doubted that such an IUS could handle the most demanding planetary launches. Boeing, however, envisioned a modular design, wherein multiple solid units could combine to give the necessary energy. "It's going to take quite a few solids, but I think we'll still be able to fly our planetary missions," a NASA official said hopefully.[29]

LaBerge had picked Boeing's approach, but this did not mean that Boeing would get to build it. Only General Dynamics could offer Centaur, and only Martin Marietta was ready to proceed with Transtage. Boeing held similar proprietary claims to Burner II, but that firm's

Table 9.1.   Capabilities of Existing Upper Stages

Upper Stage	Empty weight (lbs.)	Propellant weight (lbs.)	Payload (lbs.)
Transtage	3,242	23,472	3,230
Delta	1,666	10,292	3,283
Agena	1,250	13,950	4,450
Centaur	4,480	30,760	7,400

Source: AAS Paper 75-292, 751.

IUS was no Burner II; it was considerably larger, and it existed only on paper. Other companies had their own paper concepts, and the Air Force did not give Boeing a sole-source contract. Instead it reopened the competition, calling for bids on a solid-fueled IUS that could carry five thousand pounds to geosynchronous orbit. The new contest matched Boeing against Lockheed, General Dynamics, and Martin. Boeing won again, in August 1976, and this time it was for keeps.[30]

That firm envisioned the development of two standard solid motors, respectively with 21,400 and 6,000 pounds of propellant. The small stage, used alone, would boost Delta-class payloads from the shuttle. The large unit, again by itself, handled Atlas-class spacecraft. The two together provided a standard IUS for geosynchronous payloads. A pair of the large ones, mounted in tandem, accommodated multiple satellites. A three-stage configuration, formed by placing the small unit atop this tandem pair, gave extra energy for planetary missions. For more energy yet, this triple assembly could mount a small kick motor as a fourth stage. The two Boeing motors thus were to serve as building blocks, which could be combined to produce a family of operational IUS versions for a wide variety of applications.[31]

These IUS models promised to match the lifting power to the mission, but they did not suit the communications satellite industry, for they were to be stabilized on three axes. This industry's spacecraft were spin-stabilized, which saved money by simplifying their designs. The builders and users of comsats were accustomed to purchasing launch services from NASA, and this opened a market opportunity for a shuttle upper stage that could serve their needs.[32]

Three-axis stabilization was costly, for it demanded both attitude control and inertial guidance. The spacecraft carried the former; the latter was part of the IUS, and because this vehicle was expendable, this guidance system would fly only once. Spin stabilization dispensed with both. A turntable, mounted within the shuttle's cargo bay, would impart the necessary rotation, while the orbiter's own attitude control could point it in the proper direction. It would back away, allowing the spinning stage to fly freely, then give an ignition command. The rocket motor in the stage would do the rest, with stage and spacecraft continuing to

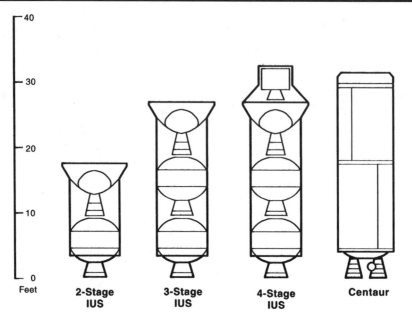

Fig. 66. *Inertial Upper Stage configurations compared to the Centaur upper stage. (Art by Dennis R. Jenkins)*

pirouette on their way to geosynchronous orbit. Once there, the satellite was to separate from the spent stage and then fire its own apogee motor to achieve a circular orbit.

Hughes Aircraft and the Aerospace Corporation conducted early studies that found promise in this approach. It offered a means to cut the launch cost for Delta- and Atlas-class payloads, for several could fit within the shuttle's cargo bay. Existing solid motors also appeared suitable for use, sparing the need for new development.[33] McDonnell Douglas picked up the concept and initially offered it to NASA as a way to launch this agency's Tracking and Data Relay Satellites, which were to provide the shuttle with its communications links to the ground. This proposal did not win acceptance. Company officials then came to appreciate that they could develop such a spinning stage as a commercial venture, selling operational units to NASA and the Air Force (which had spin-stabilized spacecraft of their own), as well as to the communications satellite industry.

The upshot was a pair of vehicles known as the Spinning Solid Upper Stage (SSUS) and, later, as the Payload Assist Module (PAM). The first of them, SSUS-D, served Delta-class payloads, with weight of as much as 2,750 pounds. Indeed, it could fly atop McDonnell Douglas's Delta launch vehicle, as well as within the shuttle. This stage used the Star 48 solid motor of Thiokol. SSUS-A stood in for the Atlas-Centaur, launching spacecraft of up to

4,400 pounds, with a version of the Intelsat V communications satellite as its designated payload. Its motor was derived from the third stage of the Minuteman III.[34]

Late in 1976 NASA and McDonnell Douglas reached agreement on terms whereby that firm was to use its own funds to develop SSUS-D. There was interest in SSUS-A at Boeing as well, and for a time there was talk that NASA might turn both stages into a federally funded program and allow the two firms to compete.[35] But in February 1977 NASA and McDonnell Douglas also reached an understanding on SSUS-A, with NASA ordering six of these stages later that year.[36]

Studies of the space tug continued, but it clearly was receding further and further into the future. At the same time, the Air Force was funding the development of the IUS, while McDonnell Douglas was ready to pay for both versions of SSUS, leaving NASA free to purchase them as off-the-shelf items once they became available. In addition, Air Force interest in the IUS was broadening, with this service slating that stage for use with the Titan 34D as well as with the shuttle.[37] It certainly now was more than a stopgap, and in December 1977, at a meeting of the Air Force Systems Acquisition Review Council, members of this review group agreed to change its name from Interim Upper Stage to Inertial Upper Stage. The new name took note of its guidance system and distinguished it clearly from the commercial SSUS.[38] But it also demonstrated anew that in Washington, interim and temporary solutions have ways of becoming permanent. The IUS indeed flew, but there would be no space tug, not then and not to this day.[39]

Budget and Management

Within Washington, NASA faced ongoing scrutiny from Congress, the White House, the Office of Management and Budget, and the General Accounting Office. Congress generally treated the agency's budgets with kindness, matching and at times even exceeding the requested amounts. But these requests originated within the OMB, which kept NASA on a tight leash. The space program rode a rising curve of expenditures during the 1970s, but inflation, measured by the Consumer Price Index, eroded their value mercilessly (table 9.2).[40] During 1971 George Low and other NASA officials sought a shuttle design that they could build within a cost ceiling of one billion 1971 dollars in annual expenditures. They met this ceiling with room to spare, not even approaching it until the Republicans were out of the White House following the 1976 election. Low also had sought an agency budget of $3.2 billion in constant dollars. After 1973, NASA never came close.[41]

Late in 1973, during preparation of the fiscal year 1975 budget, OMB officials had promised that if NASA accepted one more round of cuts and schedule delays, the OMB would impose no further slippage in the date for first orbital flight.[42] The budgeters kept their

Table 9.2.   NASA Budgets, 1972–1981
(in Billions of Current-Year Dollars)

Fiscal Year	NASA			Space Shuttle		
	Request	Appropriation	In 1971 Dollars	Request	Appropriation	In 1971 Dollars
1972	3.313	3.310	3.209	0.100	0.100	0.097
1973	3.408	3.408	3.110	0.200	0.199	0.181
1974	3.054	3.040	2.501	0.475	0.475	0.391
1975	3.267	3.231	2.436	0.800	0.798	0.601
1976	3.559	3.552	2.531	1.206	1.206	0.859
1977	3.729	3.819	2.554	1.288	1.413	0.945
1978	4.081	4.064	2.525	1.349	1.349	0.838
1979	4.372	4.559	2.551	1.439	1.638	0.917
1980	4.725	5.243	2.583	1.366	1.871	0.922
1981	5.737	5.541	2.467	1.873	1.995	0.888

Source: Data from NASA SP-4012, vol. 3, pp. 9, 69; NASA SP-4407, vol. 1, p. 586; and *Aviation Week,* 30 Jan. 1978, 28–33; 22 Jan. 1979, 16; 29 Jan. 1979, 24–26; 28 Jan. 1980, 21; 4 Feb. 1980, 27–29; 19 Jan. 1981, 25–28; and 15 Feb. 1982, 24–26.

word; subsequent delays resulted from difficulties with the SSME and the thermal protection rather than from deliberate program stretchouts. Still, the OMB did not make life easy for NASA. Every year they engaged in a stylized ritual that was as much duel as duet, with officials of each playing their roles. One may illustrate with preparations for fiscal years 1977 and 1978, wherein NASA took an unexpected cut but then regrouped to hold its own.

The budgetary gavotte began in July 1975, when James Lynn, the OMB director, sent a letter to NASA's Fletcher:

> *This letter provides Presidential policy guidance for the preparation of your fiscal year 1977 Budget submission. It also provides dollar planning targets to be used in preparing your budget estimates.*
>
> *The President's review of the fiscal and economic outlook for 1977 has confirmed the need for a balanced program that is designed to support economic growth without aggravating inflationary pressures. At the same time, the sharp growth in relatively uncontrollable programs in recent years continues to present major difficulties in managing the budget. The President expects you to develop plans for the operation of your agency in such a way that your budget submission is consistent with the enclosed targets. . . . Accordingly, the President expects your 1977 budget recommendations to be held to totals at or below the target figures.*

These target figures appeared on a separate page, in millions of dollars.[43] Budget authority was set at $3.850 million, outlays at $3.805 million. Two months later Fletcher responded with budget marks that hit the targets nearly on the nose: $3.859 million in budget authority, $3.802 million in outlays. Low wrote, "Our budget mark for FY 1977 had been $3,850,000,000, including new starts and inflation. It turned out that this was a reasonable mark and that we were able to put together a budget to meet this mark without too many drastic cuts."[44]

Still, the lion was not about to lie down with the lamb, for OMB promptly came back with a proposed cut of $305 million. This was too much even for James Mitchell, the OMB associate director who dealt with NASA. Fletcher and his budget officer, William Lilly, met with Mitchell and two of his colleagues, and Fletcher wrote, "The impression I received is that Jim Mitchell is not only sympathetic but agrees completely and will try to make this same case to Jim Lynn in the Director's Review." This was a high-level budget review that took place as an intramural OMB exercise, with no NASA officials on hand. For NASA to have a friend at court was significant indeed.[45]

Mitchell prevailed, at least to a degree, and succeeded in reducing the amount of the requested cut to between $100 million and $120 million. NASA naturally preferred the low end of this range, and came back with a proposal to trim the budget by $104 million. This was to all come out of the shuttle, but it did not mean that the date for the first flight slipped anew. Rather, it meant a one-year delay in the start of orbiter production, of assembling the vehicles that would follow Columbia. Other cutbacks reduced the capabilities of the early orbital flights. The OMB accepted this, and the budget request of President Ford, which went to Congress early in 1976, called for $3.697 billion.[46]

It was clear, however, that NASA had responded too readily to the OMB's initial target. In August 1976, as a prelude to that year's impending budget exercise, a typewritten memo with handwritten comments circulated among NASA's senior officials. These executives drew hope from the recent success of the Viking mission to Mars, which had set down a lander that had returned stunning panoramas of the surface. In this memo, Guyford Stever was Ford's science advisor, Senator Frank Moss and Congressman Olin Teague chaired the space committees in their houses of Congress, and William Lilly was the NASA comptroller:

OBJECTIVE: To achieve a minimum $250 million increase in NASA's OMB-approved budget request for FY78.

TACTICS:

Begin negotiations at a much higher figure than OMB's guideline so that we have more room to "give." [Handwritten: "Ought to go in for a 4.9 B budget."]

> *Capitalize on Viking success by aggressiveness. Remind OMB of President's publicly expressed interest in future missions (Viking 3).*

Top NASA management should consider getting more involved . . . throughout OMB negotiations.

> *Decisions made at lower OMB staff levels are difficult to overturn later.*
>
> *Try to overcome traditional adversary relationship.*
>
> *Do we have the right people talking to OMB at staff level? [Handwritten: "Lilly antagonistic to OMB, ought to change attitude."]*

Try to enlist aid of President's new science advisor.

> *Other White House staffers lack time, interest or clout to influence OMB or the President.*

After just recreating the post, Administration must pay some attention to science advisor in setting priorities.

If all else fails, go Presidential appeals route again. [Handwritten: "If he wins, bring Cong people in with me. Don't try to see him before elections—won't be able to focus."]

> *Dangerous course.*
>
> > *If President is defeated, he may leave all except major decisions to OMB. [Handwritten: "Will lean to a lower budget. Pres. will tend to do the same."]*
>
> *Don't expect help in FY78 from Carter if elected.*
>
> *Not enough time for him to revise Ford budget; Congress would do so with its new budgetary apparatus.*

A handwritten comment at the bottom read "Best bet—get Moss & Teague to support Stever—then get Stever's help."[47]

NASA and OMB were ready for a new pas de deux, and a few days later, Lynn sent his expected letter to Fletcher:

This letter provides Presidential policy guidance for your 1978 budget recommendations.

> *The President has stated that he plans to seek a balanced budget for 1979. Realization of the President's goal will require a concerted effort to hold down the 1978 budget to assure that we are on the path to a balance in 1979.*

The letter again gave the budget planning targets, this time for fiscal year 1978:[48] budget authority, $3.795 million; outlays, $3.768 million. In mid-September Fletcher responded with his own proposal: $4.107 billion in budget authority, $3.925 billion in outlays. The larger of these numbers was the proposed congressional appropriation, standing more than $300 million above Lynn's target.[49]

Fletcher did not get it, not from the OMB and not from Congress. But this time his arguments carried weight and he came close, as the appropriation came to $4.064 billion. He had aimed at a boost of $250 million over fiscal year 1977, and he nearly hit this mark, for this appropriation topped that of fiscal 1977 by $245 million. Even so, this brought no turnaround in NASA's fortunes, for inflation cut the purchasing power and left NASA breaking a little less than even. The net result of all the schemes and stratagems was to leave the agency almost exactly where it had been in fiscal year 1976 in constant dollars.[50]

As NASA wrestled with the OMB it also faced continuing scrutiny from the General Accounting Office. Analysts at the GAO had tried to critique the economics of the shuttle during 1972 and 1973 only to find themselves routed as NASA projected enough traffic to assure its cost-effectiveness.[51] But this was not the end of the matter, for the GAO returned to the shuttle during 1975.

The GAO was an arm of Congress, an investigative agency that prepared reports in response to requests from the House or Senate. *Aviation Week* stated that these reports were "forcing other government agencies to tighten control and improve management of their programs":

GAO's influence is expected to continue to grow in the coming years as congressional committees become increasingly demanding for program justification and management quality.

GAO's impact is due to the increasing scrutiny it is placing on government procurement of major military and civilian aerospace systems. In addition, GAO expects protests of government procurement decisions by losing contractors to become more frequent, particularly in aerospace.[52]

The GAO offered technical criticism, and its work was far from the best. It relied on its inhouse staff, and had tried to remedy their weaknesses by hiring more knowledgeable people. But in the words of a congressional staffer, "The technical people who can also manage programs do not want to be in GAO. They want to be running the program." A NASA shuttle official, asked to rate the GAO on a scale of 1 to 10, gave it only a 2 on its technical ability and commented, "We find ourselves having to develop our discussions with them in laymen's language to give them enough technical insight into the area being discussed to make sure they get a full-blown understanding."[53]

Better technical advice was available to members of Congress through the National Academy of Science and its National Research Council. In this fashion, Senators Stevenson and Schmitt commissioned an influential study of the SSME program, with a panel of specialists headed by MIT's Eugene Covert. Nor was NASA a stranger to the commissioning of such reviews and studies. A 1977 summary counted nineteen such assessments to date. Most dealt with specialized issues, such as subsynchronous whirl in the SSME turbopump shaft. But an important 1975 review, directed by Lockheed vice president Willis Hawkins, carried out a broad survey of technical issues and made seventy-three suggestions that covered a variety of areas.[54]

Still, in the realm of program management, the voice of the GAO would certainly be heard. It had begun to audit major systems and projects in the late 1960s, when a massive overrun on the C-5A transport came to light, and its staffers had been studying the shuttle since 1972. In 1975 the GAO assigned fifteen people to this. Since mid-1973 the GAO had been placing staffers at NASA's Kennedy, Marshall, and Johnson centers, and had scheduled weekly meetings at NASA-Johnson. Monthly meetings followed, between GAO officials and senior shuttle program directors.[55]

Management and budget were the purview of OMB, but NASA had characteristic practices in both areas, and the GAO was ready to offer some pointed critiques. In running the shuttle program, NASA relied on "success-oriented management," which sought to tailor the testing to the funds in hand. This drew on a general belief that Apollo had tested to excess, conducting many exercises that contributed little to understanding, or that had not been cost-effective. By drawing on this experience NASA could choose its shuttle tests with greater adroitness, reducing the overall number while getting the most from those performed.[56]

Conservatism in shuttle design promoted this practice. The orbiter's basic structure was that of an aluminum airplane, which offered little novelty. Many of its on-board systems drew directly on the experience of Apollo and Skylab. The ET resembled the S-II stage, and the SRBs amounted to enlarged solid boosters from the Titan III. The greatest technical risks lay within a relatively few areas: the SSME, the thermal protection, software and redundancy management. Here the testing indeed was thorough, but elsewhere this was not necessarily true.[57]

In practice, success-oriented management meant that much testing involved complete vehicles and engines rather than components. Apollo had done this, saving time in development by launching the first Saturn V as a complete stack, with all three stages live. Thus NASA bypassed the step-by-step procedure of the earlier Saturn I, which had wrung out its first stage by flying it repeatedly with dummy upper stages, only later launching it with a live second stage. The shuttle program took this approach further, for several key tests were not to take place until the early orbital flights. "We've designed our flight program so we can learn more from flight than we can from ground testing," Christopher Kraft, director of NASA-Johnson, told *Aviation Week* in 1976.[58] To a large degree this reflected the lack of suitable facilities. No wind tunnel could replicate all the conditions of airflow over the orbiter during reentry. Only flight test could do this; the best that anyone could do was to fly the vehicle on a benign trajectory during entry to subject it to the lowest possible thermal load.

Thermal vacuum testing represented another area, wherein the vehicle was to be placed within a vacuum chamber and exposed to the thermal environment of space flight. Such tests were to present the craft with the temperatures and heat loads of flight in orbit, though not of reentry. Apollo spacecraft had undergone such treatment, but no chamber was large enough to accommodate a full-sized orbiter, and the construction of such a facility would have been costly. OMS pods went into the thermal vacuum chamber; in Kraft's words, "Based on Apollo experience, we felt these were tests we could not do without." For the most part, however, the orbiter was to face its thermal environment only during flight, not on the ground and in a test facility.

"We think we understand the thermal-vacuum environment we're going to encounter in near-earth orbit better than we did in the other programs," said Harry Douglas, a manager of test operations. "We are more confident we can predict accurately the kind of heat loads the orbiter will see. If the first few space missions prove this isn't so, the orbiter crews can come back early and we can add radiator capacity or more insulation to the vehicle."[59]

Even technically feasible tests were dropped in favor of flight testing. The crew module could have fitted within a thermal vacuum chamber, but analysts preferred to trust their mathematical models of the heat loads. If these proved excessive during the first flight, that mission could be cut short. "Ground thermal tests could have been done on the crew module alone," Kraft said, "but we decided instead to instrument the orbiter and crew module heavily and see what kind of thermal loads we can get when we fly it."[60] The GAO was dubious: "A study of thermal vacuum tests for past programs showed that numerous anomalies were detected which required design, procedure, and process changes. Thermal vacuum tests for the orbiter were eliminated even though NASA officials believe the shuttle is no less complex than previous programs. Therefore, this testing approach assumes a higher degree of success than the Apollo program."[61]

Vibration and acoustic tests were also important. NASA had initially planned to conduct vibration tests in January 1978 by using a simulated orbiter mated with an ET and inert SRBs. NASA replaced the boiler-plate orbiter with Enterprise, accepting a six-month delay in the start of testing. The GAO warned in 1976 that there was "no room for slips in the scheduled availability" of Enterprise, and that the new schedule gave less time to analyze data and make changes.

The orbiter's aft structure raised particular issues in vibroacoustic testing, for it stood right next to the rocket engines. A planned test series was eliminated when it became evident that the program could not afford the $70 million price tag. Instead, most of the data came later in the program, from the cluster of three SSME engines tested in Mississippi.[62] Another vibroacoustic test program, planned for the orbiter's forward position, was also deleted. It would have verified that the crew station and payload bay did not receive excessive noise and were not subject to excessive fatigue. In the words of the GAO, "the initial operational capability date may not be met if noise levels or structural panels are found to be unacceptable during the flight test program. The Hawkins team recommended reinstatement of this ground test, but NASA rejected this recommendation because it does not believe the tests would produce a sufficient increase in confidence to warrant the expenditure."[63]

Then there were the main engines, which represented par excellence the use of complete systems for developmental testing. Prior programs had done a great deal of component testing, but the SSME program shifted at an early date to tests with complete engines. The Hawkins review had stated, "Component testing appears weak and not far enough ahead of commitment to production. Schedule should be extended and hardware brought forward for spares for test. More hardware probably required." However, the Covert committee's SSME review, three years later, did not recommend a return to component testing. Indeed, it proposed a step-up in full-scale engine testing, by reactivating a test stand owned by Rocketdyne, and this was done.[64]

Success-oriented management lent itself to criticism. It sought to correct an alleged excess of testing during Apollo, but the GAO noted, "No study has been made to show whether past programs contained unnecessary testing."[65] A NASA official told *Science* that this management approach "means you design everything to cost and then pray." Noel Hinners, a senior NASA official, said that "to take on a technological challenge with penny-pinching as the major goal was just plain stupid; if you're going to break technical ground, you can't design to cost."[66]

"If we hadn't had the dollar limitations, we probably would have built more test facilities and planned more test series," said the test manager Harry Douglas. An internal NASA report added, "The net effect of this management approach has been an absence of realistic plans, inadequate understanding of the status of the program, and the accumulation of schedule and cost deficit without adequate visibility."[67]

It is hard to fault NASA for this management approach; it had to live as best it could within a world dominated by the OMB. NASA shaped its budget accordingly and said the program was in good shape. The agency had planned in 1972 to carry through the development of the shuttle for $5.15 billion, in 1971 dollars, and to achieve first flight to orbit in March 1978. Negotiations with OMB late in 1972, on the fiscal year 1974 budget, delayed this first flight by nine months but kept the cost of development at the same level. Further negotiations on the fiscal 1975 budget, late in 1973, slipped this flight by up to six additional months and increased the program cost by $50 million, to a total of $5.2 billion. In February 1976 NASA announced a further cost growth of $20 million. The agency held to this announced level into the following year. Thus, in presenting testimony to Congress during June 1977, John Yardley, NASA's associate administrator for Space Flight, stated that the cost of development was still $5.22 billion, in 1971 dollars, with first orbital flight still set for mid-1979.[68]

The GAO had different views. It noted in 1976 that "the cost to complete individual shuttle projects has been consistently underestimated, because project managers are given predetermined ceilings which they are instructed not to exceed. NASA's top management believes providing cost goals causes project managers to search for and identify excessive or unneeded requirements. We believe it also understates project requirements and inhibits meaningful analysis of remaining reserves."[69]

The GAO also noted pointedly that development of the shuttle did not represent the whole of the shuttle program, being merely the largest and most visible line item. Separate budget categories covered construction of an operational fleet of orbiters, assembly of launch facilities, and costs of flight operations. In the words of the GAO, "Other adjustments moved funding problems into the future or out of the [development] budget into other budgets where potential cost growth will not be readily identifiable. This situation suggested that, if cost overruns are encountered, they will either not be recognized and/or not be identified until the latter stages of the program."[70]

Program stretchouts often bring such overruns. Late in 1974 NASA officials tried to determine the true cost requirements. Project estimates went up by $538 million for known program tasks and diminished by $148 million by deleting some tasks and deferring other work to later dates. The net increase was therefore $390 million. Eight months later this nascent overrun was put at $410 million (table 9.3). This estimate represented work at the project offices within NASA's field centers. Officials at NASA Headquarters assessed the validity of these estimates in the light of their own experience and judgment, and with this superior wisdom, they decided the true increase was only $210.9 million. They covered half of this by transferring $111.7 million into other budgets. The remainder, some $100 million, came within a $300 million allowance that already was part of the budget and that covered errors in estimates along with planned changes during the remaining years of the program. Presto! The overrun disappeared, and Yardley could give his assurance that the program was on track.[71]

The GAO did not agree. Its analysts rejected the numbers from Headquarters, noting that NASA had used management-to-cost techniques "to prevent estimates from exceeding funding limitations." The GAO accepted the higher numbers from the field centers, describing them as "more complete than Headquarters' estimates and are made by personnel charged with the program's day-to-day management" and hence "should be more representative of the expected outcome of the program."[72] The GAO also made its own audit of NASA's transfers of funds into other budget categories and came up with a figure that was nearly twice as high. Its 1976 report then summarized its findings in 1971 dollars: $426 million of increases were due to program stretchout, and $195 million represented funds transferred to other budgets, for a total of $621 million. Of this overrun, NASA accepted only the $70 million already admitted, over the $5.15 billion of 1972. Nor would the GAO accommodate any part of this within an allowance for errors and planned changes. The GAO stated that such reserves "cannot be reliably evaluated," but nevertheless appeared inadequate.[73]

Table 9.3. Shuttle Overrun Estimate, 1975

Project	Increase (+) or Decrease (−) (in Millions of Dollars)
Orbiter	+147.7
Main engines	+147.0
External tank	+56.5
Solid rocket boosters	+20.0
NASA-Marshall systems management	+53.2
Launch and landing	−13.5
Other	−0.4
Total	+410.5

Source: GAO Report PSAD-76-73: Staats, *Status and Issues,* 32.

In a 1977 report the GAO repeated some of its findings and offered a forecast: "In our opinion, there is a high probability that NASA will encounter major cost growth during the remaining program years. It is possible that serious technical problems will be identified because some test programs . . . have been deleted, reduced in scope, or delayed until the orbital flight test program. . . . Our past experience in reviewing major civil and defense acquisitions has shown that delaying or deleting testing can lead to costly retrofit or redesign at a later date or to deploying systems that cannot adequately fulfill their intended role."[74] During the Carter administration, NASA faced these issues.

Carter and the Shuttle

The election of Carter in 1976 brought a changing of the guard. George Low left the agency in June, even before Carter came to the White House. He had no wish to continue as deputy administrator; he saw only a fifty-fifty chance of being named as full administrator, and he wanted to take plenty of time in looking for a new position "so that I would not have to jump at the first opportunity that came along after the change of administration in 1977." He took the presidency of Rensselaer Polytechnic Institute, from which he had received his bachelor's and master's degrees. His replacement, Alan Lovelace, had been NASA's associate administrator for the Office of Aeronautics and Space Technology. Lovelace held this post through the Carter years.[75]

Fletcher stayed on until relieved in May 1977. The new administrator, Robert Frosch, had been the assistant Navy secretary for research and development, holding that service's highest research position. More recently he had been an associate director at Woods Hole

Oceanographic Institution in Massachusetts, one of the world's leading centers for exploration of the seas. His background held a touch of the exotic, for he had helped direct a United Nations environmental program in Kenya. Despite his career's heavy emphasis on ocean technologies, he was no stranger to space. He had directed Project Vela, which used satellites to detect nuclear explosions, while at the Pentagon's Advanced Research Projects Agency. He held bachelor's, master's, and doctorate degrees, all from Columbia University.[76]

Once in office Frosch faced the issue of how many orbiters were to be built. A joint NASA–Air Force position paper, dating to May 1973, anticipated 581 shuttle flights from 1979 to 1990, and derived a requirement for five orbiters in the operational fleet. The Pentagon continued to support NASA in this matter, with Low and Currie signing a position statement in January 1976: "The National Aeronautics and Space Administration and the Department of Defense agree that five Space Shuttle Orbiters are needed to meet our national traffic model requirements. Three Orbiters are funded by NASA within the DDT&E and production programs. Neither agency has budgeted funds for the remaining two Orbiters. . . . NASA and DOD agree to work together to resolve this issue as part of the FY 1978 budget cycle activities."[77] The plan at that time called for refurbishing OV-101, subsequently named Enterprise, into a flight vehicle, and to build four more in accordance with the following schedule:[78]

Manufacturing Sequence	Date	Location
2	August 1978	Kennedy
3	December 1979	Kennedy
1 (OV-101)	December 1980	Kennedy
4	December 1981	Vandenberg
5	December 1982	Vandenberg

This plan changed, as budget negotiations for fiscal year 1977 slipped production by a year. Nevertheless, Columbia was in assembly during that fiscal year, which meant it was high time to think about what would happen next.

By March 1977 plans for a five-orbiter fleet were firm. In addition to Columbia, Enterprise was still at hand, and NASA continued to anticipate refurbishing it for operational use. The Structural Test Article, STA-099, was also in the picture; in time it would fly to orbit as the Challenger, though this was not yet in the plan. Three other shuttle orbiters, including the nascent Discovery and Atlantis, now were slated for production as well.[79]

The GAO was in no mood to rush ahead and made this clear in its report issued two months later:

NASA's fiscal year 1978 budget proposal contains $141.7 million to initiate refurbishment of the development orbiters and production of three additional orbiters. . . . Based on the shuttle's performance goals and NASA's mission duration as-

sumptions, the two orbiters obtained from the development program will provide more than enough capacity to handle the number of payloads flown during the past 10 years.

NASA's five-orbiter fleet is based on the assumption that the number of payloads will more than double during the 1980s (from approximately 40 payloads a year to over 90 a year).

Analysts at the GAO reviewed NASA's assumptions, and gave several reasons to doubt that this upsurge in demand would materialize. Their report thus made its recommendation: "Production of the third orbiter could be initiated and the remaining two delayed until there are more adequate assurances regarding technical problems, space flight activity, and the cost of operations."[80]

The OMB also endorsed a three-orbiter fleet. Then in mid-December, amid preparation of the fiscal year 1979 budget, Defense Secretary Harold Brown answered with an argument that bypassed the issue of user demand. He noted that the first two orbiters, Columbia and Challenger, would be overweight and hence could not carry the heaviest reconnaissance satellites. A single orbiter with greater capability indeed might handle such payloads; however, it could be destroyed in an accident. Considerations of national security then dictated that the nation should have two such vehicles, with the second available as a backup. Brown thus insisted strongly that the proper number of orbiters was four, and this argument carried the day.[81]

The differing weights of the orbiters reflected improvements in design that had accrued in the course of development. Though people spoke of "production" of shuttle orbiters, this was not at all like the massive aircraft orders of World War II. Every orbiter amounted to an individual procurement, and there was ample opportunity to change the blueprints from one to the next. Weight reduction was essential, for a lighter orbiter could carry correspondingly heavier payloads. It was achieved through use of titanium in the primary structure, as this metal was denser than aluminum but considerably stronger. Boron-epoxy composite, particularly light in weight, also began to appear. This indicated a clear trend: early orbiters being heavy, even after refurbishment, and later ones being considerably less so:

Enterprise (as flight vehicle)	160,000 pounds
Columbia	158,000 pounds
STA-099 (as Challenger)	155,000 pounds
Discovery	151,000 pounds
Atlantis	151,000 pounds

Clearly, the latter two orbiters were the ones that would best serve the Pentagon. By comparison, Enterprise amounted to an overweight dinosaur.

NASA had expected to have Enterprise join the operational fleet in June 1981, as the next orbiter after Columbia. But late in November 1977, while Harold Brown was honing his argument, a new NASA production plan put Enterprise fifth in line. The new holder of its former position was STA-099. It was not only lighter but could be available sooner, in February 1981. This meant that Enterprise would be refurbished for flight to orbit only if there indeed was to be a five-orbiter fleet.[82]

Frosch met with Carter just before Christmas 1977 and tried anew for a fleet of five. The acting OMB director, James McIntyre, replied with a letter making it clear that Carter was no Santa Claus:

The decision was clearly to support a four orbiter option. . . . The President stated his explicit concern that no action be taken that might be interpreted as a possible commitment now by the Government to build a fifth orbiter. . . .

The President's decision on space shuttle orbiters can be summarized as follows: . . . A total fleet of four operational orbiters will meet civilian and military shuttle flight requirements and funds to proceed with production of a four-orbiter fleet are provided in the NASA budget for FY 1979.[83]

Carter's decision left Enterprise with no prospect for flight. It also amounted to an acknowledgment that the shuttle would do less work than NASA had anticipated. At the same time, it was becoming clear that this program would cost more than planned. As recently as June 1977 NASA's Yardley had told a House space subcommittee that the shuttle development cost was still at $5.22 billion in 1971 dollars. Four months later he gave new testimony, admitting that the program would experience an overrun of as much as 7 percent, or $365 million.[84]

The agency had already addressed an earlier problem by deferring $68 million in fiscal year 1977 funding to fiscal 1978. Now, at the start of fiscal 1978, NASA faced $123 million in unanticipated expenses. The agency expected to deal with this by reprogramming up to $120 million in fiscal 1978 production funding, transferring it into development and testing. This was not the budgetary shell game that had caught the GAO's attention; it was an attempt by NASA to live within its means while delaying the operational dates of future orbiters.

The $123 million represented costs incurred during fiscal year 1977; the full 7 percent overrun included projected increases for future years. A host of technical areas were involved; in Yardley's words, "These problems included the main engine turbopump, increased structural and thermal loads from wind tunnel model test data, provision for a backup flight control system, the orbiter hydraulic control system and thermal-protection system tile production. Increased efforts were also required in the manufacturing and assembly of test articles, as well as associated subcontractor work." He also pointed to extra costs for the construction of Columbia and for the three-engine SSME cluster, as well as for system software and SRB development.[85]

By shifting funds and by drawing on budget reserves, NASA managed to make it through fiscal year 1978, which ended on 30 September. Clearly, though, this could only carry the program so far. The agency soon would need cold cash to cover its overrun, and in this time of need, it found help from an unlikely patron: President Jimmy Carter. Carter was no friend of aerospace, being closely aligned with liberal Democrats such as Senators William Proxmire and Walter Mondale, who had tried to cancel the shuttle program. Indeed, Mondale now was vice president. But Carter had a strong commitment to arms control.

Prospects for the control of nuclear weapons blossomed in the wake of the 1962 Cuban Missile Crisis, which brought the superpowers to the brink of World War III. In 1963 John Kennedy and Nikita Khrushchev agreed to ban tests of nuclear weapons in the atmosphere or in outer space; such tests could be conducted only underground. These national leaders also set up a hot line, a communications link that would help them exchange messages during any future crisis. Next came the Non-Proliferation Treaty of 1968, whereby Moscow and Washington pledged to work against the spread of nuclear arms to additional nations. This was followed by SALT I, the first round of Strategic Arms Limitation Talks, which began in Helsinki in November 1969 and led to accords signed in May 1972. SALT I produced two treaties. One placed stringent limits on antiballistic missiles, which promoted massive ICBM deployments in hope of overcoming such defenses. The second froze land-based ballistic missiles at existing levels.

The early 1970s brought a flurry of new agreements. In February 1971 the superpowers prohibited the installation of nuclear weapons on the seabed. In September of that year they agreed to upgrade the hot line, and pledged as well to improve their safeguards on nuclear arms. In July 1974 they signed the Threshold Test Ban Treaty, setting a limit of 150 kilotons on their underground nuclear tests. They also pushed ahead with SALT II. This new round of negotiations began in Geneva in November 1972 and led to three summit meetings during the following two years. The third summit brought President Gerald Ford to Vladivostok, where he met with Leonid Brezhnev in November 1974. This conference set guidelines for subsequent discussions. After that, American and Soviet delegations met almost continuously in Geneva. Meanwhile, the flow of new treaties continued. In May 1976 the two nations set restrictions on nuclear explosions conducted for allegedly peaceful purposes, such as digging canals or fracturing rock formations that held reserves of natural gas.

The formal SALT II agreement was ready in June 1979; Carter and Brezhnev signed it in Vienna. It defined a set of categories, each broader than the one before, and set numerical limits for each nation (table 9.4).[86] This was not nuclear disarmament; it more nearly resembled an acceptance by each nation of the other's plans for a future buildup. The Air Force wanted the MX, a new type of multiple-warhead ICBM, and the SALT II limits gave room for its deployment. The Navy expected to phase out its Polaris submarines in favor of the new Trident class, which were harder to detect and carried more missiles. The Soviets hoped to replace their old-style armaments—standard bombers, single-warhead missiles—with new versions that could carry cruise missiles and mount multiple warheads. Yet because the Soviets would actually have to reduce their total number of vehicles, from 2,500 to 2,250, Carter argued that this treaty would serve the national interest.

He would need all the arguments he could muster, for the SALT II treaty had to be ratified by the Senate, where it required a two-thirds majority. However, it never even came to a vote. The Soviet invasion of Afghanistan late in 1979 destroyed the climate of trust that had nurtured this agreement, putting Carter's plans on hold.[87] But during 1978 and 1979, with prospects for this treaty flourishing, he was well aware that he would need satellite reconnaissance to enforce its terms.

In mid-1977, soon after taking office, Carter showed anew his commitment to arms control by canceling Air Force plans for production of the B-1 bomber. Even so, he understood that the Air Force would play an essential role in arms control. During 1978, his National Security Council conducted a review that led to a Presidential Directive, NSC-37, on the space program. It specified the roles of the secretary of defense and the director of the CIA in dealing with space reconnaissance, and gave a strong endorsement to the shuttle.[88]

Carter had scrapped plans for an operational fleet of B-1s. Yet he could not treat the shuttle so cavalierly unless he wished to weaken the means whereby the intelligence community was to verify compliance with his prized SALT II. With this, NASA reaped a long-deferred reward. The agency had shaped the shuttle, during 1969, as a logistics vehicle that could service an orbiting space station. It had limited payload and little maneuverability, and its close link to that proposed station nearly killed it before the project could get underway. It received a withering assault in Congress, as critics charged that it was merely a step toward that most extravagant of NASA's wishes: a piloted mission to Mars.

NASA responded during 1970 with a change in policy, allowing the space station to fade while bringing the shuttle to the forefront. It also gave the vehicle an extensive redesign, crafting it to serve the Air Force's needs. This allowed NASA to claim that the shuttle would serve as a national space transportation system, meeting the requirements of both military and civilian users. With political support from the Pentagon, NASA won over much of its congressional opposition.[89] Now, with Carter depending on the Air Force to turn SALT II into something more than an exercise in wishful thinking, the shuttle again would have its day.

Day and night are closely linked, and for NASA, fiscal years 1979 and 1980 brought a good deal of both. The agency made it through fiscal year 1978 by drawing on a reserve of $144 million, but by the end of that budget year, the reserve was gone and had given way to a $14 million deficit. In addition NASA committed $158 million in fiscal 1979 funds for work that was actually performed in fiscal 1978. But by late 1978 the agency was running out its string. It no longer could keep the shuttle on schedule merely by transferring funds from one pocket to another.[90]

Table 9.4. SALT II Numerical Limits, 1979

Category	U.S. Strength	Soviet Strength	SALT II Limit
ICBMs with multiple warheads	550	500	820
Above plus submarine-based missiles with multiple warheads	1,046	600	1,200
Above plus bombers carrying long-range cruise missiles	1,394	750	1,320
All strategic weapons vehicles	2,129	2,500	2,250

Source: Lanouette, "SALT II," *National Journal,* 16 June 1979, 991.

It needed a supplemental appropriation and needed it soon. The OMB offered to support a petition to Congress for $100 million. This was not enough to hold the schedule, and the OMB therefore allowed NASA to increase this to $185 million for fiscal year 1979. Carter personally took care that NASA's fiscal 1980 shuttle funding request, made to the OMB, went through without reduction. In February 1979 John Yardley warned that without this supplemental, first flight to orbit and follow-on deliveries of production orbiters would face delays of up to a year. Because the project staff would continue to draw their paychecks during this schedule slippage, the shuttle program would incur a further overrun of as much as $600 million.[91]

But during the early months of 1979 the program slipped due to problems with the SSME. The date for first flight, set hopefully for 28 September 1979, was put back to 9 November and then skidded well into 1980. Late in April 1979 Robert Frosch startled members of Congress by predicting a cost overrun of $600 million over the coming four years. He expected to cover much of this by reprogramming funds from the production budget into shuttle development, but still he knew he faced a substantial shortfall. Just then the government was midway through fiscal year 1979, with the budget for fiscal 1980 as a matter of current activity. Frosch anticipated reprogramming $70 million in fiscal 1979 and $200 million in fiscal 1980. He stated, however, that NASA would seek additional funding of up to $250 million in fiscal 1981, and "somewhat less than that" for fiscal 1982.

By now the $5.2 billion projected cost of development was all but meaningless, except as a base from which to calculate the growing overrun. This overrun stood at 16 percent, or $830 million, in May 1979, and again Jimmy Carter came riding to the rescue. His White House took quick action to obtain OMB support for a second supplemental appropriation, of $220 million for fiscal year 1980, on top of the earlier $185 million.[92]

In mid-July Carter sent a memo to Frosch, asking him to appoint "a few highly competent and independent individuals" to serve as consultants and to make recommendations on the shuttle program, both to NASA and to the White House. Frosch responded by selecting William Anders, Robert Charpie, and Vice Admiral Levering Smith. Anders had been an Apollo astronaut, had flown to the moon, and now was a general manager at General Electric. Charpie had been president of Bell and Howell, while Smith had played a major role in the Navy's Polaris program.[93]

By September the projected overrun reached 20 percent of that $5.2 billion, topping the billion-dollar mark. This reflected a further slip in the date for first flight, which now was in May or June 1980, with a fifty-fifty chance of further delay. As recently as December 1978 NASA still had hoped for a launch date of 28 September 1979; nine months on the calendar thus brought a nine-month postponement, largely due to trouble with the SSME. Administration officials met with Frosch and reaffirmed Carter's support. In a memo to Carter, the OMB's McIntyre joined Frosch in presenting estimates of additional funding: "These estimates are $200–$300 for FY 1980 and $400–$450 million for FY 1981." The fiscal 1980 funds were in addition to the two earlier supplementals; those for fiscal 1981 topped previously planned marks.[94]

In October Frosch instructed his deputy, Alan Lovelace, to devote most of his attention to the shuttle effort. Frosch himself took on Lovelace's other responsibilities, which spanned the range of NASA's activities. Problems with the thermal-protection tiles now were adding to the delays, and Lovelace anticipated a first flight in late summer of 1980. "I think July–August represents the earliest time frame that I would be willing to lay my money on, and it could slip beyond that," he told *Aviation Week.*

Frosch's consultants now were ready with their assessment; all agreed that the shuttle's problems stemmed from two sources. The program had been underfunded, while NASA had tried to do too much with too little. "Many problems in the management of the program have been cited by a host of reviewers," said Anders. "In my view, most of these have really been symptoms of the basic problem—underbudgeting by successive administrations coupled with an overoptimistic view of what work should be attempted on reduced resources."[95]

Short funding had affected even critical projects such as the SSME. Willy Wilhelm, a chief project engineer at Rocketdyne, recalls that "the budgets were tight. It was a struggle getting this done. Dividing up the pot—turbopumps weren't getting enough so let's take something from combustion devices, or my engine group was too big." *Science,* late in 1979, summarized what followed: "Rather than cautiously—and expensively—test each engine component separately, NASA's main contractor, Rockwell International, merely constructed everything to novel design, bolted it all together, and—with fingers crossed—turned on the power. At least five major fires resulted."[96]

In November 1979, with thermal-protection problems looming larger, the date for first flight slipped anew, to August or September 1980. Again there was at least a 50 percent chance of further postponement. The White House conducted a program review that observers described as the most exhaustive such analysis ever undertaken by Carter's appointees. It involved the National Security Council, the Office of Science and Technology Policy (OSTP), and the OMB, and reaffirmed the need for the shuttle as originally planned.

Carter agreed with this report, and he stated this view on 14 November at a conference within the White House's Cabinet Room. At that meeting NASA's delegation included Frosch, Lovelace, Yardley, and Lilly. McIntyre came from OMB and brought his deputy; Frank Press, heading the OSTP, did the same. Defense Secretary Harold Brown was unavailable but sent Air Force Secretary Hans Mark as the Pentagon's representative. Mark was an old shuttle man, having headed NASA's Ames Research Center. He had also served as Air Force under secretary, which made him director of the National Reconnaissance Office with responsibility for the nation's reconnaissance satellites.

Prior to the main conference, Carter met privately for fifteen minutes with Frosch in the Oval Office. In the Cabinet Room Frosch presented a status report on the shuttle, and Carter asked him to write a strong statement of support in the president's name. He also designated McIntyre as Frosch's point of contact within the White House. Discussion also turned to the billion-dollar overrun, and Carter told McIntyre: find the money.

Frosch was an old hand at shuffling funds from one budget category to another. Now McIntyre did the same, at a much higher level. He looked to Brown upon his return and prepared to transfer the required funds from the Pentagon into NASA. "Then it came back to *me*," Mark recalls. "Harold had to find a billion dollars. That's not an everyday task. So he took it out of an Air Force program." It was an ongoing effort that was installing new engines on KC-135 jet tankers. These aircraft could continue to fly with their old engines, so it was easy to slow this project to free up funds for the shuttle.[97]

Frosch shaped his own budget plans, for with OMB support, he prepared to seek a second fiscal year 1980 supplemental appropriation, this one for $300 million. It went to Congress in January 1980 along with the whole of Carter's proposed budget for fiscal 1981. NASA's total request came to $5.737 billion, topping the OMB's anticipated level of a year earlier by $1.14 billion. Some $800 million represented the increase in the shuttle program.[98]

With NASA, the Pentagon, the OMB, and the White House all standing firmly in support, this ordinarily would have ended the matter. However, 1980 was no ordinary year. It marked the second year in a row of double-digit inflation, with gasoline prices topping a dollar per gallon in the wake of Ayatollah Khomeini's revolution in Iran. The nation was slipping into recession and the deficit was blossoming. Carter responded by seeking across-the-board budget cuts.

Early in March administration officials told NASA to propose cuts totaling from $460 to $860 million. Initial discussions ran near $760 million, which amounted to inviting NASA to cover its shuttle overrun by drawing on the nonshuttle activities that constituted well over half of the agency's budget. Carter exempted the Pentagon from these cutbacks, which meant that the Defense Department could stand fast in the wake of Moscow's invasion of Afghanistan. This exemption gave Frosch an opening, as he argued that the shuttle should also be spared from the cutbacks on national-security grounds. He won this point and made headway with other arguments as well, for within a week, the requested cut dropped to $200–300 million.[99] The final budget reduction, as voted in Congress, was even less. Moreover, Congress trimmed NASA's second fiscal year 1980 shuttle supplemental only slightly, from $300 million to $285 million, and passed it just before the 4 July holiday. With this the way was open to proceed toward the first orbital flight.[100]

The shuttle had initially been planned as a six-year development program, leading to first flight in March 1978.[101] This date slipped by three years, and one may attribute more than a year of this to the OMB's budget cuts, a second year to problems with the SSME, and a third year to trouble with the thermal-protection tiles. In consequence, for fiscal years 1979, 1980, and 1981, Congress responded by appropriating $826 million more than NASA had requested in the annual budget messages from the White House. These funds included the supplemental appropriations of fiscal years 1979 and 1980, while in separate action, Congress endorsed the sharp rise in the fiscal 1981 shuttle allocation that came to it through the standard budget process and not as a supplemental.[102] This strong support reflected a general appreciation that the shuttle was something that America needed, as a program that indeed would serve the nation.

In this fashion the program received the funds it required, and the total expenditures ratcheted upward accordingly. The years of development ran from fiscal year 1971, with initial studies, to fiscal 1982, with the completion of an initial series of flights that qualified the vehicle for operational missions. During those twelve years, the total expenditure, in current-year dollars, came to $9.912 billion. On its face this appeared as a substantial overrun indeed, for NASA had promised in 1972 to develop the shuttle for as little as $5.15 billion.[103]

However, the 1970s included a number of years with severe inflation—and that $5.15 billion had been in the constant dollars of 1971. Because the dollar of 1971 had more value, the program cost in such dollars was less. There were several ways to discount the current-year dollars to take account of this. The Consumer Price Index was one. Within NASA the agency comptroller—whose formal title was assistant administrator for administration, making this official a true bureaucrat if ever there was one—maintained an independent index of inflation for internal use. Still a third index was available within the shuttle program itself, reflecting actual costs and their increases as experienced.

In this fashion, one could reduce the current-year costs to 1971 dollars, the total program cost falling substantially:

Inflation Measurement (1971 Dollars)	Program Cost (Billions of Dollars)
By Consumer Price Index	6.097
By NASA comptroller index	6.654
By actual shuttle cost increases	5.405

More specifically, there were year-to-year expenditures (which often differed from appropriations), discounted using each of these methods. The fiscal years included an interim quarter during 1976, which marked a change in the start of each fiscal year, from 1 July to 1 October (for a summary, see table 9.5).

The shuttle program office found itself experiencing an inflation rate somewhat higher than that of the Consumer Price Index, and substantially higher than that of NASA's comptroller. This certainly did not help program managers to stretch their dollars, but this high inflation discounted the current-dollar expenditures sharply. Indeed, use of this index made it possible to argue that the program had come in at $5.4 billion, which was very close to the 1972 goal![104]

Could one truly say, then, that the shuttle had indeed come in within its 1972 budget? Discounting dollars to reflect inflation represents something of an art, but there has never been any great enthusiasm in Washington for allowing every agency, and every program within that agency, to determine its own inflation rate. That would invite government officials to introduce a new excuse: "Oh, it's not really a cost overrun; we just had a problem with inflation that was higher than the official rate." Such a procedure would have introduced a new method whereby such officials could cook the books, and in this general area, NASA certainly knew its business. As noted in Chapter 1, this agency met the demanding cost-benefit requirements of the OMB simply by projecting a shuttle launch rate that was sufficiently high to address its needs. NASA's determination of its shuttle program inflation rate may well have been honest, but use of such a rate did not represent an accepted federal procedure.

When people spoke of inflation, they generally invoked the Consumer Price Index. On that basis, NASA indeed sustained an overrun on the shuttle, of just under a billion dollars. The NASA index, determined by its comptroller, put the overrun at $1.5 billion. If one accepts such estimates, then the overrun was 20 to 30 percent of the initial program cost of $5.15 billion in 1971 dollars.

Such cost escalations within a program often result from delays, and the shuttle had its share. Its first flight had originally been set for March 1978 and was three years late. However, the first year of this delay, more or less, occurred right at the start of the program, while expenditures were modest. The other two years occurred around 1980, when annual budgets

Table 9.5. DDT&E Shuttle Program Costs, Discounted to 1971 Dollars (in Millions of Dollars)

Fiscal Year	Current-Year Dollars	Consumer Price Index	NASA Comptroller	Actual Program Costs
1971	43	43.1	43.1	43.1
1972	59.5	57.5	59.5	55.5
1973	144.5	134.8	144.5	124.0
1974	502.1	428.0	469.3	400.9
1975	864.7	659.9	746.1	622.0
1976	1,165.9	833.1	920.2	763.6
Interim	331.3	233.7	256.8	212.2
1977	1,365.3	901.8	976.6	801.4
1978	1,323.6	821.0	877.1	692.1
1979	1,254.8	703.6	756.4	594.9
1980	1,079.4	527.7	583.7	456.7
1981	1,011.3	447.4	467.1	380.4
1982	767.9	305.5	333.3	258.9
Total	9,912.2	6,097.2	6,653.6	5,405.6

Source: Robert F. Thompson to Charles Donlan, letter with enclosure, 22 Apr. 1981.

were declining. Such considerations helped to keep program costs from spiraling out of control. Even so, with NASA's overall budget remaining under strict limits, the cost of the shuttle placed other programs under particularly severe pressure. Among the more noteworthy was the Space Telescope.

The Space Telescope

The shuttle was to carry satellites. There was considerable interest in crafting payloads that could take advantage of new capabilities that it offered, including opportunities to revisit an orbiting spacecraft to make repairs or to refurbish it with new instruments.[105] Such a mode of operation particularly suited the nation's astronomers. They were accustomed to operating their observatories using suites of instruments that could be changed or replaced as more capable versions became available.

Astronomers were well aware that a space telescope could overcome the limitations of ground-based versions. Such instruments look upward through the atmosphere, which causes stars to twinkle. This twinkling blurs photographic images, smearing out fine detail. In addition, the atmosphere absorbs ultraviolet starlight along with much of the infrared, forcing astronomers to view the universe through a narrow band of visible and near-infrared wave-

357

lengths. By contrast, an orbiting telescope might view the full range of wavelengths, and could form images that were vastly sharper.

Astrophysicist Lyman Spitzer was among the first and most effective advocates. As early as 1946, writing for the Rand Corporation, he anticipated that "the chief contribution of such a radically new and more powerful instrument would be, not to supplement our present ideas of the universe we live in, but rather to uncover new phenomena not yet imagined, and perhaps modify profoundly our basic concepts of space and time."[106] Spitzer became so closely identified with this concept that when it began to take form with the name Large Space Telescope, some people thought the initials LST meant Lyman Spitzer Telescope. It indeed was to be large, with a mirror having diameter of three meters or 120 inches. Bigger mirrors were feasible; the famous Hale Telescope atop Mount Palomar, near San Diego, had one of 200 inches. However, large mirrors are costly, while their telescopes grow rapidly in size and weight as the mirror becomes bigger. A three-meter instrument was about as large as might be launched into orbit.[107]

Initial concepts, late in the 1960s, envisioned launch by Titan III. However, use of the shuttle promised not only operational flexibility but lower costs. In October 1970 the shuttle was an active topic of study, though it still was more than a year away from obtaining Nixon's support. Even so, prompted by a study from Grumman, NASA's George Low gave approval to a program of studies that sought to define a suitable LST concept. Leadership in this work went to Marshall Space Flight Center, which was expanding into the field of space astronomy and had managed the Apollo Telescope Mount, a part of Skylab.[108]

Right at the start, the cost of the LST became a paramount concern. A 1972 concept called for a three-step program that was to start with an engineering model of the telescope, continue with a "precursor" flight unit, and then proceed with the definitive LST. The program cost, some $700 million, was too high. Within NASA Fletcher proposed a cost target of $300 million. Marshall responded with an LST program that was to be markedly compressed—telescoped, as it were. Out went plans for the engineering model and the precursor. Instead, a single unit was to stand at the program's focus. In addition, the new plans substantially reduced the use of on-orbit servicing. People expected that the shuttle would fly routinely and inexpensively. Marshall therefore proposed to return the orbiting telescope to Earth for major refurbishments, and then to relaunch it.[109]

The changes chopped the estimated program cost in half, to $345 million. Nevertheless, they carried risk, for program managers now planned to put their eggs in a single basket. NASA already had built smaller space telescopes; the Orbiting Astronomical Observatory program had crafted versions with mirrors as large as thirty-two inches. The agency, however, had never attempted anything on the scale of the LST. A step-by-step approach would have carried wisdom and prudence, but it was not in the budget.

Even so, this LST program failed to fly. Congress refused to fund it, not because the approach was too risky but because it still cost too much. Hence a new round of descoping was in order. Designers cut the mirror size from three to 2.4 meters, or 94 inches. This reduced the size and cost of the overall program while permitting use of a less expensive on-board system for pointing and attitude control. Instruments could also be of lesser quality, for they were to be easily replaceable. Indeed, because people still anticipated that the shuttle would provide routine access to space, such replacement was to be conducted in orbit.[110]

Allowing for inflation, a 1975 plan quoted the 94-inch version at $273 million, in contrast to $359 million for the 120-inch LST. The new concept dropped part of the earlier name, becoming simply the Space Telescope. This was the version that went forward. The program, however, saw problems long before the telescope saw stars. Part of the reason lay in its management. As *Science* reported in 1983,

> *When the telescope was approved for a new start in 1977, responsibility for scientific instrument development and post-launch operations was given to NASA's Goddard Space Flight Center in Greenbelt, Maryland, while responsibility for spacecraft engineering was given to the Marshall Space Flight Center in Huntsville, Alabama. Marshall then parceled out the work to two separate but equal "associate contractors." Perkin-Elmer would fabricate the main mirror and the rest of the optical assembly in Danbury, Connecticut, while Lockheed would build the spacecraft itself and then integrate all the components in Sunnyvale, California.*
>
> *The split between the engineering at Marshall and the science at arch-rival Goddard meant that no one was doing end-to-end systems integration. The split between Lockheed and Perkin-Elmer meant that no one was even doing a proper job of hardware integration. . . . The communications, apparently, have been nothing short of horrible. Quite apart from intercenter rivalries, the Office of Space Sciences at headquarters has had five directors in 6 years, with a similar turnover of project managers at Marshall. As [NASA administrator James] Beggs points out, "A large part of the communications breakdown was that people didn't know each other and didn't know how to ask the right questions."*[111]

The management arrangements had reasons. Marshall was lead center for the program; Goddard's role drew on its experience with scientific payloads. On paper, this arrangement tapped the strengths of both centers. However, the initiative for the LST had come from Marshall, not from Goddard, and cooperation between these centers proved difficult. Noel Hinners, who helped start the LST program at Headquarters and who became director of Goddard in 1982, stated that program leaders at this center took the attitude, "We'll do the minimum; screw it." Fred Speer, program manager at Marshall from 1979 to 1983, declared that "you can't tell another Center what to do. It tells you what it will do." Nancy Roman, chief astronomer at Headquarters, added that "I think an awful lot of the problems that the Space Telescope has had are because of the Marshall-Goddard split."[112]

The split between Lockheed and Perkin-Elmer was another approach that looked good at the time. As part of the first program descoping in 1973, Marshall proposed to save costs by taking responsibility for systems engineering and integration. Ordinarily, Lockheed would have done these things, as prime contractor, with Perkin as a subcontractor. Marshall expected to supervise both firms directly, thereby further cutting costs, but found itself facing tight restrictions on the number of personnel it could send to those companies. These staff limits came from Headquarters. They reflected a further desire to save money; they also stemmed from Pentagon concerns. Lockheed and Perkin were working on classified reconnaissance satellites, and the Defense Department wanted to minimize the number of on-site NASA people who might learn those contractors' secrets.[113]

Work with Perkin brought its own problems. The director of Marshall, William Lucas, described it as "probably the least responsive contractor we've ever dealt with. Their top management really didn't give a lot of attention, it appeared to us, to this program." Lucas attributed this to the fact that the Space Telescope "didn't constitute a sufficiently significant part of their business base." Authors Andrew Dunar and Stephen Waring add that Perkin was "not worried about NASA moving the project, because the Agency had nowhere else to take it. Perkin-Elmer had learned bad habits working on defense contracts and preferred to solve problems by spending money."[114]

Compounding these management problems was a serious underestimation of the technical difficulties. Jean Olivier, chief engineer at Marshall, noted that people initially had expected to proceed "using technology that we already fully understand." Experience showed, however, that the "technologies were much, much more demanding across the board than we ever realized when we got into it." The technologies particularly demanded tests, but amid the program's tight budget, these were not always in the plan.

For instance, Lockheed initially proposed a limited schedule of trials that left out a thermal vacuum test. This was to place the complete Space Telescope in a large vacuum chamber, amid widely varying temperatures, to simulate conditions in space. Other programs treated such tests as vital, for they could give a complete ground simulation of the satellite in orbit. Lockheed later became aware that a thermal vacuum test indeed was necessary, and this was only one of many matters that drove up costs.[115]

Perkin-Elmer encountered problems of its own, in developing a fine guidance system to point the telescope accurately at guide stars. For a time Perkin could not determine whether its sensors would work properly because it could not test them. This subsystem relied on circuitry from a contractor, Harris Electronics. Early in 1982, project cost overruns forced Harris to cut back its efforts and to lay off part of its staff. When more funds became available, Harris had to bring in a new and inexperienced group of employees.[116]

NASA had planned to place the telescope in orbit in December 1983, with the program cost standing at $575 million in 1980. Late in that year, amid escalating cost pressure, NASA slipped the launch date to October 1984 and accepted a cost of $645 million. This decision sought to cope with current money problems by delaying work and pushing higher costs into later fiscal years. This proved to offer no more than a stopgap, with the Harris Electronics problem being only one of many. At Perkin, project schedules slipped by a week per month during 1981, with the rate of slippage increasing to a month per quarter during 1982, and then to a month per month.[117]

As Program Manager Fred Speer put it, "The program was underfunded. You cannot get something like that for the money that was set aside." The budget shortfalls meant that "almost every month we found a gap. Every gap we found meant additional money was to be spent." Writing to Beggs, Lucas declared that Marshall was "not able to fully recover from the inherent problems introduced into the program as a result of those early decisions" that deleted prototype versions of the telescope and avoided placing program management at Lockheed.[118]

Early in 1983 the program stood on the brink of collapse. It was not clear that the fine guidance sensors would work properly, for they had not been tested. Precision restraints that were to hold scientific instruments in position had problems of their own, which threatened to make it impossible for astronauts to remove them in orbit. In addition, Perkin faced the problem of mounting the heavy main mirror without imposing strains that would distort its perfect optical surface.

If the nation was to have a big orbiting telescope, the program would need major change. NASA delayed the launch to the fall of 1986, to give more time for development. Time meant money, and Congress accepted a total budget of $1.175 billion. NASA Headquarters took on more authority. More tests appeared on the schedule, along with more spare parts. In the words of Dunar and Waring, the program "for the first time had resources consistent with its technical difficulty."[119]

The program encountered a further major delay, but this was not of its own making. It resulted from the loss of Challenger, which put off the launch date to April 1990 (and drove the program cost above $2 billion). By then the program had a new name: Hubble Space Telescope, named after the astronomer Edwin P. Hubble, who had discovered that the universe is expanding.[120] There was irony in the name, for the program budget had expanded as well. But the new money did not solve all the old problems, and this became clear following launch.

In orbit, it failed to focus properly. Investigation traced the problem to an error in the shaping of the main mirror, which was too flat by one-fiftieth the width of a human hair. In forming this mirror, specialists at Perkin had relied on a highly sensitive optical instrument, a "reflective null," that could detect very slight deviations from proper curvature. Unfortunately, no one checked the accuracy of the reflective null. It was off by 1.3 millimeters, more than enough to spoil the final result.

The telescope was not useless, for computer processing of its images made it possible to sharpen the focus in a number of instances. Even so, a NASA scientist said that "the Hubble is comparable to a very good ground telescope on a very good night, but it's not better than the best." The only solution was for astronauts to go up and fix it, which they did in December 1993. They installed small mirrors, shaped to precise perfection, that served as eyeglasses. These corrected the Hubble's faulty vision, enabling it to finally come into its own as an outstanding astronomical instrument.[121]

This experience showed how a cost squeeze dating to the 1970s could cast its shadow well into the 1990s, amid false economies that robbed from tomorrow to get through today. During the 1970s the shuttle program also encountered delays, which brought difficulties of their own. This experience contrasted with that of the Air Force, which was better funded.

A Tale of Two Cities

There was more to the shuttle than the shuttle itself. The Air Force's SAMSO had three associated projects: shuttle facilities at Vandenberg, the Titan 34D, and the Inertial Upper Stage. NASA, in turn, had communities of users who were awaiting the day when it would fly. These two organizations had their offices on separate coasts, with NASA Headquarters in Washington and SAMSO in Los Angeles. Both experienced schedule slips and cost overruns in their shuttle activities. Yet whereas SAMSO continued to meet its responsibilities by using the Titan III, the shuttle's delays brought a great deal of disappointment to another group of NASA clients. These were the nation's planetary scientists, who wanted to launch a mission to Jupiter.

Late in 1976 SAMSO determined that the Air Force shuttle program would cost $1.4 billion in current-year dollars through fiscal year 1983. Seven months later this estimate was close to $2 billion, with the largest part of the increase reflecting the cost of adapting the shuttle to the Pentagon's culture of secrecy. SAMSO responded in part by slipping the date for first launch from Vandenberg from December 1982 to June 1983. It also reduced the initial capability of the launch facilities, which now were to support only six flights per year, whereas earlier plans had called for twenty. This reflected an appreciation that there would not be as many missions as soon as had been anticipated.[122]

Another delay emerged during 1978 in response to an Air Force demand for more payload capacity. It wanted to place a thirty-two-thousand-pound reconnaissance satellite in polar orbit. The standard shuttle could not do this, for the orbiter had grown in weight during development. In December NASA officials elected to increase the thrust for such missions by mounting a strap-on solid motor to each of the shuttle's two SRBs. NASA soon abandoned this approach, choosing instead to pursue a lightweight SRM casing. Still, while it lasted, this strap-on concept called for changes to the design of the SLC-6 launch pad, which increased the construction time. Air Force officials slipped the target date anew, from June to December 1983.

The next delay put this date in June 1984. This happened during 1980, as the lowest bid for a major construction project came in $33 million higher than anticipated. This reflected the severe inflation of that era, which brought sharp increases in the cost of copper and electrical equipment. Air Force officials wanted to address this by reprogramming funds from other budget categories, but first they needed approval from Congress. They got it, but it took time, and this caused the program schedule to slip by several additional months.[123]

Next, during 1981, problems arose due to the Vandenberg weather. From the outset officials had expected to assemble the shuttle in the open and initially had planned to do this by using cranes on the Mobile Service Tower. That concept gave way to an alternative approach, calling for a drawbridge-like device on the mobile Payload Changeout Room, which stood on the other side of the launch pad from the MST. This device would serve as a cradle, holding the ET, swinging it upright, then placing it between the two stacked SRBs. It was to do the same with the orbiter, lifting it upright and pushing it against the ET.

As detailed design work approached completion, program officials worried that the fog and strong winds of Vandenberg would make it difficult to deal with small clearances between the orbiter and ET. They decided in September 1981 that as with NASA at KSC, assembly of the shuttle should take place indoors. They accepted a plan for a mobile Shuttle Assembly Building (SAB), an enormous enclosure to surround the components of a shuttle while mating them on the pad. It took shape as a huge windowless cube with a big American flag and the letters "USAF" painted on the side. Within the SAB assembly was to proceed using cranes; the program could dispense with the drawbridge. Yet this change in the plans boosted the projected cost from $1.9 to $2.5 billion, and delayed the anticipated date of first launch from June 1984 to October 1985.[124]

Every year, within the Defense Department budget, Congress appropriated funds for an ongoing military construction program. Under this Pentagon-wide program, each year brought the start of funding for several new construction projects at Vandenberg. The complete shuttle facilities program was extensive. Thus, in September 1982, a year after the decision to build the SAB, the list of past and future construction starts was as follows:

FY 1979 *SLC-6 launch pad, site preparation*
SLC-6 launch pad, construction
Launch Control Center

FY 1980   *Relocation of a Titan II rocket motor facility*
Orbiter Maintenance and Checkout Facility (OMCF)
Hypergolic Maintenance and Checkout Facility
Utilities

FY 1981   *Extension of existing eight-thousand-foot runway to fifteen thousand feet*
Mate/Demate Facility
SRB processing facility
ET processing facility
Seventeen-mile shuttle route from OMCF to SLC-6
Facility to test APUs used in SRB thrust vector control
Logistics

FY 1982   *Integrated Operations Support Center*
Office building for engineering and management
Small harbor for ET barges, with tow route for ETs
Flight Crew Equipment Facility

FY 1983   *Solid Rocket Booster Disassembly (located in Port Hueneme)*
Shuttle Assembly Building
Shuttle facility modifications

FY 1984   *Thermal Protection System Facility*
SLC-6 modifications to prevent ice formation on the ET
Hazardous waste management
Shuttle facility modifications
Security modifications

Changes in requirements had brought additions to this list, such as the "modifications" slated for funding in fiscal year 1983 and 1984.[125]

There certainly was far more to the Vandenberg shuttle base than SLC-6, but that launch pad remained at the center of the effort. It was the main item in the fiscal year 1979 construction package, with the work being initiated in April 1980 and reaching completion in September 1983. Projects initially funded during a particular year often took several years until they were completed. For example, this was also true of the largest project in the fiscal 1980 budget: the shuttle's hangar, or Orbiter Maintenance and Checkout Facility. Its time for construction ran from March 1980 to November 1982.[126]

The advent of plans for the SAB, in 1981, had pushed the planned date of the first shuttle launch to October 1985. During that month, the Air Force officially stated that the Vandenberg shuttle facilities were operational. Preparations went ahead for the launch of Discovery in July 1986 as the maiden flight from this center. Unfortunately, Challenger exploded in January. The shuttle was grounded, and the Air Force secretary, Edward Aldridge, abandoned those plans. No shuttle ever flew from SLC-6, for Aldridge renewed the Air Force's commitment to its trusted upgrades of the Titan III. Indeed, even these proved inadequate. Needing more lifting power, the Air Force turned in 1985 to the Titan IV, with payload capacity of as much as fifty thousand pounds.[127]

Late in the 1970s, with NASA's shuttle approaching operational status, the Air Force pursued the Titan 34D as an alternative. This represented an unwillingness to place all military eggs in the single basket of the shuttle. Until the shuttle proved itself, such a policy was certainly prudent. Indeed, Pentagon and SAMSO officials declared that the 34D would actually help smooth the transition to the shuttle.

The Air Force used three upgraded Titans as launch vehicles. The Titan III-C was the standard version, with its Transtage carrying payloads to geosynchronous orbit. The III-D was nearly identical but lacked this upper stage, serving to carry heavy payloads to low orbits. The III-B used a lengthened first stage but dropped the twin solid boosters. Its payloads were only about one-third as heavy as those of its big brothers, but with an Agena upper stage, it too could reach geosynchronous orbit. The III-E, a III-D with Centaur, was for NASA.

After 1977 the Air Force planned to consolidate this family by establishing the new Titan 34D as a standard. NASA, its eye on the shuttle, already was dropping the III-E, which flew for the last time during that year. The Air Force planned to abandon the III-C and -D once the 34D was operational, and these plans envisioned a considerable role for the 34D. When launching to geosynchronous orbit, it was to fly from Cape Canaveral and use the Inertial Upper Stage (IUS), thus achieving commonality with the shuttle. For missions to polar orbit, it would fly from Vandenberg AFB with no upper stage.

The Air Force also looked ahead to the shuttle, and therefore anticipated only limited demand for the 34D. This had consequences for the Titan III production facilities at Martin Marietta. Separate production lines had been building the two types of first stage, with their differing lengths. However, this arrangement would be excessively costly amid limited demand, and only one of these lines could survive. The survivor was to be the one that built the long first stage for the III-B. The 34D was to use this stretched first stage, with new solid boosters that grew in length to fit the longer core.

The Air Force expected to use the shuttle, but welcomed the 34D because this version of the Titan III could be available fairly quickly. Indeed, because the 34D drew on existing elements of the Titan III family, its development went forward with little difficulty. The early plans called for the Vandenberg version to use the IUS inertial guidance package with this Titan launch vehicle. That proved infeasible, for the standard Titan guidance system used analog circuitry, and Martin had to devise a new digital system that could work with IUS guidance, which was also digital. This brought difficult problems with hardware and even more so with software, and as costs began to climb, SAMSO opted instead to steer the Vandenberg 34D using a simple radio-inertial system.[128]

Development began in mid-1977, two weeks after Congress voted the initial funding.

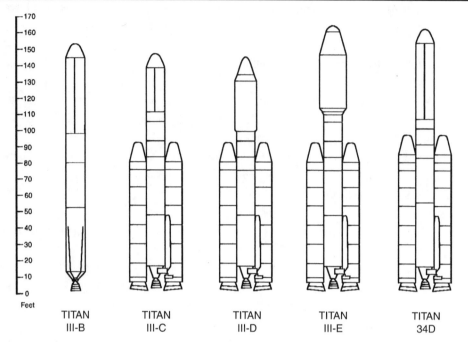

Fig. 67. *The Titan III family. The Titan III-E, with Centaur upper stage, saw limited use by NASA. The Titan 34D used the core of the Titan III-B and replaced the Air Force's Titan III-C and III-D. (Art by Dennis R. Jenkins)*

The program reached a milestone in August 1979, as United Technologies, builder of the solid boosters, ignited a test model within a tower near San Jose, California. It flamed with 1.34 million pounds of thrust, shooting a powerful plume of fire into the sky. It soon became clear that this rocket would also shoot into the sky, for eleven months later, SAMSO accepted the first three 34Ds to come off the production line.[129]

SAMSO also had the IUS on its agenda, with Boeing as the prime contractor, but this project did not go forward so smoothly. The IUS certainly was no space tug, being neither reusable nor capable of retrieving spacecraft in orbit. But the Air Force wanted reliability of 97 percent, and because this upper stage was to use inertial guidance, it needed multiple avionics packages. This quickly brought the problems of redundancy management that had marked the shuttle itself.

Boeing did not plan to use true triple redundancy, with failure isolation through voting. Rather, the IUS was to use two active computers, with a third on standby. The two active units were to continually check their own and one another's health, while running programs for inertial guidance, thrust vector control, and reaction controls. If these computers were to disagree, the third unit was to switch in and take on the computational load. The other two then would use diagnostic routines to search for the cause of the discrepancy. This approach amounted to carrying the third computer as a backup, and determining automatically when to use it. Because an IUS needed only a few hours for a flight to geosynchronous orbit, compared to days for a shuttle mission, it could achieve acceptable reliability in this manner.

The solid-fuel rocket gave an early focus for attention. It was to burn for 145 seconds, a duration sufficient to raise concern over excessive erosion of the ablative throat liner. A key test took place at Arnold Engineering Development Center in December 1977 as a full-sized IUS motor fired for 154 seconds, validating its initial design.[130] Other work dealt with the casings of the small and large IUS motors. They were of filament-wound Kevlar, strong and lighter in weight than the steel of conventional design. The IUS casings had to withstand pressures of 1,225 psi. During 1979, engineers tested them by pressurizing them with water until they burst. The large one reached 1,405 psi in May; the small casing went to 1,390 psi two months later. Full-scale firings followed, with the small and large motors attaining 25,000 and 63,800 pounds of thrust, respectively. United Technologies, builder of the motors, ran off a series of six tests between December 1979 and August 1980, and all were successful.

Software for the avionics gave a different story. The computer memory had a firm limit on its number of words, and the designers could not increase this number merely by adding more physical capacity. Each word in memory had its own address for retrieval on demand, which was stored as a 16-bit binary number. The limit for memory thus was the number of possible such binary integers: $2^{16} - 1$ words, or 65,535. During 1979 it became clear that the software was too large to fit the memory. By August it contained 70,415 words. In October the count was down to 66,629, which was still too high, though it appeared possible to reduce it to 61,144. Yet even this was too many, for the goal was a software size 10 percent below the maximum. Boeing responded by turning to its software contractor, TRW, which called for a total rewrite of the software package, thus creating substantial problems with schedules and cost.[131]

Boeing also ran into difficulty in coming up with an acceptable set of IUS designs. Its contract for full-scale development, awarded in April 1978, covered the Air Force's standard two-stage IUS as well as NASA's double-stage and three-stage vehicles. The Pentagon's version took the lion's share of Boeing's attention, and early in 1979, its managers submitted what they regarded as a final design. SAMSO officials gave it a critical review during February and told Boeing to go back to the drawing board. The design still needed work in the areas of software, rocket motors, and interfaces with the shuttle and with a ground-control facility. This project did not pass its subsequent review until November.[132]

United Technologies held the subcontract for the IUS solid rocket motors. This work suffered its own delays, as engineers dealt with issues of quality control in fabricating nozzles and tail cones, and in casting the charges of propellant. Further, in the words of the Air Force, Boeing "had handled few large space development contracts before this one" and "was consistently overoptimistic in estimating the cost and time involved in solving technical difficulties." In May 1979 the company looked ahead to first flight of an IUS as early as August 1980, little more than a year away. SAMSO did not agree, projecting first launch in July 1981 and anticipating a substantial cost overrun.[133]

In April 1978 the initial IUS development contract quoted a cost of $263 million. A renegotiated contract, late in 1979, boosted this target to $430 million, while authorized changes added another $32 million during the following months. Even this proved inadequate; in June 1981 a second renegotiation raised the price tag to a minimum of $506 million, with the government being willing to pay up to $700 million if Boeing completed the work at a financial loss.[134] By then, however, the end was in sight. TRW carried through the writing of software for the first IUS missions, which were to use the Titan 34D, and completed this work in September 1980. The programs underwent tests at Boeing, with Martin Marietta accepting software in April 1981 for the first IUS flight. Rocket-motor development culminated in a series of twelve firings, all successful, which concluded late in July. The first flight-rated IUS vehicle was already undergoing acceptance testing at Boeing, and was shipped to Cape Canaveral in November. It flew successfully in October 1982, atop a 34D, placing two military communications satellites in geosynchronous orbit.[135]

In this fashion the Air Force slogged ahead, accepting its overruns and its schedule delays but holding to its approved programs. For NASA, the shuttle's delays told a different tale, which was particularly acute within the field of planetary science. Work within this discipline was at a peak during the years of shuttle development, for in every year from 1964 to 1981, the nation launched or had a mission encounter at the moon or on some planet. At times NASA had several such encounters during a year.

We orbited Mars and mapped it with a camera, landed on its surface and looked for life. We approached Venus, probing deep into its atmosphere, and by a neat trick of celestial mechanics flew past Mercury not once but three times. We discovered the volcanoes of Mars and of Io, satellite of Jupiter; viewed the rings of Saturn close up, and were awed by the violence and immensity of those giant planets. From Venus to Saturn, every planet was the target for at least three visits by two or more distinct generations of spacecraft. We even took advantage of a rare alignment of the planets, sending the Voyager 2 probe on a planetary grand tour. It visited Jupiter in 1979 and Saturn in 1981, then swept onward toward planned encounters at Uranus in 1986 and at far-distant Neptune in 1989.[136]

These missions had vast scientific significance, for they turned the study of the moon and planets into a working discipline, drawing freely on meteorology and geology while contributing to these fields in turn. The significance was also cultural, amounting to a new age of exploration. A century earlier, the artist Albert Bierstadt had limned the West with paintings that matched this legendary land. Decades later, the photographer Ansel Adams gave portraits of his own that provided a counterpart. The planets had their own landscape artist, for the artist

Chesley Bonestell had produced brilliant color portrayals, thereby establishing space art as a new genre. The photos from NASA's missions played Adams to Bonestell's Bierstadt.[137]

Within this realm Jupiter was in the ascendancy during the 1970s. NASA started with a pair of spinning spacecraft called Pioneer 10 and 11, which flew on Atlas Centaurs respectively in March 1972 and April 1973. Each took nearly two years to reach Jupiter, sending back photos, with Pioneer 11 continuing on to a successful encounter at Saturn in 1979. In addition, 1977 brought the launch of Voyager 1 and 2, as they rode the Titan III-E Centaur on separate launches toward Jupiter. Both flew past that planet successfully, again during 1979.[138]

Their photography showed Jupiter as a phantasmagoria of vivid color. Its atmosphere appeared as a roiling mass of crimson, orange, brown, and white, deeply stirred by storms. Its Great Red Spot showed swirling detail that eclipsed even the imaginings of Bonestell and his followers. Its major satellites emerged as worlds in their own right, with Io, closest to Jupiter, ablaze with active volcanoes.[139]

The results from the Voyagers were spectacular, and whetted appetites for more. Like the Pioneers, the Voyagers executed flybys, which was like having a ship's crew sail past an island rather than land and explore. The next step called for a spacecraft that could fire a retrorocket and go into orbit around Jupiter, while dropping a probe into its atmosphere. This mission won approval in 1977. It took the name Galileo, after the scientist who discovered Jupiter's principal satellites by using his telescope and who declared that their revolutions around this planet supported Copernicus's belief that the earth circles the sun.[140]

Galileo was heavier than Voyager, for it needed propellants for its retro-fire. Even so, it had only three-fifths the weight of Viking, which had flown to Mars atop a Titan III-E Centaur. Galileo might have used this same launch vehicle, and NASA would have done itself a favor had it done this. Instead, the project found itself caught up in the difficulties of both the shuttle and the IUS, which was to serve as its upper stage.[141]

The plan initially called for launch by shuttle in 1982, and delays in that program caught Galileo's managers unaware. "There was no appreciation at the center directors' meetings that the problems would be anything this bad," Bruce Murray told *Science*. He headed the Jet Propulsion Laboratory, which served as the project's center. John Casani, the Galileo program manager, added that availability of the shuttle had been taken for granted: "We were originally scheduled to be taken up on the 26th launch, and then schedule slippages moved us up to the seventh. We sure thought we had enough padding."[142]

Weight problems exacerbated the difficulties. Midway through 1979, it became clear that the IUS and Galileo both were too heavy to fly. Galileo had an allowed maximum weight of 4,451 pounds, but could not be trimmed below that level. IUS designers were somewhat more successful, showing—on paper—that they could cut its weight, though the margin might be no more than a hundred pounds.

Even then a 1982 launch called for the shuttle to fly with its maximum launch weight as early as the fifth or sixth mission. It would need a lightweight external tank, SSMEs quali-fied for 109 percent of rated thrust at liftoff, and an orbiter stripped of unnecessary equip-ment. NASA responded by splitting Galileo into two separate spacecraft, a probe and an or-biter, each requiring its own shuttle flight. (It did not escape attention that the probe might have to be sacrificed to budget cutters, to save the rest of the mission.) The launch dates were pushed back to February and March 1984.[143]

Then during 1980 and 1981 the planetary IUS fell into mishaps. Whereas the standard Pentagon IUS used one small and one large solid motor, the planetary versions used two of the large ones, with the small unit on top for the extra energy needed to reach Jupiter. Boe-ing had been developing NASA's models alongside that of the Air Force, and when the latter encountered overruns and delays, so did the former. In November 1980 company officials quoted a development cost for the three-stage IUS of $179 million, 50 percent more than NASA had budgeted. Although the Air Force could absorb such increases, NASA could not. Within that agency, thoughts quickly turned to canceling the planetary IUS and switching to NASA's own Centaur as a shuttle upper stage.[144]

Concepts of the Centaur had blossomed briefly a year earlier, when the Air Force itself had briefly considered canceling the IUS. Within NASA, the new overrun brought these plans again to the forefront. "That was the straw that broke the camel's back," Frank Van Rensselaer, director of expendable launch vehicles, told *Aviation Week.* "When you stack the total transportation costs up and the development costs, it comes out to be significantly cheaper to go with Centaur rather than to continue with the planetary IUS development." In addition, the combination of shuttle and Centaur could launch other high-energy payloads.[145]

This left Galileo as a mission without a launch vehicle. NASA canceled both its versions of the IUS, embraced Centaur, recombined the Galileo orbiter and probe into a single space-craft, and set a new launch date of 1985. This date slipped as well, and Galileo, like Vanden-berg AFB, found itself caught up in the aftermath of Challenger. NASA then returned it to the IUS, put Galileo back on its calendar—and finally launched it in 1989, on a roundabout path that did not allow it to reach Jupiter until 1995.[146]

"It was a mistake not to go with a planetary Centaur in the first place instead of the IUS," one NASA manager declared.[147] It was also a mistake not to go with the Titan III-E Centaur, which launched the Vikings to Mars as well as the Voyagers, but which flew no more after 1977. By embracing the shuttle with unmaidenly eagerness, NASA thus damaged its plane-tary program—which certainly was not what the agency had in mind when it declared that the shuttle would be all things to all people. The abandonment of NASA's IUS, at so late a date, was far from auspicious.

Nevertheless, in early 1981 the shuttle was well past the days of dreams and designs, past the era of flaming SSMEs and of fiery cost overruns. Even the thermal-protection tiles had been tamed. Columbia was on the launch pad, and the first flight to orbit was only weeks away.

CHAPTER TEN

Preparation for Flight

Before France became a space power by crafting the Ariane launch vehicle, it built a space-port near Kourou, French Guiana, on the northern coast of South America. In an era of de-colonization, this was one of the few places where Parisians could still wave the *tricouleur.* Indeed, French Guiana counted as a *département* of France itself.

Prior to the space age, the area had been known as the site of Devil's Island. Here Alfred Dreyfus, condemned through perjury, served four years before supporters won his release. *Time* magazine writes that at Devil's Island, the worst of the *bagnards,* prisoners, "were guil-lotined and tossed to the sharks." France shut down the prisons in 1946. In consequence, Kourou, the only town, declined until its population fell to just six hundred. But the French space program brought a dramatic revival. The Ariane effort repopulated the area by attract-ing some of France's best. In doing so, this effort showed how a slumbering region in the Third World could vividly awaken by reinventing itself as a center for advanced technology.

For space shots, Kourou was ideal. It had extensive stretches of water to the north and east, allowing launches at any azimuth. Located nearly at the equator, its low latitude gave rockets the full advantage of the earth's equator. This provided an eastward speed of a thou-sand miles per hour while a booster still sat on the ground.

CNES, France's space agency, did more than simply build launch pads. It constructed a complete town for six thousand people, along with a harbor. The activity attracted a colorful crew of workers from Brazil, Martinique, other Caribbean islands, and more than a dozen other nations. Initial launches of sounding rockets took place in 1968. The successful flight of a Diamant in March 1970, which placed two satellites in orbit, demonstrated that this cen-ter was ready for use.[1]

Then in 1979, on the day before Christmas, a particularly loud thunder rumbled across the local jungles and swamplands. For the first time in eight years a large European launch vehicle was in flight and headed for space. Ariane, successor to the failed Europa rockets of earlier days, was flying its first mission. Its three stages fired in their turn, its payload fairing separated cleanly, and it was off and away. "Guidance and control was good," said Frederic d'Allest, the CNES director of launch vehicle programs. "The first and second stage looked nominal, while third stage thrust seemed somewhat higher than specified." This upper stage pushed a thirty-five-hundred-pound test satellite into a highly elliptical orbit that reached above twenty thousand miles in altitude.[2]

Ariane held a severalfold significance. It brilliantly vindicated the hopes of its advocates, who had broken with the Europa designs and had started afresh with clean sheets of drafting paper. It scored a triumph for the new European Space Agency (ESA), winning an achievement that contrasted in welcome fashion with the failures of ELDO. Its flight plan showed further contrast, for while the Europa program had struggled painfully to win success one stage at a time, Ariane flew with all three stages live.

Its development program showed nothing like the three-year delay that marked the space shuttle. In 1973, when it was called the L3S, its first launch was scheduled for 15 March 1979. Midway through 1979, with the date having slipped only to 3 November, a minor problem with engine mountings brought a delay of a month. A launch attempt on 15 December ended with engine cutoff four seconds after ignition; another attempt, eight days later, ended in a scrub. But the third time was the charm, as Ariane inaugurated a new era of serious European space capability.[3]

This era did not dawn overnight, for the second flight, five months later, splashed into the Atlantic due to failure of one of the four first-stage engines. Divers recovered it intact, along with one of its mates; the problem was diagnosed as combustion instability, and modifications to the engine's injector permitted further launches. The next two flights, both during 1981, succeeded, and although the fifth launch failed as well, eight of the next ten also flew properly.[4] Furthermore, an upgrade of the initial design boosted the chamber pressure in the main engines, stretched the third stage for extra propellant capacity and added two small solid boosters to the first stage. This brought Ariane 3, which could place 5,180 pounds on a trajectory to geosynchronous orbit. That put it on a par with the Atlas Centaur.[5]

In addition, during 1980 the backers of this Eurorocket conducted a second successful launch as they set up a commercial organization, Arianespace, to market future flights. It took responsibility from ESA for the overall program—and quickly booked customers for the first sixteen operational flights.[6] With this, and with the Air Force continuing to hold its Titan 34D, it was clear from the start that the shuttle would not have success handed to it on a silver platter. NASA had already ceased its purchases of Titan IIIs, and was looking ahead to a similar phaseout of its Delta and Atlas-class vehicles. Nevertheless, though most eyes were on the shuttle, it was not the only route to the future of space flight.

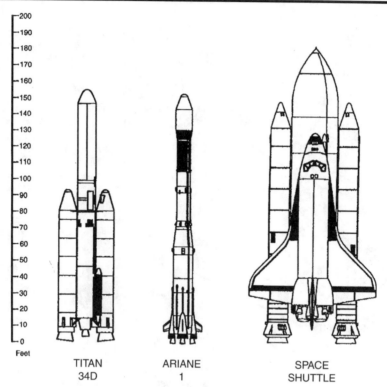

Fig. 68. *New launch vehicles of the 1980s: the Titan 34D, Ariane 1, and space shuttle. (Art by Daniel Gauthier)*

Kennedy Space Center

Preparation for first flight began with preparation of the shuttle's launch facilities at Kennedy Space Center. Many of these facilities were a direct inheritance from the Saturn V: the Vehicle Assembly Building, the huge steel platforms that would support a complete shuttle vehicle, the immense tracked transporters that were to carry shuttle and platform together to the launch pad, and the facilities of Launch Complex 39. New construction was limited, but there was much revamping of what already existed, and one may follow the planned work by tracing the flow of projected activity from the landing of an orbiter to its subsequent liftoff.

It was to touch down on a runway called the Shuttle Landing Facility, a sixteen-inch-thick slab of concrete paving with length of fifteen thousand feet and width of three hundred feet. Safety overruns added an extra thousand feet at each end for a total of well over three

373

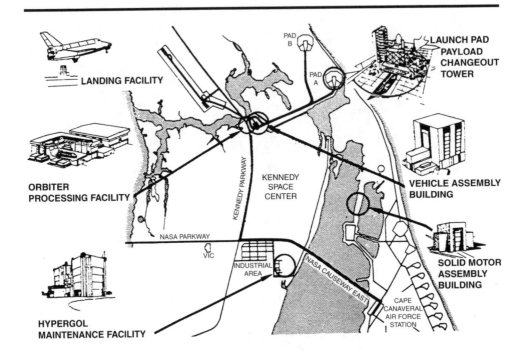

PAD B

PAD A

LAUNCH PAD PAYLOAD CHANGEOUT TOWER

LANDING FACILITY

ORBITER PROCESSING FACILITY

KENNEDY PARKWAY

KENNEDY SPACE CENTER

VEHICLE ASSEMBLY BUILDING

NASA PARKWAY

VIC

INDUSTRIAL AREA

NASA CAUSEWAY EAST

SOLID MOTOR ASSEMBLY BUILDING

HYPERGOL MAINTENANCE FACILITY

CAPE CANAVERAL AIR FORCE STATION

Fig. 69. *Shuttle facilities at the Kennedy Space Center. The Solid Motor Assembly Building, part of Cape Canaveral, supported the Titan III as well as the shuttle and belonged to the Air Force. (NASA)*

miles. The main runway was nearly twice as long as a standard commercial version and twice as wide. It also had Microwave Scanning Beam Landing System installations, again at each end, to permit landing in either direction.

Following roll-out, with the orbiter still on the runway, ground operations were to begin. These included attaching a tow vehicle, to pull the orbiter along a two-mile towway to its hangar within the Orbiter Processing Facility (OPF). This was one of the principal items of new construction at Kennedy Space Center. It was as large as a hangar for airliners, standing nine stories tall, 397 by 233 feet in dimension. Within the OPF a low bay stood between two high bays. The former included areas for electronic equipment, mechanical and electrical shops, repair of the Thermal Protection System, a communications room and supervisory control room. Each high bay accommodated an orbiter; hence two of these craft could be in the processing flow at any time. Flip-up work platforms surrounded an orbiter to provide access; two large rolling bridges, along with a fixed bridge, provided further access. Each high bay also had an emergency exhaust system in case of a hypergolic fuel spill. The OMS and reaction-control modules were not to stay within the OPF, however. They were to go into the Hypergolic Maintenance Facility, located some distance away within the Kennedy Space Center industrial area.

An orbiter would remain for months within the OPF, receiving extensive maintenance and check-out. While there, it was to obtain power and coolant from ground installations and would remain connected to the Launch Processing System. A final Orbiter Integrated Test, demonstrating its readiness, was a prelude to the next phase of work, which began by towing the orbiter out of the OPF and through a large door into the Vehicle Assembly Building. This immense structure held four high bays, each tall enough to stack a complete Saturn V with moonship. They readily accommodated the shuttle, which had only half the height. The two bays facing east, toward the Atlantic, served for assembly of complete space shuttle vehicles atop their steel Mobile Launch Platforms. The two other bays were used for the SRB segments and for the ET.

These segments arrived by rail from Utah. Upon arrival, they went into High Bay 2 for removal of transportation covers, then were moved into the adjacent High Bay 4 and placed vertically into work stands for inspection. The VAB also had a low bay, less than half the height of the main structure. It received SRB subassemblies such as nose caps, parachutes, electronics, and elements of the thrust vector control system. These supported the buildup of aft booster assemblies within High Bay 4, one for each SRB. Cranes lifted these assemblies and placed them in position on a platform in High Bay 1 or 3, a shuttle construction area. Heavy SRB segments followed, being stacked to form two complete solid boosters, standing separately as twin columns.

The ET arrived by barge from NASA's Michoud Assembly Facility, a delicate aluminum eggshell that entered High Bay 4 for its own inspections. It lay horizontally in a check-out area; a crane of 250-ton capacity hoisted it vertically and transferred it into the bay that held the SRBs, placing it between them for mating. The orbiter, entering through the VAB's north door, was towed into a broad aisle that separated the two pairs of high bays. The 250-ton crane, assisted by another of 175 tons, lifted the orbiter and swung it to the vertical. Then, for the first time since final approach prior to touchdown, the landing gear would be retracted. Members of the ground crew mated the orbiter to the ET, attached umbilicals for data and electric power, and moved platforms and work stands into position for access to the complete stack.[7]

Close to the VAB, standing literally in its shadow, was the Launch Control Center. A NASA document noted that "if the VAB is the heart of LC-39, the LCC is its brain." Within this four-story building, Firing Rooms 1 and 2 held the consoles and minicomputers of the Launch Processing System, which had supported check-out activities within the OPF and now turned to preparations for launch. The LCC had dimensions of 378 by 181 feet, with half of one floor being given over to the Central Data Subsystem that provided the LPS with its mass memory. The document added that a shuttle firing room "will be a lonely place. Launch with the Shuttle LPS will require approximately 45 operational personnel, one-tenth of the 450 needed for an Apollo launch."[8]

Within the VAB the Mobile Launch Platform, supporting the complete shuttle, showed considerable modification following Apollo. It had formerly held the Launch Umbilical Tower, a steel structure painted red and standing 398 feet tall, with a crane at the top. The shuttle dispensed with this, substituting a permanent tower, the Fixed Service Structure, at each launch pad. The shuttle also dispensed with conventional umbilicals. It relied on two Tail Service Masts that stood atop the launch platform, respectively feeding liquid oxygen and liquid hydrogen into ports at the rear of the orbiter while providing gaseous hydrogen, oxygen, nitrogen, and helium, along with coolants and electrical power. These service masts were hinged; at liftoff they were to rotate into blast-proof shelters, for protection from the engines' exhaust.

The launch platforms were substantial steel assemblies in their own right, each being 25 feet deep, 160 feet long, and 135 feet wide. Each weighed more than four thousand tons standing alone; even a fully fueled shuttle increased this weight by little more than 50 percent. During Apollo days a single hole in the platform, 45 feet square, allowed exhaust from the five first-stage engines to escape. The shuttle needed three holes, one for the SSMEs and two others for the SRBs, and crafting them was a job for heavy-duty steelworkers. These platforms were built of welded plates with thicknesses of up to six inches.

With a shuttle on top, a platform rode from the VAB to the pad on a crawler-transporter. Each of these tracked vehicles had a flat upper surface the size of a baseball diamond. It was propelled by two 2,750-horsepower diesel engines, which drove electric generators that provided power to sixteen motors, two for each tread. The general arrangement, including the power ratings, resembled that of a diesel-electric locomotive. Each transporter had a leveling system that held the shuttle within one-sixth of a degree of true vertical for a deviation of no more than about six inches. This system also kept the shuttle erect while the transporter negotiated a five-degree ramp that led to a launch pad.

To reach the pad, the ensemble of crawler, launch platform, and shuttle clanked down the crawlerway. This looked like an interstate highway, with two forty-foot lanes separated by a grassy fifty-foot median, but it accommodated the crawler alone, with each lane serving the treads on that side. (One pictures this transporter making its way along nearby Interstate 95 at its top speed when loaded of one mile per hour.) The crawlerway had three layers of fill with an average depth of seven feet and with a top surface of river gravel. When the behemoth passed, it left a flat surface of crushed rock in its wake.

At the end of this route lay Launch Pads 39-A and 39-B. These had served the Saturn V, but for the shuttle, most of their structures were removed or relocated. The S-IC, for example,

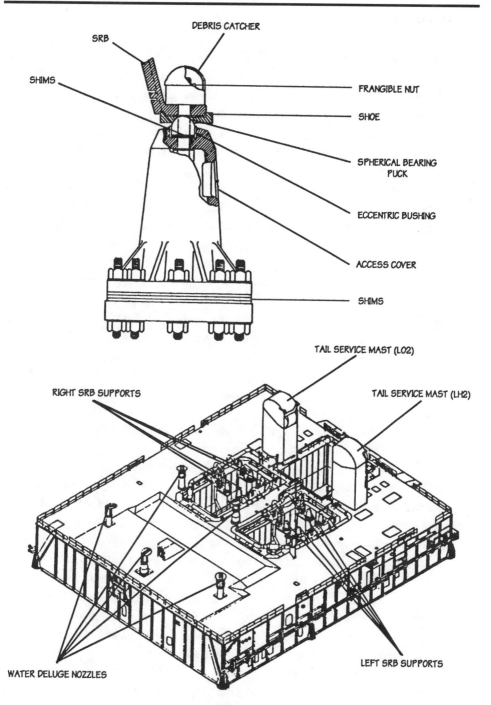

DEBRIS CATCHER

SRB

SHIMS

FRANGIBLE NUT

SHOE

SPHERICAL BEARING PUCK

ECCENTRIC BUSHING

ACCESS COVER

SHIMS

TAIL SERVICE MAST (LO2)

RIGHT SRB SUPPORTS

TAIL SERVICE MAST (LH2)

WATER DELUGE NOZZLES

LEFT SRB SUPPORTS

Fig. 70. *Hold-down posts secure the Solid Rocket Boosters prior to launch. Bottom: Mobile Launch Platform. (NASA, courtesy Dennis R. Jenkins)*

flew with more than seven hundred tons of rocket-grade kerosene, but facilities for that fuel now were obsolete. Flame deflectors also were extensively revamped. At each pad, a mobile gantry had serviced the Saturn V and its moonship, but it gave way to a completely new set of installations: the Fixed and Rotating Service Structures.

The Fixed Service Structure was a recycled Launch Umbilical Tower of Apollo, shortened to 247 feet while retaining its crane. It provided work levels at 20-foot intervals and carried three swing arms. One served as a gangplank, allowing the crew to enter the orbiter. The second arm supplied electrical power and carried a vent that collected cold gaseous hydrogen as it boiled off from within the ET. The third held a "beanie cap" that fitted the tip of the ET, collecting cold oxygen boil-off. It was necessary because this venting gas would otherwise produce a buildup of ice on the tank, which would shake loose and damage the orbiter's thermal-protection tiles.

The Rotating Service Structure was hinged to the Fixed, resting on supports that moved along a circular track. It carried its own umbilical connections, supplying liquid hydrogen and liquid oxygen for the fuel cells along with hypergolic propellants for the OMS and reaction-control systems. This structure also held a five-level enclosure that provided a clean room for payloads, allowing them to be installed at the pad. The Rotating Service Structure largely enfolded the shuttle, leaving only the top of the ET in view. When it swung away, it left the entire shuttle standing in the open. The effect amounted to an unveiling, a dramatic moment for visitors.

Pad facilities also included spherical vacuum-jacketed tanks that held the SSME propellants. The liquid oxygen was denser than water and required pumps, but the liquid hydrogen was so light that modest pressure within its tank easily pushed it through a pipeline, to fill the ET. The hydrogen system therefore used no pumps. A tall water tower also stood near each pad, with a tank of its own that held 300,000 gallons. This water sprayed copiously onto the flame deflector and the launch platform during liftoff, flowing through pipes of seven-foot diameter that could empty the tank in twenty seconds. These sprays provided sound suppression, absorbing acoustic energy that otherwise would reflect from the flat launch platform and threaten the payloads. This reduced the sound level at the rear of the orbiter from 167 decibels to 162, a cut in intensity of two-thirds. It also protected the launch pad against the rocket exhausts, which were hotter and far more severe than any blowtorch. When this water flashed into steam, it produced the dense clouds that accompanied a launch.[9]

Construction of these facilities went forward rapidly, for from 1974 until 1978, NASA officials expected to launch the first flight to orbit during 1979. The runway gave an early focus for attention, with its builders at work prior to 1975—at a time when Apollo still had one last hurrah, with a Saturn I-B scheduled for launch as part of the Apollo Soyuz Test Project. The runway project called for grading the site and then paving it with 260,000 cubic yards of concrete, a quantity that demanded a thousand railroad carloads of cement and ten thousand carloads of a limestone and sand mixture.[10]

Construction of the OPF began in June 1975. Then in 1976, workers with tall cranes began dismantling a Launch Umbilical Tower. A section only twenty feet tall weighed a hundred tons. With this tower removed, later that year, its launch platform was given its makeover. Work crews removed the steel plating that covered its top, installed new beams while cutting away some old ones, and transformed its single-hole configuration into one that had the shuttle's three holes.[11]

The overall effort at the Cape hit its stride during that year. NASA paid for it all, building toward a peak of eighteen hundred contractor employees. Construction of the runway was completed in May. The OPF also went up, initially as the low bay with a single high bay, with the latter dwarfing a mobile crane that stood nearby. The Fixed Service Structure rose above Pad 39-A, while designers prepared plans for the SRB supports on the launch platform, payload ground-handling equipment, and the Tail Service Masts. Work also began on the Hypergolic Maintenance Facility, with five existing buildings from the Gemini program being modified for this new role.

Within the VAB, High Bay 3 was the initial shuttle stacking area, with High Bay 4 serving for storage and check-out of the ET. By year's end the work stands were 80 percent complete. The height of these bays made it possible to store ETs 130 feet up, rather than on the floor, which freed considerable space for work on the SRBs. Tom Utsman, a project engineer on the KSC staff, said that this cost sixty thousand dollars and saved several million dollars.

Much work with electronics took place during 1977, as microwave landing equipment was installed on the runway. Other hardware went in at the Launch Control Center, as the first operational firing room entered service. The Rotating Service Structure at Pad 39-A was completed late that year. The modifications within the VAB were nearly finished, and existing work platforms, with a large round hole for the Saturn V, now showed three holes for the shuttle. The second OPF high bay also approached completion, with the doors of both bays displaying vertical cutouts that allowed the shuttle's tail to enter.[12]

Construction at Pad 39-A was completed in mid-1978, with 39-B being slated for 1982. There also was activity in shipbuilding, as a Florida company, Atlantic Marine, built two vessels to retrieve the SRB casings after they parachuted into the ocean. Each such ship was 176 feet long, little longer than the SRBs themselves, with displacement of 940 tons. NASA chose a subsidiary of United Technologies to operate them, and they were named UTC Liberty and UTC Freedom. They carried satellite navigation equipment along with search radars, sonar, direction finders, and gyrocompasses. Diesel engines provided propulsion, as did water-jet stern thrusters. The latter prevented the propellers from injuring skin divers, who assisted in retrieving spent SRB casings and secured them alongside the ship.[13]

Defining the Missions

Although the general design of the shuttle was in hand as early as 1972, its initial mission was not immediately clear. During 1973 and 1974, there was considerable interest in an unpiloted first flight, which would spare the lives of a crew in the event of loss. Prior experience gave strong arguments on both sides of this issue, for the Saturn I-B and Saturn V had both made unpiloted flights before they had been trusted to carry astronauts. On the other hand, all Saturn-class rockets, in a series dating to 1961, had reached orbit successfully or had flown well on suborbital launches.[14]

At a meeting at NASA Headquarters in January 1974 several senior officials and advisors debated this question. Their views were summarized in a memo from Donald Cheatham, an operations manager. Robert Gilruth, longtime director of the Manned Spacecraft Center, noted that "the crew can contribute to mission success but they cannot hold the orbiter together. If a vehicle is lost, it will not be because it was unmanned. Keep the unmanned option." This conservative recommendation was significant, for Gilruth had built his career as a leader in piloted space flight.

Holt Ashley was a professor of aeronautics at Stanford University, president of the American Institute of Aeronautics and Astronautics, and director of the National Science Foundation. He too was cautious: "There is enough time to construct a decision tree and do a better job of estimating costs and probabilities and conduct fault-tree analysis. The vehicle should be more mature to allow this to be properly done. More data are needed on risk to allow a decision. Do not be bold; retain the unmanned option."

Alexander Flax headed the Institute for Defense Analyses; he had presided over an influential study of the shuttle in 1971. He emphasized the importance of a clear decision: "If the resources can be allocated, add a vehicle and go unmanned. If the program cannot afford another vehicle then get serious because retaining the option will lead to frittering away money. If we cannot afford the impact of a dedicated unmanned vehicle, then plan only to go manned."

Allen Donovan, a senior vice president at the Aerospace Corporation, supported Flax: "If the unmanned option is retained, make it a 'hard' option and plan now to fly the first flight unmanned and allocate that vehicle as a test article. If this cannot be done, make a 'hard' decision now to go manned." Capt. Charles Conrad, a Gemini and Apollo astronaut, drew on his background: "I am concerned that the retention of the option will cause a division in the critical skills manpower and we will not be able to do a good job on either system; plan to go manned and delete the option."

Eugene Love, director of space programs at NASA-Langley, consulted his own background and agreed with Conrad: "The money allocated to the unmanned option should be diverted to improve the ground test program and the decision should be made to drop the unmanned option and go manned." Courtland Perkins, chairman of the aerospace engineering department at Princeton University, saw merit in either alternative: "If from either technical or political reasoning a failure of the first flight is deemed traumatic enough to kill the program, then decide now to go manned. If, however, a failure, while unmanned, would be considered in the same light as an experiment or aircraft flight failure such that the flight test program could continue, then we should retain the option and plan to go unmanned."

Not one of these seven participants specifically recommended an unpiloted first flight; Gilruth and Ashley merely spoke of holding an "unmanned option." Conrad and Love strongly advocated a piloted mission, while Flax, Donovan, and Perkins were prepared to go either way.[15] Nevertheless, a somewhat stronger set of views emerged a few days later, during a meeting of NASA center directors.

Christopher Kraft, head of the Johnson Space Center, was far more bullish than Gilruth, his predecessor: "JSC's position is that the crew greatly maximizes the probability of mission success and that the risks, given the improved test and analysis techniques available today, can be minimized." Rocco Petrone, director of NASA-Marshall, presented recommendations that were cautious but unmistakable: "The crew definitely increases the probability of getting back an intact orbiter. The greatest return is realized from the advantages of having the men's judgment and skills available during orbital flight. At liftoff, men are a marginal asset. MSFC's review team ended up split down the middle—a standoff, with intense views on both sides. In terms of program advantage and management judgment, would underwrite focusing resources on manned and eliminating unmanned option."

Kurt Debus, a veteran from V-2 days who now headed NASA-Kennedy, was firm: "A manned flight offers more advantages for return of an intact orbiter than the unmanned option. A certain degree of risk can be accepted. It is my opinion, assessment and intuitive judgment that the program advantage is to eliminate the unmanned option. Therefore, I would recommend strong consideration to manning the first vertical flight." Dale Myers, who chaired the meeting, was associate administrator for manned space flight and headed NASA's Office of Manned Space Flight. Even so, he did not merely root for the home team, noting that the unpiloted option might be retained "if we really enforced a minimum, simplified automated system." Nevertheless, "the baseline already calls for manning the first vertical flight."[16]

In July 1974 John Yardley, Myers's successor, presented his decision in a memo that received the signatures of Fletcher, Low, and Petrone, who now was associate administrator:

I have recently reviewed all the facts pertinent to a decision regarding manning the first Shuttle orbital flight (see enclosure). Based upon this review, I have determined that we should proceed with design, development, and testing of the shuttle considering only a manned first flight, and drop the parallel considerations of the un-

manned option. This is based on my judgment that the manned benefits of greater probability of success, less risk to overflight population, lower cost, and better operational system far outweigh the crew and program risks involved.

I have also determined that this decision can be reviewed 18 months before first orbital flight, and an unmanned option can be exercised at that time, if conditions so warrant.[17]

With the issue of the first flight settled, there was plenty of opportunity to look ahead to other early missions. The Skylab orbiting station, which had been left in space following the final departure of its astronauts in February 1974, gave an early topic of interest. Crew member Gerald Carr had used attitude-control thrusters to boost Skylab's orbit by seven miles, and mathematical studies indicated that while this orbit was slowly decaying due to atmospheric drag, Skylab would remain aloft until March 1983. Everyone knew that the shuttle would be ready by then, and initial plans called for it to revisit this station on its sixth flight.

It did not take long before people began to worry that this orbiting lab might come down considerably sooner, and the reason was sunspots. Sunspots wax and wane in number during an eleven-year solar cycle, matching an increase and decline in solar activity. This does not mean that the sun becomes hotter and cooler, but that the sun produces outpourings of charged particles, which stir the upper atmosphere and make it less rarefied. More molecules at orbital altitudes meant a shorter lifetime. A new solar cycle began in March 1976, with astronomers at the National Oceanic and Atmospheric Administration (NOAA) watching the sunspots closely. That autumn, they predicted a peak sunspot frequency that nearly doubled the previous projection. In turn, this indicated that Skylab would fall as soon as early 1980.

Solar-activity forecasting is not an exact science, and NASA officials naturally preferred predictions of longer life, which came from their own models. The first shuttle flight was set for the spring of 1979, which seemed to give leeway, and mission planners at NASA-Johnson concluded that they could rendezvous with Skylab on the fifth mission. At NASA-Marshall, attention focused on a small propulsion unit that astronauts could attach to the Skylab station, to boost it into a higher orbit.[18]

In October 1977 Robert Frosch approved this rescue mission and set the date as February 1980. This raised the question of whether Skylab would still be around. "If solar activity exceeds what we expect, it could re-enter," Robert Aller, a NASA deputy director, told *Aviation Week*. "We have that possibility. But we fully expect it to be there when we plan to visit it." Not everyone shared this expectation, for NASA officials had requested a forecast of Skylab's lifetime from NORAD, the North American Air Defense Command, which tracked satellites at a center in Colorado Springs. NORAD analysts used NOAA's sunspot predictions—and found that Skylab could make its final plunge as early as mid-1979. The reason was that its broad solar panels were increasing its drag.

NASA responded by again moving the rescue mission forward, to the third and perhaps even the second shuttle flight.[19] The agency also tried to prolong the life of Skylab by reorienting it to a low-drag attitude. Nevertheless, time continued to press. In mid-1978 the first shuttle launch could occur no earlier than June 1979, with the rescue mission definitely planned as the second flight, in October. NASA believed that Skylab could stay aloft until April 1980. NORAD and NOAA had a different date: as early as mid-November 1979, giving NASA only a few weeks to spare.[20]

NASA threw in the towel in December 1978 as Frosch advised President Carter that Skylab could not be saved. The first shuttle flight was set for 28 September 1979; nevertheless, the rescue mission could not fly before March 1980. Moreover, the propulsion module could only delay the inevitable, for Skylab weighed more than 160,000 pounds and this module lacked the energy to boost it to a safe orbit.[21] It finally reentered on 11 July 1979. Its demise drew a great deal of public attention, for the near-meltdown that March of the Three Mile Island nuclear power plant near Harrisburg, Pennsylvania, had stirred concern over the perils of technology. However, while Skylab's orbit took it over nearly all of the world's population, most of earth's surface consists of ocean or sparsely populated desert, forest, and tundra. It broke up over western Australia, producing a spectacular display of lights in the night sky accompanied by sonic booms, but its surviving pieces fell in the remote outback. No one was hurt.[22]

With Skylab off the agenda, mission planners shaped an alternate payload list. At the outset of the program, NASA officials had spoken of a two-week turnaround between flights, representing 160 hours of two-shift activity. However, this was a goal for an operational shuttle fleet, after accruing a good deal of flight experience. The early flights were to need considerably more turnaround time, and NASA had a schedule. In 1976 the plan was as follows: first mission, 29 weeks; second, 13 weeks; third, 11 weeks; fourth, 9 weeks; fifth, 34 days; and sixth, 24 days. The first four missions were to serve as test flights, with subsequent ones counting as operational missions. In practice, every launch, far into the future, was likely to amount to a test. The X-15 made some two hundred flights in the course of a decade and never ceased to be an experimental airplane, while the shuttle was more complex. Even so, flights beginning with the fifth were to carry payloads.[23]

The shuttle had been designed to provide low-cost access to orbit. NASA was highly interested in offering launch costs that were low enough to lure users away from alternatives such as the Atlas-Centaur and the Air Force's Titan 34D, as well as from Europe's Ariane. Yet the agency could not merely quote a lowball price for a launch and subsidize the users; the cost per flight was to cover the expenses of operational missions. The point of departure lay in the planned cost of $10.45 million, in 1971 dollars.

A subsequent review by the General Accounting Office stated, "This estimate equals about $13.3 million in 1976 dollars." In June 1976 NASA officials estimated that the cost per flight would come to $16.1 million, for a true increase of 21 percent. This gave a basis for costs per flight that were quoted to users. Those users included the Pentagon, other federal agencies (including NASA), foreign governments, and commercial organizations that were building communications satellites.[24]

In May 1976, concurrent with this estimate, NASA's George Low presented James Fletcher with a recommended set of prices to be charged for launch services. The era of operational flight was to begin with the fifth mission, and then divided into two phases. Phase I was the first three full fiscal years of flight activity; Phase II covered the next nine years. With the first flight to orbit anticipated during 1979, these twelve years of activity were to run from 1980 to 1992. Low anticipated that experience in launch operations would lead to a modest decrease in shuttle prices during Phase II (table 10.1).[25]

NASA began to implement its pricing policy early in 1977. A memorandum of agreement with the Defense Department, dated 7 March of that year, quoted a price per flight of "$12.2M in FY 1975 dollars," with provision for inflation. This favorable price reflected the Air Force's involvement in the shuttle program, with the Pentagon paying not only for its military spacecraft but for Vandenberg Air Force Base. For launches from Kennedy Space Center, most of them nonmilitary, the price was quoted at $19.0 to $20.9 million, again in 1975 dollars. For this expense, a user purchased not only a flight to orbit, but a guarantee of a free reflight if the launch attempt were to fail.[26]

These prices indeed succeeded in attracting users, including commercial customers. A manifest of October 1979, published in *Aviation Week,* showed thirty-seven launches projected in less than three years. Columbia was to fly first, with others following:

1981 *1 September: Tracking and Data Relay Satellite (TDRS)*
      3 November: Intelsat F5, SBS B, and Telesat E communications satellites

1982 *8 January: TDRS*
      March: Defense Department payload
      16 April: Spacelab 1
      25 May: TDRS and German scientific satellite
      July: Defense Department payload
      4 August: Intelsat F6 and Telesat F communications satellites;
         GOES F weather satellite
      14 September: Spacelab 2; first flight of Challenger
      1 October: Insat B for India; TDRS
      4 November: Long Duration Exposure Facility (LDEF); also, retrieval of
         Solar Maximum Satellite
      7 December: Spacelab 3
      December: Defense Department payload

1983   *3 February: Solar Polar out-of-ecliptic mission*
    1 March: Reflight opportunity for any earlier mission
    15 March: RCA E, Materials Processing Satellite, Telesat G, and
        Syncom 4-1; last two are communications
    7 April: Spacelab mission; Earth Radiation Budget satellite
    April: Defense Department payload
    May: Defense Department payload
    15 June: RCA F, Palapa B-1 for Indonesia, Syncom 4-2, space application
        payload
    19 July: Spacelab mission for physics and astronomy
    August: Defense Department payload
    September: Defense Department payload
    20 October: Arabsat A, Palapa B-2, Syncom 4-3
    29 November: West German Spacelab mission
    20 December: Space Telescope; first flight of Discovery

1984   *17 January: Launch Arabsat B; retrieve LDEF*
    20 February: Galileo Jupiter orbiter
    6 March: Reflight opportunity
    20 March: Galileo Jupiter probe
    March: Tiros N weather satellite plus space available; first launch from
        Vandenberg Air Force Base
    4 April: Four communications satellites: PRC 1 for China, AT&T 1,
        Telesat 1, Syncom 4-4
    April: Defense Department payload
    2 May: Spacelab physics and astronomy mission
    May: Defense Department payload
    12 June: PRC 2, AT&T 2, Syncom 4-5, Materials Processing Satellite
    21 June: Spacelab applications mission[27]

A revised manifest in March 1980 slipped the first operational mission to March 1982 and listed thirty-nine launches through number 43, in September 1984. This schedule carried four fewer Pentagon flights than the previous one, but accommodated most of the NASA and commercial payloads in spite of the delay. A new mission model, issued in June 1980, further trimmed the number of launches to twenty-seven by September 1984, but then projected a rapid expansion of activity to sixty in total by January 1986, with twenty-two during the first nine months of that year.[28]

All these schedules proved highly optimistic. The shuttle took nearly five years to make only twenty-four flights, and then stood down for nearly three years following the loss of Challenger on 28 January 1986.[29] The projected costs also were optimistic. In September

Table 10.1. Recommended Costs per Flight, 1976 (in Millions of Dollars)

User Category	Phase I	Phase II
Non–U.S. government	19.7–22.2	19.3–21.2
Civil U.S. government, Canada, European Space Agency	17.1–19.3	16.2–18.1
Department of Defense	12.3–14.0	11.7–13.2

Source: NASA Deputy Administrator to NASA Administrator, memo, 28 May 1976.

1980 NASA revised the cost per flight from $16.1 million to $27.9 million in 1975 dollars, an increase of 73 percent. It was $20.8 million in 1971 dollars, double the initial promise of $10.45 million.

This appears to reflect increasing understanding of just what it would take to prepare a shuttle for launch once operations settled down to a routine, and how much it would cost. The original figure of $10.45 million dated to 1972, when the shuttle existed only on paper. It represented a goal, something to design for, but in no way did it draw on the type of experience that builders of airliners had long since accumulated. It could not; no one had ever before built a reusable launch vehicle.

The 1980 figure, $27.6 million (in 1975 dollars), still did not draw on practical experience, for the shuttle had not yet flown. But it was only a few months away from first flight, and NASA by then had a much clearer understanding of its operational requirements. The 1980 cost per flight then reflected these insights. Subsequent estimates used 1982 dollars, with $27.6 million of 1975 coming to $49.6 million in those dollars.[30] In 1985 a White House paper stated, "The minimum acceptable bid will be $74 million (in 1982 dollars)." This higher figure took note of operational lessons, for by then the shuttle had been flying for several years.[31]

Even so, these estimates still were low, for they anticipated economies of scale due to frequent use of the shuttle. The program requirements of 1972 specified launch rates having "a maximum of 60 per year." The mission model of 1979 projected a slow buildup in flight rate during the 1980s, to stand at 55 per year from 1989 through 1993. However, there were also diseconomies of scale, which came to the forefront when the shuttle proved incapable of such use. The shuttle relied on an infrastructure that was both extensive and expensive, with a professional staff that took its paychecks whether they succeeded in launching sixty shuttles per year, or only six.[32]

The best year for shuttle operations was 1985, with nine flights. It never again came close to double digits, and with the fixed costs of its infrastructure spread over a diminished number of launches per year, the cost per flight escalated accordingly. In 1996, a decade after Challenger, this cost was quoted at $550 million—compared to $350 million for the Air Force's Titan IV, which had similar capacity.[33]

Still, in the late 1970s all eyes were on the planned first mission, STS-1. It took shape as a simple exercise wherein a crew of two was to fly to orbit and spend two days in space, work with the on-board systems, take data using instruments, reenter along a benign trajectory that promised to spare the thermal protection, and touch down on Rogers Dry Lake at Edwards AFB, close to NASA-Dryden.[34]

Astronauts for the Shuttle

To the public, NASA and its astronauts were all but synonymous. The corps of America's spacefarers had flourished. The agency hosted six selections between 1959 and 1967, choosing forty-nine pilots and seventeen scientist-astronauts. In addition, following cancellation of the Air Force's Manned Orbiting Laboratory program, seven more transferred into NASA, for a total of seventy-three. Many left for greener pastures as it became clear that they faced a long standdown between Skylab and the shuttle.[35] But as the latter program began to make headway, officials started to nurture thoughts of a new group of selectees.

They would be chosen according to new criteria, for the country had changed since the heyday of Mercury, Gemini, and Apollo. In April 1959, when the seven Mercury astronauts had faced their first press conference, they stood forth as a group of hot test pilots who had distinguished themselves in the nation's service.[36] But by the early 1970s plenty of people were ready to view them as pale, male, and stale. Of the seventy-three NASA astronauts, all had been white men; not one had been black, Hispanic, or a woman. This was unacceptable, for after 1970 public interest in space flight fell to a low ebb while concern for civil rights, defined very broadly, rose to flood tide.

James Fletcher personally raised this topic in March 1972, less than two months after obtaining President Nixon's endorsement of the shuttle. "We are working on plans to get members of minority groups into space," he told a conference on equal employment opportunity at NASA-Kennedy. "The space shuttle, which is the keystone to all our future space programs, will be an important factor in accomplishing this goal." He nevertheless was not yet ready to make a commitment, adding, "These are only plans. We don't know they'll work out."

Fletcher took another step in September as his center directors held a retreat in the Blue Ridge Mountains of Virginia. He asked that "a plan be developed for our next selection of astronauts, with full consideration being given to minority groups and women." The directors agreed that NASA-Johnson, home of the astronauts, should prepare this plan, basing it on anticipated requirements for number and frequency of shuttle flights. Later that month, Dale Myers sent a memo to Christopher Kraft, director of NASA-Johnson, inviting him to submit the plan by the following February.[37]

An early exercise sought to learn whether women had the right stuff. At NASA-Ames, twelve Air Force flight nurses took part in a five-week test program. Several of them received two weeks of complete bed rest, to simulate weightlessness, then rode a centrifuge that subjected them to 3-g acceleration, reproducing the stress of a shuttle reentry. The women were allowed to decide for themselves how long they could tolerate the g-forces, and the times were as high as seven minutes, far more than necessary for a return from orbit. Researchers concluded that no inherent problems barred women from flying aboard the shuttle, and Hans Mark, the center director, said that there was a "very high probability" that women would join the flight crews.[38]

Nevertheless, it was too early to select additional spacefarers. The review at NASA-Johnson showed that the existing corps could meet all requirements for the next several years, and in June 1973 Fletcher stated that the agency had "enough astronauts right now and doesn't plan to recruit new ones until about 1978." Two years later, John Young, head of the astronaut office, noted that NASA "took on a lot of people it shouldn't have," prematurely swelling the astronaut ranks and then leaving them with few missions to keep them in the program. "You don't pick new fellas until you need to fly," he added.

At NASA Headquarters John Yardley stayed in touch with Kraft at NASA-Johnson, where people continued to prepare for a new selection. In March 1975 in a letter to Fletcher, Kraft stated that presently active astronauts, all dating to the 1960s, were likely to retire from space flight during the next ten years, as they reached age fifty, and that an entirely new complement of people would have to replace them. Kraft also redefined the job requirements. He saw a continuing need for pilots; he also anticipated a requirement for mission specialists, holders of doctorate degrees who would work as experimental scientists.[39] This built on the early category of scientist-astronauts, of which the geologist Harrison Schmitt had flown to the moon on Apollo 17, with three other scientists spending time in space aboard Skylab.[40]

Yardley presented his astronaut-selection plan to George Low late in 1975, and Low found it satisfactory in many respects. "However," Low added, "the plan does not indicate a method for insuring application by minorities and/or women in the new astronaut group and mission specialists group. I am sure that you are aware of the importance to NASA that every opportunity be presented to these potential candidates to encourage applications and, if qualified, selection. Please let me know how you intend to proceed to solve this problem."[41]

Kraft drafted a revised plan in March 1976 and sent it to Yardley. Kraft also initiated the process of choosing new candidates. As an early step, he set up the Astronaut Selection

Board, with separate panels for pilots and mission specialists. The pilot panel had nine members, with three astronauts; these included John Young and Donald "Deke" Slayton, both of whom had headed the astronaut office. The mission-specialist group had ten members, again with three astronauts. Both panels included Joseph Atkinson, head of the Equal Opportunity Program office at NASA-Johnson, and Carolyn Huntoon, a branch chief. Atkinson was black. With this, he and Huntoon became the first of their respective race and gender to hold responsibility in choosing new astronauts.

Atkinson played a particularly strong role in recommending means whereby NASA could broaden its reach in seeking new candidates. There was no need for this in the sense of good people being hard to find, for as Slayton put it, "We pretty much knew where the expertise was in the community and generally I would have no problem in putting my hands on some four or five people, or ten or twelve if that's what we need to recruit. I would go directly to them and hire them." Fletcher had concurred, noting in 1975, "It is amazing how many people want to be astronauts. We have no trouble recruiting." However, he later declared that "blacks thought they weren't wanted so they didn't apply. It was a necessary thing for NASA to do. I don't think there is any question about it."[42]

NASA sent announcements to ninety aerospace firms, asking for help in recruiting, and to nearly a thousand colleges and universities, including all with predominantly minority enrollments. Similar announcements went to a hundred minority organizations and leaders, including the NAACP and the League of United Latin American Citizens. Women's organizations received this attention as well, including the Society of Women Engineers and the American Association of University Women. NASA reviewed a roster from the organization Blacks in Physics, examined the biographees in *Black Engineers in the U.S.* and in *Who's Who Among Black Americans,* and sent letters accordingly, following up with phone calls. The agency also placed advertisements in *Ebony, Black Enterprise, La Luz,* and the publication of the Society of Women Engineers, among others.

Nichelle Nichols helped as well. She was a black actress who played the role of Lieutenant Uhura in the television series "Star Trek," and was a member of the board of directors of the National Space Institute. She agreed to meet with members of community organizations and with minorities in universities. A NASA official stated his hope that "the word gets around that NASA really wants to get applications from qualified minorities and women. We feel Ms. Nichols is uniquely qualified to do this."[43]

Applications came from all over the world, with some of the people being military pilots stationed on aircraft carriers and as far away as Guam, Japan, and Chile. One personnel specialist recalls a man who phoned him using ham radio. He was an American scientist working at a Soviet research base in Antarctica, and he wanted to apply. This specialist said, "I don't know how he found out about it, but if that guy could find out about it, I would think anybody could."

NASA allowed a year for people to post their applications, from mid-1976 to 30 June 1977, and the final tally topped 8,000. Pilot requirements were particularly demanding, for the best candidates had two thousand hours in jets, flight experience with at least twenty aircraft, and more than 250 combat missions. They also had graduated from test-pilot school and had spent up to six years in active flight test. Few women qualified: of 659 suitable applicants, only 8 were female, and none made the final cut.

The mission-specialist requirements were less exacting, for Yardley had overruled Kraft's recommendation that applicants were to hold a doctorate degree, declaring that a bachelor's degree would do. Partly as a consequence, more than a fifth of the qualified applicants were women, and 6 percent were minorities.

NASA winnowed the applicants and came up with 208 finalists, who traveled to Houston for interviews and medical examinations between August and November 1977. Kraft made the final selection; Frosch reviewed the choices and approved 35 new astronauts in January 1978. Fifteen of them were pilots, and 20 were mission specialists—and of the latter, 6 were women. Three had backgrounds in medicine or biochemistry: Anna Fisher, Shannon Lucid, and Margaret Seddon. Kathryn Sullivan was a geologist; Judith Resnik held a doctorate in electrical engineering, and Sally Ride, who became America's first woman in space in 1983, was a physicist.

Two mission specialists were black. Guion Bluford, an Air Force major, had a doctorate degree in aerospace engineering and was a branch chief at the Air Force Flight Dynamics Laboratory at Wright-Patterson AFB. Ronald McNair held his own doctorate degree, in physics, and was a staff member at Hughes Research Laboratories. A third African American, Air Force major Frederick Gregory, was chosen as a pilot. He had been based at the Armed Forces Staff College. Another mission specialist, Air Force captain Ellison Onizuka, was of Japanese descent. He too was a branch chief, at the Air Force Test Pilot School at Edwards AFB.

The selection process had much broadening of outreach, and little if any of quotas or race-based standards. No woman qualified as a pilot-astronaut, while Hispanics still awaited a future date for selection of their own. In both groups, available candidates simply failed to make the cut. Nor did minorities win appointments as mission specialists by doing it the easy way, taking advantage of Yardley's provision whereby holders of a bachelor of science degree could qualify. One selectee, the naval flight officer Dale Gardner, held this minimal credential. But the other nineteen mission specialists all claimed higher degrees—doctor of medicine, master of science, or doctorate. In particular, both black men, Bluford and McNair, had academic backgrounds that were superlative, for both held doctorates.[44]

Some of these people became famous. The lyric "Ride, Sally, ride" was not written for the astronaut of that name; it came from a 1966 recording by Wilson Pickett, "Mustang Sally." Still it suited her as the nation cheered her flight aboard the seventh shuttle mission. In 1996 Shannon Lucid spent 188 days as a guest aboard the Russian space station Mir, setting a record for long-duration flight by an American. Others were not so fortunate. Resnik, McNair, and Onizuka all died tragically aboard Challenger.[45]

The class of 1978 did not receive flight assignments during that year, for its members had to train and to spend time in the program before any of them were ready. Plenty of veterans still were on hand, even though not all had made space flights, and during March the astronaut office chose eight of them as prime and backup flight crews for the first missions, including STS-1.

John Young, head of that office, topped the list as command pilot for this first flight. He was the world's most experienced space traveler, having already made four flights involving two different types of piloted vehicles. He came to aviation by way of Georgia Tech, where he received a bachelor's degree in aeronautical engineering in 1952, along with a commission through Navy ROTC. He became a fighter pilot, went to the Navy's Test Pilot School at Patuxent River, Maryland in 1959 and stayed there for the next three years. In 1962 he set world records for time-to-climb in the new F-4 Phantom.

In October of that year he became one of nine new astronauts, part of a group that included Neil Armstrong, Frank Borman, Charles Conrad, and James Lovell. Within this group of superstars, Young stood out, for he was the first in his group to receive a flight assignment. He flew in the right-hand seat with Gus Grissom on Gemini 3, the first piloted mission within that series. Then, during the subsequent decade, he was assigned to flight or backup crews almost continually. After Gemini 3 in March 1965 he commanded Gemini 10 in July 1966. In May 1969 he was command module pilot for Apollo 10, a flight to the moon. He reached the moon a second time in April 1972, as commander of Apollo 16, and spent three days on its surface. He also was backup pilot for Gemini 6 and Apollo 7 and backup commander for Apollo 13 and 17. He took over the astronaut office in 1974, first as acting and then as permanent chief. This made him not only the most experienced astronaut but the most senior.

His companion for STS-1 was Robert Crippen, whose career had paralleled that of Young. He too entered the Navy via ROTC, earning a bachelor's degree in aeronautical engineering at the University of Texas. He completed flight training and was assigned to carrier duty aboard USS *Independence* in 1962. Two years later he was selected for test-pilot school. He graduated—and stayed on as an instructor. In 1966 he became an astronaut for the Manned Orbiting Laboratory, transferring to NASA following cancellation of that program in 1969.

He had no space-flight experience, but Young knew that he was a good man for the shuttle: "He knows more about the computers that make this thing fly than anyone has a right to."[46] His selection continued a tradition of having rookies accompany experienced pilots on first space flights. Young himself had been such a neophyte, on Gemini 3. Donn Eisele and Walt Cunningham accompanied the veteran Wally Schirra on Apollo 7, the first flight of a piloted command module; William Anders was the new man aboard Apollo 8, which flew to the moon. Russell Schweikart played that role with Apollo 9, which exercised the lunar module in earth orbit. The Skylab program went even further. It placed two newcomers on both the first and second mission and then flew an all-rookie crew on the third, which stayed in orbit for 84 days.[47]

When Young and Crippen received their assignments for STS-1, early in 1978, other astronauts were given tickets for future shuttle flights. Joseph Engle and Richard Truly, who had flown Enterprise during the Approach and Landing Tests, were chosen as the backup crew for STS-1. This put them in the equivalent of the on-deck circle in baseball, ensuring that they soon would be at bat. Fred Haise and Gordon Fullerton, the two other ALT pilots, also found themselves selected for early missions after STS-1, with Jack Lousma and Vance Brand filling out the initial list. Of these six men, only Haise and Lousma had prior space-flight experience; the rest were also new to the trade.[48]

According to published schedules, a flight crew required 1,116 hours of training to qualify for liftoff. The training program was organized by subject, like courses at a university: orbiter electrical and mechanical equipment, avionics, life support systems, flight operations during ascent, orbit, and reentry, and extravehicular activity. Each such course began with classroom instruction and self-paced workbooks, with these studies accounting for more than one-fourth of those 1,116 hours.

Specialized training equipment, some of which dated to earlier NASA programs, dealt with specific topics. Pilots had to prepare to fly the shuttle during approach and landing, and the two Grumman Gulfstream training aircraft remained in use. They had served prior to the Approach and Landing Tests, with shuttle instruments in the cockpit and with modified aerodynamics that replicated those of an orbiter. Astronauts also wore their spacesuits and swam under water within the Neutral Buoyancy Trainer, a large and deep swimming pool. It was a legacy of Skylab, and allowed these spacefarers to float in simulated weightlessness. To learn the gracious arts of living aboard the shuttle, with its galley and zero-g toilet, flight crews spent 69 hours within a detailed replica of an orbiter's cabin.

Astronauts also honed their skills using a single systems trainer, a $1.5 million facility from Ford Aerospace. It had a minicomputer and simulated the electric power systems, environmental control, life support, OMS, and reaction control. "You can read books and ask questions about a system, and nobody has the answers yet," said Young. "But then you get in the single systems trainer, turn on the switch and it will tell you what the answer is. The learning curve when you first get into the training goes straight up."

These activities were preludes to the main form of training, which relied on flight simulators. These complemented and built on the other training aids. Thus, whereas astronauts were marked for forty hours in the Gulfstreams, the schedule called for 350 hours on reentry and landing, with much of it in the simulator. For all phases of training, these simulators called for 546 hours of work, nearly half the total.[49]

Fig. 71. *A test pilot trains for flight within a space shuttle simulator. Displays, seen through the windows, reproduce the scenes when landing at Edwards Air Force Base. (Space Division, North American Rockwell)*

The Link Division of the Singer Company, builder of the wartime Link Trainers, designed and assembled these simulators. An early version, which entered service in November 1976, replicated Enterprise with its limited array of on-board systems, and supported the Approach and Landing Tests. During 1978, Link delivered two shuttle mission simulators, which were far more complete. One had a fixed base; it remained motionless, and trained for on-orbit activities. The other had a moving base, being supported on extendable pistons. It could stand vertically during liftoff, or move in any direction while rolling and pitching, during a landing approach. To represent the g-forces produced by rocket thrust, the cabin rose from near-horizontal to vertical, placing crew members' weight on their backs. It yawed sharply to simulate a thrust imbalance between the two SRBs.

These simulators relied on mathematical models, whereby astronauts received data and observed the responses when they took action. The systems treated by these models spanned the range of on-board instruments and equipment:

Electric power: fuel cells, power distribution, distribution of fuel-cell reactants.
Mechanical power: auxiliary power units, hydraulic systems.
Propulsion: SSMEs, thrust controller, SRBs, reaction control systems, OMS.
Guidance and control: body-mounted sensors, inertial measurement, star tracker, air data such as airspeed indicator, controls and displays, control of thrusters and aerodynamic surfaces such as elevons.
Communication and navaids: TACAN, radar altimeter, microwave scanning beam, UHF and other radio bands, operational instruments, caution and warning.
Environmental control: Freon coolant, water coolant, space radiator, auxiliary power, oxygen and nitrogen supply, atmospheric circulation.
Flight computers: main events controller, master events control, engine/controller interface, formatting.
Mechanical systems: ejection seat, SRB and ET separation, payload bay doors, manipulator arm, landing and braking.
Vehicle dynamics: equations of motion for orbiter and a rendezvous target, orbiter aerodynamics, mass properties, atmosphere, wind, gravity, celestial motions, bending of fuselage, fuel slosh, ground contact at touchdown.
Tracking network: tracking, telemetry, communications.

Complementing the on-board instruments and displays, computers generated real-time color images at thirty frames per second, which appeared on television screens mounted to the windows. These were somewhat cartoonish in quality, reminiscent of computer video games of a few years later. These showed the spherical earth with its continents and landforms, the sun, moon, and stars, and views of the shuttle from the flight deck. When a crew member gave commands to open the cargo bay doors, the computer responded by showing them as they opened.[50]

Even so, simulations of the landing approach called for more than video games. NASA-Johnson had not a map but an accurate model of Edwards AFB, in color and mounted vertically on a wall. It measured twenty-four by fifty-six feet, with six television cameras scanning it to produce cockpit windshield displays. The cameras approached the runway as part of the simulation, and astronauts used this facility extensively. Prior to the Approach and Landing Tests, the pilots Haise and Fullerton made more than five hundred approaches and landings.[51]

The simulators produced sound as well as vision. No loudspeakers could reproduce the bellowing roar of liftoff, with all engines firing. Nevertheless, what the astronauts heard was loud indeed, and it complemented the instrument readings, the computer-synthesized cockpit views, and the motion from the supporting pistons. Craig Covault of *Aviation Week* rode a flight simulator in 1980, and described his experience:

> *Space shuttle main engine (SSME) ignition at T -6 sec. lurched the cockpit forward, then backward, activated main engine chamber pressure tape gauges and created a roar in the cockpit. Ground launch processing system automatically confirmed that engine thrust reached 90% within 4.6 seconds and that no malfunctions were occurring. The vehicle's distinct lurch forward and rebound backward, referred to as the twang, occurred as the three Rocketdyne main engines ignited to 1.12 million lb. of thrust, kicking the orbiter stack 19 in. in the direction of the external tank at cockpit level.*
>
> *Solid booster ignition and liftoff at zero in the countdown were simultaneous and the event was impressive as engine roar increased, cathode ray tube displays changed from a launch to ascent mode, the vehicle again lurched abruptly and immediately began intense longitudinal vibrations. The vibrations were the result of the solid propellant burn characteristics as we moved off the pad with 6.4 million lb. of total thrust.*
>
> *Vibrations during the climbout reminded me of a paint-shaker motion and often prevented me from maintaining focus on cockpit display numbers. Simulator vibrations provide a rough ride and are based on Thiokol solid-rocket-motor ground firing data.*

Sound synthesizers also reproduced the whine of motors, the aerodynamic noise of reentry, and the thump of landing-gear extension. At touchdown, a crew heard and felt the impact, with the simulator being nose-high and then rotating downward.[52]

Much flight simulation featured the active participation of the NASA-Johnson mission control center, which many of us remember from television coverage of Apollo as well as from the movie *Apollo 13.* It took responsibility for a shuttle once it was clear of the launch pad. It too showed change from Apollo days, for it had a new mainframe computer, of the IBM 370 series, which was much faster and more powerful than its predecessor. However, the number of controllers was about the same. Some 115 people had made up the mission control team for an Apollo flight, with thirty in the main control room and the rest providing contractor support. For STS-1 and the following flights, the numbers were 130 and 24, with this staff remaining high because of the large number of unknowns.[53]

Malfunctions represented an important theme in flight simulation. An instructor, working within a separate station, could make a fuel valve fail in a closed position or cause nitrogen to leak from a gear box. Some 860 such equipment problems could be specified, in avionics and other on-board systems, and flight crews had to fix them or work around them.

The instructor could also push the system beyond all acceptable limits, as by causing on-board gyros to fail, one at a time, until none were left and the vehicle fell into a wild tumble.[54]

Aborts were also part of the training. The standard ascent trajectory assumed that an engine would fail, and allowed the shuttle to gain altitude rapidly so as to place it in good position for an emergency return to the launch site. But once the guidance realized that it indeed was flying with three good SSMEs, it was to pitch the craft into a shallow dive, taking it from 85 to 72.5 miles in altitude while adding more than one Mach in velocity every thirty seconds. "It's going to be wild trying to watch the instruments because of the view outside," Crippen told *Aviation Week*'s Covault. But if an engine did fail, the crew had to turn the shuttle around, slow its speed to zero by using the good engines, accelerate westward, drop the ET, and use OMS for extra thrust to help them reach the runway.[55]

Meanwhile, as Crippen and Young used their simulators to learn what to do if anything went wrong, the ground crew at NASA-Kennedy was working with both Enterprise and Columbia, seeking to make sure that everything would go right.

Into the Countdown

When Enterprise rolled out from its Palmdale hangar and was exhibited to the press and public in September 1976, it resembled a real orbiter. Mated to an ET and a pair of dummy SRBs, it looked like part of a complete shuttle. It rode atop a launch platform at Kennedy Space Center in May 1979, the SRBs secured to this base using heavy bolts and vehicle and platform riding a crawler-transporter. Following its arrival at Pad 39-A, this assemblage looked as if it soon might be counted down and launched. It could not fly, of course; it was merely the same elaborate replica that had served in vibration tests at NASA-Marshall. But this "facilities verification vehicle" gave the ground crew a dress rehearsal for the launch of Columbia, which just then was attracting its own attentions within the nearby Orbiter Processing Facility.

The work started in the VAB, where technicians stacked their SRBs and added Enterprise to create a complete space shuttle. When the big transporter carried this shuttle as it rumbled and clanked along the crawlerway, it presented a scene that everyone hoped would be reenacted hundreds of times in the future. At the launch pad, there also was much to do. The Fixed Service Structure was to conduct fit checks, with the ET hydrogen vent line, the orbiter access room, and orbiter mid-body umbilicals being attached to the vehicle. Cryogenic propellants were to flow from storage facilities through the Mobile Launch Platform into the Tail Service Masts, though these liquefied gases did not go further, for Enterprise lacked the appropriate plumbing.

Other activities required the Rotating Service Structure (RSS). It carried the Payload Changeout Room (PCR), which accommodated installation and removal of an orbiter's pay-

Fig. 72. *Enterprise atop its Mobile Launch Platform leaves the Vehicle Assembly Building en route to its launch pad. At bottom, a Crawler-Transporter carries the entire stack. (NASA)*

load while the shuttle was stacked vertically. A twenty-thousand-pound concrete weight, representing a spacecraft, arrived within a sealed canister. With the RSS well away from the shuttle, workers hoisted the canister into the PCR, removed its dummy payload, then rotated the RSS to lie against the back of Enterprise for payload installation within the cargo bay. All this was done under strict environmental control, to prevent contamination of the "spacecraft."

Meanwhile, like a butterfly within its cocoon, Columbia was at the OPF, surrounded by work platforms. Before it could emerge from this pupal stage, it first had to grow its covering of thermal-protection tiles, and this took the next year and a half. Columbia nevertheless was a live orbiter with a full array of on-board systems, and the activities within the OPF specifically included their check-out and test. An important series of activities, beginning in mid-1979, exercised the on-board hydraulics, initially using ground-supplied power.

Fig. 73. *Enterprise, with External Tank and inert solid boosters, rides a Crawler-Transporter down the crawlerway to Launch Pad 39-A. (NASA, courtesy Dennis R. Jenkins)*

Fig. 74. *Enterprise at its Kennedy Space Center launch pad. (NASA, courtesy Dennis R. Jenkins)*

Columbia was connected to the Launch Processing System, and power-on testing ran at a twenty-four-hour pace from each Monday evening through Friday afternoon. The long weekends were reserved for additional tasks of assembly and installation of hardware, which had been shifted from Palmdale to NASA-Kennedy. These required power-off because they often called for penetration of active electrical circuits. A separate round-the-clock effort sought to verify some fifteen hundred software programs within the LPS, totaling 6.5 million lines of code. Much of this work involved playing the software against mathematical models, with the programs providing responses and engineers determining whether the system was responding appropriately. By midyear, more than two-thirds of these programs had been verified.[56]

Tests of hydraulics continued during October, this time using the orbiter's own APUs. These units used hazardous hydrazine propellant and generated poisonous ammonia within the exhaust. Hence the tests took place late at night, to avoid interfering with other work on the orbiter. Pipes carried the APU exhaust outside the OPF, and this facility was cleared of all people except for an astronaut crew aboard the vehicle. All three APUs underwent hot-fire runs for thirty minutes, with the flight crew moving the elevons, rudder, and speed brakes at their maximum rates. Other APU tests simulated complete profiles for shuttle ascent and descent. "It's a really big test because it tells you that you have an integral spacecraft as far as systems go, with the APUs' hydraulics, actuators, and the cooling that goes with it," said the astronaut Robert Overmyer.[57]

Another integrated test series, during December, gave an extensive workout of other onboard systems. "This is the one and only time that many of these systems will play together in the orbiter prior to flight," Overmyer noted. "When you've finished this test you can say you've got a system that's ready to go to launch." Rather than use ground-supplied power, Columbia relied on its own fuel cells, with Young and Crippen in the cockpit. They exercised the avionics and hydraulics along with other electrical and mechanical systems.

Five simulated flights comprised this test series, beginning with a normal launch and concluding with a forty-eight-hour mission running from ascent to descent. The latter included on-orbit activities, such as opening and closing the payload bay doors and deploying the space radiators. The activities included use of flight software in the on-board computers, fuel-cell purges, and ascent emergencies. Using ground-supplied power, controllers caused fuel-cell power buses to fail, to see how the remaining buses would take over the load.[58]

Live propellant! For the first time, ground crews worked with SRB segments that contained this highly combustible mix, as they prepared to stack the solid boosters on top of the launch platform. The work initially did not go as well, for the segments did not fit together. They had lain horizontally while being shipped and stored. The propellant charges, which surrounded a hole of five-foot diameter, slumped while bulging at the sides. The steel casings were half an inch thick, but the propellant in each segment weighed 150 tons and pushed the casings out of round.

Technicians tried to fix the problem by allowing individual segments to stand vertically, hoping that the propellant would push the casings back to proper roundness. It did not. The next plan called for repositioning the segments' lifting points in the hope that lifting them at the new points would restore the proper shape. As noted in *Aviation Week,* "It was a simple matter to unbolt and reposition the attach points. When lifted at the new points on the segment circumference, the units were found to flex back into specification, allowing proper mating." The stacking of SRBs was completed during the second week of January 1980.[59]

During the spring of 1980, with Columbia continuing to molt away its tiles nearly as rapidly as it could accumulate new ones, it underwent new power-on tests. Engineers had been pleased with the results of the earlier work, with one project manager declaring, "This is a mature set of hardware." The new exercises included the following:

1. *Pyro shock test, with live firing of the orbiter-ET separation mechanism. The firing of the pyro charges was not supposed to damage the thermal-protection tiles, or to subject the avionics to excessive shock. During the Approach and Landing Tests, one computer aboard Enterprise had dropped off line when that craft separated from its 747, and no one wanted this to recur.*

2. *Mission control center interface test, with controllers in Houston sending commands to the orbiter through a ground network. This verified the pertinent radio systems and software, and checked the ability of the spacecraft and Mission Control to work together.*

3. *Crew equipment interface test, featuring use by the flight crew of items of onboard apparatus. This test also opened and closed the cargo bay doors, for an earlier exercise with these doors had shown that crew members could not sight indicators along the hinge lines to insure proper functioning.*

4. *Cabin outgassing test. Many items of on-board equipment tended to outgas at low pressure. This test powered up all orbiter systems, with no crew inside, and with the cabin sealed. The orbiter then sat for several hours while sensors looked for toxic emissions.*

5. *OMS/reaction control integrated test. This exercised the electrical systems of the forward reaction control and aft propulsion modules, with the latter holding the OMS engines. The test opened and closed valves while gimbaling the OMS engines, but did not hot-fire any of the thrusters.*

6. *Delta orbiter integrated test, which resembled the integrated test series of the previous December. However, this time the orbiter included mated OMS and reaction control systems, which had previously been absent.[60]*

Amid persistent problems with the tiles, the launch date for STS-1 continued to slip. Midway through 1980, people were talking of May 1981, based on a three-shift operation, five days per week, with significant padding between critical tests and launch-preparation milestones.

But late in July 1980 a high-level shuttle program review set 23 November as the date for Columbia to leave the OPF and enter the VAB. The time for processing, set earlier as twenty-four weeks, now shrank to fifteen weeks as NASA officials directed a three-shift effort that was to fill the entire week. This meant launching in March.[61]

NASA managers had the custom of scheduling an early launch date, knowing that this would encourage the project staff to press the work. The specific date was 10 March 1981, which slipped 14 March during autumn due to addition of four more days of vehicle test. The realistic outlook was for no earlier than mid-April. George Page, a director of shuttle operations, declared that this round-the-clock schedule, seven days per week, represented "sixteen weeks of straight-through work except for three holidays. The only three days we are not scheduling major operations are Thanksgiving, Christmas, and New Year's. We do not have any contingency time."[62]

Activity during November transitioned into formal launch processing. On 5 November the ET was mated to the twin SRBs. The orbiter's three SSMEs, which had been removed for modifications, now were reinstalled. Then on 24 November, only one day late, Columbia rode on its own landing gear as a tow vehicle pulled it from the OPF into the VAB amid cheers from a large crowd. The scene was reminiscent of the roll-out of Enterprise at Palmdale four years earlier, but this time the craft was live.

On the following day a crane hoisted it vertically, holding it in the air like a great gaffed bird with delta wings as it stood silhouetted against the light from a tall array of windows that formed part of a VAB exterior wall. On the twenty-sixth, workers hard-mounted it to the ET, then took several days to put work stands in place and connect interfaces. They powered up the stack for the first time at 2:00 A.M. on 4 December. A shuttle now existed as a complete flight vehicle, and not as a replica built for structural or fit tests.

The LPS had seen extensive use with Columbia alone; now, also for the first time, it could work with a fully assembled live shuttle. It did this during December, amid a new set of exercises that again placed astronauts on the flight deck and transmitted data to Houston. They conducted simulated liftoffs and ascents, including both abort-once-around and abort with return to launch site, and practiced these maneuvers using both the primary and backup flight systems.[63]

The twenty-ninth of December brought roll-out, as Columbia spent the day riding from the VAB to Pad 39-A. It eased through its high-bay door around 8:00 A.M., under heavy clouds and with temperatures hovering near forty degrees. The trip covered three and a half miles and took more than ten hours, with numerous stops along the way to level the platform on the crawler-transporter. As it approached the pad, it became apparent that a cryogenic line on the mobile launch platform extended outward about six inches more than designed, and would strike an emergency escape platform on the launch pad. Someone burned off the end of the escape platform using an acetylene torch, and the vehicle moved into place. It was on

Fig. 75. *Columbia being mated to its External Tank on 25 November 1980. (NASA, courtesy Dennis R. Jenkins)*

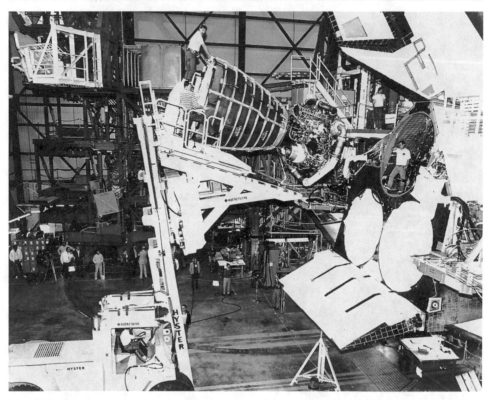

Fig. 76. *Use of a forklift truck when installing a Space Shuttle Main Engine. An adapter holds the engine in position. (NASA, courtesy Dennis R. Jenkins)*

the launch pad by 6:30 P.M., with the launch platform resting on supports an hour and a half later. It received electric power the next day.

A substantial number of major tests lay immediately ahead before STS-1 could be cleared for its countdown:

1. *Validation of interfaces between shuttle and pad.*
2. *Ground power-out test, simulating a complete loss of commercially supplied power. A NASA backup power system was to switch in, allowing ground operations to proceed.*
3. *Plugs-out test, disconnecting all launch umbilicals. This tested the shuttle's ability to maintain itself in a proper state when switched from ground to internal systems.*
4. *Loading of hydrazine fuel for the orbiter and SRB APUs.*

5. *Loading and detanking of cryogenic propellants within the ET, with first its hydrogen tank and then its oxygen tank being filled and emptied, under automatic control.*

6. *Hot-firing of APUs, with astronauts on board.*

7. *Loading of hypergolic propellants for the OMS and reaction-control systems. "Frankly, the ground servicing for the hypergolics is difficult to keep going," the operations director George Page told* Aviation Week. *"You can get little leaks here and there, and we have to do all the work in fully protective suits."*

8. *Two-day countdown leading to a simulated SSME firing, with simulated post-firing safing procedures. A flight crew was to be in the cockpit.*

9. *A "dry" countdown, without cryogenic propellants. For Young and Crippen, this was to be the most important rehearsal of preflight activities.*

10. *Mission Verification Test, a set of exercises that used the worldwide tracking and data network as well as Mission Control in Houston. Other activities took place at Edwards AFB, the prime landing site, and at White Sands Missile Range, which had an emergency site. These series of tests culminated in a "wet" countdown, leading to the Flight Readiness Firing (FRF).*

11. *FRF, with the SSMEs igniting and operating in static test for 23 seconds, and with Columbia unpiloted.*

12. *Post-FRF maintenance and test, to ensure that stresses from this firing did not change the characteristics of vehicle systems.*

13. *Launch Readiness Verification Test, with Young and Crippen repeating simulations of launch, abort, and reentry, using the on-board computers and avionics.*

14. *Refueling of the APUs and installation of ordnance for the explosively separated SRBs and ET.*

15. *Initiation of the launch countdown.*[64]

The cryogenic propellant loading tests delivered 384,000 gallons of liquid hydrogen to the ET on 22 January, followed by 140,000 gallons of liquid oxygen two days later. This brought debonding of some panels of insulation, which separated from the aluminum skin of the ET in three areas of the cold hydrogen section. The insulation had been applied using adhesive and the manufacturer, Martin Marietta, had used too much of it in places. Its coefficient of thermal expansion was five times greater than that of the aluminum, and while thin adhesive layers could stretch to accommodate this difference, thick layers simply popped loose when chilled to cryogenic temperatures. The ground crew did not attempt immediately to repair the insulation, but draped a cargo net over the damaged area. This prevented the loose insulation from shaking off during the FRF and damaging the orbiter's delicate thermal-protection tiles.

The Mission Verification Test series got under way early in February. NASA-Goddard coordinated the activities at NASA's worldwide tracking facilities. At Edwards AFB and at

White Sands, Northrop T-38 jet trainers played the role of Columbia during approach and landing. As these aircraft descended they flew the steep profile of a shuttle, with their pilots exercising NASA's communications links. Two other T-38s accompanied each pseudo-orbiter, flying chase. One carried television, while the pilot of the other flew formation and called off altitudes and airspeeds.[65]

The FRF took place on 20 February, continuing a tradition of static firings that has spanned decades. Milton Rosen, manager of the Navy's postwar Viking sounding-rocket program, wrote of a 1952 test: "When the firing button is pushed and the motor ignites, we see through the blockhouse window a chained beast, roaring ominously, straining at its fastenings. Every second is a long second, and we listen to the count, thinking it will never end."[66] Three decades later, the FRF flamed with more than fifty times the thrust of Viking.[67]

The FRF fired the three-engine cluster of flight engines, in the fashion of firings of the Main Propulsion Test Article in Mississippi, and this test did a great deal more. It used the on-board hydraulics with their APUs, to operate valves and gimbal the engines. It subjected the orbiter's computers and avionics to the acoustic and vibration loads of main-engine firing. It verified that these computers could interact properly with the LPS, and demonstrated that the orbiter could receive radio signals during launch. It showed the "twang," as Columbia bent forward by 25.5 inches, under the thrust of its engines, and then rebounded. (NASA officials had expected 19 inches, but viewed the larger value as "not out of the spectrum.") Finally, with the ET still being full of propellants, the FRF conducted an automated safe shutdown of the vehicle on the launch pad.

The engines roared. A thick cloud of steam formed from sprays of water in the sound suppression system, foreshadowing a day when Columbia was to rise amid a similar mass of vapor.[68] First, though, the ET needed to have its insulation repaired. A second such tank was in Mississippi, supporting firings of the MPTA. It served as a test bed, as technicians stripped away insulation in the pertinent areas and then made repairs that gave them practice.

At KSC, technical teams from Martin Marietta worked twelve-hour shifts around the clock for two weeks as they restored the damaged portions. The ET resembled a big aluminum balloon. Thermal contraction made it shrink when filled with supercold liquid hydrogen, but internal pressure (called backpressure) could inflate it and reduce this shrinkage. Backpressure was key to a new and severe test on 25 March.

The ET was filled with a full load of liquid hydrogen under minimal backpressure. Flight backpressure was to be 30 to 32 psi, but because it was lacking, cryogenic cooling shrank the ET structure and skin to an unusual degree. This placed particularly large stress on the insulation. However, inspection showed that the refurbished areas passed this test, while no new areas debonded. A follow-up test on 27 March, with full backpressure, confirmed these observations.[69]

Meanwhile, misfortune struck as two Rockwell employees, John Bjornstad and Forrest

Cole, died in the line of duty. They had entered the aft fuselage section of Columbia, not knowing that it was filled with a pure-nitrogen atmosphere. The nitrogen protected the shuttle from fire or explosion due to leaks of cryogenic propellants, but it asphyxiated these men, with Bjornstad dying that same day and Cole lingering for nearly two weeks before succumbing as well. Someone had failed to post a hazard sign, and George Page described the cause as "a communications problem. Now they are going to make sure everyone understands the rules and will probably announce over the intercom the specific exceptions involved in pad access."[70]

The deaths resulted from faulty safety procedures at the launch pad, and did not reflect any problem with the shuttle itself. "Frankly, I've never seen a vehicle that's so clean before its first flight," John Yardley told *Aviation Week*. With the ET insulation repairs completed and validated, the time of launch was set for shortly after dawn on 10 April. Launch processing now proceeded according to a schedule, with specific tasks set for particular dates:

28 March: Loading of data into the on-board computer's mass memories; checkout of space suits for Young and Crippen.

29 March: Installation of cables used to fire the pyrotechnics that release the SRBs from hold-down posts at liftoff. Installation of protective thermal blankets in the nozzle areas of the SRBs. Unloading of liquid oxygen into storage facilities from fifteen tankers.

30 and 31 March: Loading of hydrazine for APUs. All other work at the pad now halted, for this was a hazardous operation wherein propellant loading specialists wore protective gear with full helmets and self-contained breathing equipment.

1 April: Final installation of gap fillers within spaces between thermal-protection tiles. Initiation of preparations to load tanks on the Fixed Service Structure that would hold cryogenic oxygen and hydrogen for the orbiter's fuel cells. Installation of a camera to record separation of the ET from the orbiter. Installation of ordnance for separation and destruct systems.

2 April: Continued installation of ordnance.

3 April: Loading of liquefied fuel-cell reactants into tanks on the FSS; warmup of inertial measurement units on the orbiter.

4 April: Final tests of fuel-cell power reactants; continued warmup and calibration of inertial measurement units.

5 April: Final closeout of thermal-protection tiles in places on the orbiter where ordnance had been installed. The formal countdown, leading to the April 10 launch, began just before midnight.[71]

The count proceeded through the next four days, reaching T −20 minutes. At that point, the orbiter's four primary computers were to switch from prelaunch to launch configuration. The fifth computer, used as a backup, had remained dormant, but now was to come to life and request data from the primaries. However, there was a forty-millisecond timing error within the system, and the backup failed to receive its data according to plan. In accordance with its programming, it therefore rejected the data.

The backup computer tried again and failed, producing error messages on the flight deck and on displays within the nearby firing room and at Houston's Mission Control. Crippen, sitting in the copilot's seat, pushed a reset button, but again the computers failed to communicate. The countdown was put on hold while specialists in Houston examined the backup computer software, with the count resuming later that morning. But at T − 16 minutes, the same problem recurred, and the launch attempt was scrubbed.

The timing error proved to have arisen during initial loading of flight programs, at T − 15 hours. The cure proved to lie in dumping data from the primary computers early in the countdown. If this "time skew" reappeared, it could be eliminated by reinitializing the primaries. Computer specialists evaluated this treatment late that evening and found that it worked well.

The launch was rescheduled with a slip of two days. The count proceeded; the backup computer received its data as planned, and people cheered an announcement that "Mission Control in Houston has verified that the primary system is in synchronization." Everything now was ready. Columbia now was prepared to rise into the sky at 7:00 A.M. on Sunday, 12 April 1981, twenty years to the day since the cosmonaut Yuri Gagarin had ridden into orbit aboard his Vostok craft to become the first man in space.[72]

CODA

The Working Shuttle

With a deep booming roar, intensely bright flame leaping from both SRB exhausts, Columbia rose from amid a cloud of vapor and soared into the sky. It carried on-board instruments that measured temperatures, pressures, and vibration characteristics at various points on the orbiter. It carried no payload to deploy while in space, and it stayed aloft for only two days before returning. Still it counted as a rousing success. Paul Conrad, editorial cartoonist at the *Los Angeles Times,* caught the spirit of the venture by showing Columbia smiling broadly as it glided to its landing at Edwards Air Force Base. This played on the cheerful appearance of the aircraft of Pacific Southwest Airlines, which sported similar grins, and Conrad echoed that carrier's advertising slogan: "Our smiles aren't just painted on!"[1]

Columbia also flew the next three missions, which constituted the rest of the flight-test program. The second flight took place in November 1981, also for two days, with Joe Engle and Richard Truly in the cockpit. It carried the Canadian-built robot arm, though again it did not deploy a payload. The same was true of the third mission, which stayed up for eight days. Following the fourth flight, at midyear in 1982, NASA declared that the shuttle was operational.

To celebrate, the agency staged an extravaganza at Edwards that recalled the roll-out of Enterprise six years earlier. The event took place on 4 July, heightening the festive mood. President Reagan personally participated. A new orbiter, Challenger, was atop its Boeing 747 carrier, and Reagan himself gave the order for takeoff to the 747's flight crew. A band was playing; flags flew, and fifty thousand people were on hand. In addition, Columbia was about to make a landing, within view of the crowd.

Reagan stood with his party in the hot desert sun, with NASA administrator James Beggs as his host. Then, as Reagan later recalled, "They hurried us up on the platform, because they said it was time to get up there, the shuttle was coming in. They said it was on its approach." He asked, "Where is it now?" Someone replied, "Just over Honolulu." Hawaii

Fig. 77. *Liftoff of Columbia, 12 April 1981, at Kennedy Space Center. (NASA)*

was over two thousand miles away, and in Reagan's words, "The whole miracle was brought home to me right there."[2]

Columbia continued in service for one more flight before standing down for extensive maintenance. This fifth flight was the first operational mission, as this orbiter launched two satellites. Both were for communications: a Canadian Anik and a commercial spacecraft from Satellite Business Systems. Both rode PAM-D upper stages to geosynchronous orbit, with this mission offering a clear demonstration that the shuttle was ready to compete for commercial payloads within a booming communications-satellite industry.[3]

During the next three years the uses of the shuttle continued to expand. Spacelab flew for the first time in November 1983. It carried seventy-seven experiments, including astronomical telescopes, solar telescopes, and instruments to examine the earth and the ionosphere. *Science* noted that, ordinarily, "this would be the worst way conceivable to run a mission. Many of the experiments are utterly incompatible: Columbia will constantly be twisting down to point toward the earth, up toward the stars, and out toward the sun. No one experiment will be able to make full use of the time. But then Spacelab 1 is not a normal mission. It is an exercise in engineering exuberance."[4]

The shuttle had also been designed to support refurbishment and repair of satellites in space. A highlight of 1984 came during April of that year, as Challenger visited the orbiting Solar Maximum spacecraft. The shuttle maneuvered to within two hundred feet of Solar Max. Then, using Manned Maneuvering Units, the astronauts George Nelson and James van Hoften captured this satellite and secured it within the payload bay. Following repair, it was released into space to continue its observations of the sun.

The shuttle program also used Spacelab to promote international cooperation. Spacelab had taken shape as a West German initiative; a payload specialist from that country, Ulf Merbold, had ridden with it on that first flight in 1983. It flew three more times during 1985. On the third such mission, in November, this orbiting laboratory carried experiments for materials processing in zero-g. Mission Control in Houston controlled the orbiter, Challenger, in the usual manner. But Spacelab's technical equipment came under the control of a German space operations center located in Oberpfaffenhofen, near Munich. The work of that center complemented that of three other European payload specialists—Reinhard Furrer, Ernst Messerschmid, and Wubbo Ockels—who were aboard the shuttle in space.

Two more orbiters, Discovery and Atlantis, joined the fleet during 1984 and 1985, bringing it to full strength. Space flight seemed about to become routine, and NASA promoted this impression by opening mission opportunities to people of increasingly broad background. Senator Jake Garn flew aboard Discovery in April 1985, ostensibly as a payload specialist. A wealthy and well-connected aviation enthusiast from Saudi Arabia, Salman Abdul aziz al-Saud, followed in June. Congressman William Nelson secured a berth for a flight in January

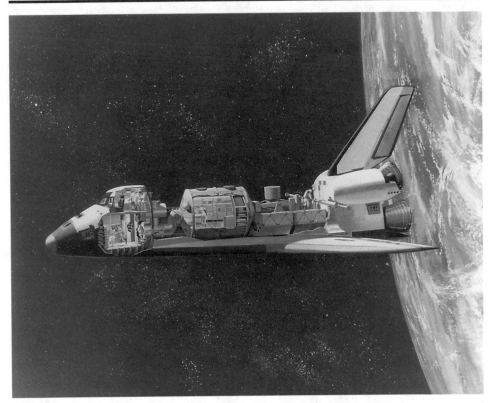

Fig. 78. *Cutaway view of Spacelab, with pallet, in the shuttle payload bay.*

1986. In addition, NASA officials decided to select a schoolteacher to fly into space, and a nationwide competition brought the selection of one Christa McAuliffe.

She was part of a crew that reflected the diversity of America in the mid-1980s. Her fellow spacefarers included Ellison Onizuka, of Japanese ancestry; Ronald McNair, a black man with a doctorate degree in physics; and Judith Resnik, who held her own doctorate. The mission patch, worn as a badge on the flight suit of each astronaut, gave the names of the seven crew members—with an apple alongside that of McAuliffe.[5]

These people were aboard the shuttle on 28 January 1986, the day of Challenger.

The image remains seared in the nation's memory: the vehicle rising on its pillar of rocket exhaust, the orange fireball, two solid boosters flying off crazily, and a shower of fragments that trailed smoke against the blue sky. The orbiter was not destroyed by explosion; instead, it was pushed broadside on to the onrushing airstream while flying at a speed above Mach 2. This vehicle could not withstand the consequent aerodynamic loads, and it broke up.

Telescopic cameras permitted identification of some fragments: a wing, the aft section with its three SSMEs still burning momentarily, the forward cabin with its flight crew. It took them more than two minutes to fall to the ocean from their initial altitude of nine miles. No parachutes were on board; when they struck the water, their deaths were instantaneous.

It certainly was no new thing for a launch vehicle to fail during ascent to orbit. However, the ensuing investigation showed that Challenger needed far more than a technical redesign. It had been destroyed due to the pressures of tight schedules, which led NASA to cut corners. The conclusion was far-reaching: the shuttle could not be all things to all people. In no way was it capable of the rapid and routine operational uses that its builders had promised. It had carried hope of frequent and low-cost space operations, but now this hope was dead. Instead, the nation would treat the shuttle as a rare and valuable commodity, to be flown only on special occasions.[6]

The Air Force had already announced plans that called for reduced reliance on the shuttle. In June 1984 it had declared that it intended to remove ten payloads from the manifest and fly them on expendables such as the Titan IV. Now the Pentagon went further, canceling plans to fly the shuttle from Vandenberg AFB. NASA scrubbed plans to use Centaur as a shuttle upper stage, for it burned volatile liquid propellants and posed safety risks that now were unacceptable. In August 1986 a statement from President Reagan stated that "NASA will no longer be in the business of launching private satellites." This decision placed the lucrative communications-satellite industry in the hands of the builders of other expendables. These launch vehicles now included Ariane, whose sponsors saw their opportunity and moved quickly to seize it.[7]

The shuttle underwent extensive engineering modifications and did not return to flight until September 1988. After that, however, it continued to make its way as a useful member of the nation's stable of space boosters. In the field of planetary science, it made valuable contributions during 1989 by launching the Magellan spacecraft toward Venus, and Galileo to Jupiter. Both missions significantly enhanced scientists' understanding of these planets.[8]

Venus lies hidden beneath a dense and hot atmosphere, amounting to an ocean of thick smog. However, radio waves pierce this atmosphere with ease. Radar observations from Earth, complemented by measurements made with a radar altimeter aboard the Pioneer Venus spacecraft, had mapped this planet's surface. The maps showed continent-sized highland regions— along with a mountain, probably a volcano, nearly as tall as Everest.

Magellan went farther, carrying on-board radar that charted the surface of Venus with a resolution of four hundred feet. This sharp detail disclosed nearly a thousand impact craters that had been formed by meteors large enough to survive passage through the atmosphere. There should have been many more, however, and scientists concluded that Venus-wide processes had erased craters more ancient than eight hundred million years. This brought the suggestion that voluminous lava flows had erupted across that planet, burying them under volcanic rock.[9]

Galileo, orbiting Jupiter, disclosed surprises of its own. The four large satellites of that planet—Io, Europa, Ganymede, and Callisto—are large enough to stand as planets in their own right. They are subject to tidal forces strong enough to flex them throughout, with these forces being strongest on the innermost moons. The flexing generates heat, and vulcanism. Photos taken by Voyager in 1979 showed that Io is the most volcanically active body in the solar system.

The Voyager spacecraft executed flybys, but Galileo made repeated close approaches and took photos that showed far more detail. These images, along with other measurements, made it highly plausible that tidal heating of Europa reduced its icy outer surface to no more than a thin shell, underlain by liquid water. Indeed, its ocean might well be far deeper and more extensive than Earth's. On Europa, a major impact failed to form a conventional crater. Instead it produced a large feature resembling the concentric ripples formed by throwing a rock into a pond, with these ripples frozen in place. The prospect of an ocean on Europa also stirred speculation that this satellite might harbor life.[10]

The shuttle played a vital role in both the Magellan and Galileo projects, for it served as their launch vehicle. For the Hubble Space Telescope, the role of the shuttle was considerably more far-reaching. After placing it in orbit in April 1990, it conducted a revisit in December 1993 that cured its faulty vision while extending its useful life. A second revisit in February 1997 further enhanced the capabilities of this orbiting observatory.[11]

Even before the initial revisit, while it still needed eyeglasses, the Hubble telescope produced enough new observations to fill two special issues of *Astrophysical Journal Letters*. In particular, it took important photographs of the remnant of a 1987 supernova that had appeared within the Large Magellanic Cloud, a satellite galaxy of our own Milky Way. Close study of this supernova gave a new measurement of the distance to this subgalaxy: 169,000 light-years, with an accuracy of 5 percent. This was three times more precise than previous distance estimates. In cosmology, it represented a valuable contribution to an ongoing effort that seeks to determine distance scales within the entire universe.

Following its on-orbit repair late in 1993, the Hubble responded with a burst of new images. A prime task involved making observations that could determine the age of the universe. The astronomer Wendy Freedman obtained clear new photos from this telescope and proposed a disturbing paradox: The universe appeared younger than the stars it contains.

Here was science at its best, raising deep issues at the foundations of astrophysics. It was not a simple matter of routine experiments and modestly improved findings. Rather, the Hubble Telescope was calling into question some of the basic methods of astronomy, casting doubt on what scientists truly know and how they can claim to know it. Yet this telescope also stood in the forefront of work aimed at resolving this paradox, by sharpening the estimated age of the cosmos.[12]

By contributing to explorations of Venus, Europa, and the universe at large, the space shuttle supported scientific studies of the highest importance. Yet such activity occupied only a small number of missions. The nation's fleet of shuttles spent far more time supporting the work of astronauts and mission specialists, with a space station as the focus of the effort.

Studies of space stations had flourished within NASA during the late 1960s, and had culminated in the Skylab program. The advent of the shuttle put plans for future space stations on hold, for there was no money in the budget. But after 1980, with development of the shuttle nearing completion, NASA officials began to look anew toward a space station as their next major program. James Beggs, Reagan's choice as NASA administrator, made it his personal goal to win presidential approval for such a station. He won his reward in January 1984, when Reagan announced his support in his State of the Union message.

This station was to serve as a center for research. Crystal growth held particular interest, not as a New Age therapy but as a basic exercise in molecular biology. Scientists knew that in this field, function follows form; the specific intricacies of a molecule's shape determine how it works within a cell. The way to learn its form was to grow a good crystal of the substance under study and then examine it by using a sharply focused beam of X-rays. This produced data that a computer could analyze, yielding the molecular shape. Molecules of interest included enzymes and other proteins, none of which formed crystals readily. The process of crystal growth was delicate, easily upset by effects due to gravity, and there was reason to believe that zero gravity would help.

These hopes grew stronger in the wake of Spacelab 1. Everyone in the field understood that crystal growth was more an art than a science—and that there were very few artists. German investigators thus stirred much interest when they described their work with proteins aboard that mission. Their crystals of beta-galactosidase were up to twenty-seven times larger than counterparts grown on Earth, whereas crystals of lysozyme were a thousand times larger.[13]

The advent of the space-station program gave new importance to Spacelab, for it now was to serve as a prelude. The demise of Challenger also brought a stand-down in flights of Spacelab, but the missions resumed in June 1991. By then, however, the space station was in trouble. Its advocates had won approval not only by touting its advantages but by quoting an unrealistically low price. When reality set in as the design matured, the price escalated, and NASA had to go back to the drawing board in hope of coming up with a new design that could cut the cost. This happened repeatedly, as the agency took an entire decade merely in deciding what it hoped to build. In mid-1993, with plans still merely on paper, the program survived a challenge in the House of Representatives by the margin of a single vote, 216 to 215.

What saved the space station was the collapse of the Soviet Union in 1991, which ended the cold war. In its wake, prospects quickly flourished for partnership between the space superpowers, as Washington and Moscow crafted plans for joint efforts. Such efforts were not new; the Apollo-Soyuz Test Program in 1975 had brought a linkup in orbit between piloted spacecraft from the two nations. However, that had been a one-time-only effort that left no immediate legacy. The new cooperation proved to be far more solidly grounded. It dated from June 1992, when President George Bush met with his Russian counterpart, Boris Yeltsin. Their agenda included renewal and expansion of a 1987 agreement on cooperation in space. The new pact provided that Russian spacefarers were to fly aboard American vehicles and vice versa.[14]

With this, for the second time in only six years, the world of space flight took a dramatic lurch. The first event, the 1986 loss of Challenger, put hopes of routine and low-cost space flight on hold. The Bush-Yeltsin agreement was the second, and was far more hopeful. It heralded an era of mutual advantage through cooperation, with America providing commercial management and capital while Russia shared its trove of rocket technology.

An early initiative created a partnership between Lockheed and the Khrunichev factory near Moscow, builder of the powerful Proton launch vehicle. Officials of the two firms set up a joint venture, International Launch Services, that went on to market the Proton to provide commercial launches. A subsequent project set out to craft a new version of NASA's Atlas-Centaur, with Russians among the bidders. Atlas had flown since the 1950s with engines from Rocketdyne, but the thrust for the new version came from Russia's firm of Trud, which formed its own partnership with Aerojet.

In the field of piloted space flight, the new internationalism initially took form amid a succession of visits by the shuttle to a Russian space station, Mir. The core of Mir, weighing 44,300 pounds, had flown to orbit atop a Proton in 1986, with subsequent missions expanding it by adding new modules. Unlike Skylab, which had hosted its crews only intermittently, Mir was continually occupied for well over a decade. It also pioneered manned orbital missions of truly long duration. The cosmonauts Vladimir Titov and Musa Masarov lived within it for an entire year, with Valery Polyakov later staying for 437 days.[15]

Visits to Mir complemented the continuing flights of Spacelab, providing a second prelude to what now was called the International Space Station (ISS). American involvement with Mir began in March 1995, when the astronaut Norman Thagard rode a Russian Soyuz rocket to fly to that station in the company of Russian crewmates. In June the shuttle Atlantis flew to orbit and docked with Mir, carrying a flight crew that included two new Russians. A crew exchange ensued, as Atlantis returned with Thagard and his comrades of March.[16] This became the first in a succession of nine shuttle missions to Mir. Seven American astronauts spent a total of thirty-two months aboard that orbiting station, with twenty-eight months of continuous occupancy commencing in March 1996. This far outpaced the cumulative totals of the shuttle program, which took sixty flights to accumulate twelve months in orbit. In addition to introducing the Yankees to Russian-style durations of spaceflight, this program pro-

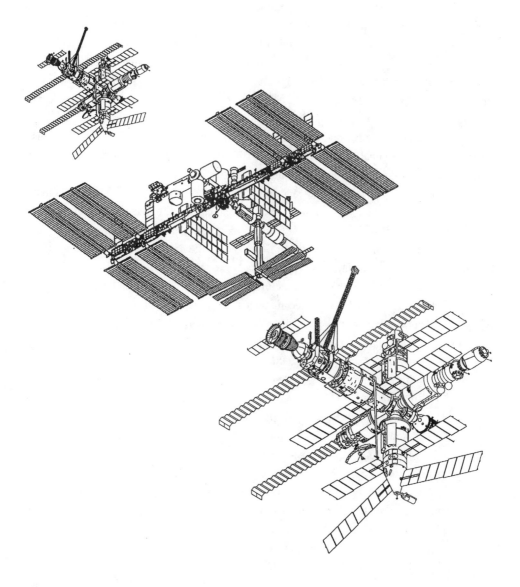

Fig. 79. *Space stations supported by the shuttle. Top to bottom: Mir, International Space Station to the same scale, and Mir in an enlarged view. (Art by Daniel Gauthier)*

vided valuable experience to both nations in crew training and in operating a true international program of piloted space flight.

Meanwhile, the ISS was taking shape. It was a concept to take the breath away, for it was to weigh more than a million pounds, making it heavier than a top-of-the-line Boeing 747. Mounting six laboratories, its design called for close to an acre of solar panels to provide electrical power. Its measurements, 290 feet long by 356 feet across, made it large enough to hold two football fields. It indeed was international, drawing support from Canada, Japan, Brazil, and the European Space Agency, as well as from Russia and the U.S. Most of the orbital equipment was to come from Moscow and Washington, but Japan expected to provide one of the six labs, while Europe planned not only to add its own lab but to provide logistical support using its capable new Ariane 5 launch vehicle. Here indeed was a program ambitious enough to suit a dazzling new century.[17]

Looking ahead to the ISS, NASA and Europe phased out Spacelab. The last flight in that series took place in April 1998. Known as Neurolab, it carried life-sciences experiments that sought to study the behavior of nervous systems in zero-g. Another phaseout came in June, as Discovery flew the final mission to Mir. It picked up the astronaut Andrew Thomas, who had been aboard since January, and brought him home.[18]

Construction of the ISS began in November, as a Proton rocket carried the initial module aloft. Called Zarya, "dawn," it mounted its own solar panels and provided communications and attitude control. Two weeks later, the shuttle Endeavour—the replacement for the lost Challenger—arrived with its own module, Unity. It displayed six docking ports, one for Zarya and the other five for sections of the ISS that were slated for arrival on subsequent flights.

Following this initial activity, little happened for a year and a half. The shuttle Discovery flew a logistics mission in May 1999, carrying a ton of supplies and equipment to the nascent station, and that was all. But activity picked up in May 2000 with a second such logistics flight. Two months later another Proton brought the module Zvezda, "star." It provided a living area with life support systems, while taking over the attitude control and communications functions from Zarya. At the end of October, the first crew was aboard, as the astronaut William Shepherd and the cosmonauts Yuri Gidzenko and Sergei Krikalev rode a Soyuz launch vehicle from central Asia.[19] A month later, another shuttle flight installed an enormous array of solar panels.

Activity continued at a high pitch during 2001, again with space shuttle missions in the forefront. In February the orbiter Atlantis delivered a large American laboratory module called Destiny. It needed supplies and equipment, and Italy was at hand to help, for it had built modules to carry instruments and experiments. The first of them, named Leonardo, reached orbit aboard Discovery in March. That same launch brought up a second crew while giving the earlier group a ride back to Earth.

Another Italian logistics module, named Raffaelo, flew aboard the orbiter Endeavour in April. This mission also installed a Canadian-built robot arm similar to that of the shuttle. A launch in July delivered an airlock for use during space walks. A month later, Discovery brought the module Leonardo along with a third crew then gave the second group its own return to Earth. By then Moscow had abandoned its Mir as well, letting it fall back into the atmosphere in the manner of Skylab. Still that nation had little cause to regret its loss, for of the first nine spacefarers to live aboard the ISS, five were Russian.

A NASA schedule projected thirty-five shuttle flights in support of the ISS from late 2001 through April 2006. These were to serve for construction as well as for operational support. With this, it was clear that the space shuttle had come full circle since its origins.[20]

The concept had sprung to life in the late 1960s, amid burgeoning hopes for the space-station concepts of the day. These orbiting facilities would need low-cost space transport, and it was clear that available launch vehicles such as the Saturn I-B would be too expensive. NASA's George Mueller, associate administrator for manned space flight, proceeded to call for a space shuttle, a reusable launch vehicle that could achieve the desired low costs.[21]

More than thirty years later, at the dawn of a new century, the working shuttle was falling well short of Mueller's hopes for low-cost operational use. Yet in another respect the shuttle indeed was fulfilling his expectations, for it was supporting the assembly and operational use of the ISS. During those decades, the shuttle program lost intended Air Force payloads to the Titan IV, while abandoning commercial communications satellites to Ariane and its competitors. But it still could fly for NASA and the international community. With this, and with its origins still fresh in the minds of people who remembered, one could say that the shuttle at last had found its home.

Notes

Prologue

1. NASA SP-4221, 59–66, 72–74, 92–94, 102–3.
2. Ibid., 110–14, 126–31, 136–50; Agnew, *Post-Apollo.* The source material for this book is on deposit at the NASA History Office at NASA Headquarters in Washington, and is available for use by researchers.
3. NASA SP-4221, 114–16, 390–92.
4. Ibid., 173–86, 189–90, 267–68; *Science,* 28 January 1972, 393.
5. *Science,* 28 January 1972, 393–94; NASA SP-4221, 199–200, 205, 233.
6. *Science,* 28 January 1972, 394; NASA SP-4221, 117, 214–15.
7. NASA SP-4221, 206–13, 216–17, 231–34.
8. Ibid., 287–89.
9. Ibid., 272–74, 356–57, 360–61, 405–6; Fletcher to Weinberger, 29 December 1971.
10. NASA SP-4221, 391–92, 408–10, 435.
11. NASA SP-4221, 143. For Hubble Space Telescope, see Jenkins, *Space Shuttle,* 288, 295.
12. *Science,* 28 January 1972, 392; NASA SP-4221, 411–13, 415–23, 434–35.

1. Launching the Program

1. NASA SP-4221, 411–13, 418, 427–32; George Shultz to James Fletcher, 17 March 1972; NASA Press Release No. 72-153, 26 July 1972.
2. NASA SP-4012, 3:69.
3. NASA SP-4221, 279, 287–88, 331, 405.
4. Ibid., 409–10.
5. NASA Fact Sheet, 5 January 1972; John Sullivan to Caspar Weinberger, memo, 18 February 1972; Dale Myers, interview with author, 12 June 1998.
6. NASA SP-4221, 420, 422.
7. For an overview of this crisis, see William Manchester, *The Glory and the Dream: A Narrative History of America, 1932–1972* (Boston: Little, Brown, 1974), 1251–54; Theodore H. White, *The Making of the President 1972* (New York: Bantam, 1973), 81–89. Manchester quotation is on 1251.
8. For Eurodollars, see Paul Johnson, *Modern Times: The World from the Twenties to the Eighties* (New York: Harper Colophon, 1985), 663–64.
9. *Statistical Abstract of the United States,* 1999 (Washington, D.C.: Department of Commerce, 1999), 493; Manchester, *Glory and the Dream,* 1253; White, *Making of the President 1972,* 88.
10. NASA SP-4012, 3:9; SP-4221, 364–66. Weinberger quotation is on 365.
11. Fletcher to Shultz, 30 September 1971; *Aviation Week,* 31 January 1972, 24–25.
12. George Low, Personal Notes No. 62, 15 January 1972; Weinberger to Fletcher, 9 February 1972.
13. Shultz to Fletcher, 16 February 1972; Fletcher to Weinberger, 6 March 1972.
14. George Low to Fletcher, memo, 22 March 1972.
15. Weinberger to Fletcher, 13 March 1972; Shultz to Fletcher, 17 March 1972.
16. Fletcher to Low, memo, 28 March 1972; Fletcher to Weinberger, 28 April 1972.
17. Weinberger to Fletcher, 25 May 1972. For inflation, see Fletcher to Weinberger, 29 September 1972.
18. Fletcher to Low, memo, 16 June 1972; Low, Personal Notes No. 72, 17 June 1972.
19. Weinberger to Fletcher, 21 July 1972; Fletcher to Low, memo, 26 July 1972.
20. Fletcher to Low, memo, 2 August 1972.
21. Myers to Fletcher, memo, 18 August 1972.
22. Low, Personal Notes No. 77, 16 September 1972.
23. *Aviation Week,* 7 August 1972, 15; 13 November 1972, 18; 15 January 1973, 17.
24. Robert Thompson to Charles Donlan, 6 September 1972.
25. Fletcher to Weinberger, 29 September 1972.

26. Weinberger to Fletcher, 2 October 1972.
27. Myers to Fletcher, memo, 11 October 1972.
28. Low, Personal Notes No. 80, 4 November 1972; Memo for Record, J. F. Malaga (NASA), 17 October 1972.
29. Low, Personal Notes No. 83, 23 December 1972.
30. Ibid. Discussion of Congress draws on comments by Roger Launius.
31. Low, Personal Notes No. 83, 23 December 1972; "NASA Program Reductions," press release dated 5 January 1973.
32. *Aviation Week,* 31 January 1972, 25; 5 February 1973, 23.
33. Ibid., 5 February 1973, 23–24.
34. NASA SP-4012, vol. 3, p. 9; *Scientific American,* January 1986, 34.
35. *Aviation Week,* 10 September 1973, 7.
36. Ibid., 11 February 1974, 23; Hechler, "Endless," 269.
37. Low to Fletcher, memo, 10 October 1973.
38. Low, Memo for Record, 19 November 1973.
39. *Aviation Week,* 11 February 1974, 23–25 (Fletcher quotation, 23).
40. *Statistical Abstract* (1999), 493.
41. *National Journal,* 23 January 1971, 163.
42. AAS History Series, vol. 4, p. 269.
43. NASA SP-4221, 329; *National Journal,* 12 May 1973, 689.
44. *National Journal,* 19 August 1972, 1331; Low, Personal Notes No. 72, 17 June 1972.
45. *National Journal,* 19 August 1972, 1328, 1332; 12 May 1973, 691–92 (Proxmire quotation, 692); *Aviation Week,* 2 April 1973, 25.
46. *Aviation Week,* 7 August 1972, 15; Fletcher to Shultz, 30 September 1971; NASA SP-4012, vol. 3, p. 69.
47. *National Journal,* 19 August 1972, 1326 (includes AFL-CIO quotation); 12 May 1973, 691 (Proxmire quotation).
48. Morgenstern and Heiss: *Analysis,* 31 May 1971; *Analysis,* 31 January 1972. See also NASA SP-4221, chap. 6.
49. Mondale to Staats, 17 February 1972; Staats, *Cost-Benefit,* 2 June 1972 (quotation, 2; table, 4).
50. Low, Personal Notes No. 70, 21 May 1972.
51. Low to Staats, 19 May 1972; Fletcher to Staats, 24 May 1972; Low, Personal Notes No. 72, 17 June 1972.
52. *National Journal,* 19 August 1972, 1331.
53. Low to Fletcher, memo, 23 April 1973 (includes table).
54. Low, Personal Notes No. 92, 28 April 1973.
55. Ibid.
56. Low, Personal Notes No. 93, 16 May 1973 (includes quotation).
57. Low, Personal Notes No. 94, 27 May 1973 (includes extended quotation). See also Low to McCurdy, memo, 21 May 1973.
58. Staats, *Analysis,* 1 June 1973 (table on 1).
59. Ibid. (quotation, 2); Low, Personal Notes No. 92, 28 April 1973.
60. Quotations: Fletcher to Staats, 24 May 1972; Mondale aide, in *National Journal,* 19 August 1972, 1331; Low, Personal Notes No. 92, 28 April 1973.
61. Philip Klass, interview with author, 11 July 1985; *Jane's Aircraft,* 706.
62. FAA: Heppenheimer, *Turbulent,* 172–83, 261–72, 283–89.
63. Highways: *American Heritage of Invention & Technology,* Fall 1991, 8–18.
64. *Science,* 14 August 1981, 741–42.
65. *U.S. News & World Report,* 28 January 1985, 14.
66. For the rationale behind Apollo, see Logsdon, *Decision* and McDougall, *Heavens.*
67. NASA SP-4221, chap. 6. For a pre-Challenger view of shuttle economics, see Roland, *Shuttle.*
68. NASA SP-4221, 427–32.
69. *Aviation Week,* 31 July 1972, 12–13 (quotation, 12); 7 August 1972, 14–15; 14 August 1972, 16–17.
70. Ibid., 31 July 1972, 12.
71. Ibid., 7 August 1972, 14.
72. Ibid., 14 August 1972, 16; 21 August 1972, 15. For Lockheed, see CASI 76A-18650, table on 3.
73. Low, Personal Notes No. 77, 16 September 1972.

74. *Aviation Week,* 14 August 1972, 16; 21 August 1972, 15 (includes Donlan quotations).
75. Ibid., 13 November 1972, 18.
76. Ibid., 14 August 1972, 17; 6 November 1972, 11; 13 November 1972, 18–19. See also Jenkins, *Space Shuttle,* 125.
77. E. W. Land, Memo for Record, 10 May 1976.
78. *Aviation Week,* 6 November 1972, 11; 13 November 1972, 18; 27 November 1972, 48.
79. For shuttle wing configurations, see NASA SP-4221, 208–17.
80. *Aviation Week,* 21 August 1972, 15.
81. Surber and Olsen, *Space Shuttle.* See also *Aviation Week,* 29 January 1973, 21.
82. Reports SV 72-19, 8 July 1972, and SV 72-2, October 1972 (North American Rockwell); *Aviation Week,* 27 November 1972, 48.
83. Report SV 72-19, 8 July 1972 (North American Rockwell); *Aviation Week,* 26 February 1973, 17–18; 16 April 1973, 18–19.
84. AIAA Paper 74-991.
85. *Aviation Week,* 16 April 1973, 18.
86. "Space Shuttle System Summary," July 1973 and January 1978 (Rockwell International); *Aviation Week,* 30 April 1973, 58–63.
87. *Aviation Week,* 14 August 1972, 16; 13 November 1972, 18; 27 November 1972, 47.
88. Ibid., 2 April 1973, 14.
89. For shuttle activities, see NASA SP-4221, 89–91, 224, 266–67, 428–31. For broader corporate histories, see Heppenheimer, *Countdown,* index references.
90. CASI 76A-18650, table, 3.
91. Ibid., text, 3; *Astronautics & Aeronautics,* January 1976, 40.
92. NASA SP-4307 covers the background of MSC.
93. The literature on Wernher von Braun is extensive. For his wartime work in Germany, see Neufeld, *Rocket;* Ordway and Sharpe, *Rocket Team.* For postwar work with the U.S. Army: Ley, *Rockets* and Medaris, *Countdown.* For work with NASA: SP-4206 and SP-4313.
94. Johnson, *Launch Vehicles,* 36.
95. For this early two-center division of responsibility, see NASA SP-4313, 179–80 for MSFC, and SP-4307, 224 for MSC.
96. NASA SP-4313, 277.
97. NASA SP-4221, 233; SP-4313, 227.
98. Leroy Day, interview with John Mauer, 17 October 1983, 60.
99. Charles Donlan, interview with John Mauer, 19 October 1983, 42–43.
100. CASI 72A-45194. See particularly figure 5.
101. NASA SP-4313, 281–82. Myers quotation is in SP-4307, 227.
102. Compare NMI 8020.18, 12 July 1971; NMI 8020.18A, 17 March 1972; NMI 8020.18B, 15 March 1973. All NMI documents are NASA.
103. NASA SP-4307, 228; NMI 8020.18; CASI 76A-18650.
104. Dale Myers, interview with author, 5 December 1996; *Who's Who in Aviation,* 198; NASA SP-4407, vol. 1, p. 760.
105. Low, "Announcement: Key NASA Personnel Change," 9 May 1974; NASA SP-4407, vol. 1, p. 768.
106. NASA SP-4307, 23, 228.
107. NASA SP-4407, vol. 4, p. 653.
108. NASA SP-4307, 228, 229; "Biographical Data: Robert F. Thompson" (JSC, Houston, June 1974).
109. NASA SP-4307, 234; SP-4407, vol. 4, p. 640.
110. NASA SP-4407, vol. 4, p. 643; SP-4221, 206–9; Turner, "Maxime Faget"; Jenkins, *Space Shuttle,* 110–17.
111. Yardley to Low and Petrone, memo, 26 December 1974. For von Braun's procedures, see Johnson, *Launch Vehicles;* also NASA SP-4206, chap. 9.
112. Yardley to MSFC, JSC, and KSC Center Directors, 30 September 1975. Compare also NMI 8020.18B with NMI 8020.18C.
113. CASI 76A-18650; NASA SP-4221, 341–46.

2. The Expanding World of the Shuttle

1. Agnew, *Post-Apollo.* Quotations are on 14–15.
2. Paine to Nixon, 7 November 1969.
3. Paine to Nixon, 12 February 1969, 12 August 1969.
4. This summarizes data from Thompson, *Space Log.*
5. ESA: HSR-2, -3, -8, -14.
6. ESA HSR-9; *Fortune:* July 1961, 156–60, 248–60; December 1962, 188–95; October 1965, 128–31, 202–12; *Science.* 18 March 1977, 1125–33.
7. Galloway, *Politics,* 80–104; Madders, *New Force,* chap. 7. Quotation is Madders, 100.
8. ESA HSR-9; dollar shares, 18.
9. Quoted text of NSAM-338 is in ESA HSR-18, 6–7.
10. Madders, *New Force,* 105.
11. Krige and Russo, *Europe,* 82; ESA HSR-18, 22. Paine quotations are in these two references.
12. *Air & Space,* May 1996, 62–63.
13. Ley, *Rockets,* 559–60; Isakowitz, *Space,* 284; Thompson, *Space Log.*
14. ESA HSR-7, -10.
15. Ley, *Rockets,* 579; Isakowitz, *Space,* 283; Baker, *Spaceflight,* 167, 194. See also *Aviation Week,* 15 June 1964, 32; 30 May 1966, 32.
16. Krige and Russo, *Europe,* 74–76; *Aviation Week,* 13 June 1966, 38. Skynet: ESA HSR-18, 7. Program cost: HSR-9, 26; HSR-10, 6.
17. ESA HSR-9, 27; Madders, *New Force,* 117–18; Krige and Russo, *Europe,* 76; *Aviation Week,* 25 July 1966, 117–19.
18. Madders, *New Force,* 130–32.
19. Krige and Russo, *Europe,* 65; Madders, *New Force,* 135–36, 139. See also note 51, 152.
20. Madders, *New Force,* 127–29. Quotations are Krige and Russo, *Europe,* 78, 81.
21. Krige and Russo, *Europe,* 78. Large Astronomy Satellite: ESA HSR-2, 38–41; HSR-8, 57–60. Europa 2: *Aviation Week,* 22 July 1968, 21–22; 14 July 1969, 23.
22. Baker, *Spaceflight,* 213, 227. See also Madders, *New Force,* 138, 140; *Aviation Week,* 16 December 1968, 21; 14 July 1969, 23.
23. Paine to Nixon, 12 February 1969.
24. Berger, Memorandum for the President, 14 March 1969.
25. NASA SP-4407, vol. 2, p. 40.
26. Paine to Nixon, 12 August 1969.
27. ESA HSR-15, Appendix 1. Paine quotations are 50, 54.
28. Madders, *New Force,* 143.
29. Krige and Russo, *Europe,* 75, 81–82; Madders, *New Force,* 140. Rocket engines: Isakowitz, *Space,* 283; ESA HSR-7, 4; HSR-10, 1; Chulick et al., *History: Thirty Years of Rocketdyne.*
30. Madders, *New Force,* 127–29, 139–40; Krige and Russo, *Europe,* 81, 82. Atlas-Centaur: NASA SP-4012, vol. 3, p. 23; *Science,* 9 March 1973, 984.
31. Krige and Russo, *Europe,* 80; Madders, *New Force,* 140; *Aviation Week,* 3 January 1972, 49.
32. Baker, *Spaceflight,* 252.
33. Madders, *New Force,* 144–45; CASI 71N-18430; Agnew, *Post-Apollo,* 15; *Aviation Week,* 5 October 1970, 17.
34. *Aviation Week,* 24 April 1972, 84, 85–91.
35. Paine to Nixon, 21 June 1970.
36. NASA SP-487, 39–48. Modular space stations: Reports SD 71-217-1, January 1972 (North American Rockwell), and MDC G2727, April 1972 (McDonnell Douglas).
37. NASA SP-487, 5–7; SP-4304, 145, 152–53; SP-4313, chap. 11.
38. NASA SP-487, 45–48. Early studies include Report MSC-04326 (NASA); AIAA Papers 71-812, -813, -814, -816, -817; Reports TM-72-1011-3 (Bellcomm), and MDC G4471, April 1973 (McDonnell Douglas).
39. Flanigan to Ehrlichman, memo, 16 February 1971.
40. Sebesta, *Blueprints,* 52.
41. *Aviation Week,* 24 April 1972, 86; Low to Fletcher, memo, 18 January 1972.

42. Walsh to Pollack, memo, 18 February 1972.
43. Elliott to Fletcher, letter with attachment, 23 March 1972.
44. Rogers to Nixon, memo, 29 April 1972; Fletcher to Kissinger, letter with attachment, 5 May 1972.
45. Low, Personal Notes No. 72, 17 June 1972.
46. NASA SP-487, 17; SP-4407, vol. 2, pp. 81–85; *Aviation Week,* 17 July 1972, 19.
47. *Spaceflight,* September 1972, 322–24; Stott and Hempsell, *European;* AIAA Paper 73-74.
48. NASA SP-4313, 433–34.
49. Krige and Russo, *Europe,* 110–11; NASA SP-487, 18.
50. NASA SP-487, 18–21, 48–51.
51. Krige and Russo, 103–8; Madders, *New Force,* 155–63.
52. ESA HSR-18, annexes 1, 4, 5. Quotations are in Annexes 1, 4.
53. Ibid., 19–23. Quotations are on 22.
54. Baker, *Spaceflight,* 269; *Aviation Week,* 15 November 1971, 20–21.
55. *Aviation Week,* 22 November 1971, 19; Krige and Russo, *Europe,* 109–10 (quotations, 110).
56. ESA HSR-18, 27–28.
57. Thompson, *Space Log.*
58. Madders, *New Force,* 146; Krige and Russo, *Europe,* 111.
59. Krige and Russo, *Europe,* 80, 115; *Aviation Week,* 11 December 1972, 19; 1 October 1973, 34–36; Madders, *New Force,* 164, 165.
60. *Aviation Week,* 1 January 1973, 14.
61. ESA HSR-14, 11–12; Thompson, *Space Log.*
62. *Aviation Week,* 1 January 1973, 14.
63. Ibid., 3 August 1970, 20–21; 20 November 1972, 23; Madders, *New Force,* 127–32.
64. *Aviation Week,* 1 January 1973, 14–15; Krige and Russo, *Europe,* 111–12; Madders, *New Force,* 164.
65. NASA SP-487, 21, 51.
66. Madders, *New Force,* 165; Krige and Russo, *Europe,* 112; *Aviation Week,* 7 May 1973, 20–21 (Ehmke quotations, 20).
67. Krige and Russo, *Europe,* 68–70; *Aviation Week,* 25 June 1973, 60–61.
68. *Aviation Week,* 6 August 1973, 15–16; Madders, *New Force,* 166–67.
69. Madders, *New Force,* 167–68; Krige and Russo, *Europe,* 112.
70. Madders, *New Force,* 177 (note 45).
71. NASA SP-487, 59; memorandum from Fletcher, 24 September 1973.
72. NASA SP-487, 27–28, 61. Memorandum of Understanding is in ESA HSR-21, Annex 3.
73. NASA SP-487, 69–75; Lord quotation, 74. *Aviation Week,* 27 May 1974, 18–19; 3 June 1974, 16; 10 June 1974, 21.
74. Heppenheimer, *Countdown,* 319, 320, 326, 335–37.
75. *Scientific American,* March 1989, 34–40.
76. Madders, *New Force,* 174.
77. This discussion draws on comments by Charles Donlan.
78. NASA SP-4206, 71–72; Serling, *Legend,* 228–29 (quotation, 228).
79. Powers, *Shuttle,* 240; Steiner, *Jet Aviation,* 27; *Spaceflight,* February 1978, 60 (includes quotation). For Saturn V logistics, see NASA SP-4206, 302–8.
80. Jenkins, *Space Shuttle,* 241, 242.
81. "Saturn V" (NASA-MSFC poster); Powers, *Shuttle,* 240; Ley, *Rockets,* 636.
82. *Aviation Week,* 18 June 1973, 22.
83. Low to Myers, memo including Pownall letter, 11 October 1972.
84. NASA SP-4206, 81–82; *Aviation Week,* 9 April 1973, 23.
85. Low, Personal Notes No. 101, 17 August 1973.
86. McDonnell Douglas, "Delta II."
87. *Aviation Week,* 16 April 1973, 18; 30 April 1973, 61, 63. For tank disposal, see James B. Odom, interview with Joe Guilmartin, 13 July 1984, 15–16.
88. *Astronautics & Aeronautics,* January 1974, 71.
89. Ibid., February 1973, 68–73; *Astronautics and Aerospace Engineering,* March 1963, 33–36.

90. Ley, *Rockets,* 498–500; NASA SP-4221, 46; *Astronautics & Aeronautics,* February 1965, 42–43.
91. Fuhrman, *Fleet,* 270, 275, 281, 286; *Astronautics & Aeronautics,* October 1972, 61.
92. NASA SP-4012, vol. 3, pp. 48, 49.
93. Fletcher, *Selection,* 2.
94. Ibid., 11–14. Quotation, 12.
95. Ibid. Quotations, 3, 4, 5, 6.
96. Ibid. Quotations, 7, 8, 9.
97. Ibid., 20.
98. Staats, *Decision,* 24 June 1974, 19.
99. Fletcher, *Selection,* 20–22. Quotation: Low, Personal Notes No. 108, 25 November 1973.
100. Investigative writers include McConnell, *Challenger,* and Shupe, *Darker Side.*
101. Launius, *Western;* McConnell, *Challenger,* 51.
102. Fletcher to Moss, 12 January 1973.
103. Fletcher to Moss, 23 February 1973.
104. McConnell, *Challenger,* 54–55.
105. Launius, *Western.*
106. *Aviation Week,* 26 November 1973, 27 (includes quotation); NASA SP-4012, vol. 3, p. 49.
107. *Aviation Week,* 14 January 1974, 22. For C-5A buy-in, see Rice, *C-5A,* 10, 18–19, 25–28.
108. "Interim Contract with Thiokol," Fletcher papers (Univ. of Utah), 12 February 1974; *Aviation Week,* 1 July 1974, 18; Acting Comptroller General to Fletcher, 9 May 1974 (includes quotation).
109. Staats, *Decision,* 24 June 1974 (quotation, 1, 22); Fletcher, *Selection,* 3, 4, 5, 6; *Aviation Week,* 1 July 1974, 18–19; Yardley et al., Memorandum for the Administrator, 26 June 1974; Low, Personal Notes No. 125, 20 July 1974.
110. Low, Personal Notes No. 125, 20 July 1974; Yardley et al., Memorandum for the Administrator, 26 June 1974.
111. Fletcher, Memo for Record, 15 May 1974; Low, Personal Notes No. 125, 20 July 1974 (quotation, 6 of attachment); Fletcher to Staats, 26 June 1974.
112. Heppenheimer, *Quake,* 45–46.
113. Ruffner, *Corona;* Heppenheimer, *Countdown,* 140–46, 278–82; Richelson, *Secret Eyes,* index references.
114. NASA SP-4221, 117, 206–15, 225–26, 231–33.
115. NASA SP-4012, vol. 3, p. 38; *Aviation Week,* 15 July 1974, 75.
116. Ruffner, *Corona,* xiv; *Quest,* Winter 1995, 40–45.
117. *Aviation Week,* 15 July 1974, 83–89.
118. *Time,* 15 December 1958, 15, 41–42.
119. *National Journal,* 24 April 1971, 869–76 (Teague quotation, 871).
120. SRB recovery: NASA SP-4012, vol. 3, p. 44.
121. *Space World,* February 1977, 30–36; *National Journal,* 22 April 1972, 706–7.
122. *Air & Space,* March 1997, 68–73.
123. Ibid.; *Aviation Week,* 30 June 1975, 32–36.
124. NASA SP-4012, vol. 3, pp. 38–42; *Aviation Week,* 17 September 1973, 95–99; 15 July 1974, 75; 15 May 1978, 48–53.

3. Odyssey of the Enterprise

1. Leonardo da Vinci, *Notebooks,* vol. 1, p. 441.
2. AIAA Paper 73-31, 4, fig. 11; Pace, "Engineering," 166–67.
3. AIAA Papers 71-662, 71-804, 71-805; Report LE 71-7 (North American Rockwell).
4. AIAA Paper 71-662; Pace, "Engineering," 168–70.
5. Personal discussion with H. D. Hunley of NASA-Dryden.
6. Pace, "Engineering," 173–74; NASA SP-4220, 117–19; Hunley, "Significance."
7. Thompson and Peebles, *Flying;* NASA SP-4220 (quotation, 116); SP-4303, chap. 8.
8. Pace, "Engineering," 172–73; AIAA Paper 73-31, fig. 11.
9. Pace, "Engineering," 174–75; *Aviation Week,* 5 October 1970, 16 (includes quotations).

10. Pace, "Engineering," 171–72.
11. *Science,* 28 January 1973, 392–96; NASA SP-4221, 117, 133.
12. Pace, "Engineering," 171–72.
13. Ibid., 170–71, 176–77.
14. Donlan, *Requirements,* 21 April 1972.
15. DeMeritte to Wetzel, memo, 14 March 1974 (includes quotations).
16. Jenkins, *Space Shuttle,* 133; *Aviation Week,* 25 September 1972, 84–86 (quotation, 85).
17. Report SV 72-19 (North American Rockwell); "McDonnell Douglas Commercial Family DC-1 Through MD-80" (Long Beach, Calif.: Douglas Aircraft, 1985).
18. Jenkins, *Space Shuttle,* 133; *Spaceflight,* June 1977, 213; *Aviation Week,* 21 January 1974, 45–46.
19. Pace, "Engineering," 177.
20. *Aviation Week,* 15 March 1976, 65; Leroy Day, interview with John Mauer, 17 October 1983, 72.
21. Discussion of Salkeld draws on personal acquaintance and on attendance at some of his presentations.
22. Malkin: *Requirements,* 12 March 1974; Memo for Record, 12 July 1974.
23. Yardley, "Reevaluation"; Aviation Week, 5 November 1973, 46; 29 April 1974, 54; 5 April 1976, 58.
24. NASA SP-4206, 309–14; *Aviation Week,* 4 February 1974, 38–41. Boeing 747 dimensions: *Pedigree,* 63.
25. Yardley, "Reevaluation"; quotations, 1, 9. See also NASA Press Release 74-160, 17 June 1974.
26. Kraft to Schneider, 10 May 1974.
27. Malkin, Memo for Record, 12 July 1974; *Aviation Week,* 24 June 1974, 21; Jenkins, *Space Shuttle,* 145; photo, 156.
28. NASA SP-4309, 149–50.
29. Jenkins, *Space Shuttle,* 144; *Aviation Week,* 5 April 1976, 58.
30. Jenkins, *Space Shuttle,* 144–45; *Aviation Week,* 8 November 1976, 150–54; 3 January 1977, 16.
31. Jenkins, *Space Shuttle,* 150.
32. NASA SP-4220, 169, 174–75 (quotation, 175); Miller, *X-Planes,* 157–60.
33. CASI 72A-45194; *COSMIC,* 173–74.
34. CASI 45194; CASI 76A-18650. Quotes from Leroy Day, interview with John Mauer, 17 October 1983, 62–63.
35. *Aviation Week,* 20 October 1975, 16–17; 1 December 1975, 20.
36. *Astronautics & Aeronautics,* January 1976, 40–43. See also *Aviation Week,* 13 January 1975, 45.
37. Low, Memo for Record, 2 January 1975; *Aviation Week,* 23 February 1976, 25.
38. *Astronautics & Aeronautics,* January 1976, 41–43.
39. Jenkins, *Space Shuttle,* 141–42, 150; Modlin and Zupp, *Structural.*
40. Jenkins, *Space Shuttle,* 150; Heppenheimer, *Colonies,* 67–68. Discussion of Hoagland draws on personal experience.
41. *Aviation Week,* 13 September 1976, 26.
42. *Los Angeles Times,* 17 September 1976, p. 1; 18 September 1976, pp. 1, 26. This discussion also draws on personal observation.
43. Jenkins, *Space Shuttle,* 150–51; *Aviation Week,* 20 September 1976, 12–14; *Spaceflight,* June 1977, 214–15.
44. CASI 72A-45194 summarizes the management of *Enterprise.*
45. AIAA Paper 71-659; NASA SP-4221, 240–42.
46. Chapter 4 of this book presents this process for the SSME.
47. Jenkins, *Space Shuttle,* 153, 159; *Aviation Week,* 27 September 1976, 14–16; 7 February 1977, 12–13.
48. *Aviation Week,* 14 February 1977, 12–15; *Spaceflight,* January 1978, 21.
49. *Aviation Week,* 1 March 1976, 22; Cassutt, *Who's Who,* 66–67; Chaikin, *Man,* 292–94, 304.
50. Cassutt, *Who's Who,* 127–28; *New York Times Biographical Service:* November 1981, 1595–96; February 1986, 266.
51. Cassutt, *Who's Who,* 50; Miller, *X-Planes,* 107–8.
52. Cassutt, *Who's Who,* 54–55; *New York Times Biographical Service,* March 1982, 332.
53. *Aviation Week,* 30 June 1975, 44; 6 October 1975, 21; 12 January 1976, 18; 29 November 1976, 42.
54. *Aviation Week,* 8 November 1976, 150 (includes quotation); 7 February 1977, 12–13.
55. Report JSC-13864 (NASA), 2–1; *Spaceflight,* January 1978, 21–22.
56. *Aviation Week,* 21 February 1977, 16–18 (quotation, 17).
57. Ibid., 14 February 1977, 12–16 (quotation, 13); *Spaceflight,* January 1978, 22.

58. *Aviation Week,* 28 February 1977, 16–21; *Spaceflight,* January 1978, 22.

59. *Aviation Week,* 28 February 1977, 20–21 (quotations, 20, 21). For Boeing 747 weights, see *Pedigree,* 63.

60. *Aviation Week,* 7 March 1977, 19–20 (quotations, 20); *Spaceflight,* January 1978, 22–23.

61. *Aviation Week,* 7 March 1977, 19–20 (quotations, 20); *Spaceflight,* January 1978, 23.

62. Report JSC-13864 (NASA), 2–1.

63. *Aviation Week,* 18 April 1977, 44–47; 27 June 1977, 14; *Spaceflight,* January 1978, 23.

64. *Aviation Week,* 27 June 1977, 12–14; *Spaceflight,* January 1978, 25; Report JSC-13045 (NASA), 2-1 and sec. 4.1.

65. *Aviation Week,* 4 July 1977, 18–19; *Spaceflight,* January 1978, 25–26; Report JSC-13045 (NASA), 2-3 and sec. 4.2.

66. *Aviation Week,* 1 August 1977, 20–21 (quotations: Fullerton, 20; Fulton, 21); *Spaceflight,* January 1978, 26–27; Report JSC-13045 (NASA), 2-5 and sec. 4.3.

67. *Spaceflight,* January 1978, 29.

68. *Aviation Week,* 22 August 1977, 12–19 (Haise quotations, 14, 18); *Spaceflight,* January 1978, 28, 40 (Peebles quotation, 29); Report JSC-13864 (NASA), 4-3 and sec. 4.3.1.

69. *Aviation Week,* 29 August 1977, 22–23 (quotation, 22).

70. *Aviation Week,* 5 September 1977, 22. For the shuttle's on-board computers, see chapter 8 of this book.

71. *Aviation Week,* 19 September 1977, 22–23 (quotations: Engle, 22; Truly, 23); Report JSC-13864 (NASA), 4–5, sec. 4.3.2.

72. *Aviation Week,* 10 June 1974, 45–47; 28 April 1975, 91–92.

73. Ibid., 3 October 1977, 24–25 (quotation, 25); Report JSC-13864 (NASA), 4–7, sec. 4.3.3.

74. *Aviation Week,* 7 March 1977, 19 (Fulton quotation).

75. Ibid., 3 October 1977, 25 (includes Slayton quotation); 10 October 1977, 27.

76. Ibid., 17 October 1977, 23–24; Report JSC-13864 (NASA), 4–9, sec. 4.3.4 (extended quotation, 4-78).

77. *Aviation Week,* 17 October 1977, 23 (includes Slayton quotation); 31 October 1977, 16–17 (Haise quotation, 16); Report JSC-13864 (NASA), 4-11, sec. 4.3.5.

78. Malkin, Memo for Record, 8 November 1977 (includes quotations).

79. NASA SP-4012, vol. 3, p. 118.

80. Modlin and Zupp, *Structural;* Jenkins, *Space Shuttle,* 158–59; *Aviation Week,* 27 March 1978, 47; 1 May 1978, 12–13; 19 June 1978, 75–80. Inert propellant: AIAA Paper 79-1136, 10.

81. Jenkins, *Space Shuttle,* 159–61, 178; *Aviation Week,* 30 April 1979, 27; 7 May 1979, 12–17; 21 May 1979, 72.

82. Jenkins, *Space Shuttle,* 161–64; Dennis R. Jenkins, private communication.

4. Propulsion I: The Space Shuttle Main Engine

1. Sam Hoffman, interview with author, 28 July 1988.

2. Report RSS-8333-1 (Rocketdyne).

3. Sutton, *Rocket,* 153–54, 196, 231.

4. Ibid., 156–57, 196; Willy Wilhelm, interview with author, 31 March 1998.

5. Henry Pohl, interview with Joe Guilmartin interview, 2 November 1983, 3.

6. Ibid., 4. For RL-10, see Sutton, *Rocket,* 156; NASA SP-4206, 138–40.

7. Pohl, interview with Guilmartin, 4. For hydrogen peroxide, see Ley, *Rockets,* 244–47, 308; Munger and Seaman, *XLR99-RM-1.* The engines of Atlas, Titan, and Thor were variants of a standard design. See *Threshold* (Rocketdyne), December 1987, 16–23; Chulick et al., *History.*

8. NASA SP-4221, 235–44.

9. Executive Summary, PWA FP 71-50 (Pratt & Whitney); Report RSS-8333-1 (Rocketdyne); Wilhelm interview, 31 March 1998.

10. Report RSS-8333-1 (Rocketdyne), III-13.

11. Dennis Jenkins, private communication.

12. Munger and Seaman, *XLR99-RM-1,* 216–17.

13. Robert Biggs, interview with author, 12 June 2000. Table and data are from Report RI/RD87-142 (Rocketdyne).

14. Biggs, "Main Engine," 89. *Titanic:* Lord, *Night,* iii, 103. J-2: Sutton, *Rocket,* 231.

15. AIAA Paper 71-659. This discussion draws on interviews with Paul Castenholz, Robert Biggs, Maynard Stangeland, and Ed Larson.
16. Report RI/RD87-142 (Rocketdyne), 1–9, 5–2.
17. AIAA Paper 71-659. For ICBM heating, see Report 1381 (NACA).
18. AIAA Paper 71-659.
19. Report RSS-8333-1 (Rocketdyne), chap. 4.
20. NASA SP-4221, 243–44, 432–34.
21. Biggs, "Main Engine," 73–76.
22. Gray, *Angle,* 170–71.
23. Biggs, "Main Engine," 76, 79; *Astronautics & Aeronautics,* January 1976, 50; J. R. Thompson, interview with author, 20 April 1998.
24. SAE Paper 730927; Report RSS-8333-1 (Rocketdyne), III-41 to III-43.
25. Low, Personal Notes No. 115, 3 March 1974; Biggs, "Main Engine," 77.
26. Low: Personal Notes No. 119, 27 April 1974; No, 121, 25 May 1974.
27. Paul Castenholz, interview with author, 18 August 1988.
28. Ibid.; Low: Personal Notes No. 121, 25 May 1974 (Low quotations); No. 123, 22 June 1974 (McNamara quotation); Low, Memo for Record, 28 June 1974.
29. Thompson interview.
30. Biggs, "Main Engine," 80–81.
31. Ibid., 77, 79.
32. NASA SP-4206, 72–73.
33. *Aviation Week,* 30 June 1975, 38–42; 8 November 1976, 61.
34. Ibid., 5 July 1976, 44; Mack Herring, interview with author, 23 April 1998.
35. Biggs, "Main Engine," 86, 87; Robert Biggs, interview with author, 17 March 1998 (includes quotation); Maynard Stangeland, interview with author, 31 March 1998.
36. Biggs, "Main Engine," 77–78.
37. Ibid., 82, 86–87.
38. Ibid., 82–87; Biggs interview (includes quotation), 17 March 1998; Stangeland interview.
39. AIAA Paper 78-1002.
40. Biggs, "Main Engine," 88–89.
41. Ibid., 89–91; AIAA Paper 78-1002.
42. Stangeland interview (includes quotation); Biggs, "Main Engine," 91. Ek quotation: AIAA Paper 78-1002, 14.
43. AIAA Paper 77-808; *Aviation Week,* 5 July 1976, 44.
44. AIAA Paper 77-808; *Aviation Week,* 8 November 1976, 59–61 (Sanchini quotation, 61).
45. AIAA Paper 77-808; *Aviation Week,* 18 April 1977, 47.
46. Biggs, "Main Engine," 76.
47. AIAA Paper 77-808.
48. Biggs interview, 17 March 1998; Report RI/RD87-142 (Rocketdyne), 5–2.
49. Biggs, "Main Engine," 79; Report RSS-8595-6 (Rocketdyne).
50. Fletcher, Statement of Determination, 4 February 1976 (includes quotation); Biggs, "Main Engine," 79–80; *Aviation Week,* 5 July 1976, 43.
51. Biggs, "Main Engine," 92–93; Biggs interview, 17 March 1998.
52. Biggs, "Main Engine," 91–92 (includes quotation).
53. Biggs interview (includes quotations), 17 March 1998. Oxygen turbopump fires occurred on 24 March 1977; 8 September 1977; 18 July 1978, and 30 July 1980. All are described in the present chapter.
54. Report RSS-8595-11 (Rocketdyne). Quotation, C-2.
55. Biggs, "Main Engine," 94–96.
56. Frosch, statement to House Appropriations Committee, 25 January 1978, 9; *Aviation Week,* 8 August 1977, 23; AIAA Paper 78-1001.
57. Biggs, "Main Engine," 96; Report RSS-8595-13 (Rocketdyne).
58. *Aviation Week,* 3 October 1977, 26; 10 October 1977, 26.
59. AIAA Paper 78-1001; Biggs, "Main Engine," 96–97.
60. Stangeland interview; Thompson interview. Quotations are from these interviews.

61. Biggs, "Main Engine," 110–12.
62. Ibid., 99–102.
63. VerSnyder and Shank, *Development;* Stangeland interview.
64. Biggs, "Main Engine," 99; Covert, *Technical Status,* 32–36.
65. AIAA Paper 78-1001; Biggs, "Main Engine," 101–2.
66. *Aviation Week,* 13 November 1972, 18; 15 January 1973, 17.
67. Covert, *Technical Status,* 9; Schmidt and Anderson, "Update," 26–32.
68. Schmidt and Anderson, "Update," 34–35.
69. Ibid., 26–28; Biggs, "Main Engine," 115.
70. Covert, *Technical Status,* iii–vi; NASA SP-4221, 243; Maynard Pennell, interview with author, 11 April 1991; Allen Donovan, interview with author, 13 January 1995.
71. Fletcher, Statement of Determination, 4 February 1976; Biggs, "Main Engine," 79–80.
72. Covert, *Technical Status,* 27–28; John Plowden, interview with author, 16 March 1998; Thompson interview.
73. *Aviation Week,* 3 October 1977, 26; Covert, *Technical Status* (quotation, 27; table, 28).
74. Covert, *Technical Status* (quotation, 24; extended quotation, 12).
75. Frosch quotation: Biggs, "Main Engine," 79.
76. Covert, *Technical Status,* 24; *Aviation Week,* 27 November 1978, 61.
77. AIAA Paper 78-1001; *Aviation Week,* 22 May 1978, 55–61; Covert, *Second Review,* 12.
78. AIAA Paper 79-1141; Covert, *Second Review,* 12.
79. SSME: AIAA Paper 73–60. For Apollo, see NASA SP-4206, 75, 206–8.
80. Jenkins, *Space Shuttle,* 166; *Aviation Week,* 29 May 1978, 49–53; Robert Biggs, interview with author, 15 November 2000.
81. AIAA Paper 78-1001; *Lagniappe* (NASA-Stennis): 11 November 1977; 23 June 1978 (includes quotation); 21 July 1978.
82. Biggs, "Main Engine," 98; Report RSS-8595-15 (Rocketdyne).
83. Covert, *Technical Status,* v; *Aviation Week,* 3 July 1978, 25; 10 July 1978, 15; 31 July 1978, 50.
84. AIAA Paper 79-1141; *Aviation Week,* 25 September 1978, 12–13.
85. AIAA Paper 79-1141; Covert, *Second Review,* 3. For abort, see *Aviation Week,* 15 October 1979, 39–45; Jenkins, *Space Shuttle,* 265–66.
86. Yardley, statement to House Committee on Science and Technology, 25 September 1978; Covert, *Second Review,* 3, 4.
87. AIAA Paper 79-1141.
88. Covert, *Second Review,* 10–11 (includes quotation).
89. AIAA Paper 79-1141; Biggs, "Main Engine," 87.
90. *Aviation Week,* 18 December 1978, 8; Covert, *Technical Status* (quotations, 12, 25); Covert, *Second Review,* 16–17.
91. Biggs, "Main Engine," 102–3; Report RSS-8595-18 (Rocketdyne); quotations, 150, 152.
92. *Aviation Week,* 8 January 1979, 12–14; *Astronautics & Aeronautics,* March 1979, 6.
93. AIAA Paper 79-1141; *Aviation Week,* 12 February 1979, 44–45; Biggs, "Main Engine," 103–5; Covert, *Second Review,* 17–18.
94. Quotations: *Astronautics & Aeronautics,* March 1979, 6; *Aviation Week,* 8 January 1979, 14.
95. Covert, *Second Review,* 16; *Aviation Week,* 16 April 1979, 19.
96. Biggs, private communication; *Aviation Week,* 16 April 1979, 19; 23 April 1979, 21.
97. AIAA Paper 79-1141.
98. David Geiger, interview with author, 13 April 1998. Description of firing draws on the author's personal observations.
99. Leroy Day, private communication; Thompson interview.
100. AIAA Papers 79-1141, 80-1129 (Sanchini quotation, 12); Biggs, "Main Engine," 115–16.
101. AIAA Paper 81-1373; *Lagniappe* (NASA-Stennis), 25 April 1980; Robert Biggs, private communication.
102. Author's personal observations.
103. Report RI/RD87-142 (Rocketdyne), 2–15.
104. Biggs, "Main Engine," 114; *Lagniappe* (NASA-Stennis), 25 April 1980.
105. Biggs, "Main Engine," 115; Robert Biggs, private communication.

106. Biggs, "Main Engine," 115; Covert, *Technical Status,* 17.
107. AIAA Paper 80-1129; Robert Biggs, private communication.
108. *Aviation Week,* 9 July 1979, 16; *Lagniappe* (NASA-Stennis), 15 June 1979.
109. Biggs, "Main Engine," 105–7; *Aviation Week,* 9 July 1979, 16; Report RSS-8595-20 (Rocketdyne).
110. Biggs, "Main Engine," 110; *Aviation Week,* 12 November 1979, 20; *Lagniappe* (NASA-Stennis), 22 October 1979.
111. Biggs, "Main Engine," 107–10; Report RSS-8595-19 (Rocketdyne).
112. AIAA Papers 80-1129, 80-1309; Biggs, "Main Engine," 110; Report RSS-8595-21 (Rocketdyne).
113. *Lagniappe* (NASA-Stennis): 17 December 1979 to 13 February 1981, passim.
114. Biggs, "Main Engine," 111–13; Report RSS-8719 (Rocketdyne). Engine 2004: *Aviation Week,* 11 August 1980, 28; *Lagniappe* (NASA-Stennis), 22 August 1980.
115. *Lagniappe* (NASA-Stennis): 18 June 1980; 29 October 1980; 26 November 1980; 17 December 1980.
116. Biggs, "Main Engine," 98–99; Report RSS-8595-22 (Rocketdyne).
117. Biggs, "Main Engine," 115–17; Robert Biggs, private communication.
118. Biggs interview, 15 November 2000 (includes extended quotation). 991 tasks: Biggs, "Main Engine," 76.
119. Biggs interview, 15 November 2000.

5. Propulsion II: SRB, ET, OMS, RCS, APU

1. Index references: Powers, *Shuttle;* Jenkins, *Space Shuttle.*
2. *Astronautics & Aeronautics,* September 1965, 74–77; AIAA Paper 65-163; NASA SP-4313, 290–92.
3. Fuhrman, *Fleet;* Report SA 44-76-01 (NASA). Fletcher quotation: NASA SP-4313, 290.
4. George Hardy, interview with Joe Guilmartin, 13 July 1984, 7–8; W. P. Horton, interview with Joe Guilmartin, 16 July 1984, 14.
5. NASA SP-4313, 289–90; Hardy, interview with Guilmartin, 9; Horton, interview with Guilmartin, 2.
6. Horton, interview with Guilmartin, 14; Dennis Jenkins, private communication.
7. Hardy, interview with Guilmartin, 15.
8. Dennis R. Jenkins, private communication with author.
9. Ibid., 2, 3–4; Horton, interview with Guilmartin, 5–6.
10. Report SA 44-76-01 (NASA); Jenkins, *Space Shuttle,* 245–51. For 1960s program, see NASA SP-4221, 46.
11. Jenkins, *Space Shuttle,* 247. See also Sutton, *Rocket,* 294–97.
12. *Thiokol's Aerospace Facts,* July–September 1973. Quotations: Thirkill, 1; Wells, 7.
13. Ibid., 7–9; *Aviation Week,* 20 February 1978, 54–57.
14. Jenkins, *Space Shuttle,* 138; *Aviation Week,* 18 June 1973, 78.
15. AIAA Paper 78-986; Report SA 44-76-01 (NASA), 4, 5.
16. "Space Shuttle Solid Rocket Motor Alternative Propellant Program," 28 January 1975. NASA Archives; no author.
17. Yardley and Hinners to Fletcher, memo, July 1976.
18. AIAA Paper 78-954 (table, 3).
19. AIAA Paper 78-952; *Aviation Week,* 8 November 1976, 84–89.
20. AIAA Paper 78-954.
21. Vaughan, *Challenger,* 5, 96–97, 121; AIAA Paper 73-62. Quotation: Horton, interview with Guilmartin, 12.
22. *Air & Space,* April/May 1996, 80.
23. Sutton, *Rocket,* 81–85, 330–32.
24. AIAA Paper 78-950. $30 per pound: IAF Paper 78-A-24, 8.
25. AIAA Papers 75-1172, 78-951.
26. Ley, *Rockets,* 496–97; *Astronautics & Aeronautics,* September 1965, 75, 77. For a general discussion, see Sutton, *Rocket,* 332–37.
27. AIAA Paper 73-62.
28. Horton, interview with Guilmartin, 5.
29. *Astronautics & Aeronautics,* February 1965, 45, 46; AIAA Paper 75-1172.
30. Horton, interview with Guilmartin, 5.

31. Sutton, *Rocket,* 317; AIAA Paper 78-987.
32. Ley, *Rockets,* 499; *Air & Space,* April/May 1996, 86; AIAA Paper 78-987.
33. AIAA Papers 75-1172 (quotation, 3); 75-1173.
34. Bacchus et al., *Aerodynamic;* IAF Paper 78-A-24.
35. Woodis and Runkle, *Structural.*
36. AIAA Papers 75-1170, 75-1172, 75-1173; Jenkins, *Space Shuttle,* 251.
37. *Aviation Week,* 14 April 1980, 52–53.
38. Biggs, "Main Engine," 115; Jenkins, *Space Shuttle,* 167–68. Burn time for SRM was some 120 seconds: AIAA Paper 75-1170.
39. Frisby, "History of Detachment Forty-Three," 1 January–30 June 1974, 29–30; 1 July–31 December 1974, 31.
40. Ibid., 1 July–31 December 1975, 42–43; AIAA Paper 75-1172.
41. Frisby, *Detachment 43,* 1 January–31 December 1976, 52–53. SRM casings: *Aviation Week,* 8 November 1976, 84–89; Jenkins, *Space Shuttle,* 245; Report SA 44-76-01 (NASA), 1, 5.
42. Frisby, *Detachment 43,* 1 January–31 December 1977, 106.
43. AIAA Paper 78-954; *Aviation Week,* 30 January 1978, 56–57.
44. AIAA Paper 78-987; *Aviation Week,* 25 July 1977, 23; 23 January 1978, 21.
45. AIAA Papers 78-951, 78-952.
46. Frisby, *Detachment 43,* 1 January–31 December 1977, 108; *Aviation Week,* 23 January 1978, 21.
47. AIAA Papers 78-951, 79-1139. Drop tests: Jenkins, *Space Shuttle,* 168.
48. Jenkins, *Space Shuttle,* 158–59; Frisby, *Detachment 43:* 1 January–31 December 1977, 108; 1 January–31 December 1978, 108.
49. Ibid., 1 January–31 December 1978, 108–115 (quotation, 110).
50. Ibid.; AIAA Paper 79-1139; *Aviation Week,* 23 January 1978, 21.
51. *Aviation Week,* 26 February 1979, 17; Jenkins, *Space Shuttle,* 167–68.
52. Frisby, *Detachment 43,* 1 January–31 December 1978, 113; AIAA Paper 79-1139.
53. Brown, *Lonesome,* 115, 146; Ambrose, *Nothing,* 337, 353.
54. AIAA Paper 79-1136; Tucker, *Detachment 43,* 1 January–31 December 1979, 63.
55. *Spaceflight,* February 1978, 60; Ley, *Rockets,* 495–96.
56. "Space Shuttle External Tank Costs" (NASA; no author or date); William Petynia, interview with Joe Guilmartin, 9 September 1985.
57. NASA SP-4206, 149, 214–15; *Spaceflight,* March 1978, 113–14. Quotes are from interviews with James Odom: by Joe Guilmartin, 13 July 1984, 15; by Jessie Whalen, 9 Feb 1988, 80.
58. Odom, interview with Guilmartin, 7. Nose cone heating: Report 1381 (NACA).
59. *Aviation Week,* 8 November 1976, 95–99; 29 May 1978, 53; James Odom, interview with Jessie Whalen, 9 February 1988, 78–79.
60. "Space Shuttle External Tank Costs" (NASA; no author or date); James Kingsbury, interview with Jessie Whalen, 17 November 1988, 452.
61. Quotes are from James Odom interviews: by Joe Guilmartin, 13 July 1984, 18–19; by Jessie Whalen, 9 February 1988, 90–91.
62. *Aviation Week,* 8 November 1976, 95–99; Odom, interview with Guilmartin, 19.
63. AIAA Papers 76-595, 78-1004.
64. *Lagniappe* (NASA-Stennis), 11 November 1977; AIAA Paper 78-1004. Norquist quotation: AIAA Paper 79-1143, 6.
65. *Aviation Week,* 8 November 1976, 95–99; 20 March 1978, 60.
66. William Barrett, interview with Joe Guilmartin, 1 and 2 October 1984; James Odom, interview with Joe Guilmartin, 13 July 1984 (extended quotation, 15–16).
67. Jenkins, *Space Shuttle,* 158–59; *Aviation Week,* 19 June 1978 (extended quotation, 77).
68. R. R. Foll, interview with Joe Guilmartin, 1 October 1984, 6.
69. Odom, interview with Jessie Whalen, 9 February 1988, 81.
70. Foll, interview with Guilmartin.
71. *Aviation Week,* 20 March 1978, 60; 29 May 1978, 53–55; "ET" (NASA chronology), 29 August 1984.
72. Sutton, *Rocket,* 175, 178, 181–82; Jenkins, *Space Shuttle,* 213–15.
73. AIAA Papers 75-1298, 79-1144.

74. AIAA Paper 76-740; Jenkins, *Space Shuttle,* 213.

75. For history of Aerojet, see Heppenheimer, *Countdown,* 37–41; Chulick et al., *History; Time,* 3 January 1969, 16 (includes Borman quotation).

76. AIAA Papers 75-1298, 76-740; *Time,* 3 January 1969, 16.

77. AIAA 76-740; Gibson and Humphries, *Orbital,* 648–49.

78. *Aviation Week,* 31 May 1976, 58–60 (quotations, 58).

79. *Time,* 3 January 1969, 16; AIAA Paper 76-740; *Aviation Week,* 31 May 1976, 59.

80. AIAA Paper 79-1142.

81. AIAA Paper 77-811.

82. Quotation: AIAA Paper 78-1005, 11.

83. *Aviation Week,* 31 May 1976, 59; AIAA Paper 79-1145.

84. AIAA Paper 79-1142 (quotations, 1); *Aviation Week,* 19 May 1980, 119.

85. AIAA Paper 79-1144; NASA SP-4307, 143, 148, 152; Taeuber et al., *Design,* 660.

86. AIAA Papers 76-740, 79-1144.

87. AIAA Papers 79-1144, 79-1145.

88. Taeuber et al., *Design,* 659–60.

89. Heppenheimer, *Turbulent,* 237; Jenkins, *Space Shuttle,* 216–17.

90. AIAA Paper 78-1007; Lance and Weary, *Auxiliary,* 680.

91. Lance and Weary, *Auxiliary* (quotation, 680).

92. Ibid.; AIAA Paper 78-1007.

93. Hughes, *Solid.*

94. AIAA Paper 79-1139; Jenkins, *Space Shuttle,* 167–68.

95. USS *Nimitz* has 260,000 horsepower: *Jane's Ships,* 793. At 100 percent of rated thrust, main turbopumps of three SSMEs total 253,410 horsepower: Report RI/RD87-142 (Rocketdyne), 2-15, 2-17.

6. Thermal Protection

1. Sutton, *Initial,* 3.

2. Report 1381 (NACA). Atlas concepts: Charles R. McLellan, memo, 6 November 1951; Smith J. DeFrance, 26 November 1952, Box 2, RG 255, National Archives, Philadelphia.

3. Hallion, *Hypersonic,* 385–86; quotation, 386.

4. CASI 81A-44344, 234; *Time,* 18 August 1958, 58.

5. Sutton, *Initial,* 6.

6. Sutton, *Rocket,* 81–83; Kreith, *Heat Transfer,* 538–45.

7. CASI 81A-44344; AIAA Paper 73-31.

8. AIAA Paper 71-443.

9. Hallion, *Hypersonic,* 344–68; Miller, *X-Planes,* 133–37.

10. Hallion, *Hypersonic,* 464–71, 510–19.

11. AIAA Papers 71-443, 73-31.

12. *Lockheed Horizons,* No. 13 (1983), 2–14. Shuttle aluminum structure is discussed within the present chapter.

13. Hallion, *Hypersonic,* 353, 357.

14. *Lockheed Horizons,* No. 13 (1983), 2–14; Schramm, *HRSI.*

15. Ley, *Rockets,* 245; Hallion, *Hypersonic,* 353, 467–68.

16. Becker, *Leading;* Korb et al., *Shuttle.*

17. *Technical* (Vought); Becker, *Leading.*

18. AIAA Paper 71-443.

19. *Astronautics & Aeronautics,* January 1970, 52–61.

20. Jenkins, *Space Shuttle,* 58–60. Specific project reports include: MDC E0056 (McDonnell Douglas); GDC-DCB69-046 (General Dynamics); SD 69-573-1 (North American Rockwell); LMSC-A959837 (Lockheed).

21. Korb et al., *Shuttle,* 1189; Schramm, *HRSI,* 1194.

22. NASA TM X-2273, 57–58; Korb et al., *Shuttle,* 1189–90; CASI 72A-10764.

23. *Lockheed Horizons,* No. 13 (1983), 11; Schramm, *HRSI,* 1195.

24. CASI 72A-10764; NASA TM X-2273, 39–93.

25. CASI 79A-17673.

26. NASA SP-4308, 133; Sutton, *Ablation.*

27. CASI 79A-17673, table 1; NASA SP-4304, 168–70. Mach 7 tunnel: NASA SP-440, 95–96.

28. NASA SP-4221, 266–67; NASA TM X-52876, vol. 3, p. 140.

29. NASA TM X-52876, vol. 3, pp. 195–200.

30. Ibid., 145–58.

31. Ibid., 175–84.

32. Ibid., 185–94; CASI 77A-35304; Becker, *Landing.*

33. NASA SP-4221, 267, 332–41.

34. Ibid. Specific project documents include: Report B35-43RP-11 (Grumman), 3–10, 3–12; SD 71-114-1 (North American Rockwell), 13; AIAA Paper 71-804 (McDonnell Douglas).

35. Reports B35-43 RP-33 (Grumman), 19; SV 71-50, 9, and SV 71-59, 3 (both from North American Rockwell); *Interim Report,* 36 (McDonnell Douglas).

36. NASA SP-4221, 341–46; Jenkins, *Space Shuttle,* 110–15.

37. Maxime Faget, interview with author, 4 March 1997; Heppenheimer, *Turbulent,* 209–10.

38. Pace, "Engineering," 179–88.

39. CASI 81A-443344; Korb et al., *Shuttle,* 1189.

40. Faget interview; Charles Donlan, interview with John Mauer, 19 October 1983, 20.

41. *Aviation Week,* 17 January 1972, 17; NASA SP-4221, 359, 408–11.

42. AIAA Paper 73-31 (Love quotation, 15).

43. Report SV 72-19 (North American Rockwell).

44. *Aviation Week,* 27 March 1972, 48; *Lockheed Horizons,* No. 13 (1983), 11–14.

45. *Aviation Week,* 26 January 1976, 125; Becker, *Leading.*

46. *Aviation Week,* 8 November 1976, 51–54; Korb et al., *Shuttle,* 1190.

47. *Aviation Week,* 8 November 1976, 51–53 (quotations, 53).

48. Ibid., 51; NASA TM X-2719, 14, 15.

49. *Aviation Week,* 8 November 1976, 51, 53; CASI 81A-44344.

50. *Aviation Week,* 8 November 1976, 51; Korb et al., *Shuttle,* 1189.

51. *Space World,* June–July 1979, 23; CASI 77A-35304.

52. Report SV 72-19 (North American Rockwell); Powers, *Shuttle,* 241; Korb et al., *Shuttle.*

53. *Aviation Week,* 24 June 1974, 68–73 (quotation, 68); 31 March 1975, 52–53.

54. Kreith, *Heat Transfer,* 534–38.

55. *Aviation Week,* 31 March 1975, 52–53; CASI 79A-17673.

56. CASI 79A-17673; CASI 81A-44344.

57. NASA SP-4304, 172–73; Korb et al., *Shuttle.*

58. AIAA Paper 82-0566; *Astronautics & Aeronautics,* January 1976, 64; *Journal of Spacecraft and Rockets:* June 1974, 437; July–August 1985, 417.

59. AIAA Papers 78-485, 83-0147.

60. *Astronautics & Aeronautics,* January 1976, 64–65; AIAA Paper 78-485, table 2.

61. *Aviation Week,* 31 March 1975, 52–53; *Astronautics & Aeronautics,* January 1976, 59–60; CASI 81A-44344.

62. *Astronautics & Aeronautics,* January 1976, 59–61.

63. Ibid., 60, 62–64.

64. *Aviation Week,* 31 March 1975, 52; AIAA Paper 78-485.

65. *Astronautics & Aeronautics,* January 1976, 60, 63–64.

66. Dotts et al., *Thermal,* 1066–69. For engineering changes compare Dotts, op. cit., fig. 2, 1064 with *Astronautics & Aeronautics,* January 1976, fig. 3, 60.

67. NASA TM X-74039 (quotation, 12).

68. AIAA Paper 83-1047; John Yardley, private communication; quotation: *Aviation Week,* 24 June 1974, 71.

69. Waldrop, *Tiles; Astronautics & Aeronautics,* January 1981, 27, 29; Schneider and Miller, *Challenging,* 403–4.

70. *Aviation Week,* 31 July 1978, 53; 27 November 1978, 64.

71. Waldrop, *Tiles; Aviation Week,* 21 May 1979, 59–63.

72. *Aviation Week,* 27 November 1978, 64.

73. 1100 workers: Waldrop, *Tiles,* 28. SSME failures and shuttle schedule are discussed in chapter 4 of this book.
74. *Aviation Week,* 5 February 1979, 13; 5 March 1979, 22–23; 19 March 1979, 21–22.
75. Ibid., 19 March 1979, 21–22; 23 April 1979, 21; Waldrop, *Tiles;* John Yardley, private communication.
76. *Aviation Week,* 23 April 1979, 21; 7 May 1979, 12–14; 21 May 1979, 59 (includes quotation); 2 July 1979, 26–27.
77. Ibid., 21 May 1979, 59–61.
78. Ibid., 17 September 1979, 22–23 (includes data for table); Waldrop, *Tiles,* 27.
79. *Astronautics & Aeronautics,* January 1981, 29.
80. Graham et al., *Nondestructive,* 693, fig. 3.
81. Ibid., 681–83; Schneider and Miller, *Challenging; Astronautics & Aeronautics,* January 1981, 29.
82. *Aviation Week,* 17 September 1979, 22; 15 October 1979, 44.
83. Ibid., 25 February 1980, 22; *Astronautics & Aeronautics,* January 1981, 29–30.
84. *Astronautics & Aeronautics,* January 1981, 30.
85. *Science,* 23 November 1979, 910–11.
86. Ibid., 911; *Aviation Week,* 7 May 1979, 14; 25 February 1980, 22–24.
87. Graham et al., *Nondestructive,* 685–87.
88. Waldrop, *Tiles.*
89. AIAA Papers 81-2468, 81-2469; *Aviation Week,* 25 February 1980, 23.
90. Schneider and Miller, *Challenging,* 408–12.
91. *Aviation Week,* 2 June 1980, 14; 9 June 1980, 20; 21 July 1980, 16–17; 11 August 1980, 28.
92. Ibid., 15 September 1980, 26; 10 November 1980, 17–18.
93. Ibid., 17 September 1979, 22; 4 August 1980, 24.
94. Ibid., 1 December 1980, 18–19 (quotation, 18).

7. The Orbiting Airplane

1. Anderson, *Aerodynamics,* 441–43.
2. *Astronautics & Aeronautics,* April 1975, 26.
3. Heppenheimer, *Hypersonic,* 128–34.
4. *Science,* 27 January 1978, 404–5.
5. Ibid., 16 January 1981, 268–69; *Astronautics & Aeronautics:* April 1975, 22–30; March 1982, 20–28. For numbers of partial derivatives, see Chapman, *Computational,* 1294.
6. *Science,* 16 January 1981, 268–69; *Aviation Week,* 3 September 1973, 14–16; *Scientific American,* February 1971, 76–87.
7. Reinhardt, *Parallel.*
8. *Aviation Week,* 24 June 1974, 68–73; 11 August 1975, 16; 18 August 1975, 16.
9. Ibid., 30 June 1975, 43–44.
10. Dill et al., *Ascent,* 151–52, 161.
11. Young et al., *Aerodynamic,* 221–22.
12. *Aviation Week,* 24 June 1974, 71; 9 December 1974, 19; Dill et al., *Ascent,* 154, 165.
13. Young et al., *Aerodynamic,* 217–18, 219–20, 232, 238.
14. Modlin and Zupp, *Structural,* 326; *COSMIC,* 175.
15. Jenkins, *Space Shuttle,* 142, 237; NASA SP-4002, 68, 76, 121; SP-4206, 360, 362–63.
16. Pinson and Leadbetter, *Results.*
17. Jenkins, *Space Shuttle,* 141–42; Emero, *Quarter;* Modlin and Zupp, *Structural,* 330–31.
18. Modlin and Zupp, *Structural* (tables, 329, 333).
19. Mackey and Gatto, *Structural,* 335–39.
20. *Aviation Week,* 10 October 1977, 26; 6 March 1978, 13; Glynn and Moser, *Structural,* 353–56.
21. Hammond et al., *Energy,* 109–10; *Scientific American,* December 1978, 70. 0.90 volts: Simon, "Electrical," fig. 4, 711.
22. Simon, "Electrical," 705–7 (extended quotation, 706).
23. Ibid., 705, 706; Jenkins, *Space Shuttle,* 215.

24. Simon, "Electrical," 708–9.
25. Ibid., 708, 709–10.
26. Ibid., 702; table, 716, 717. See also Jenkins, *Space Shuttle,* 215, 217.
27. Oberg, *Red Star,* 137–38 (includes quotations).
28. Jenkins, *Space Shuttle,* 202, 217; Powers, *Shuttle,* 79.
29. *Food Technology,* September 1977, 40–45; *Design News,* 22 October 1979, 48–50; *Space World,* August/September 1980, 6–7.
30. *Design News,* 22 October 1979, 50.
31. Jenkins, *Space Shuttle,* 217–18; Gibb et al., *Challenges,* 476.
32. Prince et al., *Challenges,* 414–17.
33. Gibb et al., *Challenges,* 478.
34. Prince et al., *Challenges,* 417–19.
35. Ibid., 419–24.
36. Gibb et al., *Challenges,* 469–70, 472–76.
37. Nason et al., *Challenges,* 451–53.
38. *Aviation Week,* 9 December 1974, 19; 26 January 1976, 126.
39. Jenkins, *Space Shuttle,* 207–8; Williams et al., *Challenges,* 480–81; Nason et al., *Challenges,* fig. 1, 452; *Aviation Week,* 23 July 1973, 61.
40. Nason et al., *Challenges,* 457, 463–64.
41. Felder, *Space Shuttle,* 859; Campbell, *Wheel,* 854; *Aviation Week,* 12 June 1989, 269.
42. Felder, *Space Shuttle,* 858; Jenkins, *Space Shuttle,* 258; Crickmore, *SR-71,* 194; also photos throughout.
43. Jenkins, *Space Shuttle,* 229; Campbell, *Wheel;* Felder, *Space Shuttle.*
44. Jenkins, *Space Shuttle,* 228–30; Carsley, *Space Shuttle.*
45. Malkin to Yardley, memo, 18 June 1974; Frutkin to Fletcher, memo, 19 June 1974. For artwork, see *Aviation Week,* 19 June 1972, 37.
46. Frutkin to Fletcher, memo, 19 June 1974 (includes quotations).
47. "Development" (NASA memo).
48. Ussher and Doetsch, *Overview,* 892; Low, Personal Notes No. 137, 1 February 1975 (includes quotations).
49. *Aviation Week,* 1 November 1976, 38–39; "Development" (NASA memo); Ussher and Doetsch, *Overview,* 901.
50. Ussher and Doetsch, *Overview;* Jenkins, *Space Shuttle,* 209.
51. *Aviation Week,* 15 May 1978, 54 (includes quotations).
52. "Development" (NASA memo; includes quote); Ussher and Doetsch, *Overview,* 892, 902.
53. NASA SP-4219, 304–5.
54. McMann and McBarron, *Challenges; Aviation Week,* 15 August 1977, 37–40.
55. *American Heritage of Invention & Technology,* Spring 1998, 55; Chaikin, *Man,* 23, 609; Heppenheimer, *Countdown,* 231–32.
56. Jenkins, *Space Shuttle,* 217.
57. *Aviation Week,* 15 August 1977, 37–40.
58. Ibid., 29 October 1979, 45; 18 August 1980, 65–68.
59. NASA SP-4219, 313–18.
60. Heppenheimer, *Countdown;* see index references.
61. Donlan, *Requirements,* 21 April 1972 (quotation, 6); NASA SP-4221, 429.
62. *Technology Review,* March/April 1977, 18, 19; *Aviation Week,* 15 October 1979, 39. Quote: "Shuttle Crew Abort/Escape" (NASA memo, MHQ-1, 12 January 1976).
63. *Aviation Week,* 15 October 1979, 39–45; Jenkins, *Space Shuttle,* 265–66.
64. Donlan, *Requirements,* 21 April 1972, 2.
65. *Technology Review,* March/April 1977, 17–18; Jenkins, *Space Shuttle,* 248; AIAA Paper 79-1139.
66. Biggs interview, 15 November 2000; Jenkins, *Space Shuttle,* 213–14; AIAA Paper 79-1144.
67. *Technology Review,* March/April 1977, 18; Jenkins, *Space Shuttle,* 215, 216.
68. Nason, *Annual,* i, 1–2; Biggs interview, 15 November 2000.
69. Quotations: Nason, *Annual,* 27, 32.
70. *IEEE Spectrum,* June 1989, 26–27, 44.
71. Report JSC-09990C (NASA), 2-2, 5-1.

72. Ibid., A-5; *Aviation Week,* 30 April 1973 (artwork, 61); Jenkins, *Space Shuttle,* 134; Baker, *Spaceflight,* 284.
73. Report JSC-09990C (NASA), 1-1, 2-1, 5-2 to 5-2a.
74. Ibid., sec. 5, passim.
75. Ibid., 5-3. Quote: Donlan, *Requirements,* 21 April 1972, 6.
76. "Product Assurance for Shuttle" (NASA memo; no author, no date).
77. *IEEE Spectrum,* June 1989, 42–43 (quotations, 42).
78. NASA SP-4201, 180 (includes quotation).
79. Biggs interview, 15 November 2000; Biggs, "Main Engine," 114–16; Report CP320R0003B (Rocketdyne), table 3, p. 67.
80. AIAA Papers 79-1141, 80-1129; Biggs interview, 15 November 2000.
81. *IEEE Spectrum,* June 1989, 45 (includes quotation).
82. Jenkins, *Space Shuttle,* 269, 275.
83. Vaughan, *Challenger,* 5. Quote: Fletcher, *Selection,* 7.
84. Rogers, *Report,* chap. 6. Reprinted in NASA SP-4407, vol. 4, p. 367.
85. *IEEE Spectrum,* June 1989, 42–44; Jenkins, *Space Shuttle,* 279–82, 286–99.

8. The Electronic Shuttle

1. *Aviation Week,* 20 November 1967, 88–94, 192–202.
2. Jenkins, *Space Shuttle,* 198–201.
3. NASA SP-504, 5–6. 1972 digital flight: Jenkins, *Space Shuttle,* 171.
4. NASA SP-504, 3, 8–9; *Aviation Week,* 28 June 1971, 58–59.
5. NASA SP-504, 3, 9; Jenkins, *Space Shuttle,* 170–71.
6. Cooper and Chow, "Development," 11, 14, 16.
7. Ibid., 15–16; Jenkins, *Space Shuttle,* 170.
8. Cooper and Chow, "Development," 6, 7, 10–11; Jenkins, *Space Shuttle,* 173.
9. Chevers, *Avionics,* 30–31; NASA SP-504, 9; Jenkins, *Space Shuttle,* 171–72, 174–75.
10. NASA SP-504, 28.
11. Ibid., 29–32 (table, 30); Cooper and Chow, "Development," 17.
12. Pfeiffer, *Thinking,* 47–49 (quotations, 49); *Scientific American,* May 1975, 37.
13. Cooper and Chow, "Development," 6; NASA SP-504, 25–26.
14. Jenkins, *Space Shuttle,* 173; Heppenheimer, *Turbulent,* 285–86.
15. NASA SP-504, 32; Cooper and Chow, "Development," 18.
16. *Scientific American:* August 1973, 4, 54; May 1975, 37, 38.
17. Cooper and Chow, "Development," 17–19; Jenkins, *Space Shuttle,* 227. Microprogramming: *Scientific American,* May 1975, 36, 37.
18. *Scientific American,* May 1975, 32–40 (quotation, 39); *Science,* 18 March 1977, 1111, 1112.
19. *Science,* 18 March 1977, 1111; Cooper and Chow, "Development," 6.
20. *Scientific American,* May 1975, 32–33, 37.
21. NASA SP-504, vi, 10.
22. *Scientific American,* May 1975, 35 (includes computer codes).
23. Ibid., 38.
24. Ibid., 38–39; Jenkins, *Space Shuttle,* 172, 173.
25. *Aviation Week,* 8 November 1976, 78; NASA SP-504, 10. Jay Forrester: Jenkins, *Space Shuttle,* 172; Pfeiffer, *Thinking,* 47.
26. Jenkins, *Space Shuttle,* 172; *Datamation,* July 1978, 128.
27. *Datamation,* July 1978, 130–36.
28. Ibid., 130; Sklaroff, *Redundancy,* 23–25.
29. Sklaroff, *Redundancy,* 26; Jenkins, *Space Shuttle,* 171–72.
30. NASA SP-504, 12–13. 38 subsystems: Jenkins, *Space Shuttle,* 226; Cooper and Chow, "Development," 16.
31. Jenkins, *Space Shuttle,* 172; Dennis Jenkins, private communication.
32. AIAA Paper 81-2135 (table: fig. 1).

33. Ibid.; NASA SP-504, 11; Jenkins, *Space Shuttle,* 214.
34. AIAA Paper 81-2135 (quotation, 8).
35. Jenkins, *Space Shuttle,* 173–74.
36. Ibid., 174; AIAA Paper 81-2135.
37. Jenkins, *Space Shuttle,* 174–76; Chevers, *Avionics* (quotation, 30).
38. *American Heritage of Invention & Technology,* Winter 1992, 21–22.
39. Ibid., 23–24; NASA SP-4303, 218–220.
40. NASA SP-504, 4, 17; *Aviation Week,* 8 November 1976, 75; Jenkins, *Space Shuttle,* 199.
41. NASA TM 81349; NASA SP-2000-4224, 111–14.
42. NASA SP-504, 6–7, 42.
43. Ibid., 11, 39–41; Jenkins, *Space Shuttle,* 224–25; *Aviation Week,* 8 November 1976, 75.
44. Jenkins, *Space Shuttle,* 198–201; NASA SP-504, 35–38.
45. Jenkins, *Space Shuttle,* 225–26; NASA SP-504, 49–54.
46. *Aviation Week,* 27 August 1973, 21; 3 January 1977, 14–16; Thompson, *Space Log,* 207, 252, 255.
47. Jenkins, *Space Shuttle,* 226; NASA SP-504, 52.
48. NASA SP-504, 16. TACAN: Nolan, *Fundamentals,* 65–68.
49. Heppenheimer, *Turbulent,* 134, 287–89; *ICAO Bulletin,* August 1974, 41.
50. *ICAO Bulletin,* August 1974, 40–44.
51. Ibid., August 1974, 40–44; August 1975, 22–23; NASA SP-504, 24–25; *Astronautics & Aeronautics,* December 1974, 52–54.
52. *Aviation Week,* 10 June 1974, 45–47; 28 April 1975, 91–92.
53. NASA SP-504, v (includes quotation). F/A-18: Dennis Jenkins, private communication.
54. Heppenheimer, *Countdown,* 112, 116–18. Quotations: Dorrenbacher, 117; Ezell, 118.
55. NASA SP-4204, 62, 347–48; SP-4206, 242, 243; Tomayko, *Evolution.*
56. CASI 75A-40626; NASA SP-4204, 348; SP-4206, 235–36.
57. NASA SP-4204, 348–52; SP-4206, 328; Tomayko, *Evolution.* 60 percent: *Space/Aeronautics,* December 1965, 66.
58. NASA SP-4204, 350, 352–56; SP-4206, 237; Tomayko, *Evolution.*
59. NASA SP-4206, 239–40.
60. Tomayko, *Evolution;* NASA SP-4204, 359–62.
61. NASA SP-4204, 347, 356.
62. *Bulletin of the Atomic Scientists,* September 1969, 84–87 (quotation, 86).
63. *Space/Aeronautics,* December 1965, 70.
64. NASA SP-4206, 239; Tomayko, *Evolution,* 7.
65. NASA SP-4206, 160, 240, 414–19.
66. Tomayko, *Evolution* (quotation, 8); *Aviation Week,* 13 October 1975, 40–43. Apollo's five systems: CASI 75A-40626.
67. Tomayko, *Evolution.*
68. Byrne et al., *Launch;* CASI 75A-40626; Dennis Jenkins, private communication. Honeywell 66/80s: Bailey, *Launch,* 535.
69. *Aviation Week,* 13 October 1975, 42; 8 November 1976, 83; Dennis Jenkins, private communication. Photo: NASA SP-4206, 236.
70. CASI 77A-35323 (includes GOAL program); Dennis Jenkins, private communication.
71. CASI 83A-27474; Dennis Jenkins, private communication.
72. Bailey, *Launch; Aviation Week,* 13 October 1975, 40–41; Dennis Jenkins, private communication.
73. Retha Hart, interview with author, 3 September 1998 (includes quotation).

9. The Program Struggles

1. CASI 76A-19330 (quotation, 2).
2. CASI 72A-45194, 8.
3. CASI 76A-18650 (quotation, 6).

4. Ibid. Quote: CASI 72A-45194, 9.

5. CASI: 72A-45194; 76A-19330; 76A-18650.

6. CASI: 76A-18650; 72A-45194; 76A-19330.

7. CASI 76A-19330.

8. CASI 76A-18650; NASA SP-4206, 286–87; *COSMIC,* 337–38.

9. NASA SP-4206, 283–88; CASI: 72A-45194; 76A-18650.

10. CASI 76A-18650. Quote: CASI 76A-19330, 6.

11. NASA SP-4221, 420–22; Mandell, *Management* (quotations, 47–48).

12. Mandell, *Management;* Fuhrman, *Fleet;* AIAA Paper 88-4516.

13. Mandell, *Management;* NASA SP-4307, 228; Dennis Jenkins, private communication.

14. Agnew, *Post-Apollo,* 15; CASI 71N-18430.

15. NASA SP-4313, 431; *Astronautics & Aeronautics,* January 1970, 38, 42.

16. AIAA Paper 73-74; *Aviation Week,* 11 September 1972, 104.

17. Low, Memo for Record, 6 September 1973 (includes quotation).

18. Low to Yardley, memo, 10 October 1973; Low to Currie, 29 October 1973. Quotations: Currie to Senator Thomas McIntyre, 21 December 1973. $100 million: Snyder et al., "History of Space and Missile Systems Organization" (hereafter cited as SAMSO), 1 July 1973 to 30 June 1975, 167–68.

19. Memo for Record, Brig. Gen. Henry Stelling, 22 February 1974; Low to Fletcher, memo, 20 February 1974 (includes quotation).

20. Low to Fletcher, memo, 20 May 1974 (includes quotations).

21. Low, Memo for Record, 17 June 1974. Shuttle as national system: Yardley to Low, memo, 27 June 1974, with enclosures.

22. Low, Memo for Record, 17 July 1974. Quote: Clements to Fletcher, 8 August 1974.

23. Thompson, *Space Log,* 51, 52, 66, 86, 153.

24. AAS Paper 75-292 (table, 751).

25. *Aviation Week,* 26 May 1975, 19; *Journal of the British Interplanetary Society* 35 (1982): 357.

26. AIAA Paper 73-74; Ley, *Rocket,* 496.

27. *Aviation Week,* 11 November 1974, 22–23; Snyder et al., SAMSO, 1 July 1973 to 30 June 1975, 169–70.

28. Quotations: Stelling: *Aviation Week,* 7 April 1975, 44; Low: Personal Notes No. 150, 9 August 1975, 7.

29. Snyder et al., SAMSO, 1 July 1975 to 31 December 1976, 106; *Aviation Week,* 15 September 1975, 20–23 (quotation, 23).

30. Snyder et al., SAMSO, 1 July 1975 to 31 December 1976, 105–13; *Aviation Week,* 27 October 1975, 14–16; 16 August 1976, 23–24.

31. *Aviation Week,* 8 November 1976, 140–41; 15 November 1976, 63.

32. For overviews of communications satellites, see *Science,* 18 March 1977, 1125–33; *Scientific American,* February 1977, 58–73.

33. Snyder et al., SAMSO, 1 July 1975 to 31 December 1976, 110; *Aviation Week,* 31 May 1976, 61; 6 September 1976, 46.

34. *Aviation Week,* 17 November 1980, 82–83.

35. Ibid., 8 November 1976, 143–44; 6 December 1976, 20.

36. Ibid., 21 February 1977, 19; 26 September 1977, 24.

37. Ibid., 15 May 1978, 48–53.

38. Bricker and Warren, "Directorate," 1 July to 31 December 1977, 16–17.

39. For a post-Challenger critique, see Report NSIAD-90-192 (General Accounting Office), July 1990.

40. Data from the following sources: NASA SP-4012, vol. 3, pp. 9, 69; NASA SP-4407, vol. 1, p. 586; *Aviation Week,* 30 January 1978, 28–33; 22 January 1979, 16; 29 January 1979, 24–26; 28 January 1980, 21; 4 February 1980, 27–29; 19 January 1981, 25–28; 15 February 1982, 24–26. Value of 1971 dollar from *Statistical Abstract* (1999), 493.

41. NASA SP-4221, 288, 331.

42. *Aviation Week,* 11 February 1974, 23–25.

43. Lynn to Fletcher, 25 July 1975.

44. Fletcher to Lynn, 30 September 1975. Quote: Low, Personal Notes No. 152, 21 September 1975.

45. Note, Fletcher to Low, 14 October 1975 (includes quotation).

46. Low to Mitchell, 30 October 1975; *Aviation Week,* 26 January 1976, 30.
47. Draft, 10 August 1976 (internal NASA paper).
48. Lynn to Fletcher, 19 August 1976.
49. Fletcher to Lynn, 14 September 1976.
50. NASA SP-4012, vol. 3, p. 9. For inflation, see *Statistical Abstract* (1999), 493.
51. Chapter 1 of this book.
52. *Aviation Week,* 1 September 1975, 46–47 (quotations, 46).
53. Ibid., 1 September 1975, 46–47 (staffer quotation, 46); 22 September 1975, 47–50 (NASA quotation, 48).
54. Covert, *Technical Status;* Yardley to Lovelace, memo, 18 August 1977.
55. *Aviation Week,* 1 September 1975, 46; 22 September 1975, 48, 50. 1972 study: Staats, *Cost-Benefit.*
56. *Aviation Week,* 8 November 1976, 39; Roland, *Shuttle,* 38; Report PSAD-76-73 (GAO), 43; *Science,* 23 November 1979, 910.
57. *Aviation Week,* 8 November 1976, 39.
58. Ibid., 64–66 (quotation, 64); NASA SP-4206, 347–60, 414, 416.
59. *Aviation Week,* 8 November 1976, 64–66. Quotations: Kraft, 65; Douglas, 66.
60. Ibid.; quotation, 64.
61. Report PSAD-76-73 (GAO), 44.
62. Ibid., 39–40 (quotation, 40); *Aviation Week,* 8 November 1976, 64–65.
63. Report PSAD-76-73 (GAO), 43 (includes quotation).
64. Yardley to Kraft, 9 April 1975 (quotation: paragraph III-7 of attachment). For Covert committee, see chapter 4 of this book.
65. Report PSAD-76-73 (GAO), 44.
66. *Science,* 23 November 1979. Quotes, 910.
67. Quotations: Douglas: *Aviation Week,* 8 November 1976, 66; NASA: *Science,* 23 November 1979, 910.
68. *Space Transportation System* (GAO staff study), 21–22; Report PSAD-77-113 (GAO), 5; *Aviation Week,* 13 June 1977, 96. See also chapter 1 of this book.
69. Report PSAD-76-73 (GAO): quotation, 30–31.
70. *Space Transportation System* (GAO staff study), 18–19; quotation, 23.
71. Report PSAD-76-73 (GAO), 7–8, 27, 31–32, 96, 98. Table, 32.
72. Ibid., 7–8 (quotations, 8).
73. Ibid., ii–iii, 37, 50–53. Quotation, 37.
74. Report PSAD-77-113 (GAO), 6.
75. Low: Personal Notes No. 163, 21 March 1976 (quotation, 1); No. 167, 4 June 1976. Lovelace: NASA SP-4012, vol. 3, p. 378; SP-4307, 228.
76. *New York Times,* 24 June 1977, p. 9; *Aviation Week,* 20 June 1977, 79–80.
77. "NASA/DOD Space Shuttle Orbiter Fleet Size Analysis," 15 May 1973. Quote: "Joint NASA/DOD Position Statement," 23 January 1976 (reprinted in NASA SP-4407, vol. 2, p. 389).
78. Richard Irwin, Memo for Record, 19 February 1975.
79. *Aviation Week,* 26 January 1976, 30; Jenkins, *Space Shuttle,* table, 178; Neal, *Bumped,* 12.
80. Report PSAD-77-113 (GAO), chap. 7. Quotes, 56–57, 63. See also *Aviation Week,* 13 June 1977, 87–89.
81. Mark, *Space Station,* 71–73.
82. Neal, *Bumped,* 13–14, 24, 27. Weights: ibid., 40; Jenkins, *Space Shuttle,* table, 178.
83. Neal, *Bumped,* 41–42; Frosch to Carter, 16 December 1977; Frosch to McIntyre, 21 December 1977. Quote: McIntyre to Frosch, 23 December 1977.
84. *Aviation Week,* 13 June 1977, 96; Yardley, "Office," 6.
85. *Aviation Week,* 19 October 1977, 26 (includes quotation).
86. *National Journal:* 31 December 1977, 1988–1989; 16 June 1979, 984 (table, 991).
87. Ibid., 16 June 1979, 986–88, 991–93. Afghanistan: ibid., 1 November 1980, 1852, Malia, *Soviet,* 379–80; Walker, *Cold War,* 251.
88. Mark, *Space Station,* 76–79. B-1 bomber: *National Journal:* 9 July 1977, 1087; 21 January 1978, 100–102.
89. See the prologue of this book.
90. *Aviation Week,* 4 June 1979, 21.
91. Ibid., 1 January 1979, 14; 22 January 1979, 16; 26 February 1979, 21.

92. Ibid., 8 January 1979, 12; 23 April 1979, 21; 7 May 1979, 18 (includes quotation); 4 June 1979, 20–21.

93. Ibid., 6 August 1979, 21; Carter to Frosch, memo, 12 July 1979 (includes quotation).

94. Frosch and McIntyre to Carter, memo, 11 September 1979 (includes quotation); *Aviation Week,* 1 January 1979, 14; 3 September 1979, 16; 17 September 1979, 22.

95. *Aviation Week,* 22 October 1979, 21–22. Quotations: Lovelace, 21; Anders, 22.

96. Quotations: Wilhelm interview; *Science,* 23 November 1979, 911.

97. *Aviation Week,* 19 November 1979, 16–18; Mark, *Space Station,* 101–3; Hans Mark, interview with author, 25 June 1998 (includes quotation).

98. *Aviation Week,* 21 January 1980, 51; 28 January 1980, 21; 4 February 1980, 27.

99. Ibid., 10 March 1980, 20; 17 March 1980, 16–17. For inflation, see *Statistical Abstract* (1999), 493. For overview of NASA budget, see *Aviation Week,* 4 February 1980, 27–29.

100. *Aviation Week,* 7 July 1980, 23.

101. Ibid., 13 November 1972, 18; 15 January 1973, 17.

102. Ibid., 30 January 1978, 28; 29 January 1979, 24; 4 February 1980, 28; 15 February 1982, 24. Numbers derived from these sources, added together, give $826 million.

103. Robert F. Thompson to Charles Donlan, letter with enclosure, 22 April 1981; communicated privately by Donlan. For $5.15 billion, see NASA SP-4221, 422.

104. Robert F. Thompson to Charles Donlan, letter with enclosure, 22 April 1981 (includes tables); communicated privately by Donlan.

105. NASA SP-4221, 260–64.

106. Smith, *Telescope,* 14–18, 30–31. Quote, v.

107. Ibid., 51; AIAA Paper 72-201; NASA SP-4313, 477.

108. Smith, *Telescope,* 74–78; NASA SP-4313, 139, 157, 476.

109. Smith, *Telescope,* 87; NASA SP-4313, 479–80.

110. Smith, *Telescope,* 43; NASA SP-4313, 481–83.

111. Smith, *Telescope,* 152–54. Quote: *Science,* 8 April 1983, 172.

112. NASA SP-4313, 488 (includes all quotations).

113. Ibid., 479–80, 490, 502.

114. Ibid. Quotations: Lucas, 498; Dunar and Waring, 503.

115. Ibid., 490 (includes quotations); Smith, *Telescope,* 264–65.

116. *Science,* 8 April 1983, 173.

117. NASA SP-4313, 495–96, 498, 500; *Science,* 8 April 1983, 172–73.

118. All quotations: NASA SP-4313, 503.

119. Ibid., 499–500, 504–5 (quotation, 504); *Science,* 8 April 1983, 173.

120. Smith, *Telescope,* 10–11; NASA SP-4313, 509.

121. NASA SP-4313, 509–16 (quotation, 511); *Science,* 17 August 1990, 735–36.

122. Hanley et al., "History of Space and Missile Systems Organization" (hereafter cited as SAMSO), 1 January to 31 December 1977, 106–7; 1 January to 31 December 1978, 112.

123. *Aviation Week,* 14 August 1978, 16–17; 4 September 1978, 104; Hanley et al., SAMSO, 1 January to 31 December 1978, 97–98, 113–55; "History of Space Division," 1 October 1979 to 30 September 1980, 92–95.

124. Hanley et al., SAMSO, 1 January to 31 December 1977, 95–96; "History of Space Division," 1 October 1980 to 30 September 1981, 127, 130–33. Illustrations: *Air & Space,* March 1997, 68–71; Jenkins, *Space Shuttle,* 164.

125. "History of Space Division," 1 October 1981 to 30 September 1982, 90–92.

126. Ibid., 1 October 1979 to 30 September 1980, 98; 1 October 1981 to 30 September 1982, 92–93, 94–95.

127. Jenkins, *Space Shuttle,* 163; *Air & Space,* March 1997, 72. Titan IV: *Aerospace America,* July 1991, 34–38.

128. Hanley et al., SAMSO, 1 January to 31 December 1977, 64–66; 1 January to 31 December 1978, 70–72, 79–81; *Aviation Week,* 15 May 1978, 48–53.

129. Hanley et al., SAMSO, 1 January to 31 December 1977, 65; "History of Space Division," 1 October 1980 to 30 September 1981, 92; *Aviation Week,* 3 September 1979, 22–23.

130. Hanley et al., SAMSO, 1 January to 31 December 1977, 82–83; *Aviation Week,* 8 November 1976, 140–41; 2 January 1978, 14.

131. Hanley et al., SAMSO, 1 January to 30 September 1979, 78–80; "History of Space Division," 1 October 1979 to 30 September 1980, 79–81; *Aviation Week,* 16 July 1979, 53–55.

132. Hanley et al., SAMSO, 1 January to 31 December 1978, 100–101; 1 January to 30 September 1979, 78, 81; "History of Space Division," 1 October 1979 to 30 September 1980, 84.

133. Hanley et al., SAMSO, 1 January to 30 September 1979, 80–81; "History of Space Division," 1 October 1979 to 30 September 1980, 81–84, 86. Quotations: ibid., 1 October 1980 to 30 September 1981, 106.

134. Hanley et al., SAMSO, 1 January to 31 December 1978, 100–101; "History of Space Division," 1 October 1979 to 30 September 1980, 87–88; 1 October 1980 to 30 September 1981, 105, 108–10.

135. "History of Space Division," 1 October 1980 to 30 September 1981, 113–16, 123; Thompson, *Space Log,* 203.

136. For an overview of planetary exploration, see Heppenheimer, *Countdown,* index references.

137. Bonestell's art: Miller, *Worlds.* NASA photos: Dixon, *Universe.*

138. Thompson, *Space Log,* 130, 136, 166; *American Heritage of Invention & Technology,* Winter 2001, 46–58.

139. *Science,* special issues: 1 June 1979, 23 November 1979.

140. *Aviation Week,* 24 January 1977, 24–25. Galileo the scientist; Durant, *Age of Reason,* 603–6.

141. Weights: Thompson, *Space Log,* 153, 166; *Aviation Week,* 3 September 1979, 53.

142. *Astronautics & Aeronautics,* December 1978, 36. Quotations: *Science,* 23 November 1979, 912.

143. *Aviation Week,* 3 September 1979, 53–57; 29 October 1979, 15; 5 November 1979, 20.

144. Ibid., 1 December 1980, 16–17; "History of Space Division," 1 October 1980 to 30 September 1981, 110–11.

145. *Aviation Week,* 26 November 1979, 18; 7 January 1980, 16–17 (includes quotations).

146. "History of Space Division," 1 October 1980 to 30 September 1981, 111–13; *Aviation Week,* 2 February 1981, 38; Thompson, *Space Log,* 260.

147. Quote: *Aviation Week,* 10 September 1979, 26.

10. Preparation for Flight

1. *Aviation Week,* 6 April 1970, 61–67; *Time:* 8 January 1965, 25; 14 March 1969, 42. Diamant: Thompson, *Space Log,* 116. Dreyfus: Tuchman, *Proud Tower,* chap. 4.

2. *Aviation Week,* 7 January 1980, 18 (includes quotations).

3. Ibid., 1 October 1973, 35; 6 August 1979, 17; 7 January 1980, 18.

4. Ibid., 2 June 1980, 17; 23 June 1980, 16; 27 October 1980, 25; "Arianespace."

5. *Aviation Week,* 25 June 1979, 91. Atlas Centaur: ibid., 8 January 1996, 116–17.

6. Ibid., 5 March 1979, 38; 9 July 1979, 18; 3 March 1980, 89; 1 December 1980, 22.

7. Beddingfield, "Facilities"; *Space World,* October 1977, 24–27.

8. Beddingfield, "Facilities," 2-1, 3-26 to 3-29 (quotations, 3-26, 3-29).

9. Ibid.; *Space World,* October 1977, 24–27. Seven hundred tons in S-IC: "Saturn V" (NASA-MSFC poster). "Beanie cap": *Aviation Week,* 14 July 1980, 112.

10. *Aviation Week,* 13 January 1975, 45.

11. Ibid., 23 February 1976, 25; 8 November 1976, 131; 22 November 1976, 63.

12. Ibid., 22 November 1976, 62–65; 5 December 1977, 43–44.

13. Ibid., 2 June 1980, 48; Beddingfield, "Facilities," 3–45.

14. NASA SP-4206, 414–19.

15. All quotations in Donald C. Cheatham to Manager, Space Shuttle Program, memo, 4 February 1974. Ashley: *Who's Who in Aviation,* 9. Flax: ibid., 89; NASA SP-4221, 362–63. Donovan, Love, Perkins: *AIAA Roster 1973.*

16. "MSF Management Council," NASA-OMSF memo, 7 February 1974.

17. Yardley to Petrone, memo, 18 July 1974.

18. NASA SP-4208, 361–63; *Science,* 7 April 1978, 28–29; *Aviation Week,* 18 October 1976, 18.

19. *Aviation Week,* 7 November 1977, 24–25 (quotation, 24); *Science,* 7 April 1978, 29, 32–33.

20. NASA SP-4208, 363–65; *Aviation Week,* 19 June 1978, 30–31.

21. *Science,* 12 January 1979, 153; NASA SP-4208, 367. Weight: Thompson, *Space Log,* 137. 28 September date: *Aviation Week,* 1 January 1979, 14.

22. NASA SP-4208, 370–71. Skylab orbit: *Science,* 7 April 1978, map, 32.

23. *Aviation Week,* 8 November 1976, 128–33 (list, 133); 31 March 1980, 54. X-15: Miller, *X-Planes,* 106–8.

24. Reports PSAD-76-73 (GAO), 58 (includes quotation); MASAD-82-15 (GAO), 15.

25. Low to Fletcher, memo, 28 May 1976 (includes table).

26. Memorandum of Agreement: NASA SP-4407, vol. 2, pp. 386–88 (includes quotation); *Astronautics & Aeronautics,* January 1977, 36.

27. *Aviation Week,* 29 October 1979, 15.

28. Ibid., 31 March 1980, 54–55; *Flight International,* 13 December 1980, 2182.

29. Jenkins, *Space Shuttle,* 268–69, 286.

30. Report MASAD-82-15 (GAO), 7. For inflation, see *Statistical Abstract* (1999), 493. Date of 1972: CASI 76A-18650, 4.

31. President Reagan, National Security Decision Directive No. 181, 30 July 1985. In NASA SP-4407, vol. 4, pp. 452–53 (quotation, 452).

32. Donlan, *Requirements,* 21 April 1972, 7 (includes quotation); "STS Schedule Baseline" (Washington, D.C.: NASA Headquarters, STS Operations, 7 June 1979).

33. Thompson, *Space Log,* 41–43; *Aviation Week,* 8 January 1996, 114–17.

34. Land and Malkin, *Flight Test,* 2-1, 2-2, A-1; Jenkins, *Space Shuttle,* 270, 271.

35. *Aviation Week,* 23 February 1976, 25; Atkinson and Shafritz, *Real Stuff,* 134.

36. Wolfe, *Right Stuff,* chap. 5.

37. Atkinson and Shafritz, *Real Stuff,* 134–38 (includes quotations).

38. *Aviation Week,* 29 October 1973, 18 (includes quotation).

39. Atkinson and Shafritz, *Real Stuff,* 138–43 (Fletcher quotation, 139); *Aviation Week,* 7 April 1975, 50 (includes Young quotations).

40. Schmitt: Chaikin, *Man,* chap. 13. Skylab: NASA SP-4208, 387–88.

41. Atkinson and Shafritz, *Real Stuff,* 143 (includes quotation).

42. Ibid., 145, 147, 150–51, 175. Quotations: Slayton, 155–56; Fletcher, 139, 156.

43. Ibid., 151, 153–55 (quotation, 155).

44. Ibid., 142–43, 146, 157, 163, 167–72 (quotation, 157); *Aviation Week,* 28 November 1977, 20; 23 January 1978, 18–19.

45. Thompson: *Space Log,* 208, 233; *Space Log 1996,* 332.

46. *Aviation Week,* 20 March 1978, 26; Cassutt, *Who's Who,* 44–45, 136–37 (quotation, 45).

47. Chaikin, *Man,* 585–94; NASA SP-4208, 387–88.

48. *Aviation Week,* 20 March 1978, 26; Cassutt, *Who's Who,* alphabetical listings.

49. CASI 81A-21726; *Aviation Week,* 14 August 1978, 50–59 (quotation, 57). Grumman Gulfstream: *Flight International,* 23 October 1976, 1277. Neutral buoyancy facility: NASA SP-4208, 170, 274–75; *Aviation Week,* 30 April 1979, 135.

50. Bird, *Design* (includes list); *Aviation Week,* 18 April 1977, 44–47.

51. *Air Line Pilot,* March 1978, 13; *Aviation Week,* 18 April 1977, 45.

52. Bird, *Design,* 250. Extended quotation: *Aviation Week,* 29 September 1980, 49.

53. *Aviation Week,* 30 April 1979, 125–35; *Flight International,* 13 December 1980, 2180–81.

54. Bird, *Design,* 250–51; *Aviation Week,* 14 May 1979, 47.

55. *Aviation Week,* 15 October 1979, 39–45; 29 September 1980, 48–62 (quotation, 56).

56. Ibid., 30 April 1979, 27; 7 May 1979, 12–17; 21 May 1979, 21; 28 May 1979, 61–65; 18 June 1979, 54–55.

57. Ibid., 29 October 1979, 16 (includes quotation).

58. Ibid., 10 December 1979, 21 (includes quotation).

59. Ibid., 5 November 1979, 18–19; 21 January 1980, 71 (includes quotation).

60. Ibid., 7 April 1980, 39–41; 19 May 1980, 119.

61. Ibid., 14 July 1980, 17; 4 August 1980, 24; 11 August 1980, 27–28.

62. Ibid., 10 November 1980, 17–20 (quotation, 19).

63. Ibid., 1 December 1980, 18–19; 8 December 1980, 23; 15 December 1980, 21.

64. Ibid., 5 January 1981, 12–13, 51–55 (quotation, 54); 19 January 1980, 99–103.

65. Ibid., 2 February 1981, 18; 9 February 1981, 24–26, 73–79; 16 February 1981, 20.

66. Quote: Rosen, *Viking,* 185.

67. Engine thrusts: Viking: Rosen, *Viking,* 245; shuttle: Report RI/RD87-142 (Rocketdyne), 1-5.

68. Riles, *Space Shuttle;* Report RSS-8656-1 (Rocketdyne); *Aviation Week,* 2 March 1981, 17–19 (quotation, 18).

69. *Aviation Week,* 9 March 1981, 253; 16 March 1981, 19; 6 April 1981, 19.
70. Ibid., 6 April 1981, 18–19 (quotation, 19).
71. Ibid., 16–20. Quote, 16; schedule, 20.
72. Ibid., 20 April 1981, 20–21 (quotation, 21). Gagarin: Thompson, *Space Log,* 55.

Coda: The Working Shuttle

1. *Aviation Week,* 20 April 1981, photos on 18, 20; Jenkins, *Space Shuttle,* 270. Conrad: *Los Angeles Times,* 15 April 1981, pt. 2, p. 13.
2. Jenkins, *Space Shuttle,* 270; Heppenheimer, *Countdown,* 315–16 (includes quotations).
3. Jenkins, *Space Shuttle,* 270; Thompson, *Space Log,* 204.
4. Heppenheimer, *Countdown,* 320. Quote: *Science,* 28 October 1983, 405.
5. Jenkins, *Space Shuttle,* 178, 268–69, 273–75. Challenger crew: ibid., 260; Atkinson and Shafritz, *Real Stuff,* 171.
6. Rogers, *Report;* Heppenheimer, *Countdown,* 323–26.
7. Heppenheimer, *Countdown,* 321, 322–23, 326–27; Statement by the President (White House), 15 August 1986 (includes quotation).
8. Thompson, *Space Log,* 252, 256, 260.
9. Dixon, *Universe,* 38–43; *Scientific American:* August 1980, 54–65; March 1999, 50–57.
10. *Scientific American:* January 1980, 88–100; December 1983, 56–67; December 1995, 44–51; October 1999, 54–63; February 2000, 40–49.
11. Jenkins, *Space Shuttle,* 288, 295; Thompson: *Space Log 1993,* 23–25; *Space Log 1997–1998,* 6–7.
12. Heppenheimer, *Countdown,* 350; *Scientific American:* June 1992, 44–51; November 1992, 54–60; January 1999, special issue. See also *Time:* 6 March 1995, 76–84; 20 November 1995, 90–99.
13. NASA SP-4221, 60–73, 131, 223, 227–30; Heppenheimer, *Countdown,* 313–19, 320; *Science,* 26 July 1985, 370–71.
14. Jenkins, *Space Shuttle,* 286, 290; Heppenheimer, *Countdown,* 335–37, 340–41.
15. Heppenheimer, *Countdown,* 341–44; Thompson: *Space Log,* 46, 234; *Space Log 1996,* 60.
16. Thompson, *Space Log 1995,* 17–18, 47, 75; Jenkins, *Space Shuttle,* 297–98; *Scientific American,* May 1998, 46–55.
17. "International Space Station." Internet address: www.shuttlepresskit.com. For Boeing 747, see *Pedigree,* 66.
18. Thompson, *Space Log 1997–1998,* 58, 61.
19. "Early Assembly Flights." Internet address: www.shuttlepresskit.com.
20. "International Space Station Assembly Sequence; Revision F (August 2000)." Internet address: www.spaceflight. nasa.gov/station/assembly/flights.chron.html.
21. NASA SP-4221, 72–73, 92–94.

Bibliography

BOOKS, REPORTS, AND PAPERS

Agnew, Spiro, Chairman. *The Post-Apollo Space Program: Directions for the Future.* Washington, D.C.: Space Task Group, September 1969.

AIAA Roster 1973. New York: AIAA.

Ambrose, Stephen E. *Nothing Like It in the World.* New York: Simon & Schuster, 2000.

Anderson, John D. *A History of Aerodynamics.* New York: Cambridge University Press, 1997.

"Arianespace Launches to Date." Washington, D.C.: Arianespace, 5 February 1996.

Atkinson, Joseph D., Jr., and Jay M. Shafritz. *The Real Stuff: A History of NASA's Astronaut Recruitment Program.* New York: Praeger, 1985.

Bacchus, D. L., D. A. Kross, and R. D. Moog. "The Aerodynamic Challenges of SRB Recovery." In Chaffee, *Conference,* 189–208.

Bailey, William W. "Launch Processing System— Concept to Reality." In Chaffee, *Conference,* 532–38.

Baker, David. *Spaceflight and Rocketry: A Chronology.* New York: Facts on File, 1996.

Becker, Paul R. "Leading-Edge Structural Material System of the Space Shuttle." *American Ceramic Society Bulletin* 60, no. 11 (1981): 1210–14.

Beddingfield, S. T. "Space Transportation System Facilities and Operations: Kennedy Space Center, Fla." Typewritten report. N.d.

Biggs, Robert E. "Space Shuttle Main Engine: The First Ten Years." In Doyle, *History,* 69–122.

Bird, J. D. "Design Concepts of the Shuttle Mission Simulator." *Aeronautical Journal,* June 1978, 247–54.

Bricker, Col. J. C., and Mrs. L. M. Warren. "History of the Directorate of Space: DCS/Research and Development." Photocopy. Headquarters, United States Air Force.

Brown, Dee. *Hear That Lonesome Whistle Blow.* New York: Bantam, 1978.

Byrne, F., G. V. Doolittle, and R. W. Hockenberger. "Launch Processing System." *IBM Journal of Research and Development* 20, no. 1 (1976): 75–83.

Campbell, Carlisle C., Jr. "Orbiter Wheel and Tire Certification." In Chaffee, *Conference,* 850–56.

Cameron, Col. R. M., and Mrs. L. M. Warren. "History of the Directorate of Space: DCS/Research and Development." Continuing series. Photocopy. N.p.: Headquarters, United States Air Force.

Carsley, Renton B. "Space Shuttle Wheels and Brakes." In Chaffee, *Conference,* 872–82.

Cassutt, Michael. *Who's Who in Space.* New York: Macmillan, 1993.

Chaffee, Norman, comp. *Space Shuttle Technical Conference.* NASA Conference Publication 2342. Houston: Lyndon B. Johnson Space Center, 28–30 June 1983.

Chaikin, Andrew. *A Man on the Moon.* New York: Penguin Books, 1994.

Chapman, Dean R. "Computational Aerodynamics Development and Outlook." *AIAA Journal* 17, no. 12 (1979): 1293–1313.

Chevers, Edward S. "Shuttle Avionics Software Development Trials, Tribulations, and Successes: The Backup Flight System." In Chaffee, *Conference,* 30–34.

Chulick, M. J., L. C. Meland, F. C. Thompson, and H. W. Williams. "History of the Titan Liquid Rocket Engines." In Doyle, *History,* 19–35.

Cooper, A. E., and W. T. Chow. "Development of On-Board Space Computer Systems." *IBM Journal of Research and Development* 20, no. 1 (1976): 5–19.

COSMIC Software Catalog. NASA CR-191005. Athens: University of Georgia, 1993.

Bibliography

Covert, Eugene E., Chairman. *Second Review—Technical Status of the Space Shuttle Main Engine.* Washington, D.C.: National Academy of Sciences, February 1979.

—. *Technical Status of the Space Shuttle Main Engine.* Washington, D.C.: National Academy of Sciences, March 1978.

Crickmore, Paul F. *Lockheed SR-71 Blackbird.* London: Osprey Publishing, 1986.

Day, Leroy, Manager. *NASA Space Shuttle Summary Report.* Washington, D.C.: Space Shuttle Task Group, 19 May 1969. See also revised version, 31 July 1969.

"Development of the Remote Manipulator System (RMS) for the Space Shuttle: Cooperative Project Between NASA and National Research Council of Canada." NASA memo, unsigned. C. 1979–80.

Dill, Charlie C., J. C. Young, B. B. Roberts, M. K. Craig, J. T. Hamilton, and W. W. Boyle. "The Space Shuttle Ascent Vehicle Aerodynamic Challenges: Configuration Design and Data Base Development." In Chaffee, *Conference,* 151–76.

Dixon, Don. *Universe.* Boston: Houghton Mifflin, 1981.

Donlan, Charles J. *Space Shuttle Program Requirements Document Level I.* Revision no. 4. Washington, D.C.: Office of Manned Space Flight, NASA, 21 April 1972.

Dotts, Robert L., Donald M. Curry, and Donald J. Tillian. "Orbiter Thermal Protection System." In Chaffee, *Conference,* 1062–81.

Doyle, Stephen E., ed. *History of Liquid Rocket Engine Development in the United States.* American Astronautical Society History Series, vol. 13. San Diego: Univelt, 1992.

Durant, Will, and Ariel Durant. *The Age of Reason Begins.* New York: Simon and Schuster, 1961.

Emero, D. H. "Quarter-Scale Space Shuttle Design, Fabrication, and Test." *Journal of Spacecraft and Rockets* 17, no. 4 (1980): 303–10.

Felder, G. L. "The Space Shuttle Program from Challenge to Achievement: Space Exploration Rolling on Tires." In Chaffee, *Conference,* 857–60.

Fletcher, James C. "Selection of Contractor for Space Shuttle Program Solid Rocket Motors." 12 December 1973. Copies in NASA History Office and in James Fletcher Papers, University of Utah, Salt Lake City.

Frisby, Joyce C. "History of Detachment Forty-Three, Thiokol Corporation." Continuing series. N.p.: United States Air Force, Air Force Contract Management Division.

Fuhrman, R. A. "The Fleet Ballistic Missile System: Polaris to Trident." *Journal of Spacecraft and Rockets* 15, no. 5 (1978): 265–86.

Galloway, Jonathan F. *The Politics and Technology of Satellite Communications.* Lexington, Mass.: Lexington Books, 1972.

Gibb, John W., M. E. McIntosh, Steven R. Heinrich, Emory Thomas, Mike Steele, Franz Schubert, E. P. Koszenski, R. A. Wynveen, R. W. Murray, J. D. Schelkopf, and J. K. Mangialardi. "Other Challenges in the Development of the Orbiter Environmental Control Hardware." In Chaffee, *Conference,* 465–79.

Gibson, C., and C. Humphries. "Orbital Maneuvering System Design Evolution." In Chaffee, *Conference,* 639–55.

Glynn, Philip C., and Thomas L. Moser. "Orbiter Structural Design and Verification." In Chaffee, *Conference,* 345–56.

Graham, L. J., F. E. Sugg, and W. Gonzalez. "Nondestructive Evaluation of Space Shuttle Tiles." *Ceramic Engineering and Science Proceedings* 3 (1982): 680–97.

Gray, Mike. *Angle of Attack: Harrison Storms and the Race to the Moon.* New York: W. W. Norton, 1992.

Hallion, Richard P., ed. *The Hypersonic Revolution.* 3 vols. Bolling AFB: Air Force History and Museums Program, 1998.

Hammond, Allen L., William D. Metz, and Thomas H. Maugh II. *Energy and the Future.* Washington, D.C.: American Association for the Advancement of Science, 1973.

Hanley, Dr. Timothy C., Dr. Harry N. Waldron, Dr. Thomas S. Snyder, and Mrs. Elizabeth J. Levack. "History of Space and Missile Systems Organization." Continuing series. N.p.: U.S. Air Force, History Office, HQ 6592 Air Base Group, Los Angeles Air Force Station, Calif.

Hechler, Ken. *The Endless Space Frontier: A History of the House Committee on Science and Astronautics, 1939–1978.* American Astronautical Society History Series, vol. 4. San Diego: Univelt, 1982.

Heppenheimer, T. A. *Colonies in Space.* Harrisburg, Pa.: Stackpole Books, 1977.

—. *The Coming Quake: Science and Trembling on the California Earthquake Frontier.* New York: Times Books, 1988.

—. *Countdown: A History of Space Flight.* New York: John Wiley, 1997.

—. *Hypersonic Technologies and the National Aerospace Plane.* Arlington, Va.: Pasha Publications, 1990.

—. *Turbulent Skies: The History of Commercial Aviation.* New York: John Wiley, 1995.

"History of Space Division." Continuing series. N.p.: U.S. Air Force, Space Division, Chief of Staff, History Office.

Hughes, Robert W. "The Solid Rocket Booster Auxiliary Power Unit—Meeting the Challenge." In Chaffee, *Conference,* 690–701.

Hunley, J. D. "The Significance of the X-15." Unpublished manuscript. NASA-Dryden.

"Interim Contract with Thiokol Chemical Corporation." 12 February 1974. James Fletcher Papers, University of Utah, Salt Lake City.

Isakowitz, Steven J. *International Reference Guide to Space Launch Systems.* Washington, D.C.: American Institute of Aeronautics and Astronautics, 1991.

Jane's All the World's Aircraft. Arlington, Va.: Jane's Information Group, 1999–2000.

Jane's Fighting Ships. Arlington, Va.: Jane's Information Group, 2000–2001.

Jenkins, Dennis. *Space Shuttle: The History of Developing the National Space Transportation System.* Marceline, Mo.: Walsworth Publishing, 1996.

Johnson, Paul. *Modern Times: The World from the Twenties to the Eighties.* New York: Harper Colophon, 1985.

Johnson, Stephen B. "Launch Vehicles and the Development of Systems Engineering." Unpublished manuscript. University of North Dakota, Grand Forks.

Kreith, Frank. *Principles of Heat Transfer.* Scranton, Pa.: International Textbook Company, 1965.

Krige, John, and Arturo Russo. *Europe in Space 1960–1973.* ESA SP-1172. Washington, D.C.: European Space Agency, September 1994.

Korb, L. J., C. A. Morant, R. M. Calland, and C. S. Thatcher. "The Shuttle Orbiter Thermal Protection System." *American Ceramic Society Bulletin* 60, no. 11 (1981): 1188–93.

Lance, Renee, and Dwayne Weary. "Space Shuttle Orbiter Auxiliary Power Unit Development Challenges." In Chaffee, *Conference,* 673–89.

Land, E. W., Jr., and M. S. Malkin. "Space Shuttle Flight Test Operations Directive." Revision C. Washington, D.C.: Space Shuttle Program Office, Office of Space Transportation Systems, NASA, January 1979.

Launius, Roger. "A Western Intellectual in Washington: James C. Fletcher, NASA, and the Final Frontier." Unpublished manuscript. NASA History Office.

Leonardo da Vinci. *Notebooks.* 2 vols. New York: Reynal and Hitchcock, 1938.

Ley, Willy. *Rockets, Missiles, and Men in Space.* New York: Signet Books, 1969.

Logsdon, John M. *The Decision to Go to the Moon.* Cambridge, Mass.: MIT Press, 1970.

Lord, Walter. *A Night to Remember.* New York: Bantam, 1963.

Bibliography

Mackey, Alden C., and Ralph E. Gatto. "Structural Load Challenges During Space Shuttle Development." In Chaffee, *Conference*, 335–44.

Madders, Kevin. *A New Force at a New Frontier.* Cambridge: Cambridge University Press, 1997.

Malia, Martin. *The Soviet Tragedy: A History of Socialism in Russia, 1917–1991.* New York: Free Press, 1994.

Malkin, Myron S. *Space Shuttle Program Requirements Document Level I.* Revision no. 6. Washington, D.C.: Office of Manned Space Flight, NASA, 12 March 1974.

Manchester, William. *The Glory and the Dream: A Narrative History of America, 1932–1972.* Boston: Little, Brown, 1974.

Mandell, Humboldt C., Jr. *Management and Budget Lessons: The Space Shuttle Program.* 1989. NASA SP-6101 (02), 44–48.

Mark, Hans. *The Space Station: A Personal Journey.* Durham, N.C.: Duke University Press, 1987.

McConnell, Malcolm. *Challenger: A Major Malfunction.* Garden City, N.Y.: Doubleday, 1987.

McDougall, Walter. . . . *The Heavens and the Earth: A Political History of the Space Age.* New York: Basic Books, 1985.

McMann, Harold J., and James W. McBarron II. "Challenges in the Development of the Shuttle Extravehicular Mobility Unit." In Chaffee, *Conference,* 435–49.

Medaris, Maj. Gen. John B. *Countdown for Decision.* New York: G. P. Putnam's Sons, 1960.

Miller, Jay. *The X-Planes, X-1 to X-29.* Marine on St. Croix, Minn.: Specialty Press, 1983.

Miller, Ron, and Fred Durant. *Worlds Beyond: The Art of Chesley Bonestell.* Norfolk, Va.: Donning, 1983.

Modlin, C. Thomas, Jr., and George A. Zupp Jr. "Shuttle Structural Dynamics Characteristics—The Analysis and Verification." In Chaffee, *Conference,* 325–34.

Morgenstern, Oskar, and Klaus P. Heiss. *Economic Analysis of New Space Transportation Systems.* Princeton, N.J.: Mathematica, 31 May 1971.

—. *Economic Analysis of the Space Shuttle System.* Princeton, N.J.: Mathematica, 31 January 1972.

Munger, William P., and Robert Seaman. "XLR99-RM-1 Rocket Engine for the X-15." In *Research-Airplane-Committee Report on Conference on the Progress of the X-15 Project,* 215–24. Langley, Va.: Langley Aeronautical Laboratory (NACA), 25–26 October 1956.

Nason, Howard K., Chairman. *Annual Report to the NASA Administrator by the Aerospace Safety Advisory Panel on the Space Shuttle Program.* Washington, D.C.: NASA, March 1977.

Nason, John R., Frederic A. Wierum, and James L. Yanosy. "Challenges in the Development of the Orbiter Active Thermal Control Subsystem." In Chaffee, *Conference,* 450–64.

Neal, Valerie. *Bumped from the Shuttle Fleet: Why Didn't "Enterprise" Fly in Space?* Washington, D.C.: National Air and Space Museum, forthcoming.

Neufeld, Jacob. *The Development of Ballistic Missiles in the United States Air Force 1945–1960.* Washington, D.C.: Office of Air Force History, 1990.

Neufeld, Michael. *The Rocket and the Reich: Peenemunde and the Coming of the Ballistic Missile Era.* New York: Free Press, 1995.

Nolan, Michael S. *Fundamentals of Air Traffic Control.* Belmont, Calif.: Wadsworth Publishing, 1990.

Oberg, James E. *Red Star in Orbit.* New York: Random House, 1981.

Ordway, Frederick, and Mitchell Sharpe. *The Rocket Team.* New York: Thomas Y. Crowell, 1979.

Pace, Scott. "Engineering Design and Political Choice: The Space Shuttle 1969–1972." Master's thesis, MIT, May 1982.

Pedigree of Champions: Boeing Since 1916. Seattle: Boeing, 1985.

Pfeiffer, John. *The Thinking Machine.* New York: J. B. Lippincott, 1962.

Pinson, Larry D., and Sumner A. Leadbetter. "Some Results from 1/8–Scale Shuttle Model Vibration Studies." *Journal of Spacecraft and Rockets* 16, no. 1 (1979): 48–54.

Powers, Robert M. *Shuttle: The World's First Spaceship.* Harrisburg, Pa.: Stackpole Books, 1979.

Reinhardt, Walter A. "Parallel Computation of Unsteady, Three-Dimensional, Chemically Reacting, Nonequilibrium Flow Using a Time-Split Finite Volume Method on the Illiac IV." *Journal of Physical Chemistry* 81, no. 25 (1977): 2427–35.

Rice, Berkeley. *The C-5A Scandal.* Boston: Houghton Mifflin, 1971.

Richelson, Jeffrey. *America's Secret Eyes in Space: The U.S. Keyhole Spy Satellite Program.* New York: Harper & Row, 1990.

Riles, Lt. Col. Warren L. "Space Shuttle Flight Readiness Firing—Dress Rehearsal for STS-1." In Chaffee, *Conference,* 510–24.

Rogers, William P., Chairman. "Report at a Glance." Washington, D.C.: Presidential Commission on the Space Shuttle *Challenger* Accident, 6 June 1986. Reprinted in NASA SP-4407, vol. 4, pp. 356–75.

Roland, Alex. "The Shuttle: Triumph or Turkey?" *Discover,* November 1985, 29–49.

Rosen, Milton. *The Viking Rocket Story.* New York: Harper, 1955.

Ruffner, Kevin C., ed. *Corona: America's First Satellite Program.* Washington, D.C.: Central Intelligence Agency, 1995.

Rutten, Maj. Thomas W. "History of the Directorate of Space: DCS/Research and Development." Continuing series. U.S. Air Force: N.p.

Schmidt, David A., and C. R. Anderson. "Update Report on the Space Shuttle Program, National Aeronautics and Space Administration." House Appropriations Committee, U.S. House of Representatives, Washington, D.C., 11 April 1978.

Schneider, William C., and Glenn J. Miller. "The Challenging 'Scales of the Bird' (Shuttle Tile Structural Integrity)." In Chaffee, *Conference,* 403–13.

Schramm, Wilson. "HRSI and LRSI—The Early Years." *American Ceramic Society Bulletin* 60, no. 11 (1981): 1194–95.

Sebesta, Lorenza. "Blueprints for the Future: Collaboration on Human Flights." Unpublished manuscript. Copy courtesy of Prof. John Logsdon, George Washington University.

Serling, Robert J. *Legend and Legacy: The Story of Boeing and Its People.* New York: St. Martin's Press, 1992.

Shupe, Anson. *The Darker Side of Virtue: Corruption, Scandal, and the Mormon Empire.* Buffalo, N.Y.: Prometheus Books, 1991.

Simon, William E. "Space Shuttle Electrical Power Generation and Reactant Supply System." In Chaffee, *Conference,* 702–19.

Sklaroff, J. R. "Redundancy Management Technique for Space Shuttle Computers." *IBM Journal of Research and Development* 20, no. 1 (1976): 20–28.

Smith, Robert W. *The Space Telescope: A Study of NASA, Science, Technology, and Politics.* Cambridge: Cambridge University Press, 1989.

Snyder, Dr. Thomas S., Timothy C. Hanley, Robert J. Smith, SSgt. Ronald Benesh, and Lt. Col. Earl Goddard. "History of Space and Missile Systems Organization." Continuing series. N.p.: U.S. Air Force, History Office, HQ 6592 Air Base Group, Los Angeles Air Force Station, Calif.

Space Transportation System. Staff study. Washington, D.C.: U.S. General Accounting Office, February 1975.

Staats, Elmer. *Analysis of Cost Estimates for the Space Shuttle and Two Alternative Programs.* Washington, D.C.: U.S. General Accounting Office, 1 June 1973.

—. *Cost-Benefit Analysis Used in Support of the Space Shuttle Program.* Washington, D.C.: U.S. General Accounting Office, 2 June 1972.

—. *Decision in the Matter of Protest by Lockheed Propulsion Company.* Washington, D.C.: U.S. General Accounting Office, 24 June 1974. Reprinted in part in NASA SP-4407, vol. 4, pp. 268–72.

Statistical Abstract of the United States. Washington, D.C.: U.S. Department of Commerce, 1999.

Steiner, John. *Jet Aviation Development: One Company's Perspective.* Seattle: Boeing, 1989.

Stott, D., and M. Hempsell. "The European Space Tug: A Reappraisal." *Journal of the British Interplanetary Society* 34 (1981): 294–98.

Surber, T. E., and D. C. Olsen. "Space Shuttle Orbiter Aerodynamic Development." *Journal of Spacecraft and Rockets* 15, no. 1 (1978): 40–47.

Sutton, George P. *Rocket Propulsion Elements.* New York: John Wiley, 1986.

Sutton, George W. "Ablation of Reinforced Plastics in Supersonic Flow." *Journal of the Aero/Space Sciences* 27, no. 5 (1960): 377–85.

—. "The Initial Development of Ablation Heat Protection: An Historical Perspective." *Journal of Spacecraft and Rockets* 19, no. 1 (1982): 3–11.

Taeuber, Ralph J., W. Karakulko, D. Blevins, C. Hohmann, and J. Henderson. "Design Evolution of the Orbiter Reaction Control Subsystem." In Chaffee, *Conference,* 656–72.

Thirty Years of Rocketdyne. Canoga Park, Calif.: Rocketdyne Division, Rockwell International, 1985.

Thompson, Milton O., and Curtis Peebles. *Flying Without Wings: NASA Lifting Bodies and the Birth of the Space Shuttle.* Washington, D.C.: Smithsonian Institution Press, 1999.

Thompson, Tina D., ed. *TRW Space Log* (cited as *Space Log*), vol. 27. Redondo Beach, Calif.: TRW, 1991. See also *TRW Space Log 1993,* vol. 29; *TRW Space Log 1995,* vol. 31; *TRW Space Log 1996,* vol. 32; *TRW Space Log 1997–1998,* vol. 33–34.

Tomayko, James E. "The Evolution of Automated Launch Processing." *Aerospace Historian,* March 1988, 2–10.

Tuchman, Barbara. *The Proud Tower.* New York: Macmillan, 1966.

Tucker, Zola. "History of Detachment Forty-Three, Thiokol Corporation." Continuing series. Photocopy. N.p.: U.S. Air Force, Air Force Contract Management Division, Air Force Systems Command.

Turner, Sarah. "Maxime Faget: Father of the Space Shuttle." Unpublished manuscript. NASA History Division, n.d.

Ussher, T. H., and K. H. Doetsch. "An Overview of the Shuttle Remote Manipulator System." In Chaffee, *Conference,* 892–903.

Vaughan, Diane. *The Challenger Launch Decision.* Chicago: University of Chicago Press, 1996.

VerSnyder, Francis L., and M. E. Shank. "The Development of Columnar Grain and Single Crystal High Temperature Materials Through Directional Solidification." *Materials Science and Engineering* 6 (1970): 213–47.

Waldrop, Mitch. "Space Shuttle Tiles: A Question of Bonding." *Chemical & Engineering News* 58 (12 May 1980): 27–29.

Walker, Martin. *The Cold War: A History.* New York: Henry Holt and Company, 1993.

White, Theodore H. *The Making of the President 1972.* New York: Bantam, 1973.

Who's Who in Aviation. New York: Harwood & Charles, 1973.

Williams, J. L., M. F. Modest, J. A. Oren, and H. R. Howell. "Challenges in the Development of the Orbiter Radiator System." In Chaffee, *Conference,* 480–89.

Wolfe, Tom. *The Right Stuff.* New York: Farrar, Straus & Giroux, 1979.

Woodis, William R., and Roy E. Runkle. "Structural and Mechanical Design Challenges of Space Shuttle Solid Rocket Boosters Separation and Recovery Subsystems." In Chaffee, *Conference,* 365–85.

Yardley, John F. "Office of Space Flight Status Summary: Statement of John F. Yardley." Photocopy. Washington, D.C.: Subcommittee on Space Science and Applications, Committee on Science and Technology, U.S. House of Representatives, 5 October 1977.

—. "Reevaluation of Space Shuttle Carrier Aircraft Selection." 25 February 1975. Typescript. Leroy Day Papers, NASA History Office.

Young, James C., Jimmy M. Underwood, Ernest R. Hillje, Arthur M. Whitnah, Paul O. Romere, Joe D. Gamble, Barney B. Roberts, George H. Ware, William I. Scallion, Bernard Spencer Jr., James P. Arrington, and Deloy C. Olsen. "The Aerodynamic Challenges of the Design and Development of the Space Shuttle Orbiter." In Chaffee, *Conference,* 209–63.

BOOKS, REPORTS, AND PAPERS, BY SOURCE

AAS (American Astronautical Society)

75-292. J. H. Guill. "Agena Interim Upper Stage." August 1976. Reprinted in *Advances in the Astronautical Sciences,* vol. 32, pt. 2, pp. 737–57.

AIAA (American Institute of Aeronautics and Astronautics)

65-163. R. G. Sampson and A. P. Peters. *Status Report on 120-Inch Motor Design and Development.* February 1965.

71-443. R. G. Helenbrook, F. M. Anthony, and R. M. Fisher. *Selection of Space Shuttle Thermal Protection Systems.* April 1971.

71-659. Paul D. Castenholz. *Rocketdyne's Space Shuttle Main Engine.* June 1971.

71-662. Warner L. Stewart, Arthur J. Glassman, and Stanley M. Nosek. *Shuttle Airbreathing Propulsion.* June 1971.

71-804. J. F. Yardley. *McDonnell Douglas Fully Reusable Shuttle.* July 1971.

71-805. Bastian Hello. *Fully Reusable Shuttle.* July 1971.

71-812. W. R. Hook and W. T. Carey. *Opportunities for Research-and-Applications Investigations in Near-Earth Orbit.* July 1971.

71-813. W. L. Breazeale and J. T. Milton. *Conceptual Studies of Research and Applications Modules.* July 1971.

71-814. E. Stuhlinger and J. A. Downey. *The Concept of a General-Purpose Laboratory in Space.* July 1971.

71-816. Fritz C. Runge. *Shuttle Utilization for Sorties and Automated Payloads.* July 1971.

71-817. J. B. Lagarde. *European Interests and Studies of Earth Orbital Systems.* July 1971.

72-201. S. L. Morrison. *120-Inch Large Space Telescope.* January 1972.

73-31. Eugene S. Love. *Advanced Technology and the Space Shuttle.* Von Kármán Lecture, January 1973.

73-60. Paul D. Castenholz and Frank M. Kirby. *Technology Applied to the Space Shuttle Main Engine.* January 1973.

73-62. E. Roberts and D. Altman. *Large Solids for Shuttle.* January 1973.

73-74. W. G. Huber. *Space Tug.* January 1973.

74-991. T. E. Surber and D. C. Olsen. *Space Shuttle Orbiter Aerodynamic Development.* August 1974.

75-1170. John Thirkill. *Solid Rocket Motor for the Space Shuttle.* September 1975.

75-1172. D. N. Counter and B. C. Brinton. *Thrust Vector Control for the Space Shuttle Solid Rocket Motor.* September 1975.

75-1173. C. D. Nevins. *Designing for Solid Rocket Booster Reusability.* September 1975.

75-1298. W. W. Regnier and T. F. Schweickert. *Space Shuttle Orbital Maneuver Subsystem.* September 1975.

Bibliography

76-595. Lawrence Norquist. *External Tank for the Space Shuttle Main Propulsion System.* July 1976.

76-740. Dan David. *Development Status of the Rocket Engine for the Space Shuttle Orbital Maneuvering Subsystem (OMS).* July 1976.

77-808. J. R. Johnson. *Update on Development of the Space Shuttle Main Engine (SSME).* July 1977.

77-811. D. David. *Space Shuttle Maneuvering Subsystem (OMS) Rocket Engine Development Status Update—July 1977.* July 1977.

78-485. M. J. Suppans and C. J. Schroeder. *Space Shuttle Orbiter Thermal Protection Development and Verification Test Program.* April 1978.

78-950. T. L. Elegante and R. R. Bowman. *Nozzle Fabrication for the Space Shuttle Solid Rocket Motor.* July 1978.

78-951. A. R. Canfield, E. E. Anderson, and G. E. Nichols. *Space Shuttle Nozzle Development.* July 1978.

78-952. C. H. Krummel and O. N. Thompson. *Space Shuttle Solid Rocket Motor Metal Case Component Fabrication.* July 1978.

78-954. J. R. Kapp. *Design, Fabrication and Test of the Space Shuttle Solid Rocket Booster Motor Case.* July 1978.

78-986. J. Baker. *Solid Rocket Motor Grain Configuration Development.* July 1978.

78-987. C. W. Bolieau, J. Baker, and S. Folkman. *Space Shuttle SRM Ignition System.* July 1978.

78-1001. J. Johnson and H. Colbo. *Update on Development of the Space Shuttle Main Engine (SSME).* July 1978.

78-1002. M. C. Ek. *Solution of the Subsynchronous Whirl Problem in the High Pressure Hydrogen Turbomachinery of the Space Shuttle Main Engine.* July 1978.

78-1004. L. Norquist. *Development Progress, External Tank for the Space Shuttle Main Propulsion System.* July 1978.

78-1005. D. David. *Space Shuttle OMS Engine System Testing.* July 1978.

78-1007. M. W. Reck and J. R. Baughman. *Space Shuttle Auxiliary Power Unit Configuration and Performance.* July 1978.

79-1045. W. C. Rochelle, H. H. Battley, and J. J. Gallegos. *Use of Arc-Jet Blunted Wedge Flows for Evaluating Performance of Orbiter TPS.* June 1979.

79-1136. Boyd C. Brinton and Joe C. Kilminster. *Space Shuttle SRM Development.* June 1979.

79-1139. A. J. Verble Jr., A. A. McCool, and J. H. Potter. *Space Transportation System Solid Rocket Booster Thrust Vector Control System.* June 1979.

79-1141. H. Colbo. *Development of the Space Shuttle Main Engine.* June 1979.

79-1142. R. Michel. *Orbit Maneuvering System Engine Qualification Testing.* June 1979.

79-1143. L. W. S. Norquist. *Preflight Status of the External Tank Portion of the Space Shuttle Main Propulsion System.* June 1979.

79-1144. D. C. Sund and C. S. Hill. *Reaction Control System Thrusters for Space Shuttle Orbiters.* June 1979.

79-1145. V. A. Blythe and G. F. Orton. *Space Shuttle APS Subsystem Static Firing Development Tests.* June 1979.

80-1129. D. J. Sanchini and H. I. Colbo. *Space Shuttle Main Engine Development.* June 1980.

80-1309. E. W. Larson. *Investigation of the Fuel Feed Line Failures on the Space Shuttle Main Engine.* June 1980.

81-1373. J. R. Johnson and H. I. Colbo. *Space Shuttle Main Engine Progress Through the First Flight.* July 1981.

81-2135. A. J. Macina. *Space Shuttle Primary Onboard Software: STS-1 to Operational Use.* October 1981.

81-2468. R. R. Meyer, Jr. and C. R. Jarvis. *In-Flight Aerodynamic Load Testing of the Shuttle Thermal Protection System.* November 1981.

81-2469. T. L. Moser and W. C. Schneider. *Strength Integrity of the Space Shuttle Orbiter Tiles.* November 1981.

82-0566. P. J. Bobbitt, C. L. W. Edwards, and R. W. Barnwell. *The Simulation of Time Varying Ascent Loads on Arrays of Shuttle Tiles in a Large Transonic Tunnel.* March 1982.

83-0147. W. C. Rochelle, H. H. Battley, J. E. Grimaud, D. L. Tillian, L. P. Murray, W. J. Lueke, and T. M. Heaton. *Orbiter TPS Development and Certification Testing at the NASA/JSC 10 MW Atmospheric Reentry Materials and Structures Evaluation Facility.* January 1983.

88-4516. Ben R. Rich. *Wright Brothers Lectureship in Aeronautics: The Skunk Works' Management Style—It's No Secret.* September 1988.

Bellcomm

TM-72-1011-3. G. T. Orrok. *Use of Shuttle in Sortie Mode.* Technical memorandum, 31 January 1972.

CASI (Center for Aerospace Information)

71N-18430. Philip E. Culbertson. *Space Tug.* ELDO/NASA Space Transportation System Briefings, Bonn, Germany, 7–8 July 1970.

72A-10764. Richard C. Thuss, Harry G. Thibault, and Dr. Arnold Hiltz. *The Utilization of Silica Based Surface Insulation for the Space Shuttle Thermal Protection System.* SAMPE National Technical Conference on Space Shuttle Materials, Huntsville, Ala., October 1971, 453–64.

72A-45194. D. D. Myers. *NASA's Management Concept for the Space Shuttle Program.* 23d International Astronautical Congress, Vienna, Austria, 12 October 1972.

75A-40626. Henry C. Paul. *Launch Processing System—A System to Support the Space Shuttle.* Canaveral Council of Technical Societies: 12th Space Congress, Cocoa Beach, Fla., April 1975, 8-3 to 8-7.

76A-18650. Frederick Peters. *NASA Management of the Space Shuttle Program.* Presented to the Project Management Institute, San Francisco, 19–22 October 1975.

76A-19330. *Space Shuttle Management Issues.* Washington, D.C.: AIIE 1975 Spring 26th Annual Conference Proceedings, 20–23 May 1975.

77A-35304. David H. Greenshields. *Orbiter Thermal Protection System Development.* Canaveral Council of Technical Societies: 14th Space Congress, Cocoa Beach, Fla., April 1977, 1-28 to 1-42.

77A-35323. Henry C. Paul. *Launch Processing System Transition from Development to Operation.* Canaveral Council of Technical Societies: 14th Space Congress, Cocoa Beach, Fla., April 1977, 7-1 to 7-3.

79A-17673. H. K. Larson and H. E. Goldstein. *Space Shuttle Orbiter Thermal Protection Material Development and Testing.* Institute of Environmental Sciences: 4th Aerospace Testing Seminar, Los Angeles, March 1978, 189–94.

81-44344. L. J. Korb and H. M. Clancy. *The Shuttle Thermal Protection System—A Material and Structural Overview.* SAMPE 26th National Symposium, Los Angeles, April 1981, 232–49.

81A-21726. James R. Bilodeau. *The Role of Simulation in Space Shuttle Training.* NASA: Summer Computer Simulation Conference, Toronto, Canada, July 1979, 8-13 to 8-18.

83A-27474. T. E. Utsman. "KSC Ground Support Operations and Equipment for the Space Transportation System." In *Shuttle Propulsion Systems.* New York: American Society of Mechanical Engineers, 1982. Proceedings of the Winter Annual Meeting, Phoenix, Ariz., 14–19 November 1982, 73–77.

ESA (European Space Agency)

HSR-2. Arturo Russo. *ESRO's First Scientific Satellite Program 1961–1966.* October 1992.

HSR-3. Arturo Russo. *Choosing ESRO's First Scientific Satellites.* November 1992.

Bibliography

HSR-7. John Krige. *The Launch of ELDO*. March 1993.

HSR-8. John Krige. *Europe Into Space: The Auger Years (1959–1967)*. May 1993.

HSR-9. Arturo Russo. *The Early Development of the Telecommunications Satellite Programme in ESRO (1965–1971)*. May 1993.

HSR-10. Michelangelo de Maria. *The History of ELDO. Part I: 1961–1964*. September 1993.

HSR-14. Lorenza Sebesta. *United States–European Cooperation in Space During the Sixties*. July 1994.

HSR-15. Lorenza Sebesta. *United States–European Space Cooperation in the Post-Apollo Programme*. February 1995.

HSR-18. Lorenza Sebesta. *The Availability of American Launchers and Europe's Decision "To Go It Alone."* September 1996.

HSR-21. Lorenza Sebesta. *Spacelab in Context*. October 1977.

GAO (General Accounting Office)

MASAD-82-15. Charles Bowsher. *NASA Must Reconsider Operations Pricing Policy to Compensate for Cost Growth on the Space Transportation System*. 23 February 1982.

NSIAD-90-192. Frank C. Conahan. *NASA Has No Firm Need for Increasingly Costly Orbital Maneuvering Vehicle*. July 1990.

PSAD-76-73. Elmer Staats. *Status and Issues Relating to the Space Transportation System*. 21 April 1976.

PSAD-77-113. Elmer Staats. *Space Transportation System: Past, Present, Future*. 27 May 1977.

General Dynamics

GDC-DCB-69-046. *Space Shuttle Final Technical Report*. 31 October 1969.

Grumman

B35-43RP-11. *Alternate Space Shuttle Concepts Study*. 6 July 1971.

B35-43RP-33. *Space Shuttle System Program Definition: Phase B Extension Final Report*. 15 March 1972.

IAF (International Astronautical Federation)

78-A-24. W. C. Rice. *Economics of the Shuttle Solid Rocket Motors' Recovery and Refurbishment*. October 1978.

Lockheed

LMSC-A959837. *Final Report: Integral Launch and Re-entry Vehicle*. 22 December 1969.

McDonnell Douglas

Interim Report to OMSF: Phase B System Study Extension. 1 September 1971.

MDC E0056. *A Two-Stage Fixed Wing Space Transportation System*. 15 December 1969.

MDC E0558. *Space Shuttle Phase B System Study Extension Final Report*. 15 March 1972.

MDC G2727. *Space Station Executive Summary*. 1 April 1972.

MDC G4471. D. C. Cramblit and F. C. Runge. *Shuttle Orbital Applications and Requirements (SOAR) Final Report—Phase II*. April 1973.

NACA (National Advisory Committee for Aeronautics)

Report 1381. H. Julian Allen and A. J. Eggers Jr. *A Study of the Motion and Aerodynamic Heating of Ballistic Missiles Entering the Earth's Atmosphere at High Supersonic Speeds*. 1958.

NASA (National Aeronautics and Space Administration)

JSC-09990C. Jerome B. Hammack and M. L. Raines. *Major Safety Concerns Space Shuttle Program.* 8 November 1976.

JSC-13045. Donald K. Slayton. *Space Shuttle Orbiter Approach and Landing Test Evaluation Report: Captive-Active Flight Test Summary.* Johnson Space Center, September 1977.

JSC-13864. Aaron Cohen and Donald K. Slayton. *Space Shuttle Orbiter Approach and Landing Test: Final Evaluation Report.* Johnson Space Center, February 1978.

MSC-04326. Jack C. Heberlig and H. R. Palaoro. *The Modularization Approach to Living and Working in Near Earth Space.* Manned Spacecraft Center, July 1971.

NMI 8020.18. Dale D. Myers. *Space Shuttle Program Management.* 12 July 1971. See also revisions: NMI 8020.18A, 17 March 1972; NMI 8020.18B, 15 March 1973; NMI 8020.18C, n.d.

SA 44-76-01. *Fact Book: Space Shuttle Solid Rocket Booster.* Marshall Space Flight Center, March 1976.

SP-440. Don D. Baals and William R. Corliss. *Wind Tunnels of NASA.* 1981.

SP-487. Douglas R. Lord. *Spacelab: An International Success Story.* 1987.

SP-504. John F. Hanaway and Robert W. Moorehead. *Space Shuttle Avionics System.* 1989.

SP-2000-4224. James E. Tomayko. *Computers Take Flight: A History of NASA's Pioneering Digital Fly-by-Wire Project.* 2000.

SP-4002. James M. Grimwood, Barton C. Hacker, and Peter J. Vorzimmer. *Project Gemini Technology and Operations: A Chronology.* 1969.

SP-4012. Linda Neuman Ezell. *NASA Historical Data Book, Volume III: Programs and Projects, 1969–1978.* 1988.

SP-4106. Robert C. Seamans Jr. *Aiming at Targets: The Autobiography of Robert C. Seamans, Jr..* 1996.

SP-4201. Loyd S. Swenson Jr., James M. Grimwood, and Charles C. Alexander. *This New Ocean: A History of Project Mercury.* 1966.

SP-4204. Charles D. Benson and William Barnaby Faherty. *Moonport: A History of Apollo Launch Facilities and Operations.* 1978.

SP-4206. Roger E. Bilstein. *Stages to Saturn: A Technological History of the Apollo/Saturn Launch Vehicles.* 1980.

SP-4208. W. David Compton and Charles D. Benson. *Living and Working in Space: A History of Skylab.* 1983.

SP-4219. Pamela E. Mack, ed. *From Engineering Science to Big Science: The NACA and NASA Collier Trophy Research Project Winners.* 1998.

SP-4220. R. Dale Reed with Darlene Lister. *Wingless Flight: The Lifting Body Story.* 1997.

SP-4221. T. A. Heppenheimer. *The Space Shuttle Decision: NASA's Search for a Reusable Space Vehicle.* 1999.

SP-4303. Richard P. Hallion. *On the Frontier: Flight Research at Dryden, 1946–1981.* 1984.

SP-4304. Elizabeth A. Muenger. *Searching the Horizon: A History of Ames Research Center, 1940–1976.* 1985.

SP-4307. Henry C. Dethloff. *Suddenly, Tomorrow Came: A History of the Johnson Space Center.* 1993.

SP-4308. James R. Hansen. *Spaceflight Revolution: NASA Langley Research Center from Sputnik to Apollo.* 1995.

SP-4309. Lane E. Wallace. *Flights of Discovery: 50 Years at the NASA Dryden Flight Research Center.* 1996.

SP-4313. Andrew J. Dunar and Stephen P. Waring. *Power to Explore: A History of Marshall Space Flight Center, 1960–1990.* 1999.

SP-4404. John L. Sloop. *Liquid Hydrogen as a Propulsion Fuel, 1945–1959.* 1978.

Bibliography

SP-4407. John M. Logsdon, gen. ed. *Exploring the Unknown: Selected Documents in the History of the U.S. Civil Space Program.* Vol. 1, *Organizing for Exploration,* 1995; vol. 2, *Relations with Other Organizations,* 1996; vol. 4, *Accessing Space,* 1999.

TM 81349. John W. Smith and John W. Edwards. *Design of a Nonlinear Adaptive Filter for Suppression of Shuttle Pilot-Induced Oscillation Tendencies.* April 1980.

TM 81899. James Wayne Sawyer and Paul A. Cooper. *Fatigue Properties of Shuttle Thermal Protection System.* Langley Research Center, November 1980.

TM X-2273. *NASA Space Shuttle Technology Conference.* Langley Research Center, March 1971.

TM X-2719. *Symposium on Reusable Surface Insulation for Space Shuttle.* Ames Research Center, November 1972.

TM X-52876. *Space Transportation System Technology Symposium.* Lewis Research Center, July 1970.

TM X-74039. Donald H. Humes. *Hypervelocity Impact Tests on Space Shuttle Orbiter Thermal Protection Material.* Langley Research Center, June 1977.

North American Rockwell

LE 71-7. *Shuttle—The Space Transporter of the 1980's.* July 1971.

SD 69-573-1. *Study of Integral Launch and Reentry Vehicle System.* December 1969.

SD 71-114-1. *Space Shuttle Phase B Final Report.* 25 June 1971.

SD 71-217-1. *Modular Space Station Phase B Extension: Preliminary Design.* 1 January 1972.

SSV 72-2. *Space Shuttle Summary Briefing.* October 1972.

SV 71-28. *Fully Reusable Shuttle.* 19 July 1971.

SV 71-50. *Space Shuttle Phase B Extension, 4th Month Review.* 3 November 1971.

SV 71-59. *Space Shuttle Program Review.* 15 December 1971.

SV 72-19. *Space Shuttle System Summary Briefing.* 8 July 1972.

Pratt and Whitney

PWA FP 71-50. *Proposal for Space Shuttle Main Engine.* 28 April 1971.

Rocketdyne

CP320R0003B. W. F. Wilhelm and P. D. Castenholz. *SSME Contract End Item Specification.* 10 May 1973.

RSS-8333-1. P. D. Castenholz. *Space Shuttle Main Engine Phase B Final Report.* Vol. 1, *Summary.* 23 June 1971.

RSS-8595-6. E. W. Larson and D. J. Sanchini. *SSME Accident/Incident Report Test 901-110 High Pressure Oxidizer Turbopump Fire.* 30 June 1977.

RSS-8595-13. M. C. Ek and D. J. Sanchini. *SSME Accident/Incident Report Test 901-136 High-Pressure Oxidizer Turbopump Fire.* 20 March 1978.

RSS-8595-15. D. J. Sanchini and P. N. Fuller. *SSME Accident/Incident Report Test 902-120 High-Pressure Oxidizer Turbopump Fire.* 12 February 1979.

RSS-8595-18. D. J. Sanchini and P. N. Fuller. *SSME Accident/Incident Report Test 901-225 Main Oxidizer Valve Fire.* 1 August 1979.

RSS-8595-19. F. B. Lary and E. W. Larson. *SSME Accident/Incident Report SSFL Test 750-041, 14 May 1979 Engine 0201 Nozzle Fuel Feed Duct Failure (Steerhorn).* 25 February 1980.

RSS-8595-20. D. J. Sanchini. *SSME Accident/Incident Report MPTA Static Firing Test SF6-01 Main Fuel Valve Failure.* 7 January 1981.

RSS-8595-22. D. J. Sanchini and P. N. Fuller. *SSME Accident/Incident Report Engine 0010 Test 901-284 High-Pressure Oxidizer Turbopump Fire.* 15 January 1981.

RSS-8656-1. R. E. Biggs. *SSME Flight Report FRF.* 1 September 1981.

RSS-8719. D. S. McAlister. *Failure Analysis Report Engine 0006 Fuel Preburner Burnthrough Incident SF10-001.* 25 March 1981.

Rockwell International Corporation

See bibliography under "North American Rockwell."

SAE (Society of Automotive Engineers)

Paper 730927. P. F. Seitz and R. F. Searle. *Space Shuttle Main Engine Control System.* October 1973.

Vought Missiles and Space Company

Technical Overview: Oxidation Resistant Carbon-Carbon for the Space Shuttle. N.d., c. 1971.

PERIODICAL LITERATURE, BY DATE

Aerospace America

Art Hanley. "Titan IV: Latest in a Family of Giants." July 1991, 34–38.

Air & Space

James R. Chiles. "Out from the Shadow." April/May 1996, 76–87.

William Triplett. "The French Succession." April/May 1996, 62–67, 74–75.

Bruce D. Berkowitz. "The Nine Lives of Slick Six." February/March 1997, 68–73.

Air Line Pilot

David C. Koch. "Space Shuttle Training." March 1978, 13–15.

American Heritage of Invention & Technology

T. A. Heppenheimer. "The Rise of the Interstates." Fall 1991, 8–18.

James E. Tomayko. "The Airplane as Computer Peripheral." Winter 1992, 19–24.

Kelly A. Giblin. "'Fire in the Cockpit!'" Spring 1998, 46–55.

Mark Wolverton. "The Spacecraft That Will Not Die." Winter 2001, 46–58.

Astronautics & Aeronautics

William Cohen. "Big Solids—Seeking a New Plateau." February 1965, 42–46.

H. L. Thackwell Jr. "Solid Rockets—A Maturing Technology." September 1965, 74–77.

George E. Mueller. "An Integrated Space Program for the Next Generation." January 1970, 30–51.

Max Faget. "Space Shuttle: A New Configuration." January 1970, 52–61.

Gen. Sam Phillips. "Launch on December 63rd." October 1972, 60–63.

Gerald W. Driggers. "Short Guide to Titan III Launch Vehicles." February 1973, 68–73.

M. S. Malkin. "Space Shuttle—The New Baseline." January 1974, 62–78.

Frederick B. Pogust. "Landing the Shuttle." December 1974, 52–54.

Dean R. Chapman, Hans Mark, and Melvin W. Pirtle. "Computers vs. Wind Tunnels for Aerodynamic Flow Simulations." April 1975, 22–30, 35.

Myron S. Malkin and George W. Jeffs. "Space Shuttle 1976: Into Mainstream Development." January 1976, 40–43.

Gregory P. McIntosh and Thomas P. Larkin. "The Space Shuttle's Testing Gauntlet." January 1976, 44–56.

George Strouhal and Donald J. Tillian. "Testing the Shuttle's Heat-Protection Armor." January 1976, 57–65.

William F. Moore and Lt. Col. Conrad Forsythe. "Buying a Shuttle Ticket." January 1977, 34–40.

Richard H. Battin. "Astrodynamics." December 1978, 36–37.

Henry Simmons. "Space Shuttle: The Month that Was." March 1979, 6.

Paul A. Cooper and Paul F. Holloway. "The Shuttle Tile Story." January 1981, 24–34, 36.

Randolph A. Graves Jr. "Computational Fluid Dynamics: The Coming Revolution." March 1982, 20–28, 62.

Astronautics and Aerospace Engineering

Col. Joseph S. Bleymaier. "ITL and Titan III." March 1963, 33–36.

***Aviation Week & Space Technology* (cited as *Aviation Week*)**

"Premature Blue Streak Shutdown Studied." 15 June 1964, 32.

Herbert J. Coleman. "U.K. Engineer Warns of Bleak ELDO Fate." 30 May 1966, 32.

—. "Britain Plans to End ELDO Participation." 13 June 1966, 38.

Warren C. Wetmore. "ELDO Management, Financing Tightened." 25 July 1966, 117–19.

Barry Miller. "747 to Set Inertial Pattern for Airlines." 20 November 1967, 89–94.

Kenneth J. Stein. "Self-Contained Avionics Broaden Scope of C-5 Missions." 20 November 1967, 193–202.

"Four Europa 1 Tests Canceled." 22 July 1968, 21–22.

"Eighth Europa Launch Scheduled." 16 December 1968, 21.

"Europa 1 Third Stage Problem Raises New ELDO Difficulties." 14 July 1969, 23.

Donald E. Fink. "French Space Center Operational." 6 April 1970, 61–67.

—. "European Meeting Moves to Unify Space Programs." 3 August 1970, 20–21.

William S. Hieronymus. "High-Speed Unpowered Landing Urged as Feasible for Shuttle." 5 October 1970, 16.

Donald E. Fink. "Europeans Back Space Cooperation." 5 October 1970, 17–18.

William S. Hieronymus. "Shuttle Avionics Concepts Studied." 28 June 1971, 58–59.

Donald E. Fink. "Failure Clouds European Launch Future." 15 November 1971, 20–21.

"Europa 2 Loss Laid to Guidance System." 22 November 1971, 19.

Donald E. Fink. "ELDO Pad Attains Operational Goal." 3 January 1972, 48–50.

"Sortie Module May Cut Experiment Cost." 17 January 1972, 17.

"NASA's Budget Seen Leveling at $3.4 Billion." 31 January 1972, 24–25.

William S. Hieronymus. "Two Reusable Materials Studied for Orbiter Thermal Protection." 27 March 1972, 48.

"Germany Plans 3-Year Space Fund Gains." 24 April 1972, 81–84.

"Many Options Hedge Diverse Space Plan." 24 April 1972, 85–94.

"North American Rockwell Alters Shuttle." 19 June 1972, 36–37.

"Europeans Delay Post-Apollo Meeting." 17 July 1972, 19.

Zack Strickland. "Shuttle Costs Remain $5 Billion." 31 July 1972, 12–13.

—. "Apollo Expertise Aimed at Shuttle." 7 August 1972, 14–15.

"Democratic Committee Attacks Shuttle Award." 7 August 1972, 15.

"Losers' Role in Shuttle Investigated." 7 August 1972, 14.

"$3.4-Billion Space Limit Passes Hurdle in Conferees Vote." 7 August 1972, 15.

Donald E. Fink. "Shuttle Subcontract Seminars Readied." 14 August 1972, 16–17.

"Shuttle Program Subcontracting Pace Aimed at Cost Control." 21 August 1972, 15.

"NASA Planners Study Interim Space Tug." 11 September 1972, 104.

Benjamin M. Elson. "NASA Gets Shuttle 'Landing' Data." 25 September 1972, 84–86.

"Cost Reduction." In "Washington Roundup," 6 November 1972, 11.

"Orbiter Construction, Flight Tests Scheduled." 13 November 1972, 18.

"Space Shuttle Design Changes Cut Cost." 13 November 1972, 18–19.

"Unified Space Agency Urged." 20 November 1972, 23.

Donald E. Fink. "Shuttle Subcontractor Briefings Begin." 27 November 1972, 47–49.

"Cape Kennedy to Be Kept Busy Despite End of Apollo Project." 11 December 1972, 17.

"French Studying New Launcher." 11 December 1972, 19.

Robert R. Ropelewski. "Europe Plans New Space Agency." 1 January 1973, 14–15.

William A. Shumann. "NASA Cuts Unmanned Effort, Research." 15 January 1973, 16–17.

Donald E. Fink. "Budget Cuts Will Slow Pace of Space Shuttle Program." 15 January 1973, 17–18.

—. "Orbiter Wing Design Change Pares Shuttle's Liftoff Weight." 29 January 1973, 21.

William A. Shumann. "No Projects Axed in NASA Budget Cuts." 5 February 1973, 23–24.

Donald E. Fink. "Lightweight Version of Shuttle Proposed." 26 February 1973, 17–18.

"Senate Unit Given Shuttle Cost Details." 2 April 1973, 25.

"Space Shuttle Orbiter Subcontractors Chosen." 2 April 1973, 14.

William A. Shumann. "NASA to Study Impact on R&D of Budget Cuts as Big as 30%." 2 April 1973, 25.

"Requests for Proposals." In "News Digest," 9 April 1973, 23.

Donald E. Fink. "Shuttle Design Faces Major Milestone." 16 April 1973, 18–19.

—. "Orbiter Weight Key Factor in Lighter Shuttle Design." 30 April 1973, 58–63.

Donald E. Fink. "Joint Space Shuttle Program Mapped by NASA and Air Force." 7 May 1973, 23.

"Europa Decision Key to German Participation." 7 May 1973, 20–21.

Robert R. Ropelewski. "Demise of Europa 2 Clears Way for L-3S." 7 May 1973, 20.

"GAO Would Eliminate Economics as Key Point in Shuttle Decision." 18 June 1973, 22.

"Lower-Thrust Solids Planned for Shuttle." 18 June 1973, 78.

"New Shuttle Forecast." 18 June 1973, 22.

William H. Gregory. "British Seeking Satellite Support for Role in Spacelab." 25 June 1973, 60–61.

"Shuttle Radiator Panels Tested." 23 July 1973, 61.

Herbert J. Coleman. "Europe Makes 79% Pledge on Spacelab." 6 August 1973, 15–16.

Philip J. Klass. "Tracking, Data Relay Satellite Lease System Studied by NASA." 27 August 1973, 21.

Benjamin M. Elson. "Computer Seen Assuming Shuttle Tasks." 3 September 1973, 14–16.

Senator Frank E. Moss. "Warning on Shuttle." 10 September 1973, 7.

Craig Covault. "Titan 3E to Fill Gap in Space Boosters." 17 September 1973, 95–99.

Robert R. Ropelewski. "Europe L3S Launcher Start Set for 1974." 1 October 1973, 34–36.

"No Bars Yet Seen to Women in Space." 29 October 1973, 18.

"C-5A Studied as Shuttle Orbiter Ferry." 5 November 1973, 46.

"Space Shuttle Solid Rocket Motor Development Awarded to Thiokol." 26 November 1973, 27.

"Lockheed Asks GAO to Overturn Space Shuttle Solid Motor Award." 14 January 1974, 22.

Donald E. Fink. "Shuttle Orbiter Nears Test Flight Decision." 21 January 1974, 45–46.

"Giant Aircraft Would Lift Shuttle Orbiter." 4 February 1974, 38–41.

"NASA Budget Increased, Shuttle Delayed." 11 February 1974, 23–25.

Donald E. Fink. "Carrier Design for Space Shuttle Orbiter Being Refined." 29 April 1974, 54–62.

Donald C. Winston. "Weight Problem Delays Contract for Spacelab." 27 May 1974, 18–19.

"ERNO Spacelab Concept Called Superior." 3 June 1974, 16.

"ERNO Chosen Spacelab Prime." 10 June 1974, 21–22.

Kenneth J. Stein. "Contract Awarded for Shuttle Microwave Beam Landing Aid." 19 June 1974, 45–47.

Erwin J. Bulban. "747 Wins Competition for Shuttle Role." 24 June 1974, 21.

Richard G. O'Lone. "Shuttle Test Pace Intensifies at Ames." 24 June 1974, 68–73.

"NASA Restudying Solid Motor Award." 1 July 1974, 18–19.

"Expendable Vehicles Fill Pre-Shuttle Role." 15 July 1974, 75.

"SAMTEC Operates Western Test Range." 15 July 1974, 83–89.

Craig Covault. "Studies Awarded on Shuttle Upper Stage." 11 November 1974, 22–23.

—. "Thermal, Weight Concerns Force Changes to Shuttle." 9 December 1974, 19–20.

"Space Shuttle Runway Takes Shape at Cape." 13 January 1975, 45.

Benjamin M. Elson. "New Unit to Test Shuttle Thermal Guard." 31 March 1975, 52–53.

"Crew Criteria for Space Shuttle Evaluated by NASA Management." 7 April 1975, 50.

"Hurdles Face Space Shuttle Upper Stage." 7 April 1975, 44–45.

Kenneth J. Stein. "Shuttle Ground Landing Aid Award Made." 28 April 1975, 91–92.

"NASA Planetary Office Plans Appeal on Interim Upper Stage." 26 May 1975, 19.

Donald E. Fink. "USAF Launch/Recovery Plan Set." 30 June 1975, 32–36.

Craig Covault. "Shuttle Engine Passes Critical Milestone." 30 June 1975, 37–42.

Richard G. O'Lone. "Tunnel Tests Yield New Orbiter Data." 30 June 1975, 43–44.

"Shuttle Orbiter/Boeing 747 Separation Methods Refined." 30 June 1975, 44.

"Explosion Damages Shuttle Wind Tunnel." 11 August 1975, 16.

"Tunnel Explosion Forces Shuttle Test Change." 18 August 1975, 16.

William A. Shumann. "Investigative Agency Impact Rises." 1 September 1975, 46–47.

Donald E. Fink. "Competition Set for Shuttle Upper Stage." 15 September 1975, 20–22.

Craig Covault. "Planetary Missions Capabilities of Upper Stage Is NASA Concern." 15 September 1975, 20–22.

—. "Space Shuttle Funding Seen at Stake." 22 September 1975, 47–50.

—. "Space Shuttle Engine Testing Delayed." 6 October 1975, 20–21.

"Modified Gulfstream 2 Begins Shuttle Checkout Flights." 6 October 1975, 21.

Craig Covault. "Cape Shuttle Capabilities Broadened." 13 October 1975, 40–43.

"Shuttle Orbiter Hardware Assembled." 20 October 1975, 16–17.

Donald E. Fink. "Upper Stage Plans for Shuttle Detailed." 27 October 1975, 14–16.

"IUS Milestones Keyed to Shuttle Program Dates." 27 October 1975, 16.

"Major Shuttle Elements Mated." 1 December 1975, 20.

Craig Covault. "GAO Report to Scrutinize Shuttle Pace." 12 January 1976, 17–18.

"Shuttle Training Aircraft Buffet Problem Studied." 12 January 1976, 18.

"Complex Process Produces Shuttle Material." 26 January 1976, 125.

"Orbiter Composite Structures Readied." 26 January 1976, 126.

"Shuttle Orbiter to Use Silica Insulation." 26 January 1976, 108–9.

"Saturn-5 Launcher Dismantled." 23 February 1976, 25.

Craig Covault. "GAO to Suggest Further Shuttle Delay." 1 March 1976, 22–23.

"Crews Chosen for Shuttle Tests." 1 March 1976, 22.

"Model Aircraft Used in Orbiter Tests." 15 March 1976, 65.

"Shuttle Carrier 747 Model Being Tested." 5 April 1976, 58.

Craig Covault. "Shuttle Maneuvering Engine Starts New Tests." 31 May 1976, 58–60.

—. "Shuttle Engine Delays Overcome." 5 July 1976, 43–49.

"USAF, Boeing Completing Shuttle IUS Negotiations." 16 August 1976, 23–24.

Barry Miller. "NASA Sets Briefing on Upper Stages." 6 September 1976, 26.

"Shuttle Orbiter Named." 13 September 1976, 26.

"Shuttle Completed on Schedule." 20 September 1976, 12–14.

Donald E. Fink. "Orbiter 101 in Final Systems Checks." 27 September 1976, 14–16.

"Shuttle Skylab Mission Scrutinized." 18 October 1976, 18.

"Shuttle Manipulator Design Reviewed." 1 November 1976, 38–39.

"Shuttle Coating Material Developed." 6 November 1976, 53.

Barry Miller. "Modular Booster Set for Varied Roles." 8 November 1976, 140–41.

"Boeing 747 Prepared for Shuttle Orbiter Ferry Role." 8 November 1976, 150–54.

"Budget Constraints Pare Test Efforts." 8 November 1976, 64–66.

Craig Covault. "Solid Rocket Booster Nears Milestones." 8 November 1976, 84–89.

Bibliography

Kenneth J. Stein. "Orbiter Avionics Aim at High Reliability." 8 November 1976, 75–78.

"Main Engine Test Program Accelerates." 8 November 1976, 59–61.

"Michoud Assembling First External Tank." 8 November 1976, 95–99.

"NASA Seeks Low-Cost Spinning Stage." 8 November 1976, 143–44.

"Post-Flight Turnaround Still Unsettled." 8 November 1976, 128–33.

"Program Language Developed for Shuttle." 8 November 1976, 78.

Richard G. O'Lone. "Thermal Tile Production Ready to Roll." 8 November 1976, 51–54.

"Shuttle Main Engine Production Set." 8 November 1976, 61.

"Upper Stage Techniques Seen Complementary." 8 November 1976, 141.

"Usable Launch Data Elements Delivered." 8 November 1976, 83.

William H. Gregory. "Shuttle Opens Door to New Space Era." 8 November 1976, 39–43.

"Shuttle Upper Stage Stresses Modularity." 15 November 1976, 63.

Craig Covault. "Cape Canaveral Prepared for Shuttle." 22 November 1976, 62–65.

"Astronauts Training for Landing Tests." 29 November 1976, 42.

"Shuttle Upper Stage Agreement Reached." 6 December 1976, 20.

Craig Covault. "TDRSS Won by Western Union." 3 January 1977, 14–16.

"Shuttle Carrier Aircraft in Flight Tests." 3 January 1977, 16.

"UTC Prepares for Shuttle Booster Role." 3 January 1977, 15.

Craig Covault. "Few Shifts Expected in NASA Request." 24 January 1977, 24–27.

"Orbiter Towed to Dryden for Test." 7 February 1977, 12–13.

Donald E. Fink. "Shuttle Orbiter, Carrier Is Readied for Flight." 14 February 1977, 12–16.

—. "Shuttle Orbiter, 747 Carrier Move Into Flight Phase." 21 February 1977, 16–18.

"Shuttle Stage Agreement Made." 21 February 1977, 19.

Donald E. Fink. "Orbiter/747 Open Flight Envelope." 28 February 1977, 20–21.

"Shuttle Orbiter Begins Flight Tests." 28 February 1977, 16–19.

—. "Orbiter Readied for Next Test Phase." 7 March 1977, 19–20.

—. "Orbiter Crews Train in Flight Simulator." 18 April 1977, 44–47.

"Shuttle Engine Passes Test Milestone." 18 April 1977, 47.

"Delay Urged on Funds for Two Orbiters." 13 June 1977, 87–89.

"Space Shuttle Sticking Closely to Cost Estimates, NASA Says." 13 June 1977, 96.

"NASA Nominee to Stress Applications." 20 June 1977, 79–80.

"Booster Recovery System Drop Tested." 27 June 1977, 15.

Donald E. Fink. "Orbiter Flight Plan Expanded." 27 June 1977, 12–14.

Jeffrey M. Lenorovitz. "Shuttle Orbiter Test Phase Trimmed." 4 July 1977, 18–19.

"Shuttle Solid Test Near Specifications." 25 July 1977, 23.

Jeffrey M. Lenorovitz. "Shuttle Orbiter Cleared for Free Flight." 1 August 1977, 20–21.

"Engine Tests Raise Confidence." 8 August 1977, 23.

Craig Covault. "Shuttle Suit Shows Advances on Apollo." 15 August 1977, 37–40.

Donald E. Fink. "Orbiter Responsive in Free Flight." 22 August 1977, 12–19.

"Orbiter Free Flight Draws Big Crowd." 22 August 1977, 13.

Donald E. Fink. "Runway Shifted for Orbiter's Next Flight." 29 August 1977, 22–23.

"Shuttle Team Corrects Computer Problem." 5 September 1977, 22.

Donald E. Fink. "Second Flight Confirms Enterprise Data." 19 September 1977, 22–23.

"NASA Orders McDonnell Douglas Spin Stage." 26 September 1977, 24.

Donald E. Fink. "Orbiter Nears Final Free Flight Tests." 3 October 1977, 24–25.

Craig Covault. "Delay in Shuttle Engine Activities Being Studied." 3 October 1977, 26.

—. "NASA Evaluating Major Shuttle Orbiter Changes." 10 October 1977, 26–27.

"Next Orbiter Free Flight Set for Oct. 12." 10 October 1977, 27.

Donald E. Fink. "Shuttle Orbiter Passes Critical Milestone." 17 October 1977, 23–24.

—. "Orbiter Experiences Control Problems." 31 October 1977, 16–17.

Craig Covault. "Early Shuttle Mission to Skylab Planned." 7 November 1977, 24–25.

Jeffrey M. Lenorovitz. "Orbiter/747 Cleared for Ferry Flight." 28 November 1977, 72–73.

"NASA Winnows Astronaut Candidates." 28 November 1977, 20.

Craig Covault. "Cape Preparations for Shuttle Pressed." 5 December 1977, 43–44.

"Interim Upper Stage Motor Test Completed." 2 January 1978, 14.

"New Astronaut Training to Start July 1." 23 January 1978, 18–19.

"Second Shuttle-Motor Firing by Thiokol Termed Success." 23 January 1978, 21.

Craig Covault. "Inflation Absorbs NASA Funding Growth." 30 January 1978, 28–33.

Bruce A. Smith. "Shuttle Motor Units Move on Air Film." 30 January 1978, 56–57.

Benjamin M. Elson. "Shuttle Booster Motor Tests Planned." 20 February 1978, 54–55.

"Space Shuttle Reusable Motor Tested." 20 February 1978, 56–59.

Photo of Structural Test Article: 6 March 1978, 13.

"Shuttle External Tank Loaded on Barge." 20 March 1978, 60.

"Shuttle Orbiter Test Crews Chosen." 20 March 1978, 26.

"Use of Shuttle Polar Orbit Jeopardized." 20 March 1978, 16.

"Shuttle Orbiter, 747 Demated." 27 March 1978, 47.

"Shuttle Prepared for Vibration Tests." 1 May 1978, 12–13.

Donald E. Fink. "Titan 34D Booster Design Completed." 15 May 1978, 48–53.

"Review Completed on Shuttle Orbiter Remote Manipulator." 15 May 1978, 54.

Craig Covault. "Propulsion Tests Provide Shuttle Data." 29 May 1978, 49–53.

"Solid-Rocket Motor Use Studied for Shuttle Tank." 29 May 1978, 53–55.

Craig Covault. "Skylab Reoriented, But Future Clouded." 19 June 1978, 30–31.

"NASA Restudies Process for Picking Shuttle Crew." 19 June 1978, 31.

"Shuttle Launch Vibrations Tested." 19 June 1978, 75–80.

"Shuttle Launch Delay." 3 July 1978, 25.

"NASA Assesses Shuttle Launch Delay." 10 July 1978, 15–16.

Craig Covault. "Shuttle Delay to Affect Few Payloads." 31 July 1978, 50–51.

"First Shuttle Launch Vehicle Being Assembled at Palmdale." 31 July 1978, 53.

Craig Covault. "Shuttle Creates New Astronaut Training Challenges." 14 August 1978, 50–59.

"New Payload Could Boost Shuttle Cost." 14 August 1978, 16–17.

"Need to Add Solids to Shuttle Sparks Defense Dept. Concern." 4 September 1978, 104.

Craig Covault. "Shuttle Engine Tests Successful." 25 September 1978, 12–13.

"Rockwell Tests Shuttle Engine Number 0201." 27 November 1978, 61.

"Thermal Tiles Applied to Orbiter 102." 27 November 1978, 64.

"Shuttle Engine Heat Exchanger Fails." 18 December 1978, 8.

Craig Covault. "Austere NASA Budget Excludes New Program Starts." 1 January 1979, 14–15.

—. "Engine Failure Threatens Shuttle's Schedule." 8 January 1979, 12–14.

Craig Covault. "NASA Budget Provides No Real Growth." 22 January 1979, 24–26.

"NASA Budget Provides No Real Growth." 22 January 1979, 16.

"Orbiter Ferry to Space Center Delayed." 5 February 1979, 13.

Craig Covault. "Further Shuttle Launch Slip Forecast." 26 February 1979, 17–19.

"Space Shuttle Motors Have Last Test." 26 February 1979, 17.

"Shuttle Supplemental Called 'Problem.'" 26 February 1979, 21.

Bruce A. Smith. "First Orbiter Ready for Florida Transfer." 5 March 1979, 22–23.

Edward W. Bassett. "ESA Evaluates Shifting Ariane Program to Commercial Group." 5 March 1979, 38.

Bruce A. Smith. "Loss of Tiles Delays Orbiter Ferry Flight." 19 March 1979, 21–22.

"Space Shuttle Engine Problem Believed Solved." 16 April 1979, 19.

"Shuttle Hardware Progress Confirmed." 23 April 1979, 21–22.

"Shuttle Orbiter, Tank Mated." 30 April 1979, 27.

Craig Covault. "Shuttle Mission Control Training Starts." 30 April 1979, 125–35.

—. "Mated Shuttle Reaches Pad 39A." 7 May 1979, 12–17.

"$600-Million Shuttle Cost Overrun Startles Congress." 7 May 1979, 18.

"First Full-Duration." In "News Digest," 21 May 1979, 23.

Craig Covault. "Thermal Tile Application Accelerated." 21 May 1979, 59–63.

"Launch Pad 39A Interfaces Verified with Enterprise." 21 May 1979, 72.

Craig Covault. "Space Center Stresses In-House Designs." 28 May 1979, 61–65.

—. "Further Shuttle Launch Delay Feared." 4 June 1979, 20–21.

—. "NASA Accelerates Checkout of Shuttle." 18 June 1979, 54–55.

"French Planning Ariane Future." 25 June 1979, 91.

Craig Covault. "First Shuttle Launch Again Postponed." 2 July 1979, 26–27.

"Europeans Organize Commercial Ariane Satellite Launch Company." 9 July 1979, 18.

"Rocketdyne Studies Engine Shutdown." 9 July 1979, 16.

Richard G. O'Lone. "IUS Propulsion System Goes Beyond Interim Status." 16 July 1979, 52–55.

"Ariane First Flight Postponed." 6 August 1979, 17.

"Carter Asks for Space Shuttle Briefing." 6 August 1979, 21.

Craig Covault. "Galileo Launch Vehicle Decision Near." 3 September 1979, 53–57.

"Space Funding Tradeoffs Loom." 3 September 1979, 16–17.

Richard G. O'Lone. "Solid Rocket Motor Tested Successfully." 3 September 1979, 22–23.

"Stage Constraints Curb Galileo Options." 10 September 1979, 25–26.

Craig Covault. "Administration Backs Shuttle Fund Rise." 17 September 1979, 22–23.

—. "Shuttle Aborts Pose New Challenges." 15 October 1979, 39–45.

—. "Closer Attention to Shuttle Needs Urged." 22 October 1979, 21–22.

"APU Firings Test Orbiter Hydraulics." 29 October 1979, 16.

"Maneuvering Unit for Shuttle Tested." 29 October 1979, 45.

"Shuttle Manifest Saturated into 1984." 29 October 1979, 15.

Craig Covault. "Kennedy Center Starts Shuttle Stacking." 5 November 1979, 18–19.

Jeffrey M. Lenorovitz "Galileo Jupiter Program Faces Rework." 5 November 1979, 20.

"Main Engine Malfunctions Cause Another Shuttle Launch Delay." 12 November 1979, 20.

Craig Covault. "Carter Backs Shuttle Fund Rise." 19 November 1979, 16–18.

"USAF Studies Inertial Stage Cancellation." 26 November 1979, 18.

Craig Covault. "Shuttle Project Faces New Problems." 10 December 1979, 20–21.

"Orbiter Integrated Test Starts This Week." 10 December 1979, 21.

"Air Force Inertial Upper Stage Delayed Year; Cost Increases." 7 January 1980, 19.

Edward W. Bassett. "ESA Successful in First Ariane Launch." 7 January 1980, 18.

Craig Covault. "Oceanic, Gamma Ray Plans Approved." 21 January 1980, 51–54.

"Problem of Propellant Slumping Delays Shuttle Solid Stacking." 21 January 1980, 71.

Craig Covault. "Shuttle Boosts NASA Budget to Record." 28 January 1980, 21.

—. "Near-Term Goals Stressed by NASA." 4 February 1980, 27–29.

"Densification Process Applied to Shuttle Tiles." 25 February 1980, 22.

"F-15 Used in Shuttle Tile Tests." 25 February 1980, 23.

"Orbiter Protective Tiles Assume Structural Role." 25 February 1980, 22–24.

Edward W. Bassett. "Europe Competes with U.S. Programs." 3 March 1980, 89–91.

"Potential Budget Cutbacks Threaten Space Programs." 10 March 1980, 20.

Bibliography

Craig Covault. "Budget Cuts Force Restructuring at NASA." 17 March 1980, 16–17.

"Defense Shuttle Flights Cut in New Schedule." 31 March 1980, 54–55.

Craig Covault. "Key Shuttle Power-on Tests Approach." 7 April 1980, 39–41.

"System Planned for Sea Recovery of Booster." 14 April 1980, 52–53.

"Orbital Maneuvering System Pod Installed." 19 May 1980, 119.

Craig Covault. "Shuttle Concerns Force Action." 2 June 1980, 14–16.

Edward W. Bassett. "More Ariane Launches Slated Despite Failure." 2 June 1980, 17.

"Shuttle Booster Recovery Ships Being Built." 2 June 1980, 48.

"Shuttle Tile Decision Set for July." 9 June 1980, 20.

"Ariane Engine Malfunction Studied." 23 June 1980, 16.

Craig Covault. "NASA Assesses Shuttle Engineering Problems." 23 June 1980, 16–17.

"Supplemental Budget Passage Averts Added Shuttle Slippage." 7 July 1980, 23.

"Shuttle Delay." In "Washington Roundup," 14 July 1980, 17.

"Space Shuttle 'Beanie' Checked Out." 14 July 1980, 112.

Craig Covault. "NASA Presses to Hold Shuttle Schedule." 21 July 1980, 16–17.

"NASA Presses to Hold Tight Shuttle Schedule." 4 August 1980, 24.

Craig Covault. "NASA Tightens Shuttle Schedule Again." 11 August 1980, 27–28.

Bruce A. Smith. "Backpack Modified for Tile Repair Use." 18 August 1980, 65–68.

"Shuttle Engine, Tile Work Proceeding on Schedule." 15 September 1980, 26.

Craig Covault. "Shuttle Launch Ascent Faces Formidable Challenges." 29 September 1980, 48–62.

"Four Shuttle Crews in Detailed Training." 29 September 1980, 55.

"Combustion Anomaly Blamed for Ariane Launch Failure." 27 October 1980, 25.

Craig Covault. "Shuttle Shifting to Launch Processing." 10 November 1980, 17–20.

"Engine Shutdown." 10 November 1980, 19.

"PAM Important in Shuttle Transition." 17 November 1980, 82–83.

Craig Covault. "Centaur Studied for Shuttle Use." 1 December 1980, 16–17.

Jeffrey M. Lenorovitz. "Injector Firing Marred by Failure Like That on Ariane 2 Launch." 1 December 1980, 22.

"The Mission Model." 1 December 1980, 22.

"NASA Finishes Shuttle Mating." 1 December 1980, 18–19.

"Mated Shuttle Begins Tests." 8 December 1980, 23.

"Simulations Begin with Complete Shuttle." 15 December 1980, 21.

Edward H. Kolcum. "Space Shuttle Moved to Pad." 5 January 1981, 12–13.

Craig Covault. "Pad Rollout Initiates New Shuttle Tests." 5 January 1981, 51–55.

—. "NASA Request Revises Upper Stage Efforts." 19 January 1981, 25–28.

"Space Shuttle Tested Plugs-Out after Software-Related Delay." 19 January 1981, 99–101.

"Space Shuttle Work Enters Final Phase." 19 January 1981, 102–3.

Edward H. Kolcum. "Shuttle Propellant Loading Test Termed Successful." 2 February 1981, 18.

"NASA Plans to Contract Centaur as Shuttle Stage." 2 February 1981, 38.

Craig Covault. Craig Covault. "Mission Verification Test Begins for Space Shuttle." 9 February 1981, 73–79.

"Firing Test Decision Tied to Orbiter Function." 9 February 1981, 24.

Craig Covault. "Shuttle Firing Test Count Starts This Week." 9 February 1981, 24–26.

—. "NASA Assesses Planning with Preliminary Budget." 16 February 1981, 19–21.

Edward B. Kolcum. "Shuttle Engine Firing Successful." 2 March 1981, 17–19.

"Launch Preparation Time Shortened." 2 March 1981, 19.

"Shuttle Engine Firing Assessed." 2 March 1981, 18.

"Shuttle Insulation Repairs Delayed." 9 March 1981, 253.

"Shuttle Tank Insulation Tests Set." 16 March 1981, 19.

Craig Covault. "Columbia Ready for First Flight." 6 April 1981, 16–20.

"Manual Reentry Control Considered." 6 April 1981, 18–19.

"Procedures Tightened as Specialists Investigate Pad Accident." 6 April 1981, 18–19.

"Columbia Exceeds Flight Goals." 20 April 1981, 18–22.

"Quick Fix Leads to Faultless Performance." 20 April 1981, 20–21.

Craig Covault. "NASA Request Rises 11%, Reflects Gains in Space Science, Shuttle." 15 February 1982, 24–26.

"Michelin Subsidiary Developing Tires for Hypervelocity Aircraft Program." 12 June 1989, 269.

"Outlook/Specifications: Launch Vehicles." 8 January 1996, 114–19.

Bulletin of the Atomic Scientists

Sidney Sternberg. "Automatic Checkout Equipment— The Apollo Hippocrates." September 1969, 84–87.

Datamation

Caroline T. Sheridan. "Space Shuttle Software." July 1978, 128–40.

Design News

R. F. Stengel. "Designing for Zero g: The Space Shuttle Galley." 22 October 1979, 48–50.

Flight International

Howard Levy. "NASA's Shuttle Trainer Delivered." 23 October 1976, 1277.

David Velupillai. "Shuttle Training: The Final Countdown." 13 December 1980, 2177–82.

Food Technology

Charles T. Bourland, Rita M. Rapp, and Malcolm C. Smith Jr. "Space Shuttle Food System." September 1977, 40–45.

Fortune

Lawrence Lessing. "Laying the Great Cable in Space." July 1961, 156–60, 248–60.

"Pie in the Sky." December 1962, 188–95.

"The Comsat Compromise Starts a Revolution." October 1965, 128–31, 202–12.

Bibliography

ICAO Bulletin

Gustav Blazek and Louis M. Carrier. "Microwave Landing System Selected for Space Shuttle Programme." August 1974, 40–44.

Frederick B. Pogust. "Space Shuttle Orbiter Landing Guidance Demands Reliability and Integrity." August 1975, 22–26.

IEEE Spectrum

Trudy E. Bell. "Managing Murphy's Law: Engineering a Minimum-Risk System." June 1989, 24–27.

Trudy E. Bell and Karl Esch. "The Space Shuttle: A Case of Subjective Engineering." June 1989, 42–46.

"How NASA Determined Shuttle Risk." June 1989, 45.

Journal of the British Interplanetary Society

Dave Dooling. "A Third Stage for Space Shuttle: What Happened to Space Tug?" Vol. 35 (1982): 553–64.

Journal of Spacecraft and Rockets

James C. Dunavant and David A. Throckmorton. "Aerodynamic Heat Transfer to RSI Tile Surfaces and Gap Intersections." Vol. 11, no. 6 (1974): 437–40.

Don E. Avery, Patricia A. Kerr, and Allan R. Wieting. "Experimental Aerodynamic Heating to Simulated Shuttle Tiles." Vol. 22, no. 4 (1985): 417–24.

Lagniappe (NASA Stennis Space Center)

"Shuttle Engine Testing Is Vital." 11 November 1977, 1, 4.

"Space Shuttle Propulsion System Successfully Fired on Longest Run." 23 June 1978, 1.

"First MPTA Series Complete After Four Successful Firings." 21 July 1978, 1.

"MPT Provides Useful Data." 15 June 1979, 1, 6.

"All Systems 'Go' for MPT Series." 22 October 1979, 1, 6.

"Shuttle's Main Propulsion Testing Scheduled to Resume This Week." 17 December 1979, 3, 4.

"First Full Duration MPT Firing Called 'Storybook' by Officials." 25 January 1980, 1, 4.

"Calibration Tests Successful; MPTA Static Firing Cut Short." 26 February 1980, 1.

"Early Morning Test Scheduled This Week." 21 March 1980, 1, 3.

"SSME Testing Milestones Met." 25 April 1980, 1, 3.

"MPTA Firing Planned May 30." 23 May 1980, 1.

"Space Shuttle Test Program Marked by Success." 18 June 1980, 1, 4, 6.

"Space Shuttle Main Engine 0009 to Complete PFC Test Series." 22 August 1980, 1, 3.

"MPTA Firing Set for November." 29 October 1980, 1.

"Test Series Under Way on Shuttle Engines." 29 October 1980, 1.

"500th SSME Test Conducted at NSTL." 26 November 1980, 1, 5.

"Two MPTA Firings Set for December." 26 November 1980, 1, 5.

"MPTA Static Firing Goes 'Exactly as Planned.'" 17 December 1980, 3.

"Engine 2008 Nears Completion of 102 Percent PFC Series." 17 December 1980, 3.

"MPTA Firing Scheduled Jan. 17." 16 January 1981, 1.

"MPTA Achieves Requirements." 13 February 1981, 1, 5.

Lockheed Horizons

Wilson B. Schramm, Ronald P. Banas, and Y. Douglas Izu. "Space Shuttle Tile—The Early Lockheed Years." Issue 13 (1983): 2–15.

Los Angeles Times

"Space Shuttle Shown Amid Palmdale Fanfare."
17 September 1976, 1.

George Alexander. "Space Shuttle put on Display;
New Era Hailed." 18 September 1976, 1, 26.

Editorial cartoon by Paul Conrad. 15 April 1981, pt. 2,
p. 13.

National Journal

Dom Bonafede. "White House Report/The Making of
the President's Budget: Politics and Influence in a New
Manner." 23 January 1971, 151–65.

Claude E. Barfield. "Space Report/NASA Feels
Pressures in Deciding on Location for Its Space
Shuttle Base." 24 April 1971, 869–76.

"Florida, California Win Space-Shuttle Bases; Air
Force Gains Major Role in the Program." 22 April
1972, 706–7.

Claude E. Barfield. "Technology Report/Intense
Debate, Cost Cutting Precede White House Decision
to Back Shuttle." 12 August 1972, 1289–99.

—. "Technology Report/NASA Broadens Defense of
Space Shuttle to Counter Critics' Attacks." 19 August
1972, 1323–32.

—. "Science Report/Space Shuttle Opponents Argue
Costs Endanger More Important NASA Programs."
12 May 1973, 689–94.

Dom Bonafede. "Going Out on a Limb." 9 July 1977,
1087.

"How the United States Got to Geneva: A Quick
History of Arms Control Treaties." 31 December 1977,
1988–89.

Daniel Rapoport. "The B-1 Bomber Keeps Coming
Back from the Dead." 21 January 1978, 100–102.

Timothy B. Clark. "SALT II Treaty—It Won't Stop
Rise in Defense Spending." 16 June 1979, 984–89.

William J. Lanouette. "SALT II—Preserving the
Balance of Terror." 16 June 1979, 990–93.

Dick Kirschten. "Rubbing SALT in the Wounds."
1 November 1980, 1852.

New York Times

John Noble Wilford. "Space Agency's New Chief:
Robert Alan Frosch." 24 June 1977, 9.

New York Times Biographical Service

Truly: Grandfather, 44, Has a Birthday in Space."
November 1981, 1595–97.

"C. Gordon Fullerton." March 1982, 332.

Malcolm W. Browne. "Veteran of Space and
Administration: Richard Harrison Truly." February
1986, 266.

Quest

Jonathan McDowell. "U.S. Reconnaissance Satellite
Programs. Part 2: Beyond Imaging." Winter 1995,
40–45.

Science

Robert Gillette. "Space Shuttle: Compromise Version
Still Paces Opposition." 28 January 1972, 392–96.

Dominique Verguèse. "European Space Program: It's
Half-Speed Ahead." 9 March 1973, 984–86.

John G. Linvill and C. Lester Hogan. "Intellectual and
Economic Fuel for the Electronics Revolution."
18 March 1977, 1107–13.

Burton I. Edelson and Louis Pollack. "Satellite
Communications." 18 March 1977, 1125–33.

William D. Metz. "Midwest Computer Architect
Struggles with Speed of Light." 27 January 1978,
404–7, 409.

R. Jeffrey Smith. "The Skylab Is Falling and Sunspots
Are Behind It All." 7 April 1978, 28–33.

—. "Shuttle Problems Compromise Space Program."
23 November 1979, 910–12, 914.

Bibliography

Gina Bari Kolata. "Who Will Build the Next Supercomputer?" 16 January 1981, 268–69.

R. Jeffrey Smith. "The Waterway that Cannot Be Stopped." 14 August 1981, 741–42, 744.

M. Mitchell Waldrop. "Space Telescope in Trouble." 8 April 1983, 172–74.

—. "Spacelab: Science on the Shuttle." 28 October 1983, 405–7.

—. "Hubble: The Case of the Single-Point Failure." 17 August 1990, 735–36.

Scientific American

D. L. Slotnick. "The Fastest Computer." February 1971, 76–87.

William C. Hittinger. "Metal-Oxide-Semiconductor Technology." August 1973, 48–57.

André G. Vacroux. "Microcomputers." May 1975, 32–40.

B. I. Edelson. "Global Satellite Communications." February 1977, 58–73.

Arnold P. Fickett. "Fuel-Cell Power Plants." December 1978, 70–76.

Laurence A. Soderblom. "The Galilean Moons of Jupiter." January 1980, 88–100.

Gordon H. Pettengill, Donald B. Campbell, and Harold Masursky. "The Surface of Venus." August 1980, 54–65.

Torrence V. Johnson and Laurence A. Soderblom. "Io." December 1983, 56–67.

James A. Van Allen. "Space Science, Space Technology and the Space Station." January 1986, 32–39.

John M. Logsdon and Ray A. Williamson. "U.S. Access to Space." March 1989, 34–40.

Index

Italic page numbers refer to illustrations